S0-CBN-734

$14 95

HUMAN GENETICS

PRENTICE-HALL BIOLOGICAL SCIENCES SERIES

William D. McElroy and Carl P. Swanson, *Editors*

Cover photo: From a sculpture by Arpi Misserlian

PRENTICE-HALL, INC. / ENGLEWOOD CLIFFS, N.J., 07632

NORMAN V. ROTHWELL

PROFESSOR OF BIOLOGY / LONG ISLAND UNIVERSITY

HUMAN GENETICS

Library of Congress Cataloging in Publication Data

Rothwell, Norman V (date)

Human genetics.

Bibliography: p.
Includes index.
1. Human genetics. I. Title.
QH431.R853 573.2'1 76-26701
ISBN 0-13-445080-9

Printed in the United States of America

10 9 8 7 6 5 4 3 2 1

Prentice-Hall International, Inc., *London*
Prentice-Hall of Australia, Pty. Limited, *Sydney*
Prentice-Hall of Canada, Ltd., *Toronto*
Prentice-Hall of India Private Limited, *New Delhi*
Prentice-Hall of Japan, Inc., *Tokyo*
Prentice-Hall of Southeast Asia Pte. Ltd., *Singapore*
Whitehall Books Limited, *Wellington, New Zealand*

To Robert F. Lewis / friend and colleague

CONTENTS

chapter 1:

SOME FUNDAMENTAL GENETIC PRINCIPLES, 1

The Human and Other Species in Genetic Analysis, 1
The Particulate Nature of the Hereditary Material, 5
Mendelian Inheritance in Mice and the Human, 8
Application of Mendel's First Law to Human Genetics, 17
Variation in Gene Expression in the Human, 18
The Environment and Gene Expression in the Human, 24

chapter 2:

THE PHYSICAL BASIS OF INHERITANCE, 30

The Distribution of the Genetic Material, 30
The Human Chromosome Complement, 36

chapter 3:

CHROMOSOME BEHAVIOR AND SEXUAL REPRODUCTION, 47

The Need for a Reduction Division, 47
Meiosis in the Male, 48
Meiosis in the Female, 56
Meiosis and the Generation of Variation, 59
Meiosis and Mendel's Laws, 60

chapter 4:

SEX CHROMOSOMES AND THE GENES THEY CARRY, 68

Sex Determination, 68
Anomalies of the Sex Chromosomes, 71
Genes on the X Chromosome, 80
Blood Clotting and Genes Which Influence It, 86
Sex-linked Recessives and Autosomal Recessives, 90
Sex Chromosomes and Mendel's Laws, 92

chapter 5:

SEX AS A COMPLEX CHARACTERISTIC, 97

Sex-limited Genes, 97
Sex-influenced Genes, 99
Aspects of Hermaphroditism, 103
Sexual Preferences, 105
The Female and Mosaicism, 106
Evidence for the Lyon Hypothesis, 109
Practical Implications of a Knowledge of Sex Chromosome Constitution, 113

chapter 6:

LINKAGE, CROSSING OVER, AND HUMAN VARIABILITY, 117

Linked Genes Tend to Stay Together, 117
Linked Genes May Separate, 119
Gene Locations Can Be Mapped, 122
Assignment of Human Genes to Chromosomes, 127
Sexual Reproduction and the Generation of Variation, 130
The Biological Significance of Crossing Over, 135
Types of Twins and the Variation in a Twin Pair, 138
Effects of Inbreeding on Variation, 142

chapter 7:

GENES AND THE IMMUNE SYSTEM, 148

Antigens and Antibodies, 148
Inheritance of ABO Blood Types, 150
The MN Blood Groupings, 154
The Rh Types, 156
Blood Groups and Associated Effects, 161
Transplants and Grafting, 163

chapter 8:

SINGLE GENES, POLYGENES, AND POPULATIONS, 170

Types of Variation, 170
Inheritance of Skin Pigmentation, 173
Problems in Analysis of Polygenic Inheritance, 177
Populations, Species, and Gene Flow, 184
Factors in Variation Among Populations, 186
Differences Among Human Populations, 192
Human Races and Their Distinction, 194
Skin Pigment and Natural Selection, 197
Detrimental Genes in Human Populations, 198
The Role of Chance in Gene Frequency, 201

chapter 9:

CHEMICAL ASPECTS OF GENETICS, 207

Composition of the Hereditary Material, 207
The Watson-Crick Model and Some of Its Implications, 209
Genes and Protein Formation, 216
The Role of Messenger RNA, 219
Amino Acids and Steps in Protein Construction, 221
The Genetic Code, 227
Colinear Molecules and Their Significance for the Human, 230

chapter 10:

GENES, MOLECULES, AND DISEASE, 235

Genetic Blocks and Human Disorders, 235
Mutant Genes and Hemoglobins, 243
Genetic Disease and Early Detection, 251
Heredity and Environment in Expression of Disease, 256

chapter 11:

CHROMOSOME ANOMALIES AND APPROACHES TO THE PROBLEM OF GENETIC DISORDERS, 262

Inherited and Non-inherited Genetic Disorders, 262
Aneuploidy Involving Sex Chromosomes, 264
Aneuploidy Involving Autosomes, 265
Parental Age and the Origin of Chromosome Anomalies, 266
Polyploidy, 268

The Deletion, 269
The Translocation and the Inheritance of Chromosome Anomalies, 271
The Inversion, 275
Complications of Some Unusual Genetic Conditions, 277
The Increasing Challenge of Genetic Impairments, 280
Detection of Genetic Defects, 281
Genetic Counseling, 282

chapter 12:

PEDIGREES AND PROBABILITY, 287

The Geneticist and the Application of Probability, 287
Combining Simple Probabilities, 288
Probability and Degrees of Relationship, 292
Use of the Binomial Expansion in Genetics, 295
Representative Pedigrees and Genetic Counseling, 300
Linkage, Probability, and Prenatal Diagnosis, 305

chapter 13:

THE ORIGIN AND SIGNIFICANCE OF GENE MUTATION, 313

General Effects of Gene Mutation, 313
Spontaneous Mutation Rate and Some of Its Effects, 315
Mutation and the Degenerate Code, 319
Harmful Genes and the Calculation of Their Mutation Rate, 320
Some Mutagenic Factors, 322
Some Properties and Genetic Effects of Radiations, 323
Ultraviolet Light as a Mutagen, 325
Action of Radiation in Genetic Change, 329
Chemical Mutagens, 330
Mutant Genes in Populations, 332

chapter 14:

CONTROL MECHANISMS, 339

Cell Specialization and Gene Expression, 339
Control of Gene Expression, 342
The Jacob-Monod Model, 344
Control of Gene Activity in Higher Life Forms, 349
Gene Control and Virus Activity, 355
Viruses, Cancer, and Controlling Mechanisms, 358
A Consideration of Various Cancer Causing Factors, 364
Factors in Aging, 367

chapter 15:

HUMAN BEHAVIOR AND OTHER PROBLEMS, 373

Genetic and Environmental Components in Species Specific Behavior, 373
Human Sex Roles in Society, 376
Intelligence and Problems in Its Measurement, 379
The Use of Twin Studies in Genetic Analysis, 381
Heritability and the Intelligence Controversy, 384
Genes, Environment, and Mental Capacity, 387
Heredity and Environment in Mental Illness, 389
Ethical Considerations in Genetic Investigations, 391
Genetic Engineering and Problems for Society, 393

GLOSSARY, 401

REVIEW QUESTION ANSWERS, 419

INDEX, 423

PREFACE

The basic objective of this book is to present the major concepts and problems of human genetics in a manner that can be readily understood by a reader with little or no background in the subject. Specific topics and examples were selected with two major audiences in mind: First, for "non-science" college students who wish to become familiar with biological principles that will remain pertinent to their everyday affairs. To satisfy that need, several schools offer a variety of courses in specific areas of the life sciences, among them human genetics intended for the non-science undergraduate. The number of such courses is increasing because of a growing awareness of human genetics problems. This book is also written for persons preparing for careers as nurses, physicians' associates, clinical technicians, and social workers concerned with communicating genetic information to community groups. The demand for physicians' associates and other allied medical personnel continues to increase with advances in human genetics, particularly as related to prenatal diagnosis, detection of carriers in high-risk groups, and the manipulation of environmental factors.

In fact, any person who desires to understand the genetic and environmental components that interact to produce the human being will find this book of value. No attempt is made to satisfy the requirements for a course in general genetics; instead we deal almost completely with the human organism. Examples from other species are given only when they best serve to make a major point. No plant genetics is included nor are microorganisms explored as genetic systems. However, references to bacteria and viruses are made throughout the text because of the insights these groups provide as to the nature of the gene and molecular interactions. Phage replication and the prophage concept are also included to explain their relation to environmental and genetic interactions and the triggering of diseases.

Technical nomenclature is minimized but the reader is not "talked down to" by substituting awkward phrases in place of common genetic expressions. All the elementary genetic terms are given such as homozygous, heterozygous, genotype, allele. All terms are explained thoroughly and used continually so as to integrate them with a reader's vocabulary without requiring rote memorization.

Certain elementary mathematical principles are stressed, mainly those applying to the concepts of probability. Nowhere are these concepts more important than in their application to problems in human genetics. Failure to understand the principles of probability as they relate to simple genetic ratios and gene frequencies in populations can lead to naive and unfortunate interpretations of the facts. The person who encounters genetic problems in daily life must understand the simple fundamentals of probability. The necessary, elementary mathematical concepts are carefully developed here and applied to specific examples of the human organism.

Genetic counseling is reviewed within the framework of recent ideas that attempt to define the role of counseling teams. The material on counseling is intended for the reader with a medical orientation or anyone who may be unaware of the role of this essential service. The chapter on pedigree analysis shows the dependence of the genetic counselor on probability.

Although the specific genetic disorders included are rather extensive this book is not a clinical text. These disorders are afflictions that are familiar to some persons (such as cystic fibrosis, sickle cell anemia, Tay-Sachs disorder). Others which may be less well known (such as glycogen storage diseases) are explained in regard to the information they provide on gene expression and the manipulation of the environment. Certain subjects are treated (homosexuality, for example) even though no genetic bases may have been established for them. The controversial topic of sex-role inheritance is also included. Such material is intended to answer questions which occur to a reader who often finds such topics omitted from basic genetics texts.

The presentation assumes no familiarity with biology or other sciences beyond that encountered in most high school curricula. Chemistry is kept to the essentials needed to appreciate the nature of gene action and of molecular disorders, their detection, and diagnosis. Mitosis and meiosis are stressed in relation to the transmission and maintenance of balanced sets of genetic information. Names of meiotic substages are not given nor are any other terms that are unnecessary for an appreciation of the biological significance of nuclear divisions. The concepts of linkage and crossing over, difficult ones for many students, cannot be eliminated from any text on human genetics without the danger of presenting a distorted viewpoint and an incomplete picture of the relationship between genes and chromosomes. Moreover, our knowledge of human linkage groups and chromosome maps is rapidly growing and will continue to be applied more routinely to the prenatal diagnosis of human afflictions. The treatment of these topics here is kept as simple as possible and is always tied to familiar situations and practical applications.

A glossary, which follows the last chapter, includes all terms defined and used in the text, as well as elementary biological expressions that may not be familiar to the student with little science background.

Norman V. Rothwell

Brooklyn, New York

ACKNOWLEDGMENTS

The author expresses his appreciation to the following individuals and sources for permission to reproduce the material cited below:

Figures

1-3 A. M. Winchester, *Human Genetics,* Charles E. Merrill Publishing Co., Columbus, Ohio, 1971.

1-10 Carolina Biological Supply Company.

2-4(A,B) Victor A. McKusick, *Human Genetics,* 2nd ed., Prentice-Hall, Inc., Englewood Cliffs, N.J., 1969.

2-5(A) Schreck, et al., *Proc. Nat. Acad. Sci.* (U.S.) 70: 804–807.

2-5(B) Frank H. Ruddle and Raju S. Kucherlapati, "Hybrid Cells and Human Genes," pp 40–41, *Scientific American,* July, 1974. Copyright © 1975 by *Scientific American.* All rights reserved.

2-6(A) Victor A. McKusick, *Human Genetics,* 2nd ed., Prentice-Hall, Inc., Englewood Cliffs, N.J., 1969.

2-6(B) The National Foundation: March of Dimes.

3-5 Dr. John Melnyk, City of Hope National Medical Center.

4-3(A,B) Malcolm A. Ferguson-Smith, "Chromosomal Abnormalities II: Sex Chromosome Defects," *Medical Genetics* (eds., McKusick and Claiborne) HP Publishing Co., Inc., New York, 1973.

4-3(C) The National Foundation: March of Dimes.

4-6(A) The National Foundation: March of Dimes.

5-6(A) Malcolm A. Ferguson-Smith, "Chromosomal Abnormalities II: Sex Chromosome Defects," *Medical Genetics* (eds., McKusick and Claiborne) HP Publishing Co., Inc., New York, 1973.

5-6(B) Carolina Biological Supply Company.

6-13 Kathleen and Eileen Murphy.

6-14 Elaine and Steven Scavelli.

6-16 N. V. Rothwell, *Understanding Genetics,* The Williams and Wilkins Co., Baltimore, Md., 1976.

Table 8-2 T. Dobzhansky, *Mankind Evolving,* Yale University Press, New Haven, Conn., 1962.

8-12 L.S. Penrose.

9-12(A) James D. Watson, *Molecular Biology of the Gene,* 2nd ed., W. A. Benjamin, Inc., Menlo Park, Ca., 1970. Copyright © by James D. Watson.

10-4 Dr. Christian B. Anfinsen.

10-5 Vernon M. Ingram, *Nature* 180: 362, 1957.

Table 14-2 M. Green, "Oncogenic Viruses," *Ann. Rev. Biochem.* 39: 701, 1970.

14-1 J. Cairns, *Cold Spring Harbor Symposium Quantitative Biology* 28: 44, 1963.

14-2 N. V. Rothwell, *Understanding Genetics,* The Williams and Wilkins Co., Baltimore, Md., 1976.

14-3 N. V. Rothwell, *Understanding Genetics,* The Williams and Wilkins Co., Baltimore, Md., 1976.

15-2 Monroe W. Strickberger, *Genetics,* Macmillan Publishing Co., Inc., New York, 1968.

15-4 Shapiro, et al., "Isolation of Pure *lac operon* DNA," *Nature* 224: 768–774, 1969.

chapter 1

SOME FUNDAMENTAL GENETIC PRINCIPLES

In some families, one child may combine features of both parents, whereas another resembles only the mother or the father. And then, there are those cases in which the offspring bear no physical resemblance to any family member. What accounts for such variations as these? Obviously, parents must transmit hereditary material to their children, but exactly what is being passed down? It must be something which carries the information required to guide the formation of a human being. All living things must carry specific information which can be transmitted to the next generation. This transfer of information insures that cats give rise to cats, dogs to dogs, oak trees to oak trees, and similarly for all forms of life. The science of *genetics* encompasses a study of the nature of the hereditary material, how it is transmitted, and how it interacts with the environment to bring about an effect on a cell or an individual. A knowledge of genetics helps us to explain why an individual may resemble both parents, one of them, or neither.

The Human and Other Species in Genetic Analysis

Most of the information on patterns of inheritance and the action of the hereditary material has been assembled from forms of life lower than the human. It is true that well before the turn of the century certain simple patterns of inheritance in humans were known. While it was appreciated that such traits as the presence of extra fingers (polydactyly) and colorblindness occur in a characteristic fashion in human pedigrees, the basis for the transmission was not well understood. Pioneer investigators utilized both plants and animals in their genetic studies. More recently, elegant analyses employing bacteria and viruses as genetic tools have provided an insight into the very nature of the hereditary material and the way in which it acts in the cell. The basic principles established in these lower groups have been found to apply to the human. Since it is the human species which concerns us most, why is it that biologists have not concentrated solely on the human to obtain fundamental genetic information? The human is not the ideal creature for such research, and the reasons become clear when we consider several factors which are essential to a detailed study of inheritance patterns.

Extremely important to such research is the need to follow not just one or two, but several generations in order to determine the way in which a particular trait occurs among members of a family line. The span of a human generation is long, in the vicinity of 25 years. Consequently, not many generations of one family can be followed in the lifetime of an investigator. Moreover, the average person can contribute little to a study of his family, since he knows little about his ancestors.

Another requirement for genetic analysis is the existence of a number of clear-cut traits in the organism being used as a genetic tool. Suppose we wanted to study the inheritance of eye color, a characteristic of all humans which occurs in many forms: blue, gray, brown, hazel, etc. Each of these alternatives is considered a *trait,* a variant form of a characteristic [Fig. 1-1(A)]. To gain information on the inheritance of any characteristic, it must exist in variant forms. If everyone had brown eyes, it might be appreciated that brown eye color has a hereditary basis; but lacking an alternative to brown, not much could be surmised about the transmission of information for the color characteristic. Moreover, the best traits for genetic study are those which are easy to recognize and to describe. Eye color in humans is a characteristic whose traits meet this requirement. In contrast, a characteristic such as height presents difficulties [Fig. 1-1(B)]. While persons may be classified into general categories such as tall, short, and medium, such groups are separated by ill-defined boundaries. Many persons would not fit precisely into any one of them. A further complication entails the influence of the environment on height. Poor diet can stunt the growth of any individual, who might otherwise be able to grow taller. A complex characteristic such as

Figure 1-1. Characteristics and traits. (A) A characteristic is a general attribute of an individual which may occur in 2 or more well-defined forms. Characteristics such as the first six noted above are obvious ones whose traits can be recognized by casual inspection. Many characteristics such as color vision or blood grouping demand closer inspection or special techniques before their variant forms can be recognized. Nonetheless, for each of these characteristics, rather sharply defined categories are apparent. (B) Height is a characteristic in which the variation is not distinct. No clear-cut traits exist. Instead, the variation describes a continuum from one extreme to the other. Such characteristics present more difficulties in genetic analysis than those in which the variant forms are clearly defined.

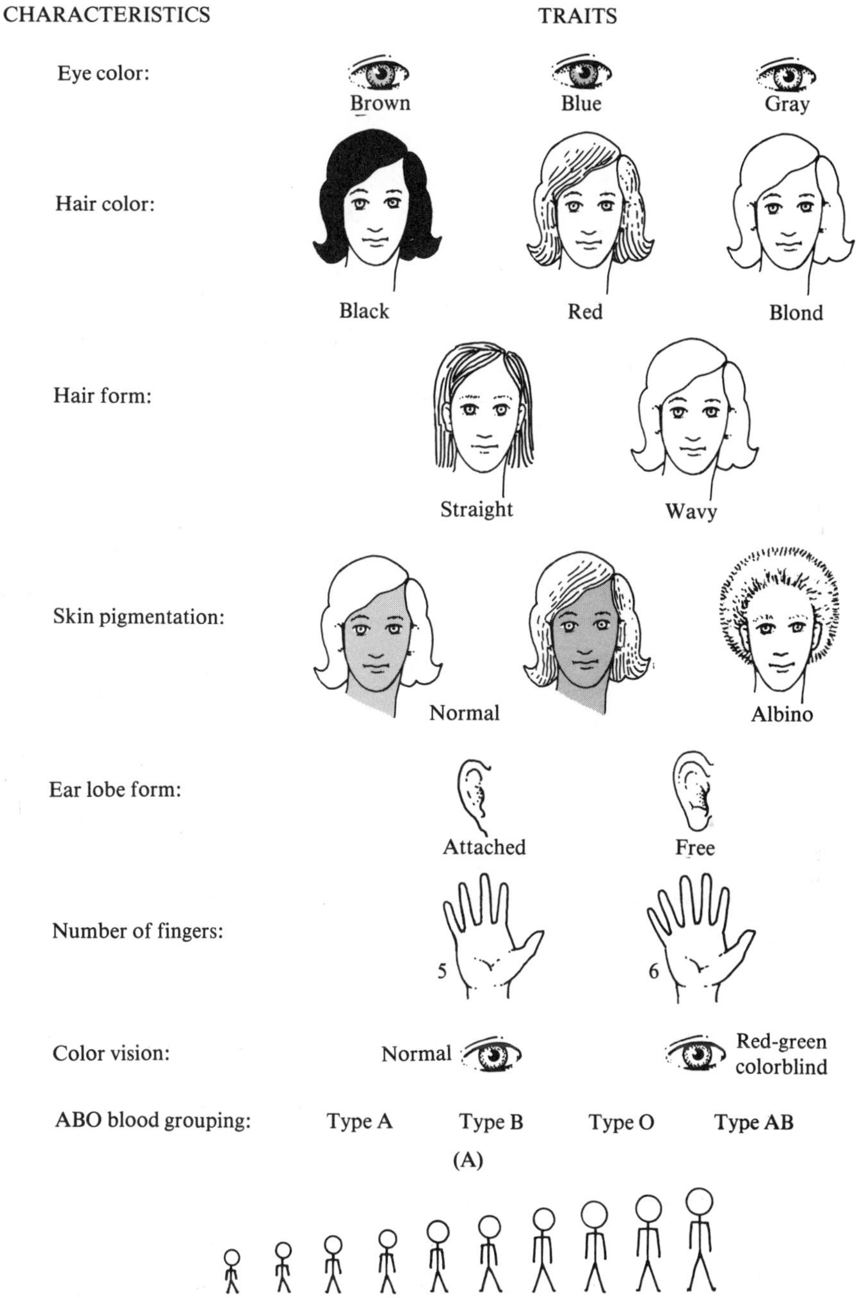

CHARACTERISTICS | TRAITS

Eye color: Brown, Blue, Gray

Hair color: Black, Red, Blond

Hair form: Straight, Wavy

Skin pigmentation: Normal, Albino

Ear lobe form: Attached, Free

Number of fingers: 5, 6

Color vision: Normal, Red-green colorblind

ABO blood grouping: Type A, Type B, Type O, Type AB

(A)

(B)

Figure 1-1

height introduces many problems and would be a poor choice for a study designed to clarify elementary genetic principles. Many nineteenth century biologists failed to obtain conclusive results because of the nature of the characteristics they chose to follow. Several plant and animal species provide an assortment of well-defined traits which lend themselves admirably to such investigations. In addition, many species present another advantage, one certainly lacking in the human. This feature is the large number of offspring which can be produced in a relatively short period of time by just a single pair of parents. A family of 8 or 9 is considered by human standards to be a good sized one, but these numbers do not compare to the hundreds of individuals which may be produced by a pair of flies or from a cross between two corn plants. Compared to the millions and millions of offspring which can arise in a matter of hours from bacterial cells and virus particles, the number of children in an average human family is insignificant.

One might ask why it is important for the geneticist to work with large numbers of individuals over the course of several generations. Suppose we wanted to determine whether or not the number of boys produced by the human is about the same as the number of girls. After observing just two families over the course of three generations, it might be found that during this time 16 girls and 4 boys were born. It would be erroneous to conclude from this sampling that humans produce four times as many girls as boys. Anyone would realize that too small a sample had been taken and that the numbers observed probably reflected the operation of chance. To minimize the role of chance factors, we must study larger numbers. To find 4 girls in a family of 4 is no different than recording 4 heads in a row after 4 tosses of a coin. In both cases, the sampling is too small to say anything about the true probability of a girl versus a boy or of a head versus a tail.

A further advantage which many other life forms hold is the ease with which a mating or cross may be manipulated. If so desired, a white-eyed fruit fly may be easily crossed to a red-eyed one. Brothers and sisters may be mated, and offspring can even be crossed with their parents in order to observe how a trait is transmitted. Such manipulations also enable the investigator to determine whether or not an individual is pure breeding for a specific trait. The student of human genetics cannot control specific matings in order to learn more about the inheritance of any trait. Instead, he must study the histories of many families over as many generations as possible, a more time-consuming and less direct approach than is possible with lower species.

There are, therefore, many drawbacks to the use of the human in genetic studies, and so it is not surprising that simpler organisms (the fruit fly, the mouse, corn and pea plants, bacteria and viruses) have played the most important role in the elucidation of elementary genetic principles. While not the species to provide the data from which the basic laws of genetics have been formulated, the human has nevertheless been the source of information which has established certain fundamental biochemical principles. This knowledge has contributed significantly to an appreciation of the activity of the genetic material in all living things. Sophisticated techniques are now making it possible to utilize the human as a tool for the study of the hereditary material on the molecular level. Refined chemical analyses of humans are enabling us to probe more deeply into the structure of all kinds of biological populations. Many exciting discoveries continue to be made concerning the human, and these hold significance not only for mankind itself but for all life forms. Now that the principles of genetics, clarified in lower species, have been shown to apply to the human, the human in turn is rapidly becoming an organism supplying information applicable to the genetics of other living things.

The Particulate Nature of the Hereditary Material

Over 100 years ago, the Austrian monk, Gregor Mendel, recognized the garden pea plant as a species which possesses that combination of features valuable to experimental genetic studies. Unlike those biologists studying human populations or those concentrating on the inheritance of complex characteristics in a variety of plant and animal species, Mendel obtained highly significant results from his work with simple traits in the pea plant. From his studies, he was able to formulate the laws upon which the science of genetics rests. While they were reported in 1865, Mendel's discoveries were not called to the attention of the general scientific world until 1900. A host of other biologists then demonstrated that Mendel's laws apply to various other living things as well as to the garden pea. Before 1910, it was suspected that the inheritance of certain human traits could also be explained on the basis of Mendel's principles. It has now been firmly established that the human is no exception to Mendelian inheritance. While peas, flies, chickens, and humans may seem to be completely unrelated, the transmission of the hereditary material in these diverse groups nevertheless conforms to the same basic laws. The knowledge gained from the lower forms has gone a long way to clarify our understanding of the genetics of the human species.

In the nineteenth century, the prevailing concept of inheritance was a blending theory in which the hereditary substance was viewed as some sort of fluid which was passed down by each parent to the offspring. These hereditary fluids, possibly even blood, would blend together and in some way give rise to the attributes of the next generation (Fig. 1-2). Even today such an idea persists among some who have no acquaintance with elementary genetics. The very expression "blood relative" is rooted in this premise. One of the great contributions of Mendel's work to biological thought was its demonstration that blending does not occur and that the hereditary material behaves as if it were particulate rather than fluid in nature. Even before the discovery of Mendel's

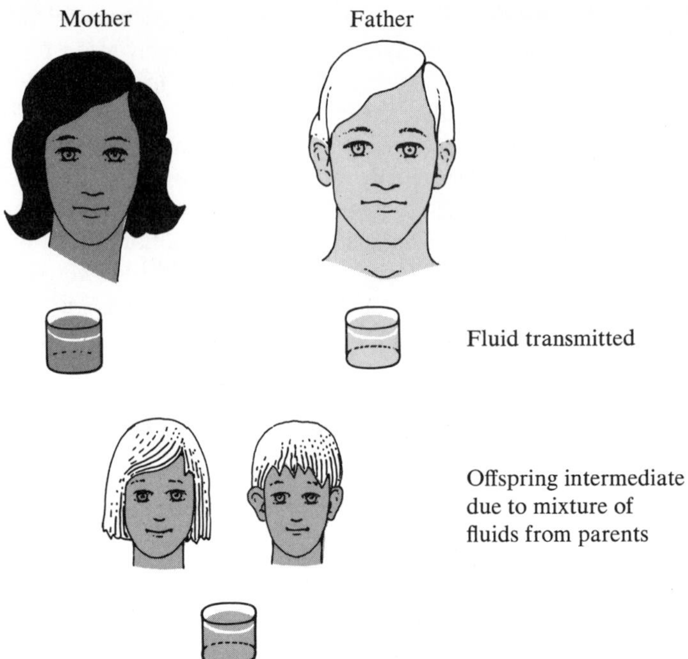

Figure 1-2. Concept of blending inheritance prevalent in the nineteenth century.

work, there was a school of biologists who supported the particulate theory of heredity and who suspected that certain human traits were inherited in a particulate manner. For example, some human pedigrees can be used to follow the albino trait in a family history. Albinism is recognized by a reduction in

Figure 1-3. Persons showing oculocutaneous albinism in which melanin pigment is greatly reduced in the skin and the iris of the eye.

the amount of melanin pigment, the main pigment which imparts coloration to the skin, hair, and iris of the eyes. As will be noted later in this chapter and in Chapter 4, several distinctly different genetic types of albinism exist. In one of the most extreme forms considered here, the cells cannot synthesize melanin at all; hence, this pigment is absent in both the skin and the eyes. Consequently, these persons are usually fair-skinned and sensitive to the sun. The reduced amount of eye pigment causes discomfort in bright daylight (Fig. 1-3). Figure 1-4 shows a pedigree for the pigment characteristic and explains some of the conventions used in representing pedigrees. The pedigree shows that the albino trait, present among members of a generation, may skip another one, only to reappear later. When it does so, the albino persons in the later generations are just as much albino as were the original ones. There has been no dilution of the albino trait. No intermediate albinos appear, as one would

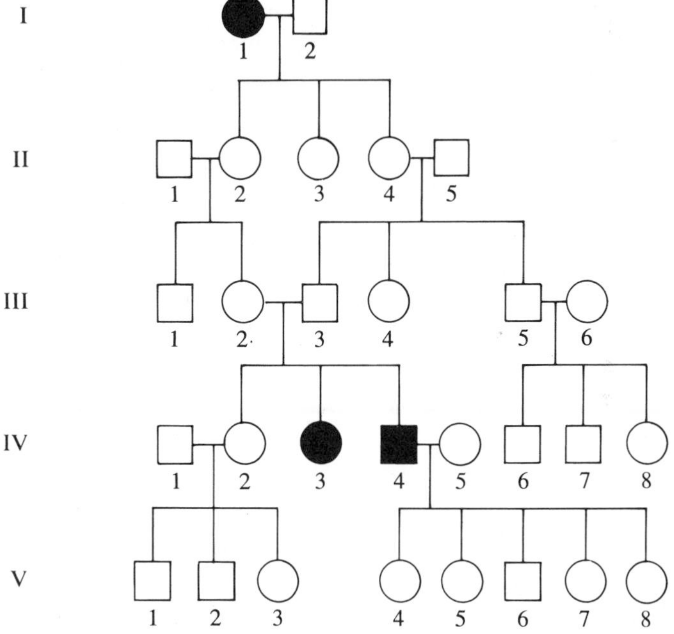

Figure 1-4. Pedigree following albinism. In the construction of a pedigree, circles represent females and squares are reserved for males. A bar between two individuals indicates a mating between them. Their offspring are represented by circles and squares suspended from this bar. Roman numerals designate successive generations in the pedigree, and Arabic numbers point to specific individuals in a given generation. A shading denotes an individual expressing the specific trait under consideration. We see here that person I-1 is an albino. The trait does not appear in generations II and III, but it reappears in generation IV and is expressed in persons IV-3 and IV-4. This type of inheritance pattern cannot be explained by the concept of blending as shown in Fig. 1-2.

predict if the responsible hereditary material were some kind of fluid. If we picture the albino substance as a white fluid and the normal hereditary substance as a red one, then we might expect some of the persons in the pedigree to exhibit a kind of intermediate albino trait. The trait should become less and less pronounced as more and more normal persons marry into the family and contribute their "normal" substance which should dilute the white. Yet nothing of this sort happens. The normal persons continue to appear normal and not less deeply pigmented due to any "albino fluid" in the family line. Then suddenly the albino trait appears, as if it had been unaffected by passing through normally pigmented persons. Certainly, from a mixture of red and white fluids in varying proportions, pure red (normal) and pure white (albino) would not be expected to persist or to separate out as if no mixing had ever taken place.

Many observations of this type were made around the turn of the century by biologists working with a variety of plant species and such diverse animal groups as insects, poultry, and mice. All the data confirmed Mendel's experiments with the garden pea which had demonstrated the particulate nature of the hereditary factors and which also showed that the transmission of traits in that species occurs in a characteristic, predictable fashion. Since the science of genetics rests on Mendelian principles which are incorporated into Mendel's first and second laws, these concepts must be firmly grasped and their implications fully appreciated by any student of genetics. The first law and its applications will be the main subject of the rest of this chapter. Since the human cannot be used in experimental matings, examples from other species will at times be given to introduce and illustrate several basic points. The significance of these will then be related to inheritance in the human.

Mendelian Inheritance in Mice and the Human

Among laboratory mice, fur color may be the typical gray in some strains, but in others it may be white (albino). Under controlled conditions, gray mice can be selectively bred only to other gray ones, and likewise white only to white, for many generations. Such selective breeding enables us to establish that certain lines are pure breeding, some for gray fur and others for white. Only the gray fur color appears within the gray line, and only the white within the white strain. Pure breeding gray animals can be mated to white ones, and from the original pairs of parents subsequent generations can be observed for the coat color characteristic (follow Fig. 1-5 in this discussion).

The first pair of parents in any genetic study is referred to as the *first parental generation,* or the P_1. The offspring of the P_1 make up the *first filial generation,* the F_1 [Fig. 1-5(A)]. In the case of the mice, all members of the F_1 will be gray, exactly the same in appearance as the original P_1 grays. We say that all the animals are expressing a gray *phenotype.* The word "phenotype" refers to the physical appearance of any living thing with regard to one or more characteristics. (We will learn later that phenotype also applies to any detectable characteristic of an individual; some of these may not be evident by casual inspection but require sensitive procedures for their detection.) However, the gray F_1's, unlike the P_1's, have albino parents. Consequently, we might suspect that with regard to fur color the genetic constitution, the *genotype,* of the F_1 animals is different from that of the P_1. This is shown to be the case when the F_1's are bred among themselves [Fig. 1-5(B)]. It is found that any one pair of F_1's may give rise to both gray offspring and albinos. Evidently, the genotype of the F_1 gray animals is different from that of their P_1 gray parents, even though they display the same phenotype. From matings such as these, it becomes evident that the inheritance of the albino and gray traits is behaving in a particulate fashion. Among members of the *second filial generation,* or F_2 (the offspring of the F_1's), the grays are as gray as the original P_1 grays and the albinos as white in fur color as the original albino parents.

The F_2 animals can in turn be used in further crosses [Fig. 1-5(C)]. When F_2 albino animals are mated to each other, only albino offspring are produced. A cross between two F_2 gray mice may produce all gray litters or litters in which both coat colors are represented. F_2 gray animals may also be crossed to albinos. When this is done, some of the pairs produce only gray offspring, whereas others may give rise to both gray and white.

All of these observations can be explained in terms of Mendel's first law, the *law of segregation.* According to this concept, the hereditary factors are particulate entities which occur in pairs in the body cells of an individual. We may refer to these factors as *genes,* a term not used by Mendel. (In later pages, more information will be presented on the nature of the gene.) There must be two factors or genes for fur color in each mouse. We can represent the gene responsible for gray fur color as *C* and that for albino as *c*. A pure breeding animal with a gray phenotype would have the genotype *CC*. It would be pure breeding, since it carries only the gene *C*. The same reasoning applies to any albino animal which is pure breeding, since it carries only *c* in double dose and therefore has the genotype *cc.*

The genes *C* and *c* are responsible for producing contrasting traits (colored fur versus white). These genes can be considered contrasting or alternative forms of one gene, the form *C* required for colored fur and the gene form *c* associated with the albino condition. Any such pair of contrasting gene forms which is related to a particular characteristic is known as a pair of *alleles.* When it is said that *C* and *c* are alleles, there is the implication that they are alternative forms of a gene and that they influence the same specific characteristic, in this case fur color. They may be referred to as the genes *C* and *c* or as the alleles *C* and *c.*

A pure breeding gray mouse possesses the genotype *CC* and is said to be

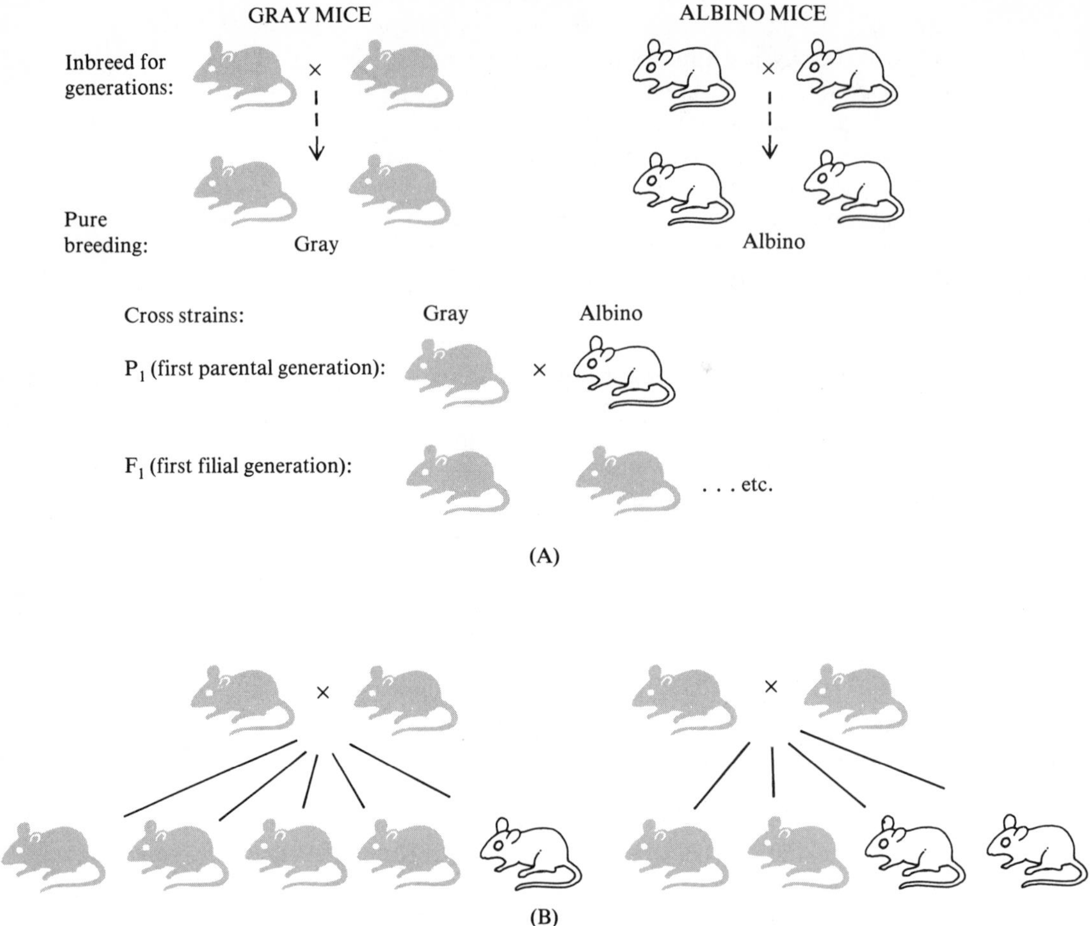

Figure 1-5. Fur color inheritance in mice. (A) Pure breeding strains of mice can be established by continued inbreeding. In this example, mice with gray fur give rise only to gray animals, and the albinos only to other albinos. When the two strains are crossed in a genetic study, the first pair of parents is designated the P_1 and their offspring are designated the F_1. We see here that the cross of the pure breeding gray and albino mice gives rise to F_1 animals, all of which have a gray phenotype. (B) When the F_1 mice are bred to each other, any one pair of gray parents may give rise to two classes of offspring: gray phenotype and white phenotype. This shows that the genotype of the F_1 gray mice is different from that of the P_1 gray mice. (C) The F_2 mice may be bred in various combinations. In some cases, 2 F_2 mice will produce only gray offspring. Other gray pairs will give rise to both gray and albino mice. When some gray mice are bred to albinos, only gray results, whereas in other such crosses, both gray and albino appear. Albinos, however, produce only albinos when they are mated to each other.

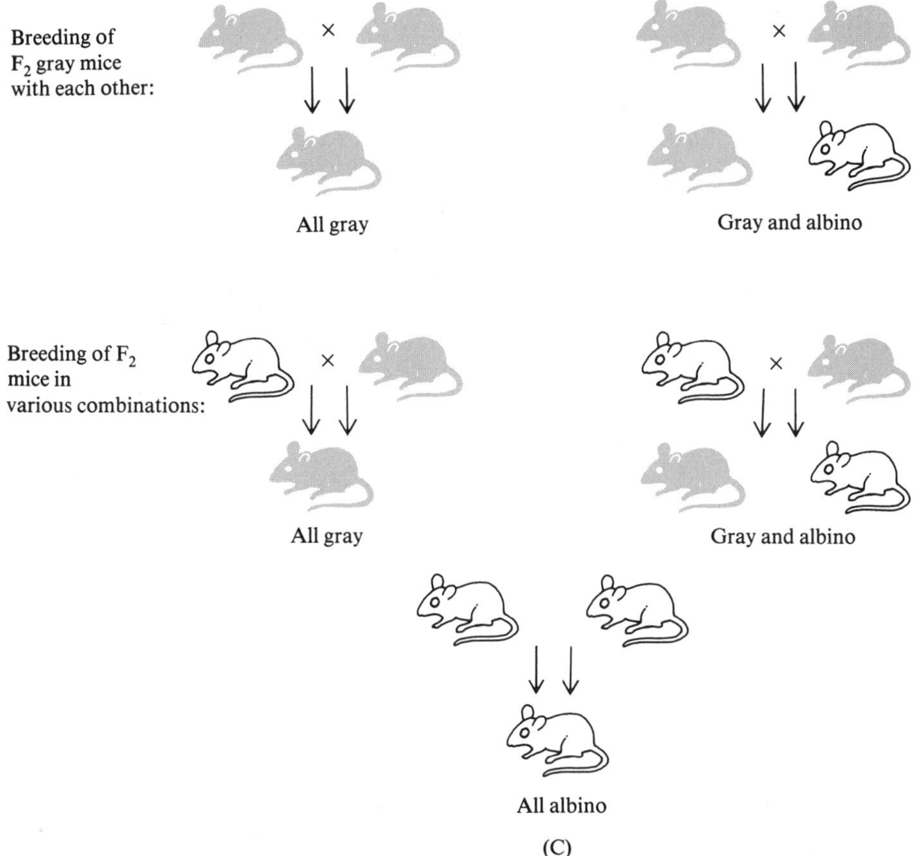

(C)

Figure 1-5. Concluded

homozygous for coat color, meaning that the alternative form of the gene is *not* present. Likewise, the pure breeding albino is homozygous, *cc,* and lacks the gene required for gray fur color, *C.* Indeed, if *both C* and *c* are present in an animal, as is true in an F_1 hybrid (Fig. 1-6), the animal has a gray phenotype. Even though the genotype of the hybrid is *Cc,* the phenotype is the same as a homozygous *CC* animal. The hybrid *Cc* is also called a *heterozygote,* meaning that the individual carries the two alternative forms of the gene. Since the animal possesses the gray trait, even though the contrasting gene form for albino is present, we say that the gene for gray is *dominant* to that for albino. Any gene such as the one responsible for albino fur, which fails to express itself in the presence of its contrasting form (its allele), is said to be *recessive.*

As we will see later in this chapter, the distinction between dominance and

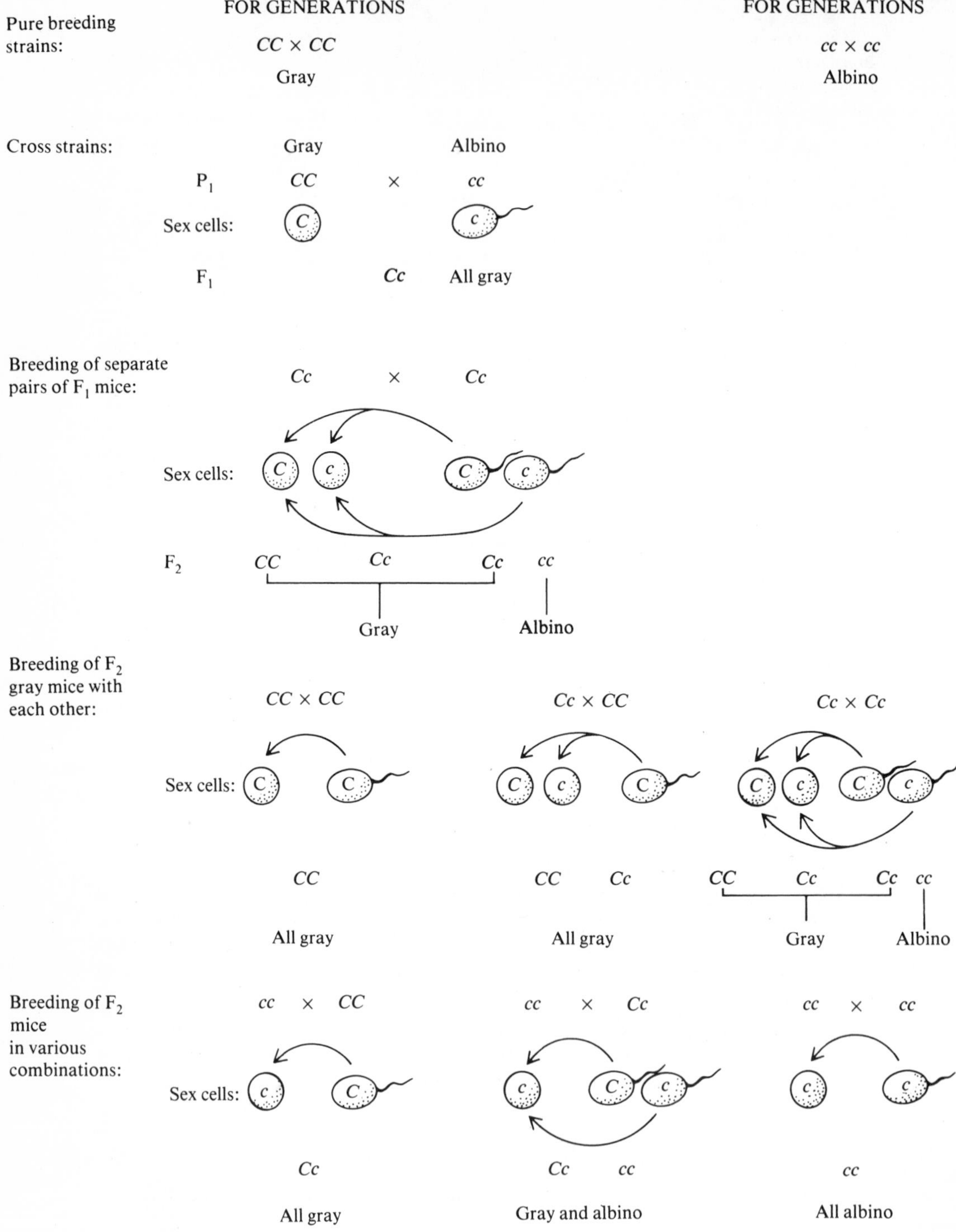

GRAY MICE INBRED FOR GENERATIONS
ALBINO MICE INBRED FOR GENERATIONS
Pure breeding strains:
CC × CC
Gray
cc × cc
Albino
Cross strains:
Gray
Albino
P1
CC
×
cc
Sex cells:
C
c
F1
Cc
All gray
Breeding of separate pairs of F1 mice:
Cc
×
Cc
Sex cells:
C
c
C
c
F2
CC
Cc
Cc
cc
Gray
Albino
Breeding of F2 gray mice with each other:
CC × CC
Cc × CC
Cc × Cc
Sex cells:
C
C
C
c
C
C
c
C
c
CC
CC
Cc
CC
Cc
Cc
cc
All gray
All gray
Gray
Albino
Breeding of F2 mice in various combinations:
cc
×
CC
cc
×
Cc
cc
×
cc
Sex cells:
c
C
c
C
c
c
c
Cc
Cc
cc
cc
All gray
Gray and albino
All albino

Figure 1-6.

recessiveness is not at all clear-cut. Absolute dominance probably does not exist, and the designation of a gene as dominant or recessive is frequently artificial and used for convenience. Actually, Mendel coined the expressions "dominant" and "recessive" to refer to traits, *not* to genes or genetic factors at all. He called a trait dominant when it appeared in the hybrid or heterozygote. In contrast, a trait which was not expressed in the hybrid was designated recessive. Throughout the years, the terms "dominant" and "recessive" have become extended as a convenient way to refer to genes as well as to traits. Since this usage is so widespread, it will be employed in our discussions.

Figure 1-6 summarizes the crosses made with the mice on the basis of Mendel's first law. This law holds that, while the genes occur in pairs in an individual, the two members of a pair separate or segregate at the time the sex cells (the gametes) are formed. Consequently, the gametes carry only the half number of genes. Therefore, any sex cell, either a sperm or an egg, can carry only one member of any pair of alleles. In the case of the mice, a gamete may carry either *C* or *c,* not both and not more than one dose of either one of them. In the P_1, the homozygous gray parent donates one *C,* and the albino donates one *c.* These contrasting gene forms or alleles, *C* and *c,* come together in the F_1 hybrid, and the double number of genes is thus restored in the body cells of these offspring. All F_1 animals possess a gray phenotype, since the gene for gray, *C,* is dominant to its recessive allele, *c.*

The F_1 gray animals are *monohybrids,* meaning that they are heterozygous with reference to one pair of contrasting genes which is being followed in the cross. Any gamete of a monohybrid will contain *either one* of the two contrasting gene forms. Therefore, a heterozygous gray animal produces two types of gametes in equal numbers, half of them carrying gene *C* and half the allele, *c.* This is so because there is an equal chance that either the dominant or the recessive allele will enter a sperm or an egg. As Figure 1-6 shows, there are two types of sperms produced by the male (*C* and *c*) as well as the same two

Figure 1-6. Genotypes and fur color in mice. This figure gives the genotypes of the various mice used in the crosses in Fig. 1-5. Note that each mouse carries 2 genes for coat color in its body cells. At the time of gamete formation, only 1 gene of the pair enters a gamete. The figure shows why crosses of some F_2 gray mice produce only gray mice. In these cases, at least one parent was homozygous for the dominant gene for coat color. Therefore, no offspring can receive a recessive from both parents in these cases, and the albino trait cannot appear. If both parents are heterozygotes, the recessive trait may appear among some offspring, since each parent contributed a recessive gene. Since the albino phenotype results from the double-recessive condition, breeding albinos together can only produce albino litters. In all of these crosses, the sex of the parent is of no importance in the inheritance of coat color. The same results arise from the cross of an albino female (*cc*) with a heterozygous male (*Cc*) and the cross of a heterozygous female (*Cc*) with an albino male (*cc*), and similarly for all the other possibilities.

types of eggs produced by the female. These two classes of gametes can unite in several ways. A *C*-carrying sperm may fertilize a *C*-carrying egg. However, there is just as much chance that a *C*-carrying sperm will unite with a *c*-carrying egg. Similarly, a sperm with a *c* may fertilize an egg of either constitution, *C* or *c*. The Punnett square illustrated in Fig. 1-7 is one method used to bring together in all possible combinations the different kinds of gametes produced by each parent. The result of crossing two monohybrids is an expected phenotype ratio of 3:1 in the next generation. Keep in mind that the 3:1 ratio is approached only after crossing several pairs of monohybrids. With just a few offspring, numbers close to 3:1 cannot be expected, since chance factors may operate to distort the ideal figures. Note in Fig. 1-5(B) that a 3:1 ratio does not necessarily occur among the F_2 of any one pair of F_1 hybrid mice. This point illustrates again the advantage in basic genetic studies provided by an organism with great reproductive capacities.

Returning to the F_2 mice (Fig. 1-7), note that while only two phenotypes are present (gray and albino in a proportion of 3:1), there are actually three genotypes: *CC, Cc,* and *cc* in a ratio of 1:2:1. While the monohybrid phenotypic ratio is 3:1, the monohybrid genotypic ratio is 1:2:1. We can now understand why some crosses between gray F_2 mice give only gray (*CC* × *CC* or *CC* × *Cc*), whereas others may give both gray and albino (*Cc* × *Cc*). Although we cannot distinguish the homozygous *CC* gray types from the heterozygous *Cc* animals by inspection, there is a rather direct genetic method which may be used to recognize them. This entails crossing the grays with albinos (Fig. 1-8). While the genotypes of the gray F_2 animals are uncertain (either *CC* or *Cc*), we do know that the genotype of any albino must be *cc*, since the recessive gene can express itself only if its dominant allele is absent.

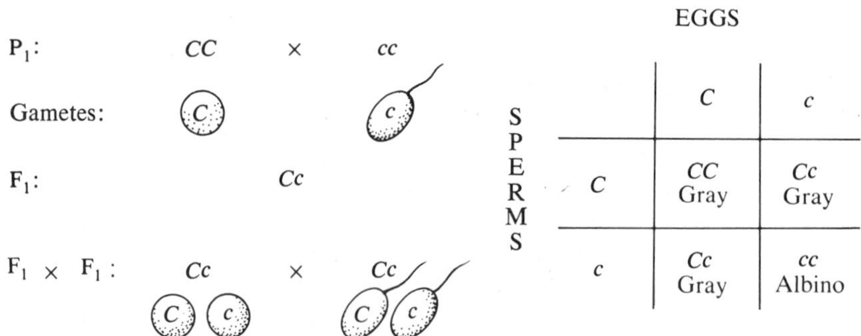

Figure 1-7. Diagram of a monohybrid cross. This is the cross through the F_2 of the gray and albino mice shown in Figs. 1-5 and 1-6. We see here that the different kinds of gametes from the hybrids can be easily combined in all possible combinations using a Punnett square. All the gametes from one parent are placed on the top. All those from the other one are placed on the side. The combinations are then shown in the appropriate boxes where the gametes from each come together.

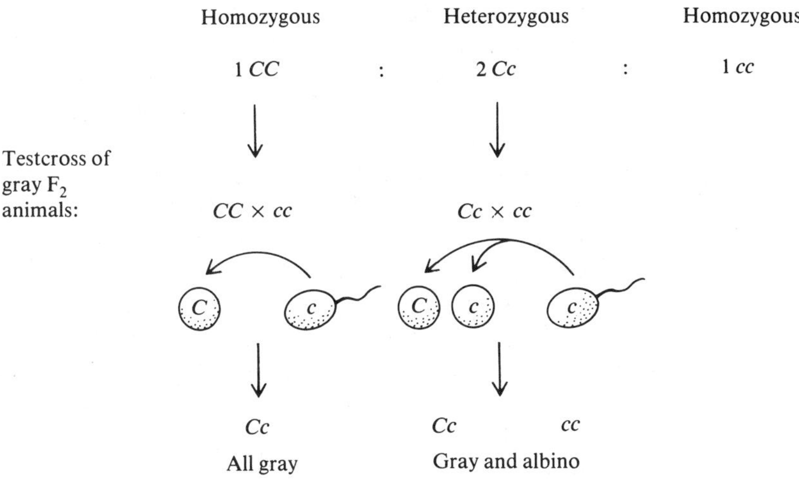

Figure 1-8. Testcross of F_2 animals. When enough F_1 monohybrids are mated together as diagrammed in Fig. 1-7, the F_2 results should approach a phenotypic ratio of 3:1 and a genotypic ratio of 1:2:1. The F_2 gray animals which are heterozygotes may be distinguished from the homozygotes by crossing grays to albinos. When heterozygotes are crossed, albinos will appear among the offspring, since the albino parent contributes only recessives in a cross. Therefore 2 out of 3 of the F_2 individuals in a monohybrid cross which express the dominant trait can produce offspring expressing the recessive when crossed to an albino. Only 1/3 of those with the dominant trait will produce offspring which express only the dominant gene.

A cross between an albino and a homozygous gray mouse will produce only gray offspring, but a cross between a heterozygous gray animal and an albino will produce both gray and albino in a ratio of 1:1. Therefore, valuable genetic information may be obtained by crossing an individual of uncertain genotype to a recessive. Such a cross is known as a *testcross.* Since the testcross parent carries only recessive genes for the characteristic which is being studied (coat color here), any recessives which are hidden by dominants in the other parent may come to expression. Testcrosses of the F_2 gray mice show that 1/3 of them are homozygotes and 2/3 are heterozygotes (Fig. 1-8).

Crosses such as these with the mouse clearly demonstrate that fur color in this animal is inherited according to the tenets of Mendel's first law: an organism carries particulate factors (genes) in pairs, and these segregate at the time of sex-cell formation in such a way that only one member of any pair enters a gamete. Along with the concept of dominance, recognized by Mendel and several earlier hybridizers, the first law provides valuable insights into human genetics. The inheritance of albinism in humans (Fig. 1-9) is similar to the transmission of white coat color in the mouse. The gene for normal skin pigmentation behaves as a dominant to its recessive allele for the albino condition. The pairs of alleles can be represented as *A* and *a* for normal and albino, respectively. (Note that the same letter is used to represent a pair of alleles. This convention prevents confusion when more than one pair of genes are being followed in a cross.) About 1 normally pigmented person in 50 is a heterozygote carrying the recessive gene for albinism. Humans, therefore, can be any one of three possible genotypes with regard to this gene pair: *AA* (homozygous for the normal allele), *Aa* (heterozygous carriers of the recessive), and *aa* (albinos who are homozygous for the recessive).

A mating between an albino (whose genotype must be *aa*) and an unrelated person of normal phenotype is most likely to result in normal offspring. This is so because the chance is only 1/50 that the unrelated mate will be a carrier of the recessive gene. Therefore the chances are high that the mating will be *AA* × *aa,* a cross which produces offspring only of genotype *Aa* who are normal phenotypically. If by the unexpected chance the albino *does* mate with a carrier, there would be a 50:50 chance that any one offspring will be albino. In such a case, the mating *Aa* × *aa* can produce offspring of either genotype,

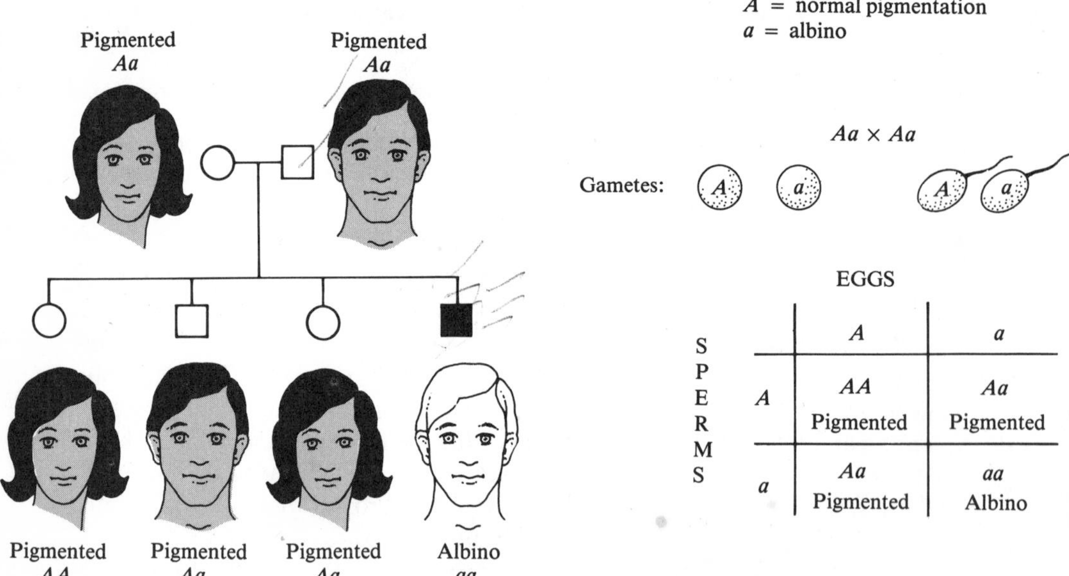

Figure 1-9. Monohybrid cross in the human. A common form of albinism is inherited as a simple Mendelian recessive. At left is a representation of a simple mating and pedigree in which the parents are both heterozygotes. At right is a Punnett square illustrating the same cross. It must be appreciated that in such a cross the *chance* is 3 to 1 for a normal offspring at any one birth. By no means does the figure mean that 1 out of 4 offspring *must* be albino in such a family.

Aa or *aa.* A reexamination of the pedigree in Fig. 1-4 shows that it is possible to write something about all the persons represented. Note that the parents of the albinos in the third generation are first cousins and *must* be heterozygotes. The genotypes of several of the normally pigmented persons in the pedigree cannot be written completely (II-1, III-5, for example). However, we do know that the dominant gene *A* *must* be present, and so these people can be represented as *A*_ (as can the genotype of any normally pigmented person about whom nothing else is known).

Application of Mendel's First Law to Human Genetics

Just from the information embodied in Mendel's first law, we are in a position to answer certain questions concerning a family in which albinism has occurred. Two parents with an albino child might wish to know the chances of their having another such child. Since each of these parents must have the genotype *Aa,* we know that a mating between them represents a monohybrid cross (Fig. 1-9). If a large number of such heterozygous parents were studied, we would expect to find approximately 3 normal children to 1 albino among all the offspring tabulated. For any *one set* of parents, however, we cannot expect to find the 3:1 ratio, even if 4 or more children are produced. Mendel's first law tells us that in a cross between two monohybrids, the chance is 3:1 that an offspring expressing the dominant trait will arise at any one birth. The chance remains only 1 in 4 for the appearance of the recessive trait. If 2 heterozygous individuals have 3 normal children, the chance is still 3 out of 4 that the next offspring will be normal, and it remains so no matter how many children they produce. The reason is exactly the same as the reason why we expect a head to appear with a 50% probability each time we toss a coin. It doesn't matter if a head appears at one throw, since the occurrence doesn't influence the next throw. The chance remains 50:50 or 1:1 for a head or tail at each toss of the coin. Only after a large number of throws do we expect to obtain approximately 50:50. Similarly, if the chance is 3/4 for a normally pigmented offspring and 1/4 for an albino at each birth, that probability remains for each successive birth and does not change. In families of 4 children where both parents are heterozygous, any combination of normal and albino can arise. It would not be surprising to find 4 normal children and no albinos in one such family and 4 albinos in another. Among any small number of offspring, departures from the expected will be common. Only when the results are compiled from a large number of crosses will the figures approach those of a theoretical ratio. Misunderstanding of the correct meaning of genetic ratios results in their incorrect application and leads to unfortunate misinterpretations. In Chapter 12 ratios will be related to the counseling of families concerned about the transmission of a genetic disease.

Several human disorders are known whose inheritance follows the simple monohybrid pattern described for albinism. The very first disease to be associated with a Mendelian pattern was alkaptonuria, a defect characterized by a deposit of pigment in the connective tissue and the onset of a severe type of arthritis later in life. In 1902, this rare condition was recognized as a simple recessive trait. It was also the first genetic disease suspected to be linked to an inherited biochemical defect in the metabolism of the cell. After the recognition of alkaptonuria as a recessively inherited disease, other disorders were soon associated with a Mendelian inheritance pattern. Thus, the study of human genetics was launched early in the century. The information which has accumulated on the inheritance of human traits, both pathological and normal, has continued to increase to the present day. A great deal has been learned about the molecular basis of some of these, and techniques are available in some cases to aid in the detection of afflicted offspring while they are still *in utero* (Chapter 5). For some afflictions, methods are available to permit the recognition of carriers among persons of normal phenotype. This enables a genetic counselor to advise two prospective parents on the chance of producing an afflicted child. The application of these procedures to various inherited human disorders will be discussed in some detail throughout this text.

Variation in Gene Expression in the Human

The story of sickle cell anemia, a fatal blood disorder, can be used to demonstrate several important points about inheritance. The gene responsible for the anemia may be designated Hb^S in contrast to the normal gene form, Hb^A. These symbols represent hemoglobin S, the abnormal hemoglobin in sickle cell victims, and hemoglobin A, the normal form of the blood protein. When hemoglobin S is the only type of hemoglobin present, the red blood cells have a tendency to assume abnormal shapes, many of them like sickles (Fig. 1-10). This is especially so under lower oxygen tension. As a result of the distortions, capillaries become clogged, and fatal consequences ensue. Homozygotes for the sickle cell gene (Hb^S Hb^S) can produce only the abnormal hemoglobin, and they manifest a collection of symptoms, among them severe physical weakness, abnormal spleen, circulatory disturbances, and brain damage. Such a group of associated symptoms which characterize a particular condition is known as a *syndrome.* Therefore, the sickle cell gene, Hb^S, is responsible for the appearance of not just one change in the normal phenotype, but several. The term *pleiotropy* refers to the collection of effects associated with a specific gene. Actually, when most genes are studied in detail, they are found to be pleiotropic. Several seemingly unrelated effects may be ascribed to a specific gene. In some cases, one of these may be more pronounced than the others, and some may even remain hidden in the absence of the refined analyses needed for their detection.

The heterozygote, the normal-appearing carrier of the sickle cell gene (Hb^A Hb^S), is generally healthy and may be unaware that he is carrying the defective gene. However, while free of the anemia, the heterozygote is *not* identical to the person homozygous for the normal hemoglobin (Hb^A Hb^A). The heterozygote is said to have the *sickle cell trait.* Both the normal hemoglobin A and the sickle cell form, hemoglobin S, are present in the red blood cells of the heterozygote. This difference from the normal homozygote (Hb^A Hb^A) may become evident at high altitudes or in other situations where oxygen is lower than in the average environment. Effects of the anemia in the heterozygote may then become manifested; spleen damage may even be encountered. Moreover, the heterozygote may be distinguished from the normal homozygote by a simple blood test. This is possible since a small proportion of the blood cells of the carrier assumes abnormal shapes under reduced oxygen tensions both in the blood vessels and in the test tube. Strictly speaking, the terms "dominant" and "recessive" cannot be used to describe this pair of alleles, Hb^A and Hb^S, which affects the formation of hemoglobin. Under scrutiny, the heterozygote is found to be somewhat different from the homozygote. When both genes are present, both types of hemoglobin are produced. Since he has a sufficient amount of the normal protein to be healthy under the average environmental conditions, the heterozygote may remain undetected. However, both alleles *do* express themselves in the heterozygote, and they cause the production of detectable products (the two types of hemoglobin).

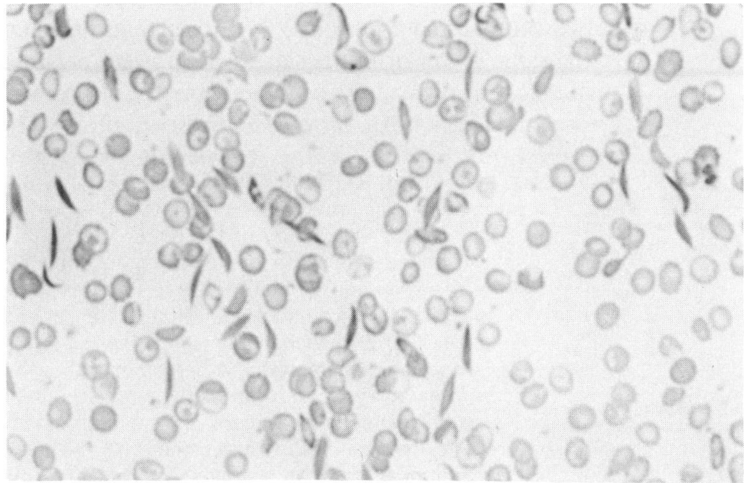

Figure 1-10. Red blood cells from a victim of sickle cell anemia, showing cells of abnormal shape among disc-shaped ones. Courtesy of Carolina Biological Supply Company.

Genes such as these are said to be *codominant.* The term *incomplete dominance* is often used to describe situations where the effects of one of the alleles in the heterozygote is more pronounced than the other. No matter which term is applied in a case such as this, both alleles are expressing themselves. Actually, a lack of complete dominance between two members of a gene pair appears to be the rule rather than the exception. Heterozygotes in the human and other species, when examined closely, frequently display a difference from both homozygotes. When the homozygote for the normal allele and the heterozygote appear identical, the proper detailed examination often reveals distinctions. Perhaps most, if not all, heterozygotes are different in some way from their normal homozygous counterparts; the demonstration of some difference may depend on the techniques available to the investigator. In Chapters 11 and 12, we will return to this point, as it holds significance for the detection of several human disorders.

Another clinical condition in the human illustrates further aspects of gene inheritance and expression. In approximately 1 birth in 10,000, an individual is born with a condition known as *osteogenesis imperfecta.* The responsible gene, *O,* behaves as a dominant in relation to its normal allele, *o.* A syndrome of clinical conditions can be recognized in the presence of gene *O*: extreme fragility of the bones, weakness of ligaments and tendons, deafness, and a blue coloration of the sclera (the white of the eye). Thus, the gene for the condition is pleiotropic [Fig. 1-11(A)]. However, when those known to be carrying the gene are studied, it becomes evident that an affected person does not necessarily exhibit all the clinical defects. While one person may manifest all the symptoms, another may be free of one or more of the defects. However, a parent who shows only the blue sclera may pass the gene to an offspring who may exhibit all the symptoms. It is clear that the gene does not express itself to the same extent or in the same way in every person who carries it. A gene such as this whose expression is not constant but varies from one individual to the next is said to be of *variable expressivity* [Fig. 1-11(B)]. While the expres-

Figure 1-11. Variability in gene expression. (A) The gene responsible for *osteogenesis imperfecta* is inherited as a dominant. When present, it can cause extreme fragility of the bones, deafness, and discoloration of the eye whites. (B) Since the gene is uncommon in the population, most persons carrying the gene would be heterozygotes (*Oo*). The normal mate of such a person would be homozygous for the normal condition (*oo*). There is thus a 50:50 chance that any offspring will be affected. The gene, however, is of variable expressivity. A person may show only one abnormality, such as affected eyes. An offspring who receives the gene from such a person may have affected eyes, ears, and skeleton. Another may have brittle bones but no abnormalities of the eye or ear. (C) The dominant gene, *O*, has reduced penetrance. A person may possess the gene and exhibit no abnormalities whatsoever. However, an offspring of this individual who receives the gene could show one, all, or any combination of the abnormalities which are part of the syndrome.

sivity of some genes, such as the one responsible for *osteogenesis imperfecta,* differs from one individual to the next, other genes may be very constant in their expression, showing little or no variation at all. For example, those genes responsible for the human ABO blood types are so constant that a person

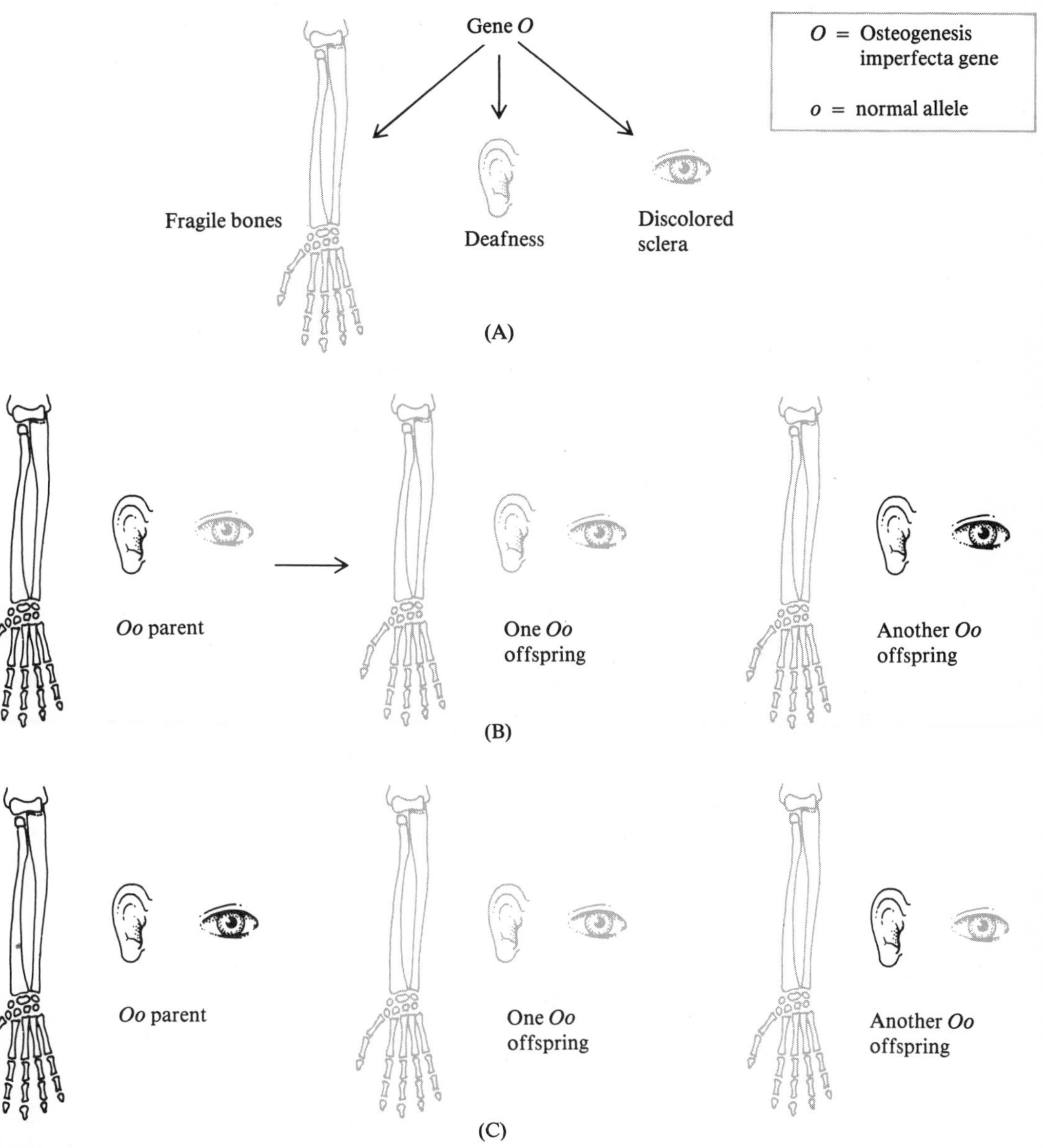

Figure 1-11

carrying the gene for A-type substance is almost certain to express it in his blood cells (Chapter 7).

The gene for *osteogenesis imperfecta* exhibits still another type of variation. Certain individuals carrying this dominant may actually show no defective symptoms of any kind! At times, certain genes, a dominant one or a recessive in the homozygous condition, may remain unexpressed when present in the genotype. When a gene expresses itself in *any way* at all, we say that it is *penetrant*. The genes for the ABO blood types are 100% penetrant, meaning that when they are present they always express themselves. (A rare exception is discussed on p. 153). The gene for *osteogenesis imperfecta* is a dominant which has *reduced penetrance*; when present, it does not bring about a detectable effect in all individuals. Nine out of 10 persons carrying this dominant will show one or more of the symptoms, whereas the one remaining will appear completely normal [Fig. 1-11(C)]. The gene has a penetrance of 90%, expressing itself in *some way,* in 90% of those carrying it. However, the gene is present in unaltered form in those carriers who show no effects whatsoever, since they may pass the dominant gene down to offspring who may exhibit one, some, or all of the symptoms associated with the disorder. Thus, this gene is pleiotropic (has multiple effects), shows reduced penetrance (does not express itself in all those with the appropriate genotype), and is of variable expressivity (its expression varies greatly among those who do express it).

Reduced penetrance and variable expressivity may at first seem to contradict the basic principles of heredity. However, the basis for each phenomenon becomes understandable when we consider several important points. In the first place, no gene or pair of alleles acts independently of other genes or gene pairs. The expression of any one gene may be greatly influenced by the expression of some other which is present in the genotype. The presence of one gene may completely prevent the expression of another. It is very important to realize that there is *no one* gene for any characteristic. Eye color, for example, involves not just one pair of alleles but the interaction of many. The pair of genes involved in *osteogenesis imperfecta, O* and *o,* is certainly not the only pair which affects the bones or the eyes. At least three different genes have been identified which can reduce the amount of melanin pigment and cause albinism. Two of these, while different genes, cause lack of pigment in both skin and eyes. The third gene causes a lack of pigment only in the eye and is inherited as a sex-linked recessive (Chapter 4).

In any specific case which is being studied, the geneticist may happen to be concerned with just one pair of genes; however, in most species, many genes interact in the normal development of a character. Probably no gene acts completely by itself without interacting with some other part of the genotype. The molecular basis for some of these interactions is known and will be the topic of Chapter 10, but it is important to realize at the outset that the effect of any one gene is not independent of others. The normal development of any characteristic, whether it be pigmentation, sight, hearing, etc., requires the proper genotype, which includes many genes which influence that character (Fig. 1-12).

Certain genes, known as *modifiers,* have just a slight effect on the expression of some other gene, altering that expression in a quantitative or measurable way. An obvious example is seen in the case of various spotted animals (dogs, mice) where the presence of any spotting at all depends on a recessive, *s.* Nevertheless, in those animals of genotype *ss,* the size of the spots varies from large to tiny. The variation has been shown to result from the presence of different modifiers among animals with the same recessive genotype (*ss*) so that spotting size varies from one individual to the next.

An abnormal effect may be completely suppressed by the presence of some other gene. Such a gene which prevents the expression of some variant (mutant) gene is called a *suppressor.* The existence of modifiers and suppressors is well known in experimental organisms, and no doubt they operate in humans as

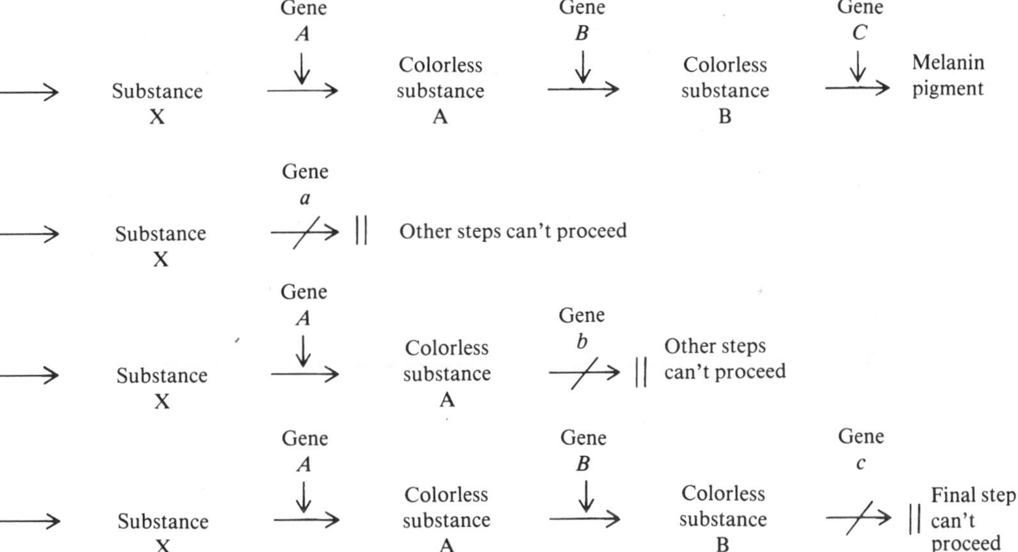

Figure 1-12. Genic interaction. This figure is greatly simplified to demonstrate that genes may interact and influence the same characteristic, such as pigment deposition. Genes may interact by controlling chemical steps in the formation of an essential product. For normal pigment development, all steps in pigment formation must proceed. Most of the steps are under genetic control. If a defective allele replaces the normal gene, one of the steps is blocked, and the others cannot proceed, even if the rest of the genotype is normal. Therefore, substitutions of gene *a, b,* or *c* will reduce pigment level. (The figure does not mean that each step in pigment formation is controlled in the exact sequence shown here.)

well. Variable expressivity in cases of genes associated with certain clinical pictures, as well as the phenomenon of reduced penetrance, can be readily explained in part on the basis of genic interaction involving modifiers and suppressor genes.

The Environment and Gene Expression in the Human

The important role of the environment in gene expression is recognized by every geneticist. Interaction of genes is greatly influenced by factors in the environmental setting. This includes the environment external to the organism as well as the internal one of the body. There are many examples in the human which indicate environmental influence as a factor in variable expressivity. One of these is polydactyly, a condition in which extra digits may be present on the hands and/or feet. The responsible gene is a dominant which has a penetrance of 90%. However, among those 9 out of 10 persons who express the gene, there is great variation. In some, there may be a well-developed extra toe on each foot and an extra finger on each hand. In others, only one extra digit, a finger or a toe, may be present. However, any extra digit may vary from complete to just a slight protuberance, even if more than one occurs on a person. In the case of polydactyly, therefore, variable expressivity is found not only among individuals; any one person with the trait may exhibit extreme variation in the development of the extra fingers or toes. Such variation implicates the environment in the expression of the responsible dominant gene.

The importance of nutrition as an environmental factor in gene expressivity is dramatically illustrated by the recessive gene, *p,* responsible for phenylketonuria (PKU), a human disorder for which the molecular basis is well understood (Chapter 10). An infant born without the normal allele, *P,* and who thus has the double recessive genotype, *pp,* may be doomed to mental retardation if fed a normal diet. However, if the condition is recognized shortly after birth, a diet can be instituted which eliminates certain common foods. This prevents the accumulation in PKU babies of toxic substances which hinder the normal development of cells of the central nervous system. This example shows not only that the environment can influence the expression of the genotype but that the *time* at which a particular environment operates may be critical. Placing an affected PKU child on the proper diet will be to no avail once the brain has been damaged.

The environment is capable of so altering gene expression in the case of certain disorders such as PKU that a person carrying defective genes may in most respects resemble an individual with a normal genotype. Thus, a person of genotype *pp* who has received the proper diet at the crucial period in infancy may appear no different from the individual who is homozygous for the normal allele (*PP*). However, it is most important to bear in mind that while the environment has prevented full expression of the genotype, the environment has not altered the genes themselves in any way. The genotype remains *pp* in the normal-appearing person who escaped the dire effects of the disorder. Such a person, unlike the normal homozygote, *PP*, will transmit the defective recessive gene to each offspring [Fig. 1-13(A)]. An individual like this is called a ***phenocopy,*** one whose phenotype has been environmentally

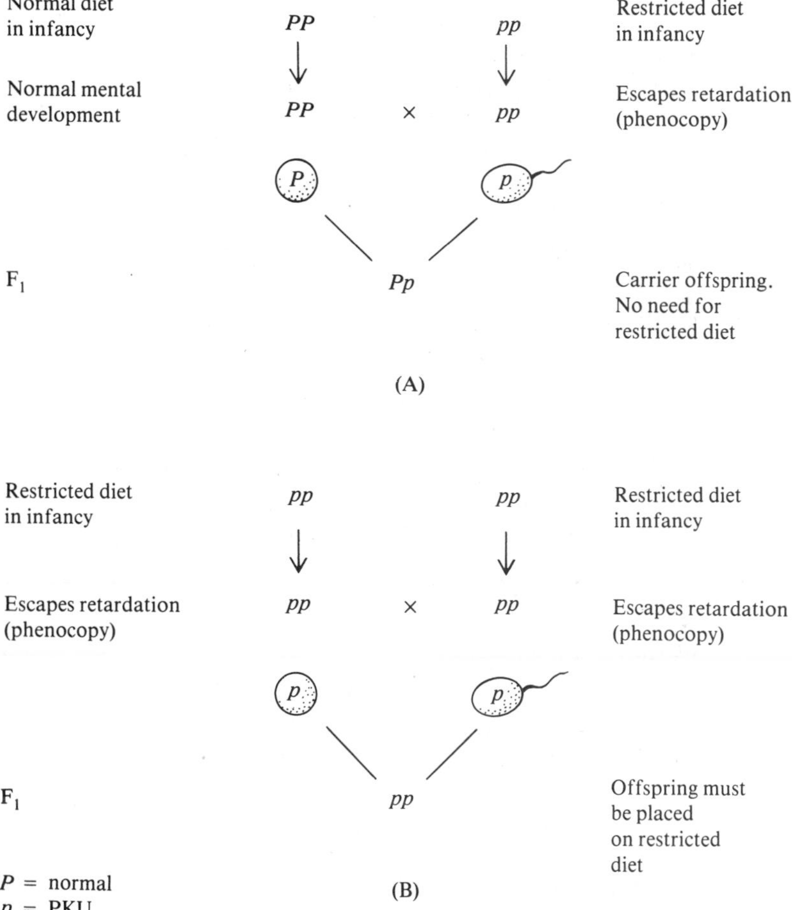

Figure 1-13. Environment and gene expression. (A) Due to restricted diet, a person may escape mental retardation even though the double-recessive genotype is present for PKU. Such persons, however, will transmit the recessive gene to their offspring, since the environment has not changed the gene. It has altered its expression. (B) Two phenocopies who have escaped the mental deterioration due to PKU can produce only children who are homozygous for the recessive gene. These children must, in turn, be placed on restricted diets if mental retardation is to be avoided.

altered so that it mimics the phenotype usually associated with a specific genotype. In essence, the phenocopy is an imitation of a phenotype typical of another particular genetic constitution. In the case of PKU, the normal-appearing phenocopy still carries the genes for the expression of the disorder. Two such phenocopies will produce offspring who also have the genotype *pp*. If these children are to be normal, they in turn must receive the proper diet during their early years [Fig. 1-13(B)]. And so the story will repeat itself. The genes of the phenocopies are not being changed; it is only the expression of the genes which is being altered.

There is a reverse side to the origin of phenocopies, and this can be illustrated by further aspects of phenylketonuria. A PKU woman of genotype *pp* who has escaped brain damage will bear children who are heterozygotes (*Pp*) if her husband carries the dominant allele in the homozygous condition (*PP*). Actually, most persons would *not* be carrying the recessive, since it is uncommon in the population. Therefore, chances are great that such a woman would encounter such a man and thus produce offspring who carry the gene for the dominant, normal trait. However, a PKU woman *must* be placed on a special diet which eliminates certain foods during her pregnancy. Otherwise, any fetus she carries will be born brain-damaged regardless of the genotype. Due to the high concentration of toxic substances in the uterine enrivonment, a child who is heterozygous and thus possesses the genetic constitution for normal brain development will suffer the effects of phenylketonuria, just as would the homozygous child, genotype *pp*. We see here an example of a phenocopy in the other direction: a heterozygote who would otherwise be normal suffers mental retardation due to the intrauterine environment. The need to recognize such a possibility in the case of a PKU mother is increasing in importance today as more PKU women escape mental retardation and bear children (Chapter 15).

Perhaps the most familiar example of a reverse phenocopy is that caused by the drug thalidomide, an ingredient which was present in certain sleeping pills. The tragic consequences following the taking of this drug by pregnant women became evident in Germany and other European countries with an increase in the frequency of birth of babies with abnormal limbs, usually greatly reduced and in some cases little more than vestiges which somewhat resemble the flippers of a seal. The drug was withdrawn once the association was made between the ingestion of thalidomide by pregnant women and the incidence of this abnormal condition, known as *phocomelia*. However, it was too late to change the phenotype of the unfortunate children, for the drug acts early during embryonic development and exerts an inhibitory effect on the formation of the limb buds, which otherwise have the potential to grow into normal arms and legs. The victims of phocomelia phenotypically resemble those individuals who have reduced appendages due to a rare hereditary defect, whose exact mode of inheritance is still unknown but which may in some cases be due to a dominant gene. However, the unfortunate thalidomide victims possess a completely normal genotype for limb development. No matter how deformed the appendages, such an individual will transmit only normal genes for the upper and lower limbs. This person is a phenocopy of a victim of a disorder which can result from a rare defective genotype.

The relationship of the genetic endowment and the environment is often so interwoven, as in the case of very complex characteristics such as intelligence and personality, that it becomes impossible and meaningless to debate which is the more important. Rarely, if ever, is either heredity or environment the sole determining factor in the expression of a trait or a characteristic. Realizing the influence of both, the geneticist regards as non-issues those arguments which debate the relative importance of nature versus nurture. It *is* important, however, to understand that any characteristic develops as a result of the interaction of many genes and that the expression of that genic interaction is influenced by the environmental setting, both external and internal. An individual with the best complement of genes may fail to express his potential if the environment is unfavorable at a critical period, which could be well before birth in the case of many genes. A person with a hereditary defect may be spared suffering if the environment is altered at a crucial time so that the expression of the defective genotype is prevented. And conversely, even in the most favorable environment, the expression of certain defective genes cannot be prevented, as seen in the case of albinism and many others which will be encountered throughout later chapters. In the discussion of human genetics which follows, the influence of the environment must never be ignored. At the same time, it must always be remembered that the environment cannot directly change the genetic material in a given direction. Some of the reasons for this will become evident in the very next chapter which deals with the nature of the hereditary material and the manner in which it is transmitted.

REFERENCES

Jenkins, J. B., *Genetics.* Houghton Mifflin, Boston, 1975.

McKusick, V. A., *Mendelian Inheritance in Man,* Third edition. Johns Hopkins Press, Baltimore, 1971.

Mendel, G., "Experiments in Plant Hybridization." Translated in *Classic Papers in Genetics,* Peters, J. A. (Ed.). Prentice-Hall, Englewood Cliffs, N. J., 1959.

Merrell, D. J., *An Introduction to Genetics.* W. W. Norton, N. Y., 1975.

Moody, P. A., *Genetics of Man.* W. W. Norton, N. Y., 1967.

Srb, A. M., Owen, R. D., and Edgar, R. S., *General Genetics,* Second edition. W. H. Freeman, San Francisco, 1965.

Stern, C., *Principles of Human Genetics,* Third edition. W. H. Freeman, San Francisco, 1973.

Strickberger, M. W., *Genetics.* Macmillan, N. Y., 1976.

Sutton, H. E., *An Introduction to Human Genetics,* Second edition. Holt, Rinehart, and Winston, N. Y., 1975.

Whittinghill, M., *Human Genetics and Its Foundations.* Reinhold, N. Y., 1965.

REVIEW QUESTIONS

1. In the human, a particular dominant trait is associated with the formation of projections on certain bones of the body. However, only 6 persons out of 10 who carry the responsible gene express the trait. Name the genetic phenomenon involved.

2. A certain gene in the human results in anemia, kidney failure, spleen damage, and impaired mental functions. Name the genetic phenomenon illustrated by the expression of this gene.

3. In certain breeds of dogs, black fur (*B*) is dominant to the brown fur (*b*) trait. Following a testcross, 1 black female gives rise to both black and brown puppies. A second black female is testcrossed several times and produces 5 black puppies in the first litter, 5 in another, and 4 in a third. Give the most probable genotypes of female 1 and her offspring and of female 2 and all her offspring.

4. Polydactyly (extra digits) behaves in the human as a dominant (*E*) to the normal condition (*e*). A person with an extra toe marries an unrelated person with the normal number of digits. Their first child has an extra finger and the second child is normal. The normal offspring eventually marries a normal, unrelated person and produces a child with an almost complete extra toe on one foot. Explain these observations.

5. In chickens, a specific gene is associated with brittle, twisted feathers. Birds homozygous for the gene appear very abnormal and are said to be *frizzled.* The gene responsible for the frizzled condition (*F*) is incompletely dominant to its allele for normal feathers (*f*). *A.* Write genotypes of (1) birds which are mildly frizzled, and (2) a normally feathered bird whose parents were mildly frizzled. *B.* What would be the results of a cross between (1) a mildly frizzled bird and a normal one? (2) a mildly frizzled bird and a frizzled one?

6. The gene responsible for sickle cell hemoglobin (Hb^S) behaves as a codominant to its allele for normal hemoglobin (Hb^A). A blood test reveals that two prospective parents each have both types of hemoglobin in their red blood cells. *A.* What are the chances that they will have a child with sickle cell anemia? *B.* Suppose the couple has 3 children who are healthy. What is the chance that the next one will have sickle cell anemia? *C.* Suppose that an apparently healthy person learns that both of his parents are heterozygous for the sickle cell alleles. What is the probability that he too is heterozygous?

7. Assume that treatment of sickle cell anemia leads to prevention of the fatal consequences of the disorder so that individuals of genotype Hb^S Hb^S live a normal life span. If two such persons become parents, what would be the genotype of their children? Would the children require treatment?

8. Select the proper answer or answers: The environment: (1) cannot affect the expression of genes; (2) can cause a genetic effect in the next generation by its influence on the body of a parent; (3) can be so altered that a gene may not express itself at all when present; (4) can influence expressivity of a gene; (5) sets the potential for the development of an individual.

chapter 2

THE PHYSICAL BASIS OF INHERITANCE

The Distribution of the Genetic Material

Before 1900, several biologists proposed that the hereditary material must be in the chromosomes of the cell nucleus, and there were many good reasons for this suspicion. Cytologists, students of cell structure, knew that most cells contain a nucleus and cytoplasm. They also knew that the gametes, the sex cells which unite at fertilization, often differ greatly in size, as is true in the case of the sperm and the egg. The latter may be thousands of times larger than the sperm (Fig. 2-1). This pronounced difference is due to the larger amount of cytoplasm present in the egg. The cytoplasmic content of the mature sperm is quite small, confined mainly to the neck and tail regions. However, nuclei of the female and male gametes are comparable in size and chromosome number. Since inheritance in sexual species appeared to be equal from both parents, it was logical to assume that the hereditary material is in the nucleus, specifically in the chromosomes.

The chromosomes can usually be visualized only at the time the nucleus is undergoing division, even though they are present intact at all times. Cell division is one very critical aspect of growth in any many-celled organism. It is also needed to replace or repair old and worn out tissues. Without cell division, growth would come to a stop; replacement of damaged or defective cells would cease, and death of the organism would be likely.

When a cell divides, the distribution of its cytoplasmic contents may not be equal. In contrast to this, the chromosomes are very accurately distributed to the two new cells, a fact which also suggested to cytologists that the hereditary material is in the chromosomes. The changes which take place at the time of division are very similar from one cell type to another, both animal and plant.

Typical stages can be recognized, and these have been given specific names. The events taking place throughout the stages of division guarantee that the chromosome material will be equal in the two new cells which arise. If the hereditary material is indeed in the chromosomes, it is essential for them to be distributed accurately, since loss of hereditary material in one cell and an excess in another would tend to upset the balance of information which is coded in the hereditary material. If the genetic material were distributed in a haphazard fashion at the time of cell division, the result would be confusion in the genetic instructions needed to carry out basic cell processes and the critical activities required for the division and the repair of the cell. The process responsible for the orderly and accurate distribution of genetic material at cell division is called *mitosis.* It is essential for growth and the maintenance of orderly development. Strictly speaking, mitosis refers only to the division of the nucleus. In certain special cell types, the cytoplasm does not divide, with the result that cells arise containing more than one nucleus. However, the cell divisions which take place in the body of any plant or animal depend upon

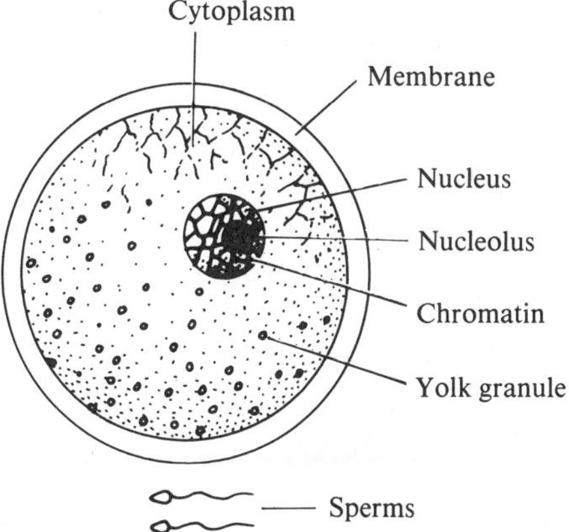

Figure 2-1. Diagram of human egg and sperm (approx. × 250). In a typical nondividing cell, observation with the light microscope usually reveals a distinct nucleus containing chromatin which stains with various dyes. The genetic material is contained in the chromatin. A nucleolus or several nucleoli can usually be seen in the nucleus as one or more small discrete bodies. The cytoplasm which surrounds the nucleus may contain an assortment of substances, many of which vary with the cell type. This diagram of an egg cell indicates the presence of yolk, which is not as abundant in the mammalian egg as it is in that of an amphibian, bird, or reptile. The cytoplasm may contain many other features which require the electron microscope for resolution. As seen here, the size of the egg contrasts with that of the sperm. The volume of the female gamete is approximately 50,000 times that of the male gamete. Despite this size difference, the amount of chromatin contributed to the offspring is the same. All the chromatin, and hence the genetic factors, is confined to the head piece of the sperm.

mitosis so that the nucleus or nuclei of any cells all carry the same genetic information.

Figure 2-2(A)-(D) summarizes the main events of the mitotic process. In the nondividing nucleus, the chromosomes are so greatly elongated that they cannot be recognized as discrete bodies. Staining the nondividing nucleus with certain dyes reveals little more than the *chromatin,* which appears as granular material or in the form of a network, and the nucleolus, a conspicuous body inside the nucleus whose importance to the cell will be noted in Chapter 9 (p. 227). At *prophase,* the first stage of mitosis [Fig. 2-2(A)], the chromatin appears to condense as the bodies known as the *chromosomes* can be recognized. Although the chromosomes are present all the time in the nondividing cell, they are usually so stretched out that they cannot be seen as independent entities. All that can be made out is the chromatin which composes them. At the end of prophase, the chromosomes are very evident, and it is clear that each one is actually composed of two halves. Each half-chromosome is called a *chromatid.* Although the chromosome is double throughout its length, there is actually one small region which appears constricted and which is not double. This is the *centromere,* a most important portion of the chromosome, since it is responsible for chromosome movements at the time of nuclear divisions. The term *kinetochore* is used by many geneticists and cytologists and is generally considered to be synonymous with the centromere. If the centromere (or kinetochore) is lost, a chromosome will be unable to move, and unequal

Figure 2-2. Stages of mitosis in whitefish. (A) Prophase. The chromosomes become more and more evident throughout this stage. Meanwhile the spindle fibers begin to appear. The photo and diagram show a dense region, the centrosome, from which fibers emerge in many animal cells. The centrosome divides, and two centrosomes become evident as prophase progresses. By the end of prophase, a spindle has formed. The fibers which compose it emanate from the centrosomes which come to mark the poles of the nucleus. By the end of prophase, the nuclear membrane and the nucleolus have disappeared. Note (right) that each chromosome at prophase is composed of two halves, each called a chromatid. (B) Metaphase. All the chromosomes, each clearly composed of two chromatids, become oriented at the middle of the spindle, the fibers of which run from one pole to the other. The centromere of each chromosome becomes attached to a spindle fiber. Other fibers form no attachment with the chromosomes. (C) Anaphase. The centromere of each chromosome divides, with the result that the two identical chromatids composing a chromosome move to opposite poles of the spindle. (D) Telophase. Nuclear membranes reform during this last mitotic stage. Nucleoli reappear, and the chromosomes become less and less distinct. The changes occur in reverse to those of prophase. Since the cytoplasm has divided by the end of telophase (division may start in anaphase), two cells will usually be present at the completion of this stage. The nucleus of each of the two cells carries the identical genetic material, and this is identical to that of the original cell which gave rise to them.

chromosome distribution will follow. This leads to unbalance of the genetic material and various types of abnormalities.

Throughout prophase, as the chromosomes become shorter and more obvious, other changes are taking place in the cell. In the cytoplasm, a system

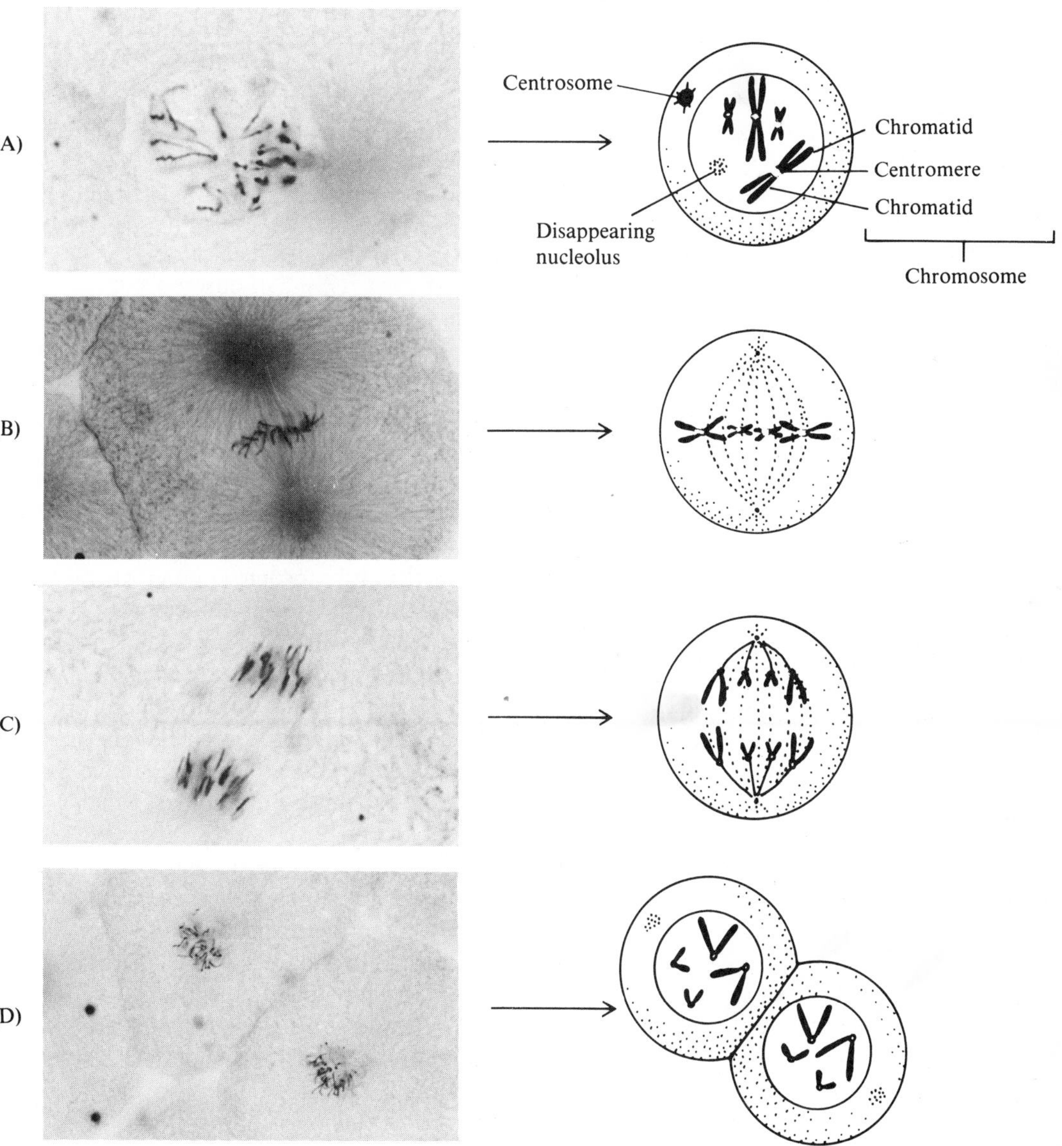

Figure 2-2

of fibers arises which compose the *spindle.* The spindle fibers are really constructed of microtubules which are protein in nature. By the end of prophase, the spindle is fully formed, the nuclear membrane has disappeared, and so has the nucleolus.

At *metaphase,* the next mitotic stage, the centromere of each chromosome arranges itself at the midregion of the spindle. Some of the spindle fibers pass from one pole to the other, but some attach to the centromeres of the chromosomes [Fig. 2-2(B)].

The next important event is the division of each centromere. This initiates *anaphase,* the stage of mitosis at which the chromosome halves, the chromatids, separate and move to opposite poles of the spindle. At anaphase, each chromosome is thus composed of only one chromatid. The number of chromosomes (or chromatids) at each pole is identical. Since the two identical halves of each chromatid separated at anaphase, the chromosome constitution at each pole is exactly the same [Fig. 2-2(C)].

At the following and last stage of mitosis, *telophase,* new nuclear membranes form around each group of chromosomes. Nucleoli reappear, and the spindle gradually fades as the fibers which compose it become dissembled. Typically, the cytoplasm has divided by the end of telophase [Fig. 2-2(D)]. In animals, this is achieved by the formation of a constriction which develops midway between the poles of the cell. The constriction becomes more and more pronounced as the two newly forming cells are pinched apart. From an examination of Fig. 2(A)–(D), it can be seen that the events of mitosis, especially those at metaphase and anaphase, insure that the chromosome content of the two new cells will be identical. The number and kinds of chromosomes in both new nuclei are the same. The figure also shows that a chromosome at some stages of the cell cycle may be double (as in prophase and metaphase) and at other times single, composed of only one chromatid (as at anaphase and telophase). As one of the newly formed cells goes on to divide again, it will in turn enter another prophase, and the identical events which have been described will take place. This means that the same chromosomes, which were single at telophase, the last mitotic stage at which we can see them, will each consist of two parts, two chromatids, when we see them again at the next prophase. Therefore each chromosome did not simply split at anaphase, for this would mean that it would become thinner with each new division. The reason for the constant double appearance at prophase is due to the fact that replication of each chromosome takes place sometime in the interval between telophase and the following prophase. *Replication* involves the construction of a chromatid identical to the single chromatid of each telophase chromosome. The genetic material of the chromosome is thus able to build exact copies of itself. Details of this process will be found in Chapter 9.

Today we know a great deal about the chemistry of the hereditary material.

The chromatin composing the chromosome is really an assortment of substances. It includes different kinds of proteins which are associated with two other kinds of molecules, DNA (deoxyribonucleic acid) and RNA (ribonucleic acid). For many years, the chemical nature of the gene was unknown, but protein was believed to be the major component, since proteins are the most complex of all cellular ingredients. Starting in the 1940s, investigations with bacteria and viruses cast doubt on this idea and indicated that the hereditary material is actually the DNA. Today we know that the gene in all cells, whether those of bacteria, trees, cats, dogs, or the human, is composed of DNA. The genetic material of many viruses (which are not cellular entities) is also DNA; however, in certain viral groups, RNA rather than DNA is the hereditary substance.

In all kinds of cells, with the exception of bacteria and blue-green algae, the DNA of the chromosome is complexed with proteins. When the chromosome replicates between cell divisions, the DNA of the chromosome directs the formation of new DNA in such a way that the DNA content of each chromatid of a chromosome is identical. There is also the formation of additional protein which is closely synchronized with the formation of the new DNA. In essence, mitosis is a process which guarantees that the DNA content of one cell is transferred equally to the two cells which are derived from it. All the activities which we observe at the stages of mitosis are concerned with this orderly distribution. Consequently, even after thousands of mitoses, the DNA content should be identical in all the cells which trace back to an original one (Fig. 2-3). In this way, even though a human begins life as just a single cell, a fertilized egg, the nuclei of all the cells of the body of an adult will carry the same DNA. This is the same as saying that they will have the same genetic information. Any one cell will not use all the information which it carries. A liver cell, for example, has different functions from that of a cell of the skin. The two will use different portions of that information at different times. How they do this is another important topic (Chapter 14), but it must be understood at this point that all cells must carry the same information. While a large part of this information may not be used by any one cell type, all cells undoubtedly require some common basic information for cell growth, repair, and the maintenance of regulated cellular activity. If there were no orderly process to distribute the genetic material, cells would arise lacking some basic critical information needed to conduct normal cell affairs. Moreover, at the time that liver cells, skin cells, etc. should normally arise in the development of the body, certain information needed to direct the changes required to produce these specific kinds of cells would be lacking in some of the cells which should give rise to specialized tissues. The outcome of any departure from an orderly chromosome distribution would be disruption in any many-celled organism.

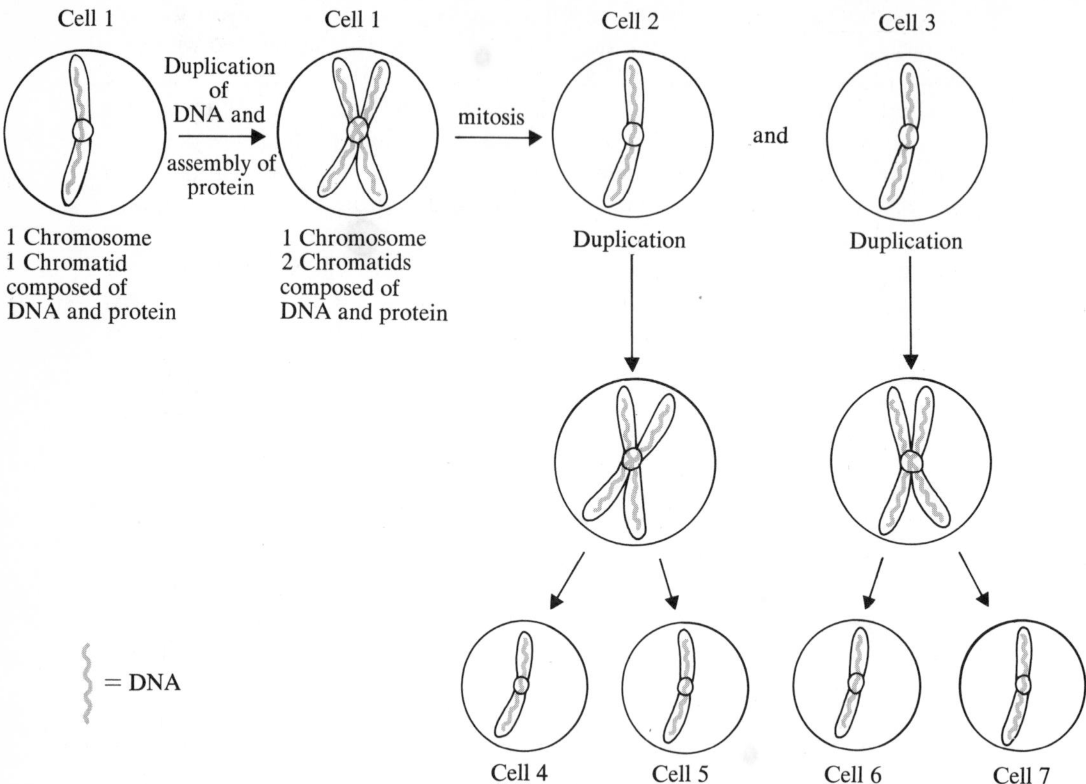

Figure 2-3. Cell division and DNA distribution. It is actually the DNA of the chromatin which carries the genetic information. Mitosis is responsible for the accurate distribution of the DNA and insures that all cells tracing back to an original cell will have the same amount of genetic material. This simplified illustration follows just one chromosome. It can be seen that the four cells (cells 4, 5, 6, and 7) which trace back to cell 1 carry the same DNA as the original.

The Human Chromosome Complement

Occasionally mistakes do take place spontaneously at mitosis. The incidence of these increases with the application of radiations and certain kinds of chemicals (Chapter 13). If one cell arises with extra chromosomes and another with one or more less, some kind of abnormality typically ensues. Depending on where and when they occur, such errors adversely affect the development of certain groups of cells or cause severe upsets in all parts of the individual. So it is essential that all the cells of an individual carry the same kinds of chromosomes in the same amount. When the many plant and animal groups are examined cytologically, their cells are found to contain a chromosome number which is typical of their particular species. All cats carry 38 chromosomes in their body cells; all dogs 78. At times in all creatures, departures from the typical number are found, and these are almost always associated with some kind of deviation from the normal phenotype. Humans are no exception. The normal chromosome number for *Homo sapiens* is 46. Figure 2-4(A)–(C) shows the metaphase chromosomes of a male and a female and briefly describes the procedure used to obtain such preparations. Human chromosomes are seen to vary greatly in size and shape. Several major groups can be easily recognized. From a photograph of a human cell, the chromosomes may be cut out and arranged according to descending size as shown in Fig. 2-4. When this is done, it is seen that the chromosomes of a body cell occur in pairs. For any one chromosome of a particular size and shape, there is usually another to match it. Chromosomes which correspond exactly in size and shape and which carry genes which influence the same characteristics are said to be *homologous.* We would expect the cells of a member of any sexual species to contain homologous chromosomes, since two parents contributed chromosomes to the formation of the individual. A human receives one set of chromosomes from the male parent and a corresponding or homologous set from the female.

When the human chromosome set is inspected (Fig. 2-4), it is seen that the lengths of the chromosome arms vary from one group of chromosomes to another. This is due to variation in the location of the centromere, the constricted region which holds the two chromatids together at prophase and metaphase. In some chromosomes, the centromere is approximately in the center, so that the two arms of the chromosome are of equal length. Such a chromosome is called *metacentric,* and the centromere is said to be *median* in position. If the centromere is slightly off center, it is said to be *submedian.* The chromosome is then classified as *submetacentric* and will have one arm somewhat longer than the other. If the centromere is located way off to one end, then one arm of the chromosome is distinctly larger in relation to the other, which may appear tiny. The chromosome is called *acrocentric.* The term *telocentric* is often used to describe a chromosome with only one apparent arm and a centromere which seems to be at the extreme end.

When the human chromosomes are paired and systematically arranged according to size and shape, 7 groups (A–G) can be recognized (Table 2-1). The chromosome complement of any cell is called its *karyotype.* A normal human karyotype includes 46 chromosomes or 23 pairs. In the female [Fig. 2-4(B)], every chromosome has a homologue which corresponds to it exactly in size and shape. The same is true for the chromosomes of the male except for one pair, the sex chromosomes [Fig. 2-4(A)]. These two chromosomes of the male are distinctly different in appearance: a large X and a small Y. In

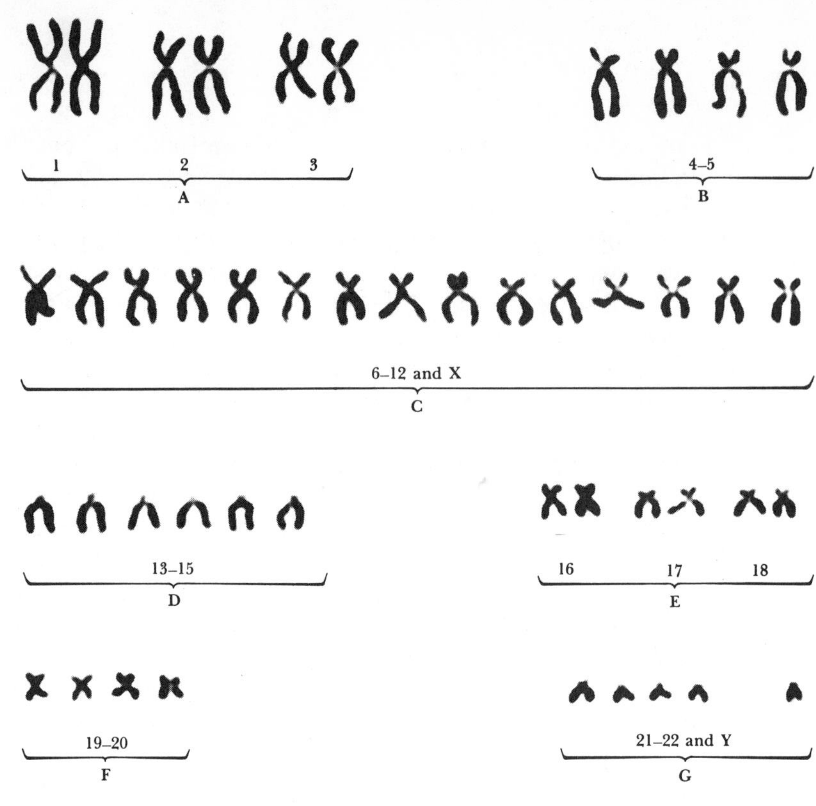

(A)

Figure 2-4. Studying the human chromosome complement. (A) Chromosomes of a male arranged in pairs. (B) Chromosomes of a female arranged in pairs. (C) Photo of human chromosomes in a white blood cell. Accurate counting and examination of human chromosomes has been made possible through perfection of a simple procedure in which cells may be stimulated to divide after exposure to a mitotic stimulator, such as an extract from the castor bean. A tiny sample of blood contains a large number of white blood cells which may be stimulated. Following 2–3 days of incubation in the presence of the stimulator, mitosis is arrested by exposure of the cultured cells to the drug colchicine, a chemical which prevents formation of the spindle. Since anaphase cannot occur, the chromosomes of the cell remain in metaphase for a longer than normal time period. In effect, the drug traps metaphases. After a few hours' exposure, the cells are quickly killed by an appropriate fixative and exposed to a low concentration of a salt solution. This treatment greatly swells the cells and causes the chromosomes to become untangled and so dispersed that each chromosome may become separated from any other. The cells are then affixed to slides and stained for examination with the microscope. A photo such as that shown here can then be taken. The chromosomes may be cut out from an enlarged photo. Their centromere positions and arm lengths are then easily determined. They can be paired and arranged systematically in order of decreasing size as shown in (A) and (B) above. Note that the centromeres of all the chromosomes are placed at the same level in the arrangement. This brings out clearly any differences in arm lengths from one chromosome to the next.

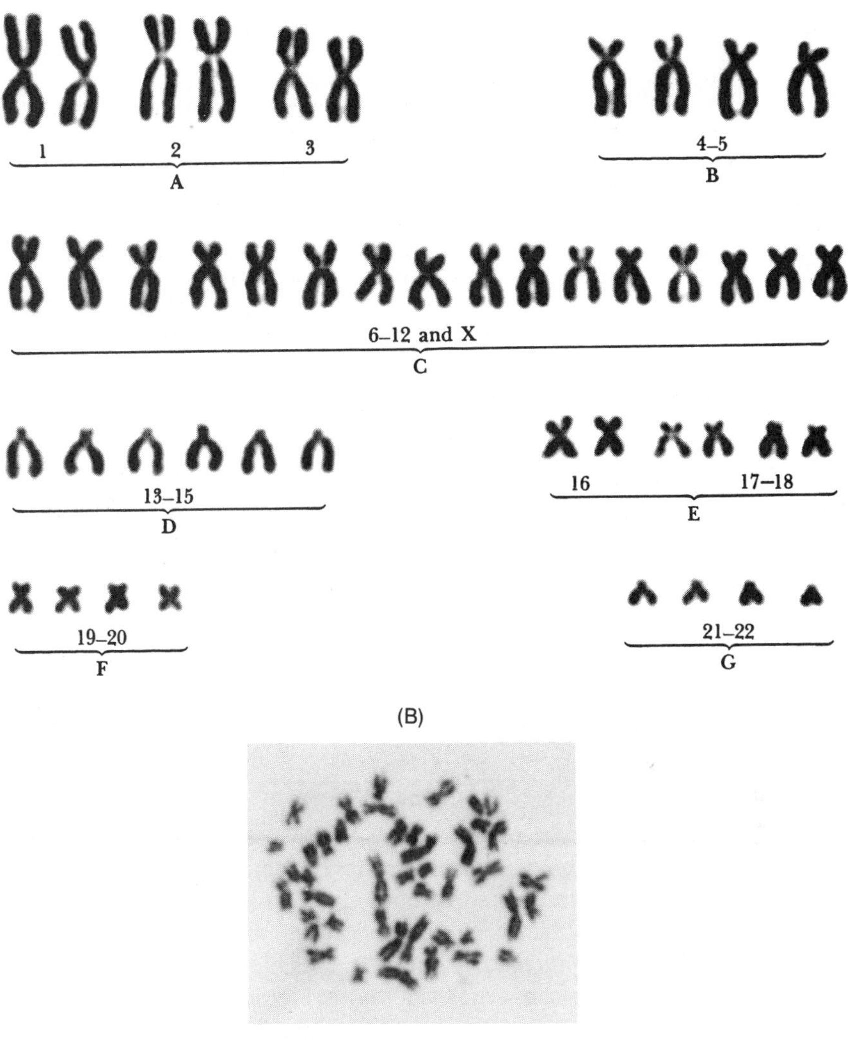

(B)

(C)

Figure 2-4. Concluded

the female, 2 X chromosomes are present, and the Y is absent. All the chromosomes in a cell other than the sex chromosomes are called the *autosomes.* The autosomes are numbered 1–22 on the basis of decreasing size. A human female thus has 22 pairs of autosomes plus 2 X chromosomes. In contrast, a human male carries in his body cells 22 pairs of autosomes plus 1 X chromosome and 1 Y. On the basis of size, the X chromosome is placed in group C and the Y in group G.

Table 2-1. Conspectus of human chromosomes

Group A, chromosomes 1-3:	The three largest chromosomes. Centromeres approximately median. Readily distinguished from one another on basis of length and position of centromere.
Group B, chromosomes 4-5:	Large chromosomes but shorter than those in Group A. Centromere position subterminal, giving distinctly unequal arms. Difficult to tell apart. Chromosome 4 slightly longer than chromosome 5.
Group C, chromosomes 6-12:	Chromosomes of medium size. Centromere position submedian, slightly off center. The X chromosome resembles the larger chromosomes, especially chromosome 6. The members of this group are difficult to identify by ordinary visual inspection.
Group D, chromosomes 13-15:	Chromosomes of medium size. Centromere position acrocentric, giving one very short arm. Members cannot be distinguished by ordinary visual inspection. A small appendage, a satellite, is present on the shorter arm of each member, but it may not be evident in a preparation due to problems in the preparation of the material.
Group E, chromosomes 16-18:	Chromosomes rather short in length. Number 16 possesses a centromere which is approximately median. Chromosomes 17 and 18 cannot be distinguished by ordinary visual inspection.
Group F, chromosomes 19-20:	Short chromosomes. Centromeres approximately median. Cannot be distinguished by ordinary visual inspection.
Group G, chromosomes 21-22:	Very short chromosomes. Centromere position acrocentric. A satellite is present on the shorter arm of each. The two members are not readily distinguishable by ordinary visual inspection. The Y chromosome is placed in this group and can often be distinguished from 21 and 22 due to its greater variability in appearance.

As the karyotypes in Fig. 2-4(A)–(B) suggest, it is often impossible to distinguish among the chromosomes of some of the groups by ordinary inspection. The chromosomes in groups C and G are especially difficult. Fortunately, recent techniques utilizing certain kinds of stains (fluorescent ones) and special lenses for the microscope make it possible to tell differences among all the chromosomes in a cell, even those in the difficult groups. These refined procedures bring out characteristic banding patterns along the lengths of the chromosomes [Fig. 2-5(A)–(B)]. The banding of the 22 kinds of autosomes and the X and the Y differs enough to permit distinctions; the identical banding between members of a pair of chromosomes in a cell makes it possible to identify the homologues.

Just like any other creature, the human will suffer some sort of abnormality if there is a departure from the karyotype which is typical for the male and female of the human species. An unusual karyotype with extra or fewer chromosomes in a cell would mean that the amount of DNA, and hence the balance of genetic information, is atypical. The consequences in the human of departures from the normal chromosome complement or karyotype present

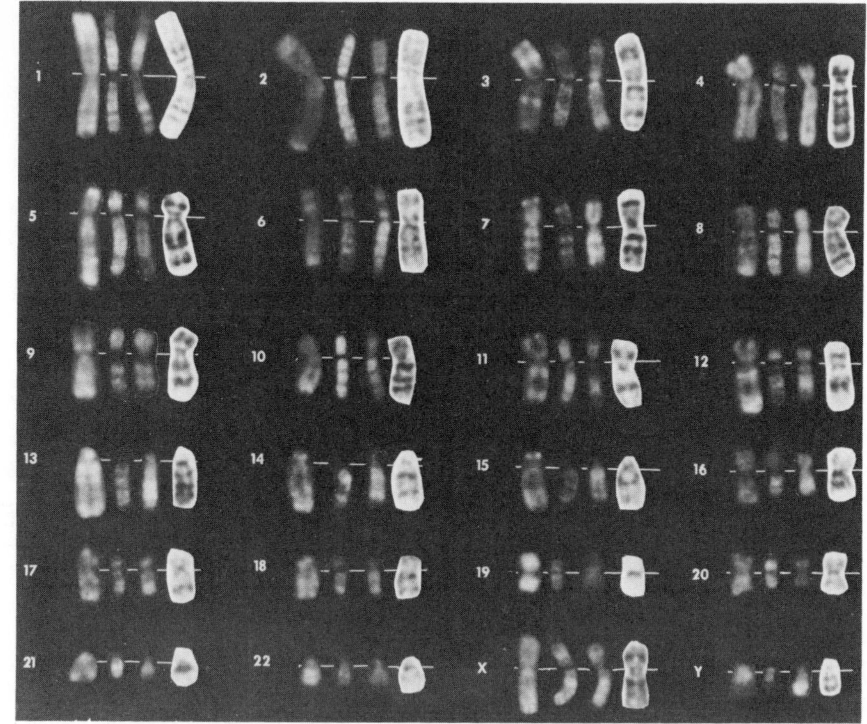

(A)

Figure 2-5. Identification of each human chromosome. (A) Each of the human chromosomes specifically identified after the application of procedures which bring out characteristic bands. A banding pattern is consistent for a given treatment. Shown in this photo are all the human chromosomes and the resulting patterns seen for each kind of chromosome when four different techniques are applied. In each specific chromosome grouping, the two on the left have been exposed to certain antibodies which have combined with a fluorescent dye. The third chromosome shows that particular chromosome after treatment with the fluorescent stain, quinacrine mustard. On the right, the chromosome is seen stained with the Giemsa stain following a chemical treatment which alters the DNA. The banding patterns produced by the last three techniques are quite similar, whereas the first is approximately the reverse of the others. The banding pattern does not reflect the DNA content of the chromosomes which is quite uniform from one end of any chromosome to the other. It is suspected that the bands are a reflection of the arrangement of the building units within the DNA, not the quantity of DNA present. (B) This is a diagrammatic summary of the banding patterns of the human chromosome complement. The enlargement of chromosome 1 (left) illustrates the meaning of the shading and designations which are shown. The dark bands indicate positive bands, those which stain brightly with certain fluorescent dyes such as quinacrine mustard. The white indicates those bands which contrast with the bright ones and are negative bands. The crosshatched regions are variable bands which react differently in chromosomes of different persons. The initials indicate a particular gene and refer to some product which the specific gene controls. For example, PPH designates a certain enzyme, phosphopyruvate hydrolase. The genes are listed under the specific chromosomes on which they are known to reside. If a certain gene has been associated with a particular band, it is placed opposite that band. The question mark indicates that it is still uncertain which of the two chromosomes is associated with that gene. Procedures used for identifying specific genes with specific chromosomes are discussed in Chapter 6.

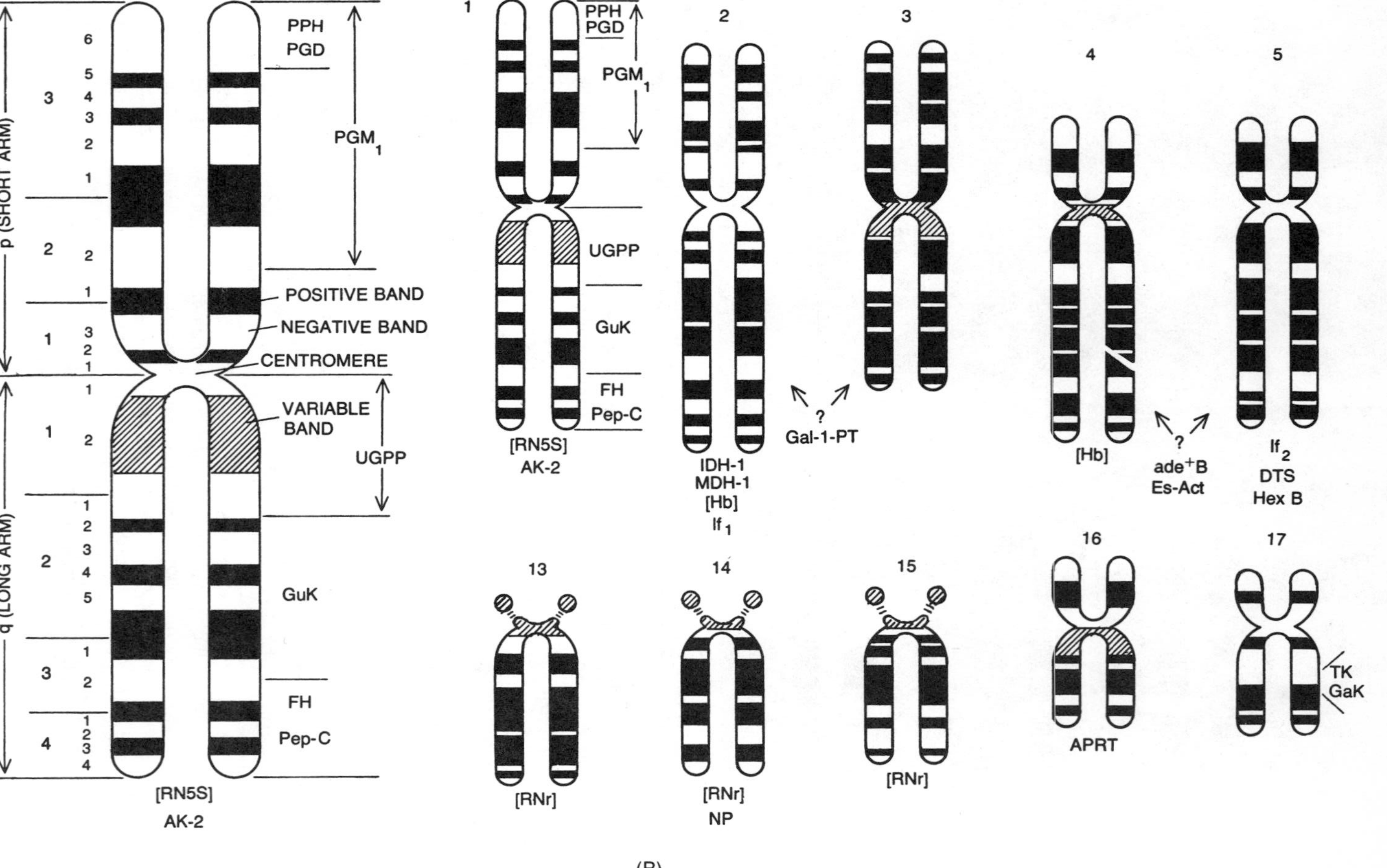

Figure 2-5. Continued

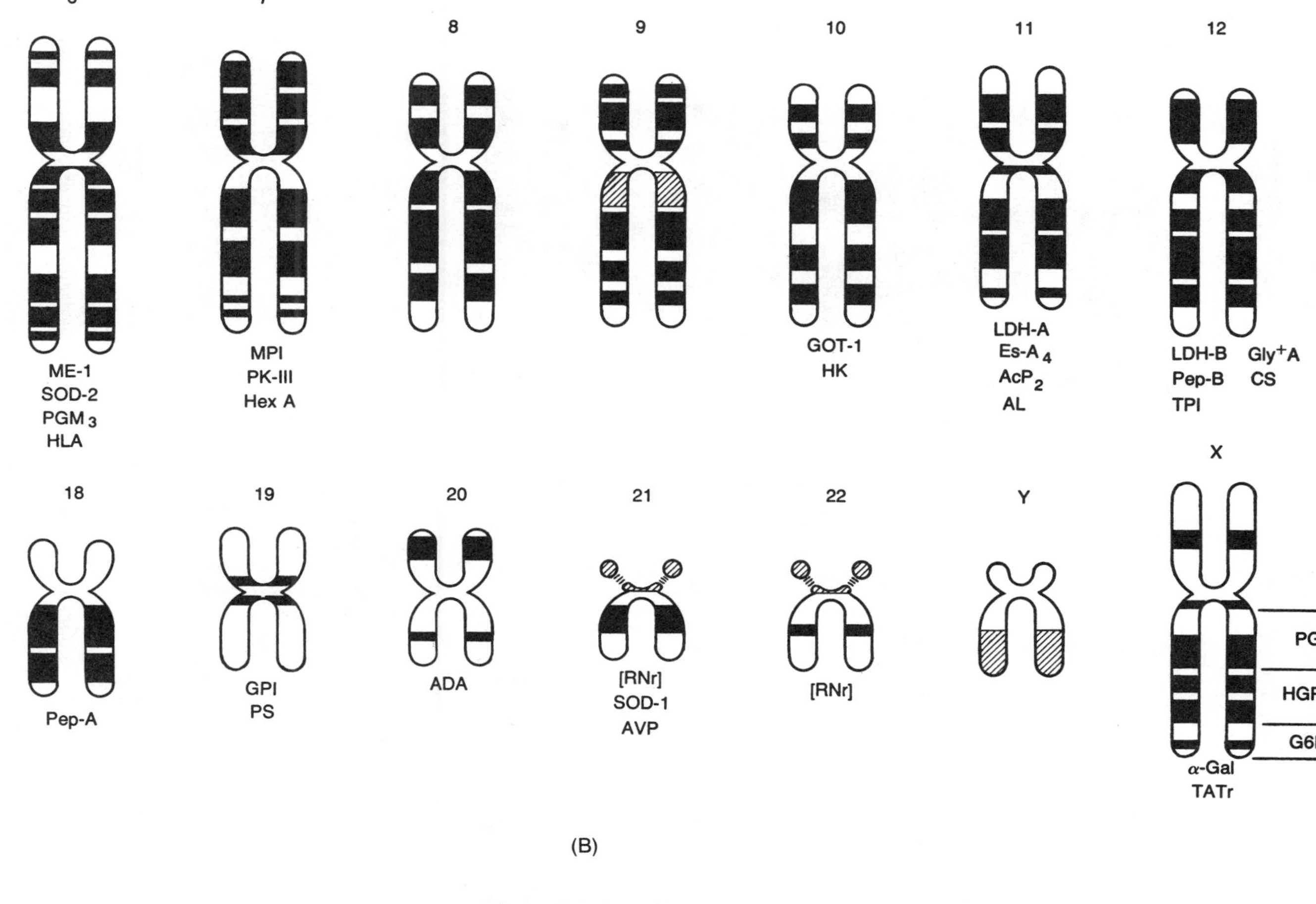

Figure 2-5 Concluded

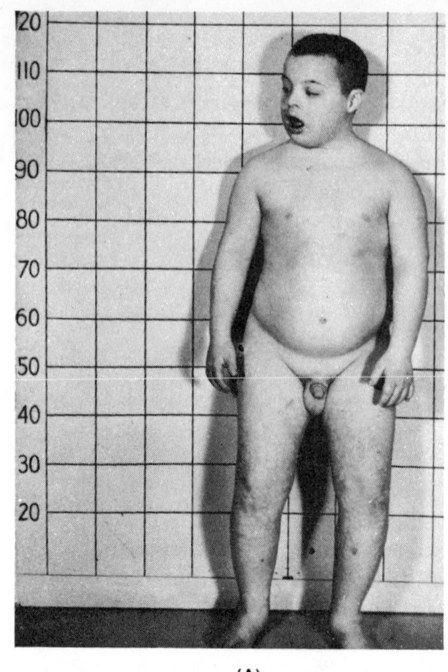

(A)

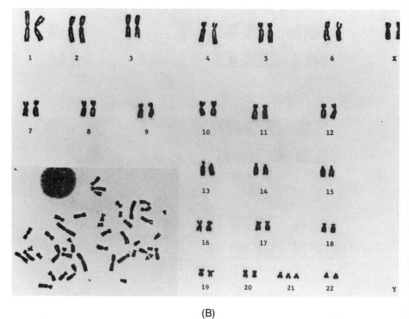

(B)

Figure 2-6. Down's syndrome. (A) Individual showing characteristic features of Down's syndrome. (B) Karyotype of an individual showing a chromosome number of 47. Down's syndrome will arise if the cells of an individual possess just one extra dose of chromosome 21, the tiniest of all the autosomes.

a real and significant problem not only to the geneticist and the medical person, but to society as a whole. To make this point clear, mention need be made at this point only of Down's syndrome. This unfortunate condition occurs in approximately 1 out of 500 births. Victims of this disorder have a characteristic appearance [Fig. 2-6(A)] which includes a folding of the skin of the upper eyelid, flattened face, small ears, and decrease in stature. An assortment of abnormalities affecting internal organs such as the heart, pituitary, and thyroid is typical. Severe mental retardation is also an accompaniment, and persons with Down's syndrome account for about 15% of the cases of mental deficiency in U.S. institutions. All the features of the syndrome stem from a general retardation in development. Most afflicted individuals do not survive beyond the age of puberty, although a few Down's females have actually lived to bear children.

The serious disorders which are part of Down's syndrome are related to an extra dosage of the chromosome designated #21, one of the smallest in the human set. (See Chapter 11 for more details on this chromosome.) Figure 2-6(B) shows a karyotype typical of a Down's patient. All of the unfortunate effects of this syndrome arise just from the addition of an extra dose of this small chromosome. This fact underscores the importance of the orderly distribution of the chromosomes at the time of nuclear division. In Chapters 4 and 11 we will encounter several disorders which are associated with abnormalities in the chromosome constitution. First, it is necessary to appreciate a few additional facts about chromosome behavior and distribution so that the basis of various human afflictions such as Down's syndrome may be thoroughly appreciated.

REFERENCES

Caspersson, T., Lomakha, G., and Zech, L., "The 24 Fluorescence Patterns of the Human Metaphase Chromosomes: Distinguishing Characters and Variability." *Hereditas,* 67:1, 1971.

Conference in Standardization in Human Cytogenetics, Paris, 1971. "Birth Defects." Original Articles Series, Vol. VIII, No. 7, October, 1972. The National Foundation-March of Dimes.

Du Praw, F. J., *DNA and Chromosomes.* Molecular and Cellular Biology Series. Holt, Rinehart and Winston, N. Y., 1970.

Hsu, T. C., "Longitudinal Differentiation of Chromosomes." *Annu. Rev. Genet.,* 7:153, 1973.

Mazia, D., "How Cells Divide." *Sci. Amer.,* Sept: 205, 1961.

Mazia, D., "Mitosis and the Physiology of Cell Division." In *The Cell,* Vol. 3:77, Brachet, J., and Mirsky A. E. (Eds.). Academic Press, N. Y., 1961.

Patil, S. R., Merrick, S., and Lubs, H. A., "Identification of Each Chromosome With a Modified Giemsa Stain." *Science,* 173:821, 1971.

Pearson, P., "The Use of New Staining Techniques for Human Chromosome Identification." *J. Med. Genet.,* 9:264, 1972.

Schreck, R. R., Warburton, D., Miller, O. J., Beiser, S. M., and Erlanger, B. F., "Chromosome Structure as Revealed by a Combined Chemical and Immunological Procedure." *Proc. Nat. Acad. Sci.,* 70:804, 1973.

Swanson, C. P., Merz, T., and Young, W. J., *Cytogenetics* (Chapters 1–2). Prentice-Hall, Englewood Cliffs, N. J., 1967.

Thomas, C. A., Jr., "The Genetic Organization of Chromosomes." *Annu. Rev. Genet.,* 5:237, 1971.

REVIEW QUESTIONS

1. Name the part of the chromosome required for chromosome movement.

2. How many chromatids compose each chromosome at mitotic prophase? at telophase?

3. Name the stage of mitosis at which (1) chromosomes are arranged at the midregion of the spindle; (2) the spindle fibers are forming; (3) nuclear membranes are reconstituted; (4) the nucleolus disappears; (5) chromatids separate and move to opposite poles of the spindle.

4. *A.* How many chromatids would be present in a human cell nucleus at (1) prophase? (2) telophase? *B.* How many chromosomes at (1) prophase? (2) telophase? *C.* How many autosomes are present in a human cell nucleus?

5. What is the chromosome number typical of the karyotype of a person with Down's syndrome?

6. Assume that the two chromatids comprising one of the human chromosomes fail to separate at mitosis and move to the same pole. How many chromosomes would be present at the next prophase in the two resulting cells, assuming each survives?

7. When does chromosome replication occur in a cell?

chapter 3

CHROMOSOME BEHAVIOR AND SEXUAL REPRODUCTION

The Need for a Reduction Division

Since the transmission of the hereditary material from one cell generation to the next is very exact in mitosis, those cells which trace back to an original one would not vary in their genetic content. The last chapter stressed the point that this is essential for normal development in a many-celled organism. However, in addition to the transfer of DNA from one cell to the next in one individual, there must be some sort of orderly transmission of DNA from one generation of individuals to their offspring. In higher animals, information must be transmitted by way of the sperm and egg so that the orderly development of an offspring will take place. Without the proper amount of necessary information to guide the formation of a human, a fertilized egg (a zygote) will either die, result in an abnormal embryo which will abort, or go on to produce an offspring suffering from some defect. An example of the last is the birth of a child with Down's syndrome as a consequence of an imbalance of genetic information which disrupts normal development. There must therefore be some process which guarantees that the gametes carry the proper amount of genetic information.

Before the formation of sex cells, there must be some kind of nuclear division other than the mitotic one. A moment's thought indicates why this must be so, for the sperm and egg cannot carry the same number of chromosomes as does a cell of the body. If human sperms or eggs were to arise by mitosis from another body cell, they too would carry the 23 pairs of chromosomes found in the nuclei of the body. A zygote arising at fertilization from the union of such sperms and eggs would contain 46 pairs of chromosomes, twice the number associated with the human. If gametes continued to arise by mitosis, the chromosome number would continue to double with each new generation. To prevent this, there must be another kind of nuclear division which takes place prior to gamete formation in all sexual forms of life. This process is necessary to maintain constancy of chromosome number from parents to their offspring. To do this, it must result in a reduction of the chromosome number before gamete formation so that a sex cell will carry exactly one-half the number of chromosomes found in the cells of the body. The nuclear division which achieves this reduction is called *meiosis.* Unlike mitosis, meiosis produces cells whose chromosome content is quite different from that of the cell which gave rise to them. A human cell with 46 chromosomes is said to carry the *diploid,* or 2*n,* number of chromosomes, meaning that the chromosomes are found in corresponding pairs. After a meiotic division, the cells derived from a diploid cell will contain 23 chromosomes. Only 1 member of each chromosome pair will be present. This half-number of chromosomes, typical of gametes, is known as the *haploid,* or *n,* number in contrast to the 2*n* number of the body cells. Each haploid cell will carry only 1 of every kind of autosome and either an X or a Y chromosome. A human egg, therefore, contains the *n* number of 23, which includes 22 autosomes plus 1 X. Any sperm will also have the haploid number of 23, again including 22 autosomes plus 1 sex chromosome, but this may be either the X or the Y. To understand how reduction in chromosome number is accomplished, we will examine aspects of meiosis in the male and female which terminate in the formation of sperms and eggs.

Meiosis in the Male

When a cross section of the testis is examined (Fig. 3-1), it is found to contain tubules, the seminiferous tubules. These are greatly convoluted, so that in any 1 section it is possible to see more than 1 part of the same tubule. The walls of the tubules include several types of cells. Among them is the *spermatogonium.* This kind of cell contains the diploid chromosome number, 46. It is a cell which is capable of undergoing mitosis to produce others identical to it. In some of the spermatogonia, however, certain events take place which set the cell on the course of a meiotic division which will lead to sperm formation. The most important of the early changes which take place is the formation of chromosome pairs in early prophase of meiosis. This formation of chromosome pairs does not take place in prophase of any mitotic cell. Moreover, as Fig. 3-2 indicates, it is not a simple matter of pairing between any 2 chromosomes. Rather, the pairing is always between corresponding or homologous chromosomes. So each kind of chromosome pairs with its homologue. The pairing involves the entire length of the chromosomes. The X and the Y in the male also form a kind of pair, but the association is confined to the end of the sex chromosomes (see below). The pairing of homologous chromosomes

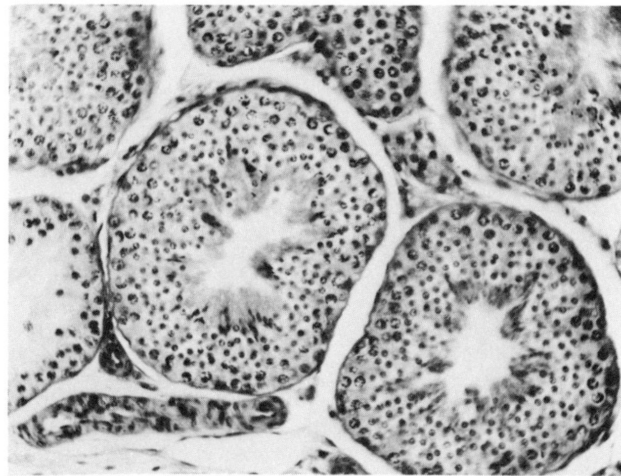

Figure 3-1. Cross section of human testis showing seminiferous tubules within which meiosis takes place.

is referred to as *synapsis.* When meiosis begins, the cell can no longer be called a spermatogonium. It is now termed a *primary spermatocyte* (Fig. 3-3). Figure 3-4(A)–(B) shows spermatogonia and primary spermatocytes in seminiferous tubules of a young man.

As pairing continues, the chromosomes become conspicuously thicker. Each chromosome is actually double, consisting of 2 chromatids as is true in prophase of mitosis. But at meiotic prophase, unlike the mitotic, each double chromosome is paired with its homologue, which is also double. Each of these

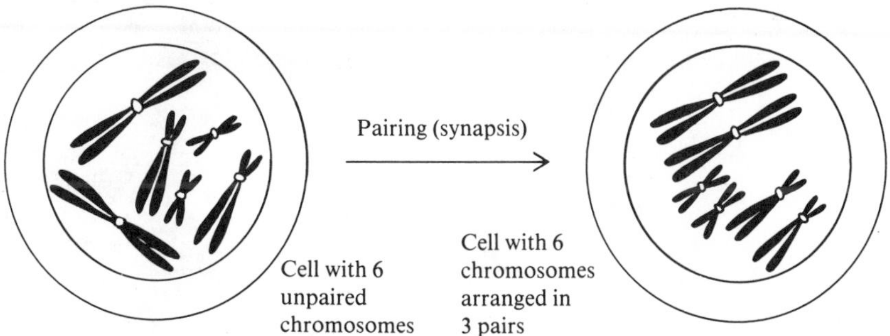

Figure 3-2. Pairing of homologous chromosomes. For the sake of clarity, only six chromosomes are represented. As seen in this diagram, there are three different kinds of chromosomes in the cell, and each chromosome has a homologue. When synapsis takes place in early prophase of meiosis, the homologous chromosomes associate to form pairs, each of which is composed of two homologues.

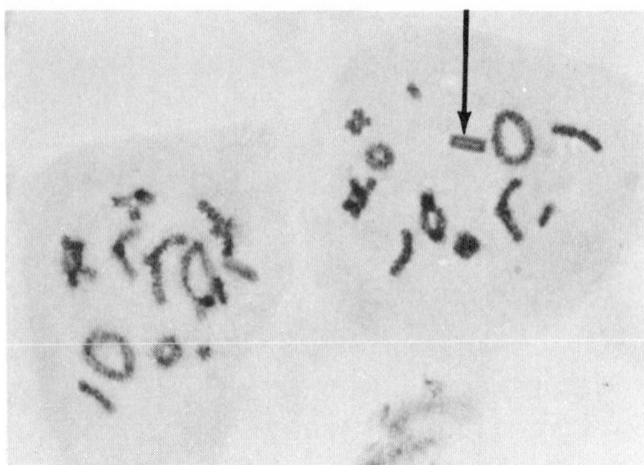

Figure 3-3. Primary spermatocytes in grasshopper. The chromosome number of the grasshopper is low in comparison to that of the human, a feature which makes counting of chromosomes quite simple in routine squashes of the testis. Early meiosis in the male grasshopper shows 11 associations of paired chromosomes. The X chromosome (arrow) remains unpaired since the male grasshopper, like the male of certain other lower organisms including some insect species, has 1 less chromosome than the female. It has 1 X but no Y.

pairs is known as a *bivalent,* an association of 2 homologous chromosomes, each of which consists of 2 chromatids. Thus, by the end of meiotic prophase in the human, 23 pairs of chromosomes can be counted. The X-Y association is often referred to as the *sex bivalent,* and it can be recognized due to the pairing which is confined to the ends (Fig. 3-5).

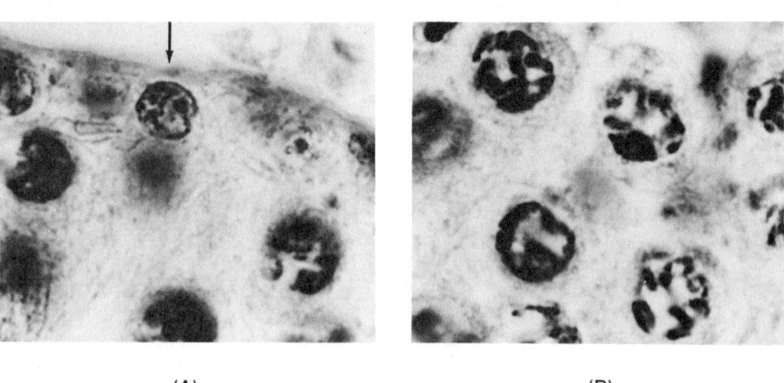

(A) (B)

Figure 3-4. Some contents of human seminiferous tubules. (A) The spermatogonium (arrow) is the cell type from which the male gametes are derived. Spermatogonia can be found just inside the outer membrane of the tubule. Some of them continue to undergo mitosis and so replenish those spermatogonia which become committed to meiosis. When prophase of meiosis commences, the cell is called a primary spermatocyte (cells lower in Figure). (B) Primary spermatocytes containing paired chromosomes.

By the end of prophase of meiosis, the nuclear membrane and the nucleolus have disappeared and a spindle has formed just as in mitosis. Metaphase of meiosis then takes place, and it can be recognized by the arrangement of the bivalents (23 in the human) at the equator of the spindle. Metaphase of mitosis and metaphase of meiosis have certain similarities. At both, chromosomes are at the midregion of the spindle. At mitosis, single chromosomes are distributed around the midregion; however, at meiosis, paired chromosomes (bivalents) are arranged at the equator [compare Figs. 3-6 and 2-2(B)].

One very important point must be grasped about the metaphase arrangement at meiosis. Remember that a bivalent is composed of a chromosome which the individual has received from the male parent and the homologous chromosome received from the female parent. Figure 3-6 shows that nothing dictates that the maternal chromosomes must all face the same pole while the paternal ones orient themselves in the direction of the other pole of the spindle. Instead, any arrangement is possible from one bivalent to the next. In some bivalents, the maternal chromosomes are directed to the upper pole, and so are some of the paternal ones. This fact is the basis for many of the genetic

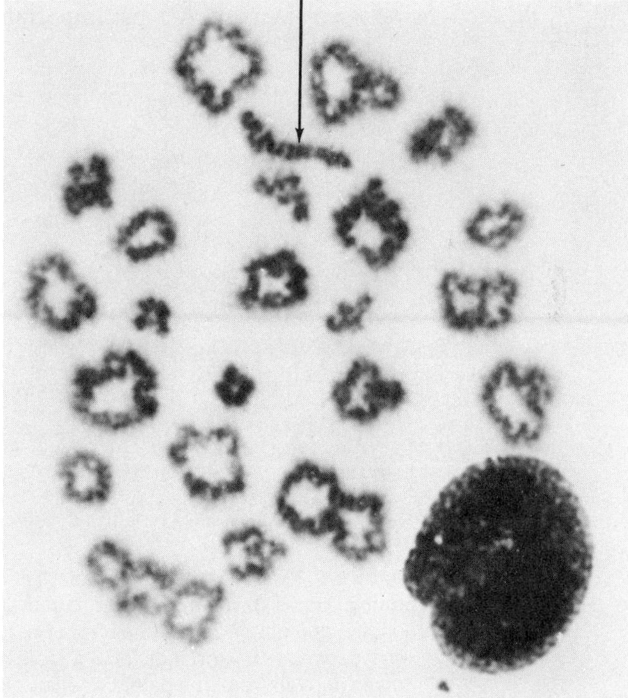

Figure 3-5. Primary spermatocyte of human. Twenty-three bivalents are present. The sex bivalent, composed of the X and Y chromosomes, is indicated.

results which occur in all sexual species. It must be fully understood in order to appreciate the reasons for some very common patterns of inheritance.

Following metaphase, anaphase takes place, and the chromosomes move to the poles of the spindle (Fig. 3-7). However, there is an important distinction between mitotic and meiotic anaphases. At this anaphase of meiosis, whole chromosomes, each composed of 2 chromatids, separate from each other and travel to opposite poles. Unlike the picture at mitosis, the centromere does not divide at this point. If we count the number of chromosomes at each pole at meiotic anaphase, or the ensuing telophase, we find the *n* number, 23 in the human. However, each chromosome is composed of 2 chromatids. At anaphase and telophase of *mitosis,* there would be 46 chromosomes at each pole, and each chromosome would consist of only 1 chromatid.

In the male, there are thus 2 cells which are formed at the telophase which follows a meiotic division of a primary spermatocyte. These cells are called *secondary spermatocytes,* and each carries the haploid number of chromosomes, 23. However, each chromosome is double. This double condition does not persist, for a second division of meiosis rapidly follows the first. Superficially, this second meiotic division resembles a mitotic one; however, there are several distinctions between the two. A most important one is that no new DNA is

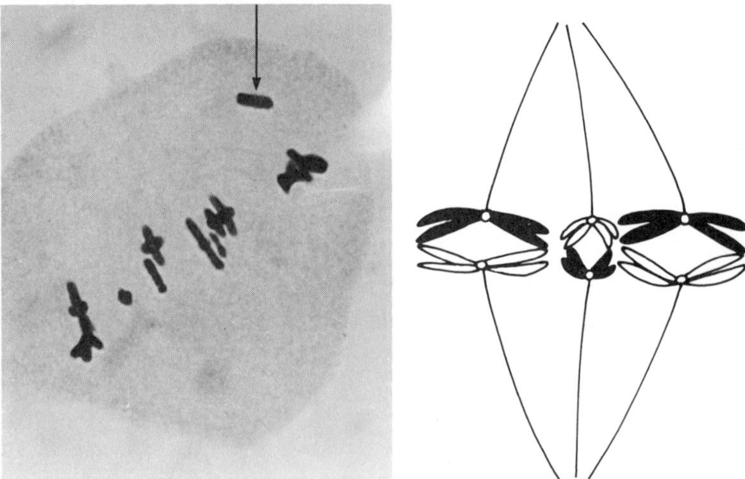

Figure 3-6. First meiotic metaphase. The diagram (right) shows that paired chromosomes (bivalents) are arranged at the middle of the spindle. The shading depicts chromosomes from the female parent. The unshaded ones were derived from the individual's male parent. Note that all the chromosomes derived from one parent need not face the same pole. This indicates just one of the possible arrangements. The photo (left) shows the bivalents at meiotic metaphase in a male grasshopper. The unpaired X (arrow) is off to one side and stains less intensely than the other chromosomes at this stage.

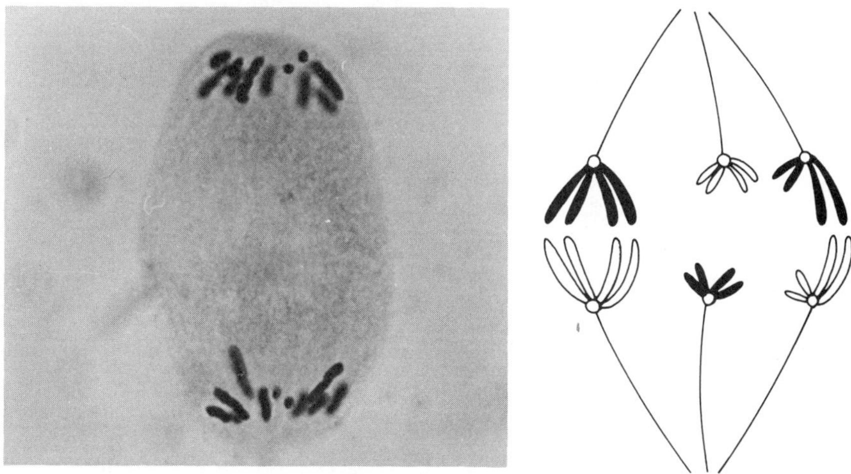

Figure 3-7. First anaphase of meiosis. The diagram (right) shows the separation of whole chromosomes, the movement of homologues to opposite poles of the spindle. Note that the arrangement of chromosomes shown in Fig. 3-6 will result in this separation: 2 maternally derived chromosomes will move to the upper pole with 1 chromosome which is paternally derived; the lower pole will receive 1 maternal and 2 paternals. The photo (left) shows anaphase in a meiotic cell in the testis of the grasshopper. The upper pole has received 11 chromosomes and the X. The lower one has just 11 chromosomes, since the X is unpaired in this insect.

made before the onset of the second prophase of meiosis. Since no replication of chromosomes takes place, this division should not be called mitotic, as is erroneously done. Additional differences also exist, but these need not concern us here. At the second meiotic division, as at a mitotic one, the chromosomes enter a prophase during which no pairing takes place. Obviously, homologous pairing is impossible, since the homologues separated during the first meiotic division. At second metaphase, the single chromosomes, each composed of 2 chromatids, arrange themselves at the middle of the spindle. And then, just as at mitotic anaphase, the centromere of each chromosome divides and the 2 chromatids move to opposite poles. A telophase follows, and from each of the 2 secondary spermatocytes, 2 cells have formed: 2 *spermatids* [Fig. 3-8(A)–(B)]. This means that from 1 original spermatogonium, 4 cells eventually arise. Each spermatid carries the *n* number of 23 chromosomes, each composed of but a single chromatid. Therefore, after the entire meiotic process, which includes the first and second divisions, the chromosome number is completely reduced by one-half. So starting with 1 spermatogonium containing 46 double chromosomes, meiosis produces 4 cells in the male (spermatids) each with 23 single chromosomes. Each spermatid will develop into a sperm. The entire process in the male, which produces a true reduction of chromosome number

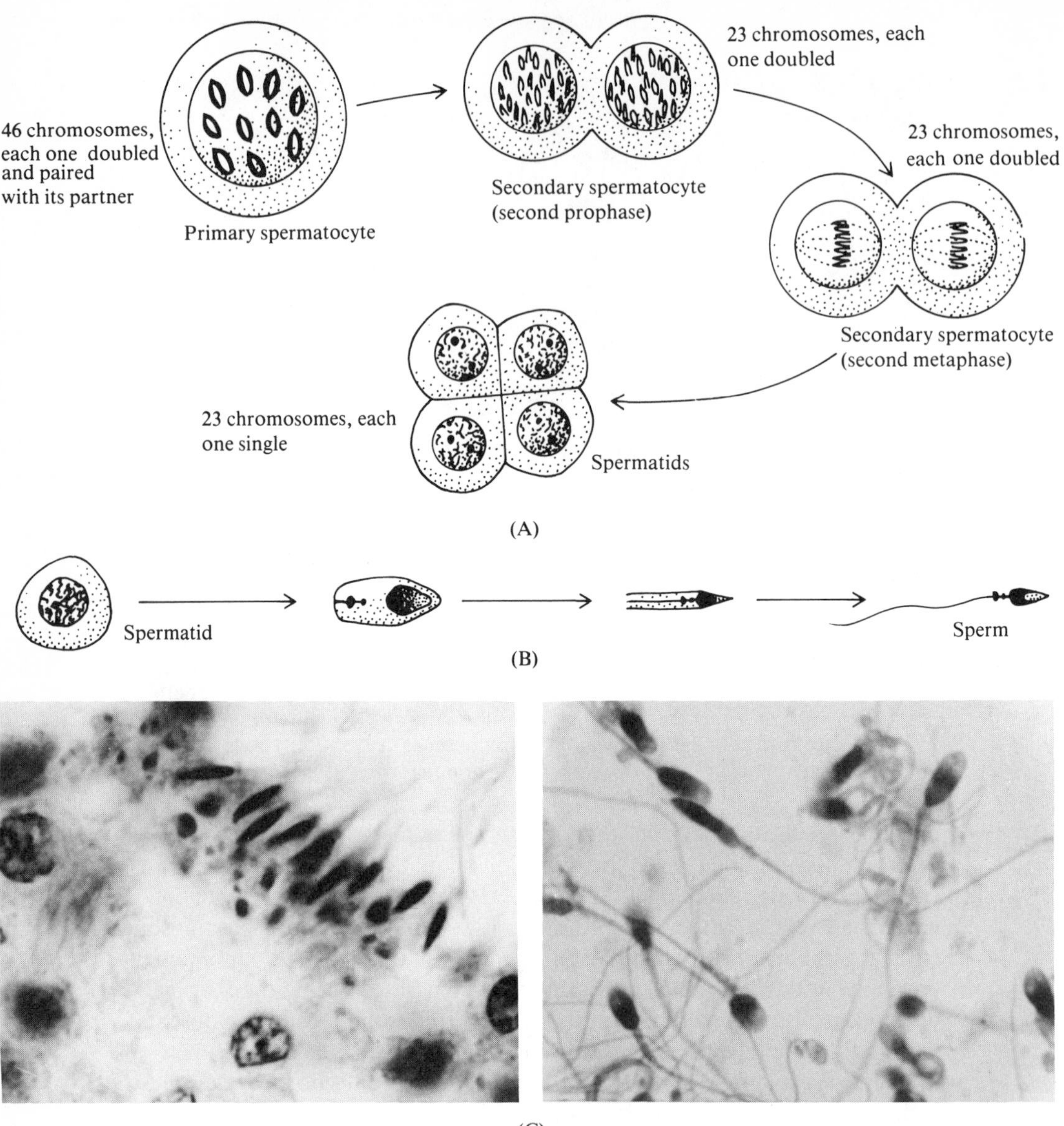

Figure 3-8. Spermatogenesis in the human. (A) In the male, from one original diploid cell which undergoes meiosis (the spermatogonium which becomes a primary spermatocyte) four functional cells are formed with the haploid number. These are the spermatids. (B) Each spermatid undergoes marked transformations during which most of its cytoplasm is cast off and the chromatin becomes packed into the head piece. No further change in chromosome number or in genetic constitution is involved in spermiogenesis, that part of spermatogenesis entailing the changes from spermatid to sperm. (C) Seminiferous tubules of a young man showing (left) transforming spermatids. Cast-off cytoplasm can be seen as darkly staining masses in the vicinity of the very deeply stained developing sperm heads. From the transformation of the spermatids, mature sperms arise (right), each with a conspicuous head, neck, and tail region.

and the formation of sex cells, is known as *spermatogenesis* [Fig. 3-8(A)–(B)]. The time required for spermatogenesis in the human, starting from a spermatogonium through the development of mature sperms, has been estimated to be about 64 days. This has been determined from studies involving injection of radioactive materials and the study of biopsies taken at regular intervals after these injections.

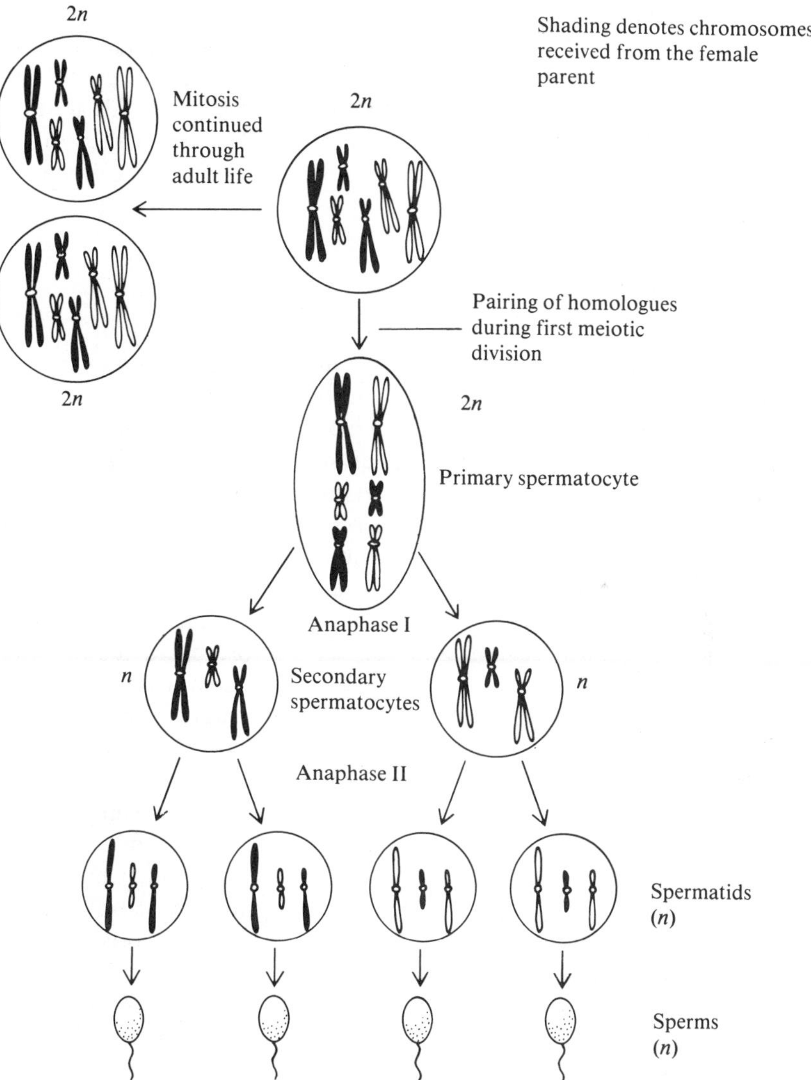

Figure 3-9. Spermatogenesis. This arrangement of the chromosome pairs at metaphase in the primary spermatocyte is just one of several, each having the same chance of occurrence.

Following formation of the spermatids, very striking changes take place in these cells as they are transformed into sperms. Most of the cytoplasmic content of a spermatid is actually eliminated, and that which remains contributes mainly to the development of the neck and the tail [Fig. 3-8(B)–(C)]. These parts function in the generation of energy and the movements typical of healthy sperm cells. Except for a small rim of cytoplasm, the head piece consists of nuclear material, primarily the chromosomes which become very condensed and compacted. The transformation of spermatids into sperms, the process of *spermiogenesis,* entails extremely marked reorganization of the cell, but it does not involve any further changes in the composition of the chromosomes. The 23 chromosomes of the spermatids are found in the sperms; the genetic constitution of a sperm cell is identical to that of the spermatid from which it arose. Figure 3-9 summarizes the main cellular events of spermatogenesis.

Meiosis in the Female

In the mammal, including the human, egg cells arise through the process of *oogenesis,* which is basically the same as spermatogenesis. Two meiotic divisions are also involved, and there is a complete reduction from the diploid to the haploid chromosome number. However, unlike the male, in whom meiosis takes place continuously from puberty to an indefinite age, the meiotic process in the female is much more restricted. In the ovary, there are types of cells which correspond to those which are found in the testis. Among these are the *oogonia* which can undergo mitotic divisions. However, these mitotic divisions, unlike those of the spermatogonia, occur only during the development of the female embryo. Well before birth, the mitotic divisions of the oogonia have ceased, and they become *primary oocytes,* cells in early prophase of meiosis. This means that a human female is born with all of the potential egg-forming cells that she will ever have in her ovaries. No more cell divisions take place after birth to produce more primary oocytes. This does not mean that there is any danger of a shortage of female gametes, since the number of oocytes derived from the oogonia is about 1 million at birth. There is a decrease to about 300,000 by the age of 7 years. Childbearing covers a span of about 30–40 years, and during this time only 350–450 cells from the original number will go on to mature. The others will never resume meiosis but will degenerate.

The chromosomes of the primary oocytes remain in the early meiotic prophase stage for years, each one of them contained in a cavity of the ovary called a *follicle* (Fig. 3-10). Meiotic division remains arrested until the age of sexual maturity, roughly 11–15 years of age. At this time, a primary oocyte in a follicle of the ovary may enlarge and start to complete meiosis. Usually only 1 such oocyte at a time each month achieves completion of the process. Maturation of other oocytes is usually suppressed once an oocyte has resumed meiosis. During the first meiotic division, which includes metaphase and anaphase, the chromosome behavior is the same as that described for the male. However, as Fig. 3-11 shows, the division of the primary oocyte produces 2 cells of unequal size, since the division of the cytoplasm is not midway between the 2 new nuclei. The larger of the two is the *secondary oocyte* and contains most of the cytoplasm. The tiny cell, called a *polar body,* will not survive to play any further significant role. The secondary oocyte is the cell in which the second division of meiosis may take place. However, this second division will not occur unless a sperm penetrates the secondary oocyte. The sperm nucleus will remain in the cytoplasm until the meiotic process of the secondary oocyte is completed. The oocyte produces 2 cells which are again unequal in size, another tiny polar body which will degenerate, and an egg with a large amount of cytoplasm. The egg nucleus will then fuse with that of the sperm which has already penetrated the cell.

Therefore, in contrast to the male, who continuously produces a tremendous number of gametes throughout his reproductive life, the female produces a small number. Of the 400 or so oocytes which ever resume meiosis, the only ones to complete it will be those which have been stimulated by penetration of the sperm. This means that a female will, in a strict sense, never produce a gamete or a true egg unless fertilization has taken place. So although it is common to speak of an egg or an ovum being produced each month during the reproductive years of a woman, this is actually not so. Rather, it is a secondary oocyte which is the so-called "unfertilized egg." And it is this oocyte which is released by the ovary for possible fertilization.

Despite these differences in gamete formation between the sexes, the overall result is the same. Only 1 functional gamete or egg is derived from an original primary oocyte, but it will carry the true haploid number of chromosomes, 23. So in both sexes, meiosis brings about the formation of sex cells with a chromo-

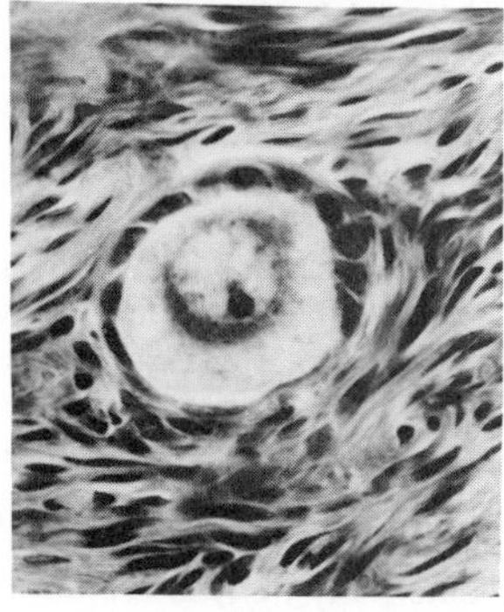

Figure 3-10. Human ovary with follicles containing primary oocytes.

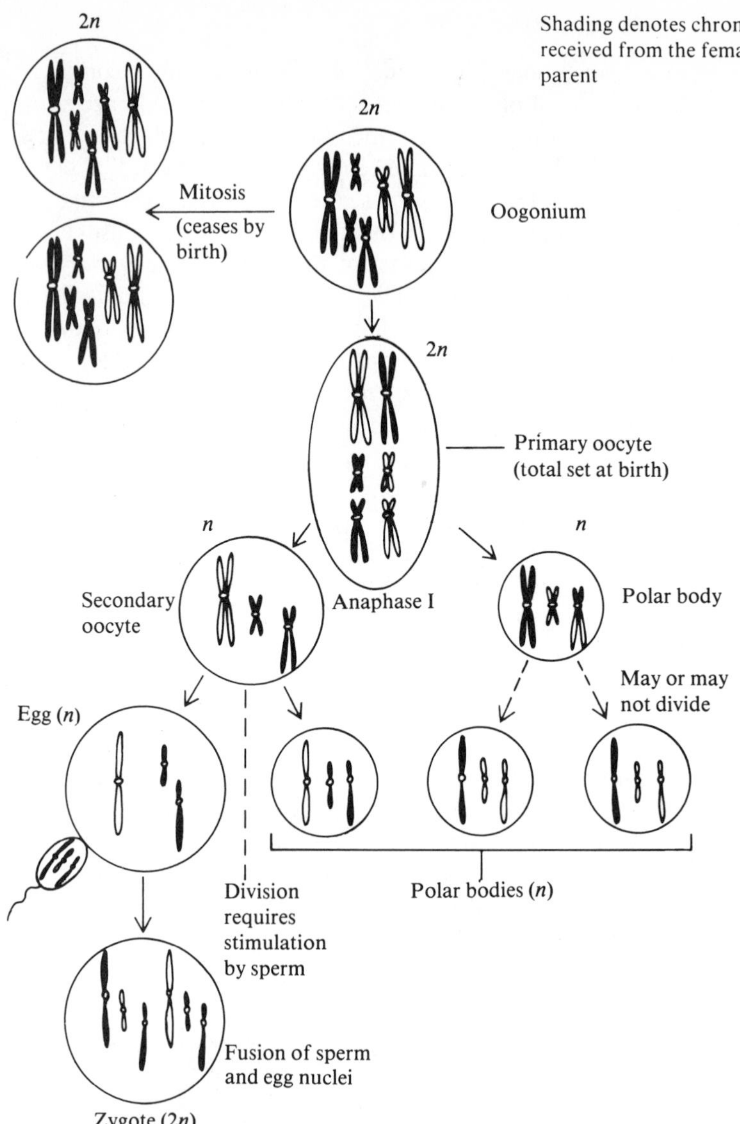

Figure 3-11. Oogenesis. This arrangement of the chromosome pairs at metaphase in the primary oocyte is just one of several possibilities. It happens to be a different one from that shown in Fig. 3-9. Any of the possible arrangements can occur in a primary spermatocyte or an oocyte. Note that following fertilization, the diploid ($2n$) number is restored. The zygote has a chromosome content different from that of either parent. The male gamete in this figure has contributed two maternally derived and one paternally derived chromosomes. The egg here also contributes two maternally derived chromosomes, but note that the combination of maternal and paternal chromosomes is not the same as that contributed by the sperm. The processes of spermatogenesis and oogenesis make possible an almost endless number of gene combinations among the offspring as a result of the chance arrangement of chromosomes at metaphase in the first meiotic division.

some number which is half that of the body cells. The diploid number of 46 is restored at the time of fusion of egg and sperm nuclei (Fig. 3-11), and this remains the number in the body cells. In the cells of the germ line, meiotic division then intervenes in both sexes at a certain time. Without this crucial division, the proper number and kinds of chromosomes could not be transmitted to the next generation.

Meiosis and the Generation of Variation

Meiosis produces another very significant effect which is extremely important to the evolution of all species including the human. It has been emphasized many times in our discussion that mitosis is responsible for maintaining the identical kind of hereditary material from 1 cell generation to the next. Mitosis results in no genetic variation. In contrast to this, meiosis does bring about variation. Not only do the meiotic products have half the number of chromosomes of the original cell, they also have an assortment of different chromosome combinations. Figure 3-9 shows that after first metaphase and anaphase in the male, the nuclei of the 2 secondary spermatocytes have different combinations of hereditary material. Therefore, there will be 2 classes of sperms derived from them. While every sperm will have 1 of each kind of human chromosome present, the nuclei carry different combinations of the chromosomes received from the male parent and those from the female parent of the individual. Figure 3-9 shows just 1 possible arrangement of the chromosomes at meiosis. At another meiotic division, some other arrangement of maternal and paternal chromosomes could face the same pole, producing gametes whose genetic constitution differs greatly from these. It must be kept in mind that the figures do not show all 23 pairs of chromosomes. With 23 pairs, there are 2^{23} or about 10^7 different combinations which can occur, an almost limitless number.

In the female, the story is the same (Fig. 3-11). A primary oocyte which goes on to complete meiosis can give rise to an egg whose combination of chromosomes is different from that of any other which will ever be produced. Therefore, meiosis generates new combinations of chromosomes and, hence, new combinations of the genes. These different combinations of genes in the sperms and eggs will result in offspring which are different, even though they have the same parents. Just by the independent arrangement of the bivalents at meiosis, chances are that no two offspring of any set of parents will ever be identical (except for identical twins; but as we will see in Chapter 6, even this is no exception to the statement). This variation produced by meiosis is essential to the evolution of the species. Certain combinations of genes will enable the individuals carrying them to cope better with the environment. These better-adapted individuals will tend to leave more offspring than those who lack the gene combinations which gave the advantage. The good combination received by the offspring can in turn undergo further combinations with other genes, and some of these may prove superior to the original one. In short, the sexual process, in which meiosis is the central feature, generates variation. Variation makes possible the establishment of individuals better and better adapted to a set of environmental factors. And since the environment itself changes, the greater the variation among members of a species, the greater the chance that some of them will be well adapted to a new set of conditions when it arises. Variation, therefore, is essential to evolutionary progress (Chapter 8).

Meiosis and Mendel's Laws

With an appreciation of meiosis as a mechanism which generates variation, we can now understand more fully the basis of Mendel's observations. Mendel realized that prior to the formation of gametes in a hybrid, contrasting factors (we now say alleles) must somehow separate so that only one goes to a gamete. This was expressed in Mendel's law of segregation which states that hereditary factors occur in pairs and that the members of a pair separate prior to gamete formation so that only one member of a pair enters a sex cell. Another glance at Figs. 3-9 and 3-11 shows that at anaphase I of meiosis the homologous chromosomes segregate. This means that a gene (let us say *A* on the maternal chromosome) separates from its allele (*a* on the paternal) at this time. Only one of the allelic members, *A* or *a,* will be found in any one sperm or egg. It is therefore the behavior of the chromosomes at the meiotic division, specifically at first metaphase and anaphase, which is responsible for the segregation of paired genetic factors (alleles).

In addition to studying monohybrid crosses, those in which only one pair of genes is being followed, Mendel performed dihybrid crosses to observe the inheritance of two pairs of alleles at the same time. From his observations with the pea plant, he formulated his second law, the *law of independent assortment,* which applies to the inheritance of two or more pairs of traits being studied simultaneously. While Mendel knew nothing of chromosomes or nuclear divisions, we now appreciate the fact that the behavior of the chromosomes at meiosis also accounts for the observations embodied in his second law: when two or more contrasting pairs of factors (genes) segregate, they segregate independently of one another. Figures 3-9 and 3-11 show that the chromosomes do exactly this. Different combinations of maternal and paternal chromosomes move to the poles of the spindle, and this results in variation in the genetic constitution of the gametes. A hypothetical mating in the human can serve to illustrate Mendel's law of independent assortment.

Many people have ears with lobes that hang free; the lobes of others are attached to the sides of the head so that no pronounced lobes are evident

(Fig. 1-1). These traits are of little significance, other than being curiosities. The gene responsible for free lobes appears to be inherited as a simple Mendelian dominant gene (*F*); the contrasting allele (*f*) for attached lobes behaves as a recessive. Another characteristic that is of no known consequence is the ability to taste a certain chemical, PTC (phenylthiocarbamide). Paper treated with this harmless substance is readily available from biological supply houses. Most persons are able to detect some kind of taste, usually bitter, when treated paper is placed on the tongue. Tasters, however, vary in the degree to which they taste the chemical. Some respond to very low levels; others detect a taste only at high levels. Such persons are called *tasters,* and they carry the dominant gene, *T*. Those people homozygous for the recessive allele, *t,* are non-tasters, unable to experience any kind of taste sensation at all from the PTC.

Let us assume that two parents in a certain study have the following genotypes: *FFtt* and *ffTT*. One parent has free ear lobes and is a non-taster; the other has attached lobes and can taste PTC. According to Mendel's law of segregation, we would expect one gene for the lobe characteristic and one for taste ability to enter a gamete so that the first generation of offspring are dihybrids, individuals with free ear lobes who are tasters [Fig. 3-12(A)]. Figure 3-12(B) shows that such a dihybrid will form four different classes of gametes. While it is unknown which chromosomes carry the genes for the lobe and taste characteristics, there is reason to suspect that they are found on different chromosomes, that is, chromosomes which are not homologous. This means that in a dihybrid individual, *FfTt,* the pair of alleles for the one characteristic (*F, f*) is associated with one pair of homologues, whereas the other allelic pair (*T, t*) is found on a different homologous pair.

It has been well established that any particular gene occupies a specific site on a particular chromosome. For example, the gene *F* or its allelic form *f* would have a specific location on a certain chromosome. The same is true for the gene *T* and its allele *t*. This is simply a consequence of the fact that any gene is a part of the DNA which is assembled in the chromosome. A particular portion of the DNA, and hence a particular gene, would be a fixed region of the chromosome. A segment of DNA or a gene is not located in one part of a chromosome at one time and then in a different location some other time. The genes do not shift about in such a way that their locations along the length of the chromosome are always changing. The characteristic location of any gene is definitely fixed, and it is known as the *locus* of the gene. Since an allelic pair represents alternative forms of the same gene, both members of the pair would have the same location or locus in reference to a specific chromosome. That is why we would expect the locus for *F* and *f* to be the same, and similarly for *T* and *t* or any other gene pair [Fig. 3-12(C)].

With regard to the earlobes, one specific chromosome may carry *either F* or *f,* never both, at the locus which influences the earlobe characteristic. The same is true for the two alleles, *T* and *t,* for taste ability. The locus associated

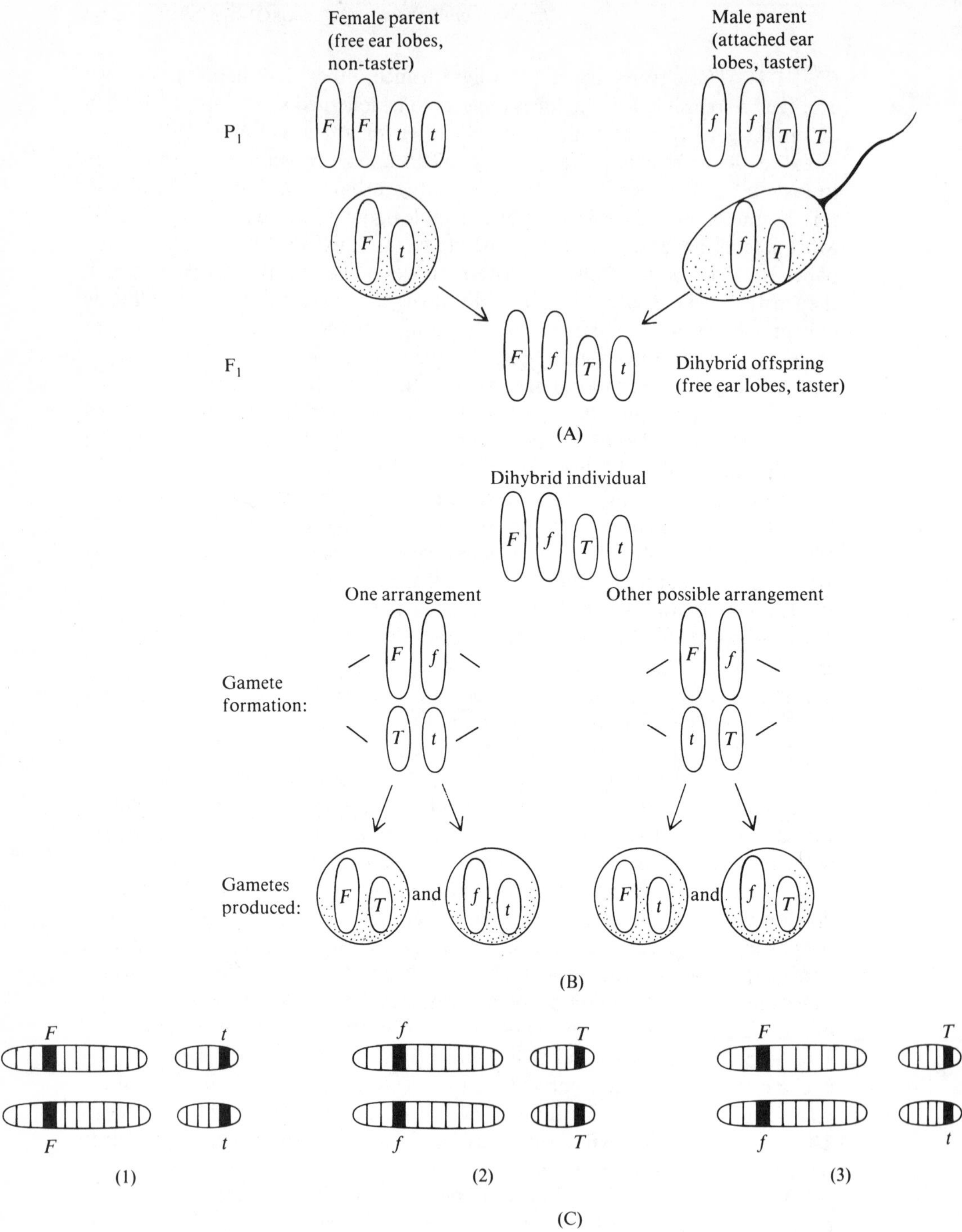
Female parent
(free ear lobes,
non-taster)
Male parent
(attached ear
lobes, taster)
P1
F
F
t
t
f
f
T
T
F
t
f
T
F1
F
f
T
t
Dihybrid offspring
(free ear lobes, taster)
(A)
Dihybrid individual
F
f
T
t
One arrangement
Other possible arrangement
Gamete
formation:
F
f
T
t
F
f
t
T
Gametes
produced:
F
T
and
f
t
F
t
and
f
T
(B)
F
t
F
t
f
T
f
T
F
T
f
t
(1)
(2)
(3)
(C)

Figure 3-12.

with the taste traits is located on another chromosome; so in the case of the earlobe and taste traits, two different loci are involved, each associated with a different chromosome. Since the chromosomes occur in homologous pairs in the body cells, the locus for the earlobe character is represented twice, one on each homologue, and similarly for the ability to taste the chemical. In the dihybrid, the dominant allele will be found at one locus and its recessive allele at the same locus on the homologue [Fig. 3-12(C)].

At meiosis, the alleles segregate in the dihybrid, and when they do, they segregate independently of the alleles associated with any other chromosomes [Fig. 3-12(B)]. This, we have seen, happens at spermatogenesis and oogenesis when the maternal chromosomes segregate independently of the paternal ones (Figs. 3-9 and 3-11). The independent assortment of the two pairs of alleles comes about because two pairs of chromosomes are involved. There is nothing which ties one chromosome to another nonhomologous one. Any one pair of homologous chromosomes arranges itself on the spindle at meiosis independently of any other pair. Since the arrangement will vary from one meiotic cell to the next, four different kinds of gametes can be formed from any one dihybrid.

In the cross of the 2 dihybrids, *FfTt* × *FfTt* (Fig. 3-13), the 4 different kinds of gametes come together in several ways. The Punnett square shows that 9 different genotypes are possible. Because of dominance, only 4 phenotypes occur, and these arise in a characteristic ratio, 9:3:3:1. This means that at each birth in this case there are 9 chances out of 16 that a child will be born with free earlobes and will also be a taster. There are 3 chances out of 16 for a child with free lobes who is a non-taster, 3 out of 16 for an offspring with attached lobes who is a taster, and only 1 chance out of 16 for one who has attached lobes and cannot taste PTC.

Figure 3-12. Two pairs of alleles. (A) With respect to two different pairs of genes associated with two different pairs of chromosomes, any one gamete will carry only one member of a homologous chromosome pair. As shown here, following segregation each gamete will contain one chromosome which carries a gene for an earlobe trait and one chromosome with a gene for a taste trait. The diploid number is restored at fertilization, and the offspring in a cross such as this would be a dihybrid with pairs of homologous chromosomes and two pairs of alleles for the characteristics being followed here. (B) Any dihybrid will produce four classes of gametes with respect to two pairs of alleles. This is simply a direct result of the two different ways in which the two pairs of chromosomes can orient themselves during meiosis. (C) This diagram indicates that there is a definite, fixed location on a chromosome for any gene and its alternative forms (alleles). This is the gene locus. There is thus a locus for the earlobe characteristic and a different locus for the taste characteristic. These happen to be on different chromosomes. An individual can be homozygous at both loci (1, 2). Either the dominant or the recessive allele can occur on any one chromosome. The body cells will contain two loci of each kind. In a dihybrid, a pair of alleles is represented at each locus in the body cells (3).

Mendel's laws of segregation and independent assortment are also demonstrated in a testcross between a dihybrid and a double recessive who has attached lobes and is a non-taster (Fig. 3-14). This is a dihybrid testcross, and the figure shows that four types of offspring are possible. The chance for any one is the same as that of any other one. Since we cannot control human

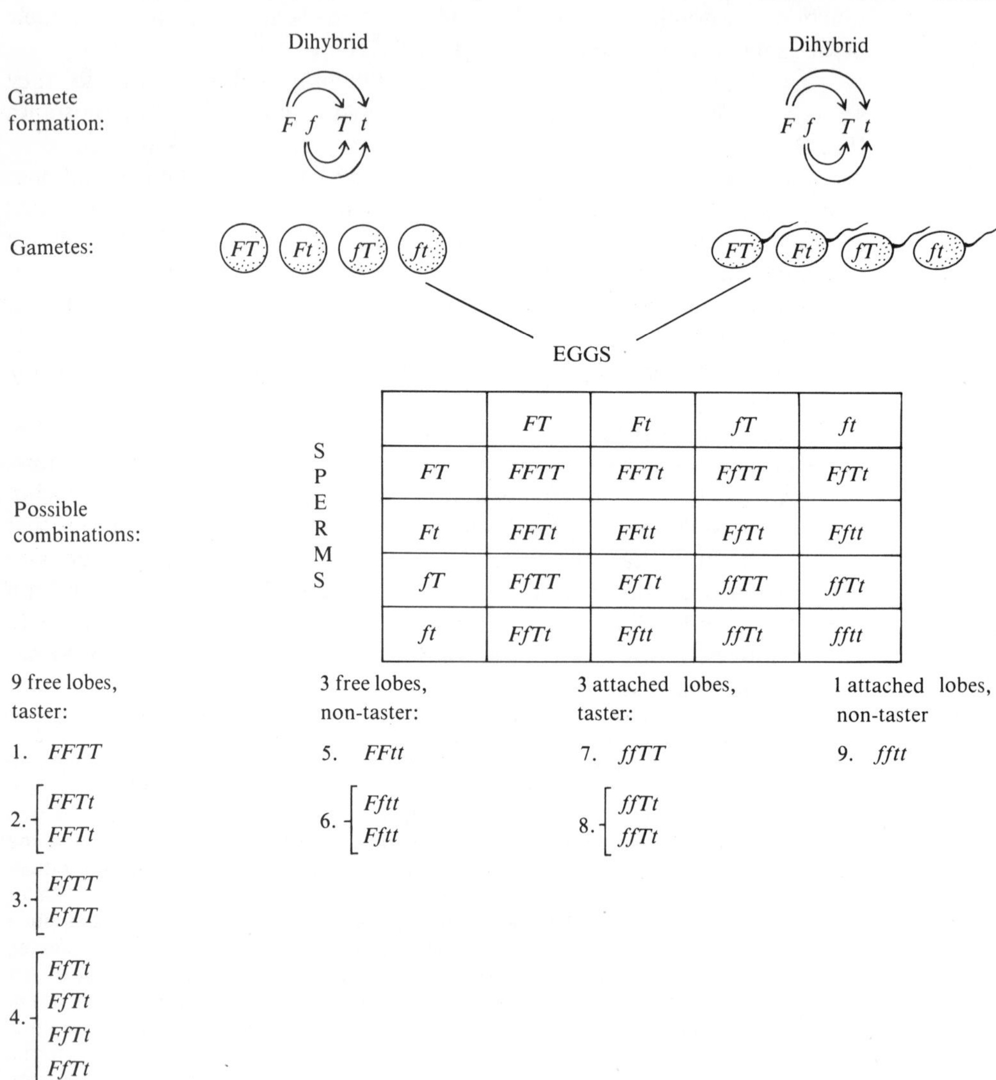

	FT	*Ft*	*fT*	*ft*
FT	*FFTT*	*FFTt*	*FfTT*	*FfTt*
Ft	*FFTt*	*FFtt*	*FfTt*	*Fftt*
fT	*FfTT*	*FfTt*	*ffTT*	*ffTt*
ft	*FfTt*	*Fftt*	*ffTt*	*fftt*

Figure 3-13. Dihybrid cross. Each dihybrid parent produces 4 different classes of gametes in equal proportions. The chromosome combinations of these fall into 9 different genotypes. When dominance is operating in both pairs of alleles, the number of phenotypes will be only 4 in a ratio of 9:3:3:1.

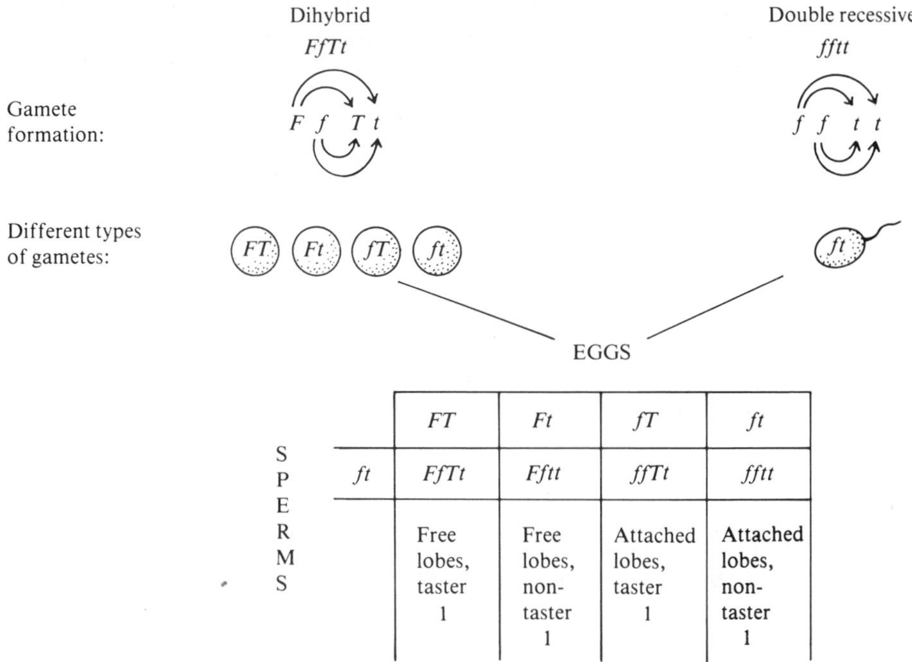

	FT	*Ft*	*fT*	*ft*
ft	*FfTt*	*Fftt*	*ffTt*	*fftt*
	Free lobes, taster 1	Free lobes, non-taster 1	Attached lobes, taster 1	Attached lobes, non-taster 1

Figure 3-14. Dihybrid testcross. When a dihybrid mates with any individual carrying the double-recessive condition at both loci, 4 different combinations are possible among the offspring, each having an equal chance of occurrence. This kind of cross clearly demonstrates segregation and independent assortment. The 1:1:1:1 ratio indicates the different kinds of gametes produced by the dihybrid, as well as the proportion in which they are formed. This comes about since the double-recessive parent can form only 1 class of gamete and contributes only recessive genes which cannot mask any recessives donated by the dihybrid parent.

matings as we can with experimental plants and animals, we often do not know the exact genotype of a person when two pairs of genes are being followed. However, the study of many human pedigrees makes it possible to analyze all kinds of matings, monohybrid and dihybrid crosses, and testcrosses. The data show that independent assortment occurs in the human as it does in other life forms. This is so even though we do not know the precise chromosome locations (loci) of most human genes.

We will learn later (Chapter 6) that independent assortment does not take place between two or more pairs of genes whose positions (loci) are on the same chromosome. However, this is no exception to Mendel's second law which applies only to genes whose locations are on separate nonhomologous chromosomes. We say that genes are *linked* when their loci are on the same chromosome. Evidence for linkage in humans and the importance of this information in diagnosis will be a major topic of Chapters 6 and 12. Before the topic of linkage is pursued, attention must be turned to the sex chromosomes, the genes found on them, and the effects of sex on the expression of certain genes.

REFERENCES

Clermont, Y., "The Cycle of the Seminiferous Epithelium in Man." *Amer. Journ. Anat.,* 112:35, 1963.

Copenhaver, W. M., Bunge, R. P., and Bunge, M. B., *Bailey's Textbook of Histology,* Sixteenth edition. Williams and Wilkins, Baltimore, 1971.

De Robertis, E. D. P., Saez, F. A., and De Robertis, E. M. F., Jr., *Cell Biology,* Sixth edition. W. B. Saunders, Philadelphia, 1975.

Rhoades, M. M., "Meiosis." In *The cell,* Vol. 3, *Mitosis* and *Meiosis,* 1, Brachet, J., and Mirsky, E. (Eds.). Academic Press, N. Y., 1961.

Segal, S. J., "The Physiology of Human Reproduction." *Sci. Amer.,* Sept: 52, 1974.

Volpe, E. P., *Man, Nature, and Society.* William C. Brown, Dubuque, Iowa, 1975.

REVIEW QUESTIONS

1. Name the exact stage of meiosis (1) during which chromosome pairing takes place, and (2) at which the two chromatids composing each chromosome move to opposite poles.

2. What term refers to (1) the actual process of pairing of homologous chromosomes, and (2) the unit composed of two intimately paired homologues?

3. How can one distinguish a cell in mitotic metaphase from one in first meiotic metaphase?

4. In the human, give the chromosome number of the following cells types: (1) A cell resulting from the mitotic division of a spermatogonium. (2) A primary oocyte. (3) A secondary oocyte. (4) The polar body formed along with a secondary oocyte. (5) A spermatid. (6) A cell formed following spermiogenesis.

5. In which of the cell types in Question 4 are bivalents present?

6. *A.* In the male, how many functional gametes arise from one germ cell which undergoes meiosis? *B.* Answer the same question for the female.

7. Normal skin pigmentation is a dominant trait (*A*); albinism is recessive (*a*). The free earlobe condition (*F*) is dominant to attached lobes (*f*). Give as much as possible of the genotypes of each of the following persons: (1) A person with normal pigmentation and free lobes whose father was an albino with free lobes. (2) A person with normal skin pigmentation and attached lobes whose mother was an albino

with free lobes. (3) An albino with attached lobes whose parents have free lobes and are normally pigmented. (4) A normally pigmented person with free lobes whose father was an albino with attached lobes.

8. Give the different kinds of gametes which can be formed by persons of the following genotypes: (1) *AaFf,* (2) *AAFf,* (3) *AAFF,* and (4) *AaFF.*

9. A person who is dihybrid for both the skin pigmentation and earlobe characteristics marries an albino with attached lobes. What kinds of children can be expected and in what proportion?

10. Suppose the two persons described above have a child who is an albino with attached lobes. What is the chance that the next child will have this same combination of traits?

chapter 4

SEX CHROMOSOMES AND THE GENES THEY CARRY

Sex Determination

In the human and other mammals, a distinct difference can be seen between the chromosome complement of a cell from a female and one taken from a male (Fig. 2-4). All the chromosomes of the female can be arranged in pairs, each composed of two chromosomes which match exactly in size and shape. However, in the male there is one pair in which the two chromosomes differ markedly. A distinctly large one, the X, and a very small one, the Y, can be recognized. The female carries in her body cells two X chromosomes and no Y. The X and the Y are the *sex chromosomes,* so-called since they compose the pair which differs between the sexes and as we shall see, cause the embryo to develop into a female (two X chromosomes) or a male (one X and one Y). All the other chromosomes are the *autosomes,* and these do not vary according to the sex of the individual.

The X chromosome is placed in the C group and the tiny Y is with the other smallest chromosomes, those of the G group (Table 2-1). In the female, the 2 X chromosomes pair during prophase of meiosis and behave just as any of the autosomes throughout the meiotic divisions. This means that all the egg cells will carry 1 X chromosome in addition to 1 set of 22 autosomes. The X and the Y, as was noted in Fig. 3-5, also form a pair at prophase of meiosis, although the association is only at the chromosome ends. Nevertheless, the X and Y behave much as the other chromosomes. The outcome is that the X separates from the Y at first anaphase, resulting in 2 classes of sperms: those carrying 1 X plus a set of 22 autosomes and those with 1 Y plus a set of 22 autosomes (Fig. 4-1). When the sex chromosomes alone are being discussed, a female may be represented as XX and a male as XY. From Fig. 4-1 it can be seen that this chromosome difference will produce female and male offspring in a ratio of 1:1. The X-Y mechanism, common to all mammals and present in many other species, guarantees that both sexes will be produced in about equal numbers.

Genes on the sex chromosomes determine whether the gonads of the embryo will develop into normal testes or ovaries. The early embryo possesses the forerunners of the sex organs, but at the earliest stages of embryology these are not differentiated. Up to the sixth week, there are pairs of rudimentary gonads in the embryo which are neither those of a female nor those of a male, but which are capable of developing into either ovaries or testes. About the sixth week, the presence of a Y chromosome triggers the embryonic gonads in the male pathway as differentiation is then initiated in the direction of the testes. One arm of the tiny Y is much shorter than the other. There is reason to believe that genes in the short one control the formation of a substance (or substances) which acts as an inducer to start differentiation in the direction of a testis. In the absence of the Y chromosome, the gonads continue to remain undifferentiated at this time, since there is no corresponding female-inducing substance produced by the XX condition. In the absence of male

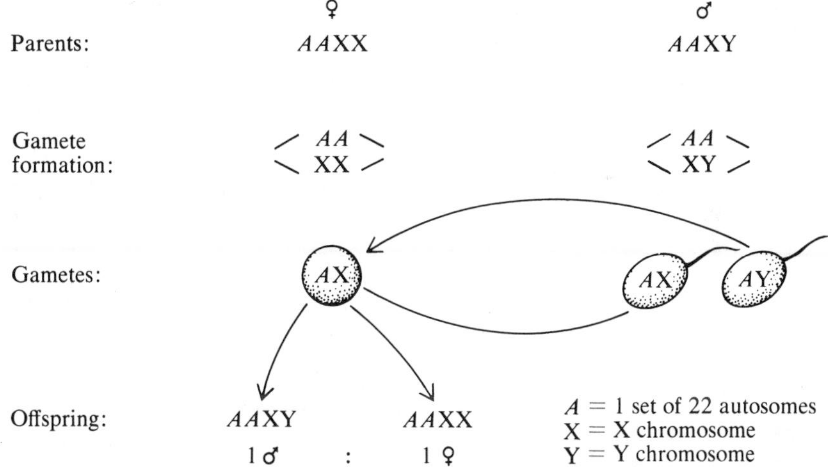

Figure 4-1. A human female possesses 2 sets of autosomes, each composed of 22 chromosomes, plus 2 X chromosomes. The male also carries 2 sets of autosomes, but his sex chromosomes consist of 1 X and 1 Y. At meiosis, each pair of homologous chromosomes in the sets of autosomes undergoes synapsis or pairing. So do the sex chromosomes. The female can produce only 1 kind of egg with respect to the kind of sex chromosome content. The male, however, produces 2 kinds of gametes, X-bearing and Y-bearing. This mechanism insures an approximately equal number of male and female offspring.

inducer, the gonads will proceed to develop into ovaries about the twelfth week. In the human, two X chromosomes are required for normal ovarian development.

Once testis development has been established, male hormones (mainly testosterone) are produced which in turn guide the differentiation of other embryonic structures into the typical parts of the male reproductive system. In addition, the testes of the fetus elaborate a substance which prevents the growth of any rudimentary structures in the female direction (Fig. 4-2). In the absence of this inhibitor, these embryonic parts can develop into uterus and fallopian tubes. It is thus the absence of male substances in the XX embryo which permits development of certain portions of the internal female reproductive system. For example, if the gonads are removed early in embryology so that there is a complete absence of male sex products at the critical stages,

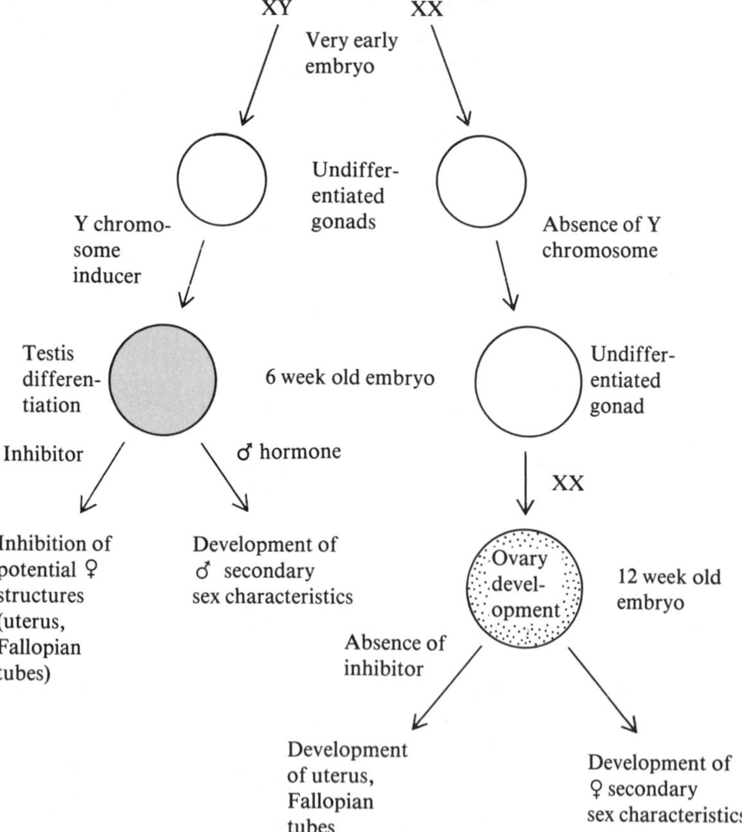

Figure 4-2. Summary of some interactions between the chromosome constitution and hormones in sex determination.

internal reproductive structures will arise which are female. That two different groups of substances are involved is indicated by the fact that injection of male hormone into a castrated fetus will cause development of the majority of the male characteristics. However, fallopian tubes and uterus will still arise, showing the need for the presence of the inhibitor. Some male secretions, then, are essential at a certain time to cause differentiation of particular internal embryonic structures in the male direction, and other secretions are essential to suppress growth of any which have a female potential. In the absence of the products of the testis, differentiation will proceed along the female pathway. The XX constitution will then exert an influence on normal ovary development as well as the formation of other female reproductive structures.

Anomalies of the Sex Chromosomes

The importance of an XX or XY chromosome constitution to normal sex determination is seen in those cases where the number of sex chromosomes departs from the typical. At times in the human as well as in other species, individuals arise with a karyotype which possesses some alteration in the number of chromosomes or in chromosome structure. Such abnormalities are generally referred to as either *chromosome anomalies* or *chromosome aberrations.* They may involve either the autosomes or the sex chromosomes. Down's syndrome, noted in Chapter 2 and discussed at length in Chapter 11, is an example of a disorder due to a chromosome anomaly: the presence of one extra autosome, chromosome #21. The full significance of chromosome aberrations can be appreciated when it is realized that they may be responsible for about 20% of the half million spontaneous abortions which occur each year in the U.S. alone. Among American babies born each year, approximately 20,000 will incur some kind of disorder or effect due directly to some type of chromosome aberration. At this point, we will examine a few of those anomalies which are related primarily to the sex chromosomes. Details on other kinds of chromosome abnormalities will be given in Chapter 11.

Approximately 1 out of 500–600 male babies is born carrying a Y chromosome but more than 1 X. Most of such individuals have an XXY constitution, giving a total of 47 chromosomes instead of the normal 46. As infants, they may appear normal, but as they mature, they will exhibit a collection of phenotypic features which describe the *Klinefelter syndrome.* Persons with Klinefelter syndrome are definite males, but typically, the penis and testes are small, the body hair sparse, and the breasts somewhat enlarged (gynecomastia) [Fig. 4-3(A)]. In addition, the arms and legs are quite long in proportion to the rest of the frame. These males are usually sterile, since the testes are abnormal and spermatogenesis does not usually take place. Some of these persons are

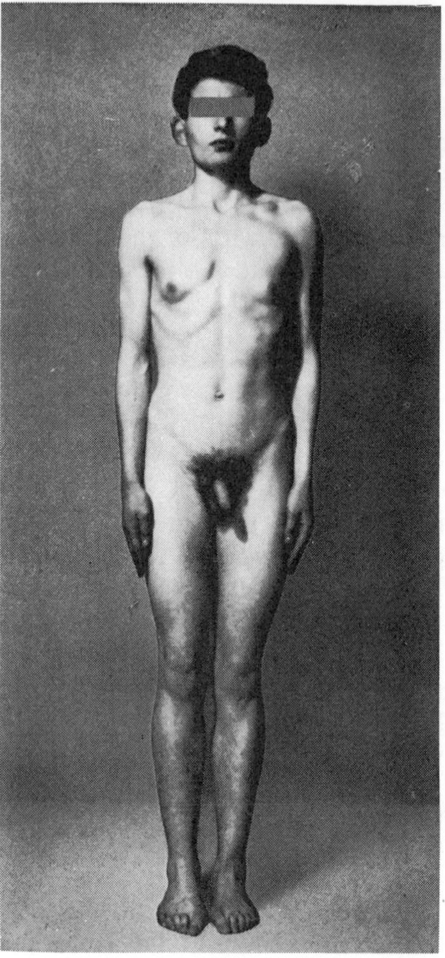

(A)

Figure 4-3. (A) Patient with Klinefelter syndrome. The karyotype of this individual is most unusual. The chromosome complement is apparently that of a normal female with 2 X chromosomes! Rare cases of XX Klinefelter males are believed to result from an exchange of genetic material between the X and Y chromosomes at meiosis in the paternal parent of the affected person. In this case, a portion of the Y chromosome is believed to have become inserted into one of the X chromosomes. (B) Karyotype of a Klinefelter male with an XXY constitution, the one most frequently associated with the syndrome. (C) Karyotype of a Klinefelter male having 3 X chromosomes and a Y.

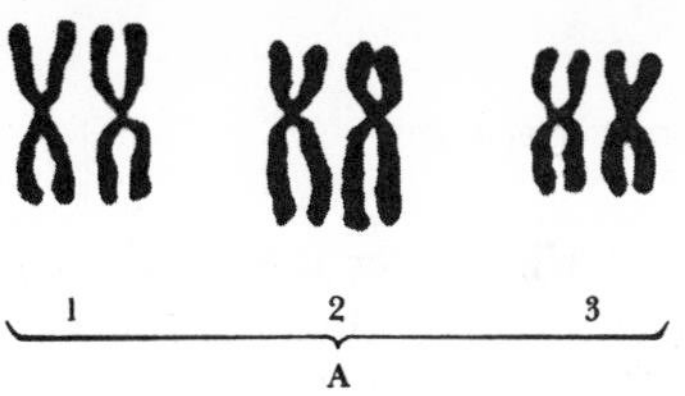

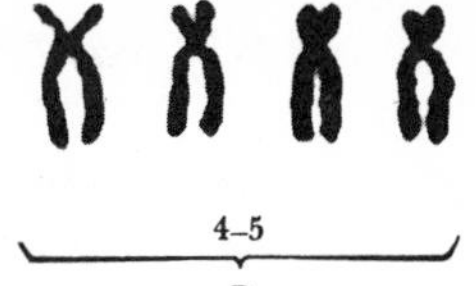

(B)

(C)

Figure 4-3. Concluded

also mentally retarded. While most Klinefelter males have 47 chromosomes and an XXY karyotype [Fig. 4-3(B)], other chromosome constitutions also occur among Klinefelter males: XXXY, XXXXY, XXYY, and XXXYY (Fig. 4-3(A), (C)]. The amount of mental retardation and the other features of the syndrome appear to be more pronounced in those individuals with three or four X chromosomes.

We have just noted that normal male sex development requires the action

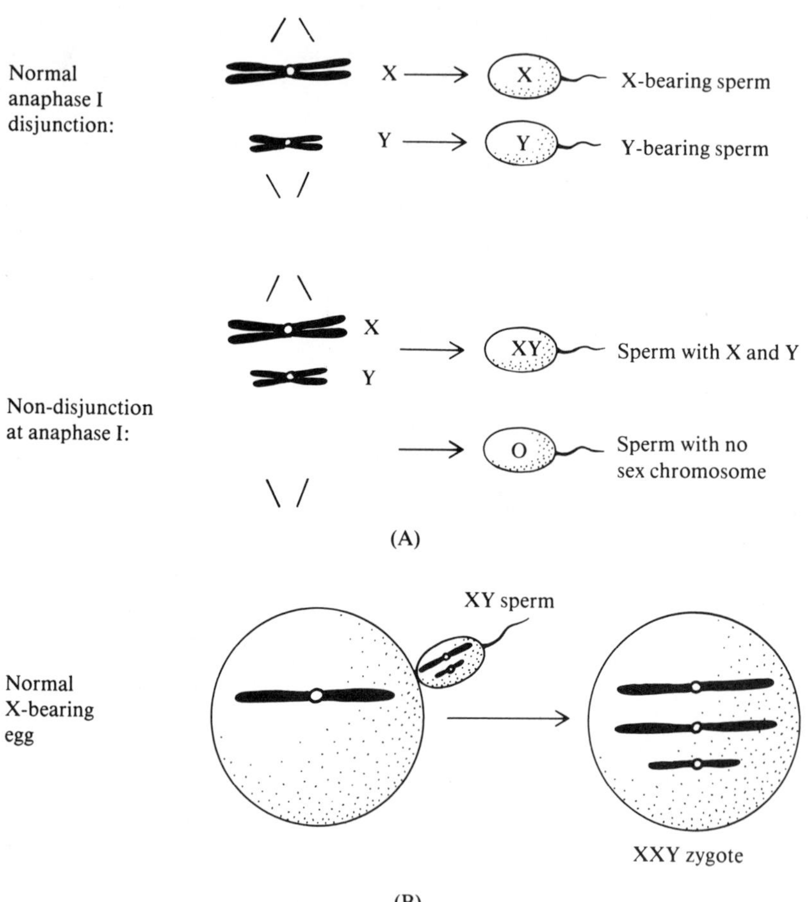

Figure 4-4. Origin of Klinefelter male due to error at spermatogenesis. (A) Normal anaphase separation (above) results in normal X-bearing sperm and Y-bearing sperm. Nondisjunction of the X and Y chromosomes at first anaphase of meiosis (below) will produce sperms with both the X and the Y chromosomes and sperms with no sex chromosomes at all. (B) Fertilization of a normal egg by an XY sperm will produce a zygote with an XXY constitution.

of the Y chromosome at an early stage in embryo development so that formation of the testes will be initiated. In a Klinefelter individual, the Y apparently acts in this way, but the presence of 1 or more additional X's, by causing an increase in the level of female sex hormone, may interfere with the development and maturation of the normal male phenotype. What factors are responsible for the origin of individuals with extra chromosomes? Since the early part of this century, cytologists have known that while mitosis and meiosis are extremely accurate processes, errors do sometimes arise. One such phenomenon, known as *nondisjunction,* takes place on rare occasions at nuclear divisions. It refers to the failure of proper chromatid separation at anaphase of a mitotic or meiotic division. As Fig. 4-4(A) shows, this can involve the sex chromosomes during spermatogenesis. If the X and the Y fail to separate at first anaphase, exceptional types of sperms will be produced: those which carry both the X and the Y and those which carry neither. If the XY sperm [Fig. 4-4(B)] unites with a normal egg bearing 1 X chromosome, then an XXY zygote will be produced. This can then develop into a Klinefelter male with 47 chromosomes. The sperm which carries no X or Y but only the set of 22 autosomes may also fertilize an egg. The result in this case is a zygote which has only 1 X and no Y (symbolized XO) and which may develop into an abnormal female type with the *Turner syndrome,* which will be described shortly.

Nondisjunction at spermatogenesis may also take place at the second meiotic division. Figure 4-5 shows the consequence when *both* chromatids of the Y chromosome pass to the same pole at anaphase II. Again exceptional sperms arise: those with 2 Y's and no X and those with neither an X nor a Y. If a YY sperm succeeds in fertilizing a normal egg, the result is an individual who is XYY. Persons of this chromosome constitution have been the subject of continuing investigations concerning the role of the genetic material in certain types of behavior patterns. The XYY individual is a male who is usually much taller than average. He may be normal in intelligence and otherwise no different from the typical XY male. However, a significant number of XYY males have been found among mentally retarded and highly aggressive men in state hospitals and penal institutions. The evidence is compatible with the hypothesis that the presence of an extra Y may predispose the individual to aggressive and hostile behavior. However, the exact relationship is not clear, since some apparently normal, well-adjusted men carry an extra Y. The frequency of XYY persons has been estimated to be of the order 1/2000 males, but a recent report from Harvard Medical School places the figure much higher, about 1/1000 births. This would make the XYY condition about as frequent in the population as the XXY Klinefelter syndrome and also as Down's syndrome. Unfortunately, many glib, dramatic statements have been made concerning the XYY chromosome anomaly. Any conclusions drawn at this time about the direct relationship between criminal tendencies or antisocial

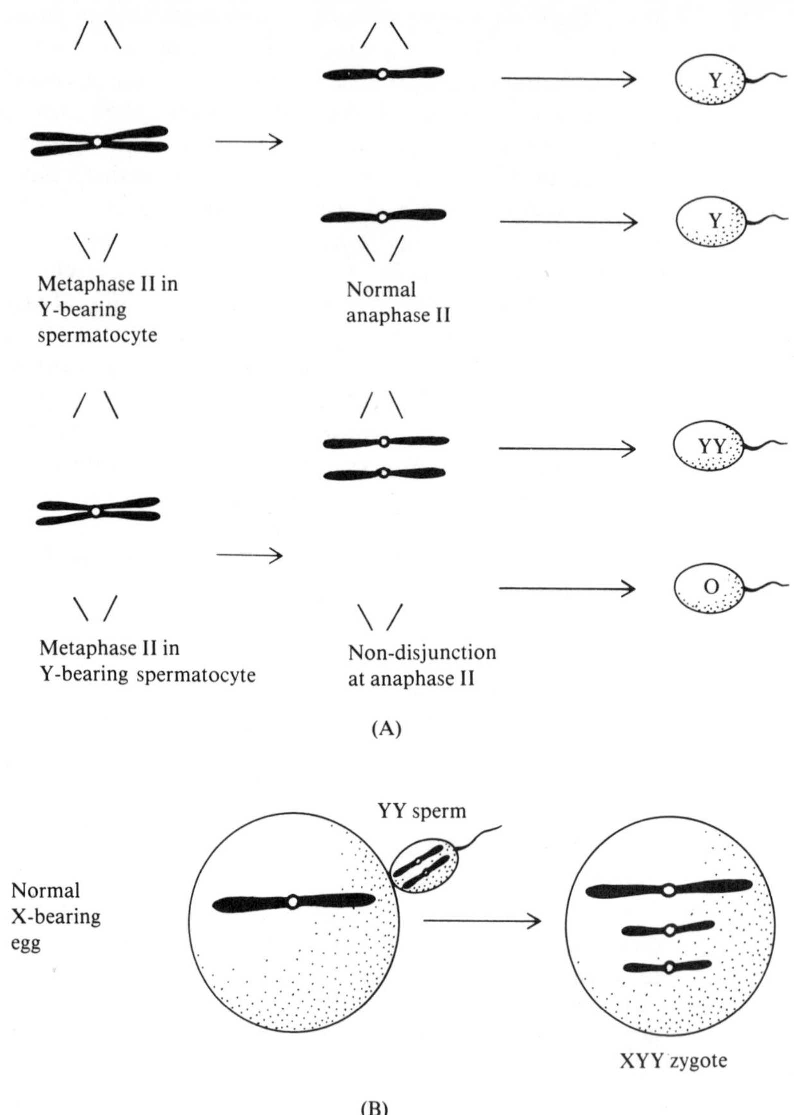

Figure 4-5. Origin of XYY condition. (A) The secondary spermatocytes formed following the first division of meiosis are of two classes: X-bearing and Y-bearing. When normal disjunction occurs in the latter at anaphase II, normal Y-bearing sperms are produced (above). If nondisjunction arises at second anaphase (below), exceptional sperms are produced, those which have two Y chromosomes and those with no sex chromosomes at all. (B) If a YY sperm fertilizes a normal egg, the result is an XYY zygote.

behavior and the presence of an extra Y chromosome would be premature. The role of the environment has not as yet been properly evaluated. It is possible that an XYY person may be born with certain metabolic or neural abnormalities which directly produce a behavior deviation, but it is just as possible that the environment may trigger the behavior problem in the XYY type. An unusually tall person or one who is somewhat below average in intelligence as well could be subjected to unpleasant experiences during his formative years. These could elicit the antisocial attitudes by inflicting psychological damage. Further investigations are in progress on this important subject, which relates genetics and human behavior and which also raises serious ethical problems for society (Chapter 15). No matter what the exact relationship may prove to be between the XYY karyotype and behavior, it is very evident that the XYY male is by far more normal phenotypically as a male than is the XXY Klinefelter person who also has 47 chromosomes. The XYY male appears to suffer no reduction in fertility, a contrast to the usually sterile XXY. The importance of the Y in the initiation of maleness during embryological development is again illustrated. An extra Y obviously permits the initiation of masculine development and does not interfere with its completion. An extra X along with a Y, however, will result in feminization, probably by causing in some way a reduction in the effectiveness of the male hormone.

Figures 4-4(A) and 4-5(A) show that nondisjunction at first or second anaphase of spermatogenesis can result in the origin of sperms which are of the constitution O, meaning that they carry no kind of sex chromosome at all. If such a gamete succeeds in fertilizing an egg, a zygote will be formed which possesses an XO karyotype. Some of these survive and develop into individuals who have 45 chromosomes, 1 less than the normal number, since only 1 sex chromosome, the X, is present. As might be suspected from what has been said about the role of the Y chromosome, these persons are females. However, they express traits which collectively describe the Turner syndrome [Fig. 4-6(A)–(B)]. Typically, the Turner female is short in stature, has a webbed neck, broad chest with undeveloped breasts, and an assortment of cardiovascular defects. Except for shortness in stature, 1 or more of the associated clinical features may be absent from case to case. While the external genitalia of these sterile persons are female, the internal reproductive organs fail to mature, the uterus remaining small and the ovaries rudimentary. A Turner female is not necessarily mentally retarded, but special procedures indicate that though such persons tend to score about normal on verbal IQ tests, their scores on performance IQ are below average. However, the XO chromosome constitution, in which there is 1 less chromosome than normal, apparently causes more serious imbalance than does the presence of an extra X, as in the Klinefelter syndrome. The bodily abnormalities of the Turner female are often serious enough to demand medical correction. The severity of the absence of a chromosome is indicated by the fact that only about 1 in 3500 female births

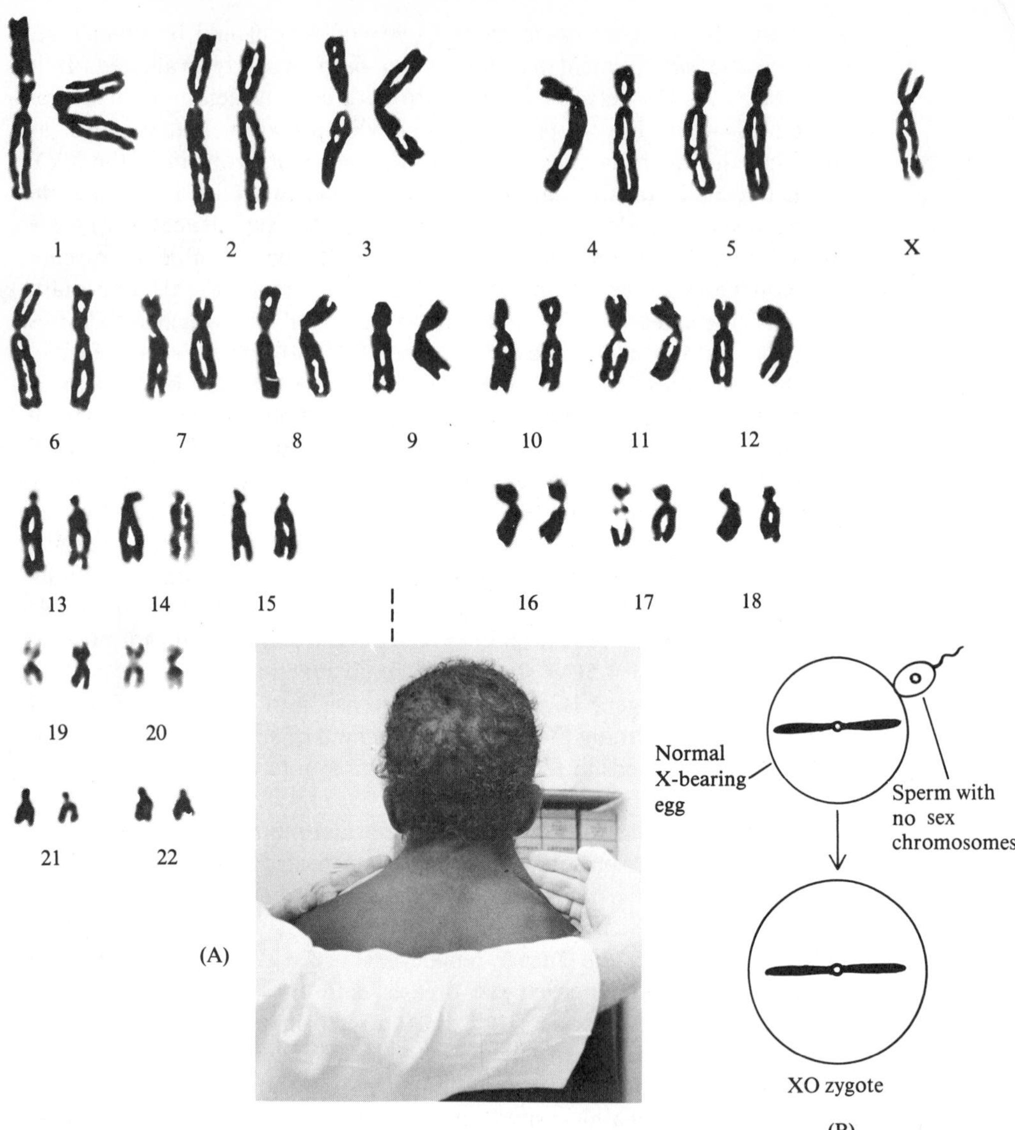

Figure 4-6. The Turner syndrome. (A) A Turner female showing typical webbed neck and the associated karyotype. (B) One way in which a Turner female can arise. A sperm with no sex chromosome fertilizes a normal egg. The resulting zygote carries only one sex chromosome, an X, and can develop into a Turner female.

is a Turner child. Studies of spontaneously aborted embryos and fetuses indicate that most of the XO types are aborted and that only 2% of the zygotes which arise go on to full development. The loss of a single chromosome, giving 1 less than normal, is thus seen to be serious enough to have lethal consequences. We will see later that the loss of just a single autosome may produce effects which are more severe than the loss of an X or a Y.

Another type of aberration of the sex chromosomes in females is the presence of one or more extra X's, producing chromosome complements which are XXX or XXXX. Even females with five X's have been reported. The XXX female arises in about 1 birth in 1400, the others being less frequent. Women with 1 extra X do not exhibit the severe bodily defects associated with the XO condition. Some of them show no evident anatomical abnormalities whatsoever. One extra X chromosome appears to have little effect on fertility. Apparently, the extra X may be lost during development of oocytes, so that only normal eggs with 1 X are produced. No XXY or XXX offspring have been reported from XXX mothers. The presence of extra X chromosomes has, however, been associated with mental retardation, about 7 per 1000 mentally defective females.

The association between extra sex chromosomes and degree of mental retardation has been a point of debate. It would appear that the incidence of persons with sex chromosome anomalies among mentally defective groups is about four times greater than one would expect by chance, indicating some relationship between these chromosome aberrations and reduced mental capacity. However, studies of the general, so-called "normal population" indicate that many of these anomalies of the sex chromosomes are compatible with the development of normal mental ability. Indeed, those mentally defective persons with aberrations of the sex chromosomes are not at all as mentally deficient as unfortunate PKU victims or those who have some abnormality of an autosome.

However, it does appear in the case of sex chromosome anomalies that the more extreme the chromosome picture (XXXX opposed to XXX or XXXY vs. XXY), the more likely there will be a detrimental effect on both mental and physical development. One conclusive point is that we must guard against sweeping generalizations in reference to any sex chromosome aberrations and their effects on behavior of any sort when we are dealing with rare disorders which have provided only small samples for analysis.

Normal meiotic division may fail at oogenesis, and this can also be responsible for the production of exceptional gametes which may result in anomalies that involve the sex chromosomes. If nondisjunction occurs at first or second anaphase of oogenesis (Fig. 4-7), exceptional eggs will be produced which have the constitutions XX and O. Fertilization of these by either normal X-bearing or Y-bearing sperms will result in zygotes of the following types: XXX, XO, XXY, and YO. The YO condition has never been found. The X chromosome,

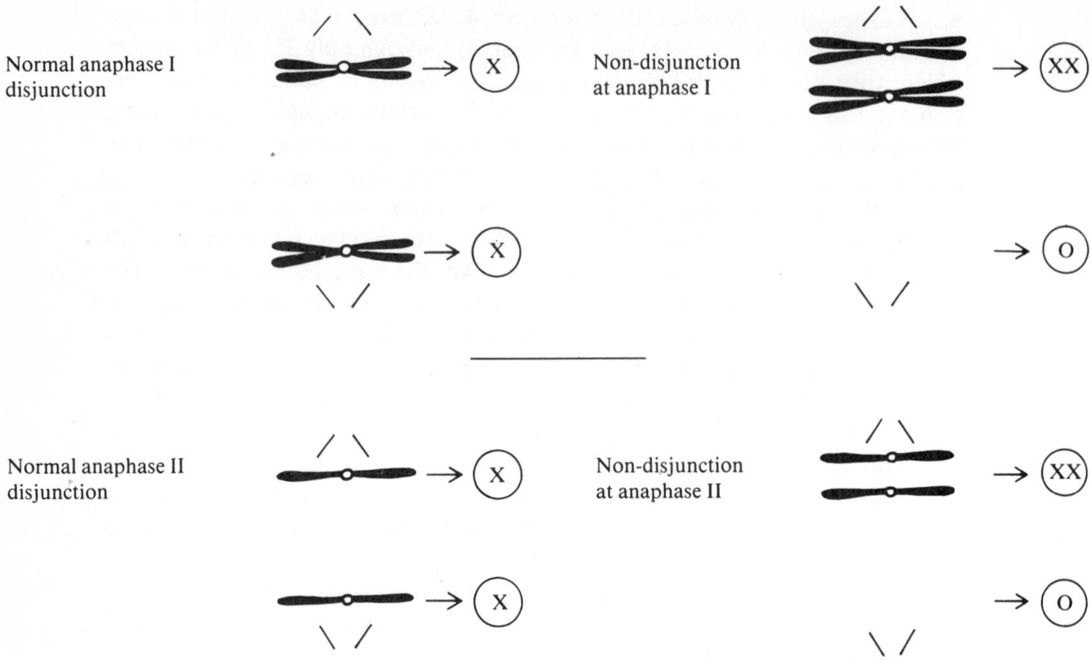

Normal anaphase I disjunction → X → X

Non-disjunction at anaphase I → XX → O

Normal anaphase II disjunction → X → X

Non-disjunction at anaphase II → XX → O

Figure 4-7. Nondisjunction at oogenesis. Normally the two X chromosomes separate at anaphase I (left) and the two chromatids of the X separate at anaphase II (below). Failure of separation at either anaphase will lead to the formation of eggs with exceptional chromosome constitutions (right).

one of the larger of the human chromosomes, carries many important genes other than those related to sex, and its presence is required for the zygote to complete development.

Klinefelter, Turner, and other individuals who have departures from the normal chromosome number are referred to as *aneuploids. Aneuploidy* indicates a chromosome constitution in which the number of chromosomes in the body cells departs from the normal by less than a whole set. The addition of just 1 or a few chromosomes from a human set of 23 constitutes an aneuploid condition, as in the cases of XXY or XYY ($2n = 47$) and XXXY ($2n = 48$). Aneuploidy also pertains to those cases in which 1 or a few chromosomes is missing from the complement, as in the XO Turner syndrome ($2n = 45$).

Genes on the X Chromosome

Since the X is one of the larger chromosomes, it is not surprising that it carries the loci of many different kinds of genes in addition to those having a role in sex determination. Genes found on the X are said to be *sex-linked* or X-linked.

X-linked genes are never found on the Y chromosome, which in the human appears to carry only genes for male-determining features. Genes found only on the Y are said to be *holandric.* In mammals, therefore, the male-determining genes are holandric, and there is no conclusive evidence that there are any holandric genes which affect characteristics other than the male sexual ones. The main argument presented today for the existence of any such gene is that for the *hairy ear trait.* This feature is found mainly in southern India and Sri Lanka where men may be seen with a luxurious growth of hair bordering the pinnae of the external ears. The trait is not found among any of the women. The possibility exists that it could be holandric, but other interpretations are also possible.

This is in marked contrast to the X, which has been shown to carry the loci of over 100 sex-linked genes known to influence the expression of a vast assortment of characteristics other than sexual ones. Among the most familiar are genes affecting color vision, blood clotting, and muscle development. The first conditions in the human to be recognized as sex-linked were red-green colorblindness and hemophilia. The inheritance pattern of genes on the X chromosome is so distinctive that it can be readily distinguished from that of genes on the autosomes. However, there is no exception to any of the basic principles discussed so far. The differences which do arise simply result from the fact that females carry 2 X chromosomes, whereas males have only 1. A normal male will always produce two classes of sperms in relation to the kind of sex chromosome; half of them carry an X and the rest carry the Y. The male is said to be *heterogametic* or the *heterogametic sex,* implying that his gametes differ in respect to the kind of sex chromosome they carry. The female is *homogametic,* meaning that all her gametes are alike insofar as they all carry the same type of sex chromosome, the X. Having only 1 X, a male can never be homozygous or heterozygous for any gene on the X chromosome, since those terms imply that 2 gene doses are present. And so, it is said that males are *hemizygous* for their X-linked genes; they carry them only in a single dose.

Let us examine the inheritance of red-green colorblindness to illustrate the consequence of these facts. As the common name implies, such colorblind persons have difficulty distinguishing red color from green. Approximately 75% of colorblind persons have a form of colorblindness known as the *deutan* type. Since it is inherited as a sex-linked recessive condition, most individuals affected with this colorblindness are males. Some facts are known about the basis of the deutan condition. The retina of the eye, it may be recalled, it composed of several layers of cells. The receptor cells of the retina are required for the sense of sight and are of 2 types, the rods and the cones. The former are extremely light-sensitive and are responsible for vision in dim light. However, they are not sensitive to color. The cones are the receptor cells which enable us to detect color and which provide us with sharp images in bright light. In the deutan type of colorblindness, the presence of the sex-linked recessive *d* results in the production of defective green-sensitive cone receptors in the retina. The presence of the dominant allele D is required for the normal development of those cones sensitive to green. Persons with the deutan type of colorblindness have defective green-sensitive cells, so that green color is poorly perceived and in some cases not recognized at all.

If a male carries the recessive gene d, he must be red-green colorblind, since the gene is on his X chromosome, and he carries no other X which could bear the normal allele (Klinefelter males will be ignored in our discussion here). Any male can have only the following two genotypes in relation to the sex-linked allelic pair for deutan colorblindness (D and d): $X^{-D}Y$ and $X^{-d}Y$. A female, however, may be one of the following three: $X^{-D}X^{-D}$, $X^{-D}X^{-d}$, or $X^{-d}X^{-d}$. When designating genotypes, a sex linked allele is often written as a superscript to the letter X (for X chromosome) as has been done here. However, many prefer to eliminate the X. Following this system, the possible genotypes in respect to the alleles for deutan colorblindness would be DY and dY for the male and DD, Dd, and dd for the female. Note that when sex-linked genes are being considered, the Y chromosome must be represented in the genotype of the male. This is necessary to take into account that he is heterogametic and forms two classes of sperms: X-bearing and Y-bearing.

It should be obvious why deutan colorblind females are much less common than colorblind males (less than one-half of 1% for females compared to approximately 6% in males.). In order to express deutan colorblindness, a female must be homozygous for the colorblind allele, since she carries 2 X's and the trait is recessive in females. The male, on the other hand, will be colorblind if the colorblind allele is present on his single X. There cannot be a

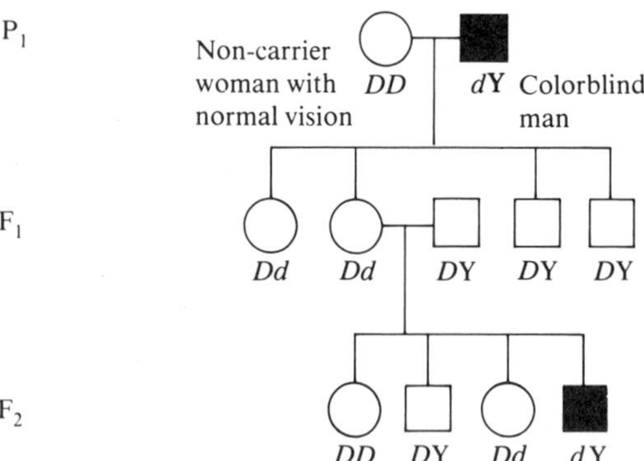

Figure 4-8. Pedigree of deutan colorblindness. Any X-linked gene, such as that for deutan colorblindness, will be transmitted from an affected male to his grandsons through carrier daughters. Since any male receives a Y chromosome only and no X from his father, an affected male cannot transmit any X-linked genes to a son.

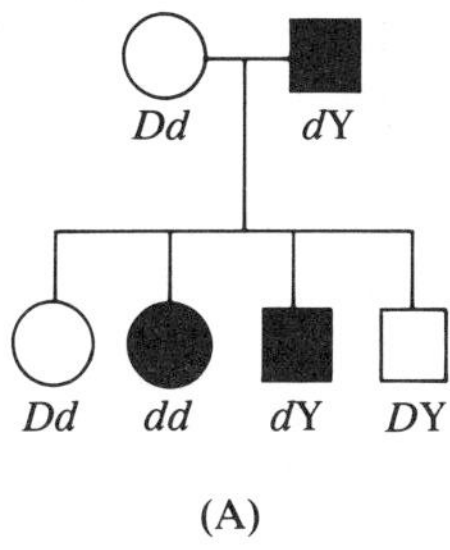

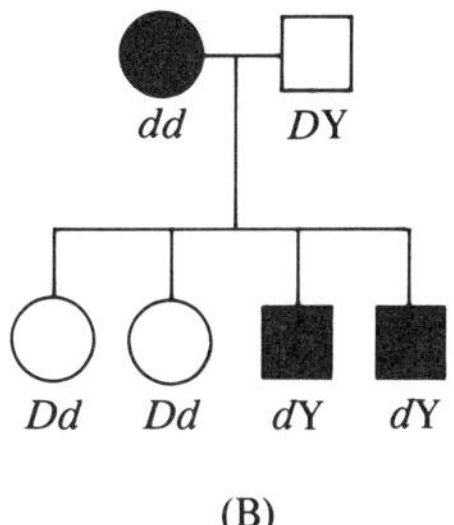

D = normal vision
d = deutan colorblindness

Figure 4-9. Colorblind females. (A) In order for a colorblind female to arise, the male parent must be colorblind and the female parent must be a carrier (or be colorblind herself). This is so because any female who expresses a sex-linked recessive trait must be homozygous for the responsible gene and must receive one X chromosome with the recessive gene from each parent. (B) All the sons of a colorblind woman *must* be colorblind. A male receives his one X chromosome from his mother; therefore, he must express any sex-linked recessive trait which the mother expresses, since she must be homozygous and can only contribute an X with the recessive gene.

normal allele present at the same time, since the Y is present instead of another X. Figure 4-8 illustrates a typical pattern of inheritance of deutan colorblindness in a family. One important point to note is that a male never passes an X-linked gene to his sons. This is so because transmission of a Y chromosome produces a son. Any X-linked genes of the male parent are passed to daughters. In the case of an X-linked recessive, all the daughters of an affected male will be heterozygous carriers who will not express the defective gene, if their mothers were homozygous for the normal allele. These carrier daughters, however, when mated to normal men, will pass the sex-linked recessive to half of their sons, who *must* express it, and in this case would be colorblind. Therefore, any X-linked recessive is passed from grandfather to grandson through a carrier daughter who usually will not express the recessive trait (since her mother will usually be homozygous for the normal gene). The pedigree (Fig. 4-8) shows why more males than females would be expected to show the trait. In order for any female to be colorblind, her mother must be heterozygous (or colorblind herself), and her father must be colorblind. Figure 4-9(A)–(B) shows how a colorblind woman can arise and the result of a mating between such a woman and a normal male. All the sons of the colorblind woman must be colorblind, and all her daughters will be carriers, normal in vision if their father is normal. This is an example of what is known as *crisscross inheritance:* the trait which appears in the female parent is expressed only in her sons, whereas the trait shown by the male parent occurs only among the daughters. This crisscross pattern does not occur when genes on the autosomes are being followed. It is so typical of sex-linked inheritance that sex linkage should always be suspected when such an observation is made.

We see from the above, therefore, that reciprocal matings produce different results in the case of sex linkage: when a colorblind male and a homozygous normal female produce offspring, all of them are normal phenotypically; however, when a normal male and a colorblind female have children, a criss-cross pattern is seen. This difference between reciprocal crosses does not occur in the case of any autosomal gene. In albinism, for example, it doesn't matter which parent is the albino; all the offspring will be phenotypically normal if the other one is homozygous for the dominant gene required for melanin deposition in the eyes and skin (Fig. 1-4). Nor does the sex of the parent matter if one of them is a heterozygote (*Aa*) and the other an albino. There is just as much chance in either case for a daughter or a son with oculocutaneous albinism. In contrast to this in the case of sex linkage, the sex of the parent who is carrying the recessive genes, as seen in deutan colorblindness, *does* cause a difference in the chances of producing a son or a daughter with the specific trait (Figs. 4-8 and 4-9).

Although there are differences in inheritance patterns between autosomal traits and sex-linked ones, there is actually no true exception to Mendel's laws. Segregation of sex chromosomes takes place and hence the genes on them segregate, as is true of any of the other chromosomes and their genes. The difference seen in the patterns is simply due to the fact that in the male the X chromosome segregates from the Y, and the two chromosomes are not completely homologous. There is also no contradiction to the concept of dominance. A sex-linked recessive can only occur in a single dose in a male and it will be expressed, but it will not be fully expressed in the presence of its dominant allele in a heterozygous female.

It was stated at the beginning of the discussion of colorblindness that 75% of colorblind persons have the deutan type. This implies that other genes exist which affect color vision, and this is indeed the case. Color perception is a complex characteristic and, as stated in Chapter 1, there is no *one* gene for any character. At least 3 other genes are known which influence the ability to detect colors. Approximately 2% of the males in the population carry a different kind of sex-linked recessive gene, *p,* which causes the *protan* type of color vision defect. The dominant allele, *P,* is required for normal development of those cones of the retina which are sensitive to red light. A male of the genotype $X^{-p}Y$ is therefore unable to recognize red color and is likely to confuse red with green. Colorblind females of the protan type, who must have the genotype $X^{-p}X^{-p}$, are exceedingly rare and are found at a frequency well under 0.05%. Since both the protan and the deutan types of color defect result in poor distinction between red and green, both forms were originally thought to be the same and to result from the presence of the same recessive X-linked gene. The 2 types can, however, be recognized with the use of special kinds of color recognition charts and by refined analysis of the retina.

A very uncommon sex-linked recessive gene, which affects well under 1% of colorblind persons, causes defects in the development of those cones sensitive to blue, as well as to those sensitive to red and to green. This results in the tritan type of colorblindness in which there is difficulty in the recognition of all colors. An exceedingly rare *autosomal* recessive gene is known which somehow prevents the development of any cones at all, resulting in total colorblindness. Since this gene is not sex-linked, we would expect just as many females as males to have this very rare type of vision defect. As noted in Chapter 1, there is a form of albinism which causes complete absence of melanin pigment in the iris of the eye but does not affect the skin. The responsible gene for ocular albinism is carried on the X chromosome.

Since we speak of sex-linked recessive genes, there must be sex-linked dominants. Obviously the alleles, X^{-D} or X^{-P}, for normal color vision are sex-linked dominants. A few sex-linked dominants are known which produce abnormal phenotypes. An example is the X-linked dominant which may bring about a type of skeletal defect, a vitamin D resistant rickets, and which typically causes a lowering of the phosphate level in the blood. Any X-linked dominant will be expressed more frequently among females, as illustrated in this case [Fig. 4-10(A)–(B)]. When a male carries the X-linked dominant, he passes it to all of his daughters who will express the trait, and to *none* of his sons. When a female shows the trait, she passes it to one-half of her sons and to one-half of her daughters. The reason why the female parent in Fig. 4-10(B) is represented as $X^{-R}X^{-r}$ is that the gene for the rickets is very rare, as in the case of most X-linked dominants which are associated with pathological defects. Therefore, almost all females carrying a defective sex-linked dominant will be carriers of

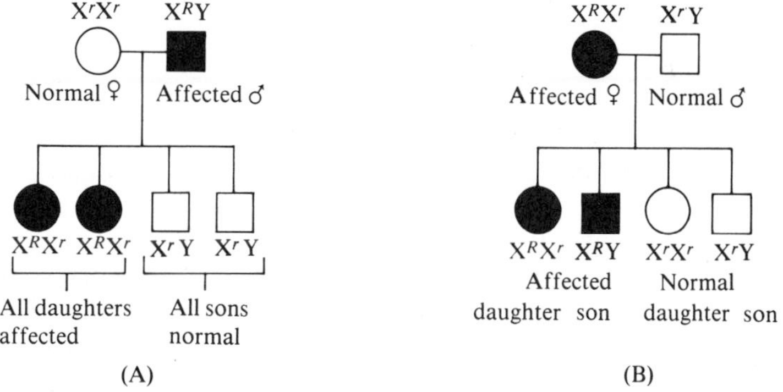

Figure 4-10. Pedigree of a sex-linked dominant. (A) If a male expresses a defect due to a sex-linked dominant gene, all his daughters must be affected, since each one would receive from him an X with the dominant gene. No sons will be affected, since they receive an X with the normal recessive allele from the female parent. (B) If a female expresses a dominant trait due to a rare sex-linked gene, she is undoubtedly heterozygous and will pass the defective dominant gene to half of her sons and half of her daughters.

the normal recessive allele. And so, on the basis of what is known about sex chromosomes and meiosis, more females than males would be expected to express a sex-linked dominant trait, whereas more males than females should express any sex-linked recessive. This is exactly what is found.

Blood Clotting and Genes Which Influence It

No discussion of sex linkage would be complete without mention of the sex-linked recessive gene, *h,* responsible for hemophilia. This gene, which afflicts about 1/10,000 males, has been of special interest because of its historical impact. Victims of hemophilia are almost always males, who tend to bruise very easily and who experience hemorrhages from relatively trivial injuries. Serious blood loss into body cavities and joints may occur. Even the trauma of birth may cause fatal bleeding inside the cranium. The basis of the disorder is an impairment of the blood clotting mechanism, a very complex process which involves many different blood components. Specific genes are known which control the production of these particular blood clotting elements. Some of the genes are sex-linked, and others are autosomal. The gene responsible for the more common form of hemophilia, hemophilia A, has a locus on the X chromosome and accounts for about 80% of all hemophilia cases. The normal allele is responsible for the production of a substance known as the *anti-hemophilia factor* (AHF). A deficiency of this substance results in an interruption of one of the many steps in the chain of reactions leading to blood clotting (Fig. 4-11). Consequently, the blood fails to coagulate at the normal rate. Clotting time may be prolonged from the normal by 5 minutes to 2 hours.

Defective blood clotting was recognized in biblical times, and the pattern of its inheritance was noted long before the recognition of Mendel's laws. Circumcision was not required of a male baby if any of his mother's brothers had died from uncontrolled bleeding or if his mother had borne previous victims of hemophilia. However, attention became focused on the disease in the mid-nineteenth century due to its appearance among the royal houses of Europe. While most human families lack extensive pedigrees, the family histories of royal personages are well recorded. Hemophilia was unknown among European royalty until the rule of Queen Victoria. One of her sons suffered from hemophilia, and several of her daughters were carriers (Fig. 4-12). There is little doubt that the normal gene, *H,* underwent a mutation to its recessive form, *h,* either in an egg of Queen Victoria's mother or in a sperm of the Queen's father, since it is almost certain that Queen Victoria was a heterozygote, *Hh.* (Note that the X may be left out when genotypes of sex-linked conditions are represented. The Y chromosome, however, must always be shown for the male.)

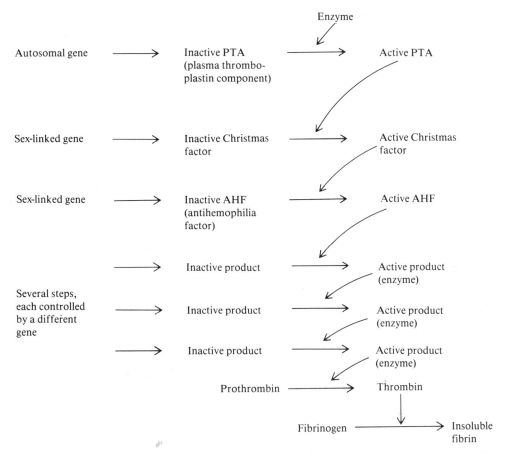

Enzyme

Autosomal gene → Inactive PTA (plasma thrombo-plastin component) → Active PTA

Sex-linked gene → Inactive Christmas factor → Active Christmas factor

Sex-linked gene → Inactive AHF (antihemophilia factor) → Active AHF

Several steps, each controlled by a different gene

→ Inactive product → Active product (enzyme)

→ Inactive product → Active product (enzyme)

→ Inactive product → Active product (enzyme)

Prothrombin → Thrombin

Fibrinogen → Insoluble fibrin

Figure 4-11. Summary of steps in blood clotting. The formation of a blood clot depends on the ultimate conversion of the protein fibrinogen to the insoluble fibrin. This requires the activity of thrombin, which acts as an enzyme in the conversion. Active thrombin itself must be derived from the inactive protein, prothrombin. This step also requires an active catalyst. The overall scheme of blood clotting entails the conversion of one inactive substance to its active, enzymatic form. In its active form, the enzyme can effect the conversion of another inactive product to its active enzyme form, and this in turn catalyzes another reaction. Each step is controlled by a different gene, some sex-linked, others autosomal. AHF is just one substance needed in the chain leading to blood clotting. Lack of AHF due to a sex-linked recessive gene is the basis of the commonest form of hemophilia.

Figure 4-12 shows how the actual inheritance of the condition can be explained in the royal pedigree. One of the Queen's daughters to receive the gene was Alice. She in turn transmitted it to a son who became a hemophiliac and to two daughters who were also carriers. One of these, Alix, became the wife of Tsar Nicholas II of Russia. Alix passed the gene to her son, the

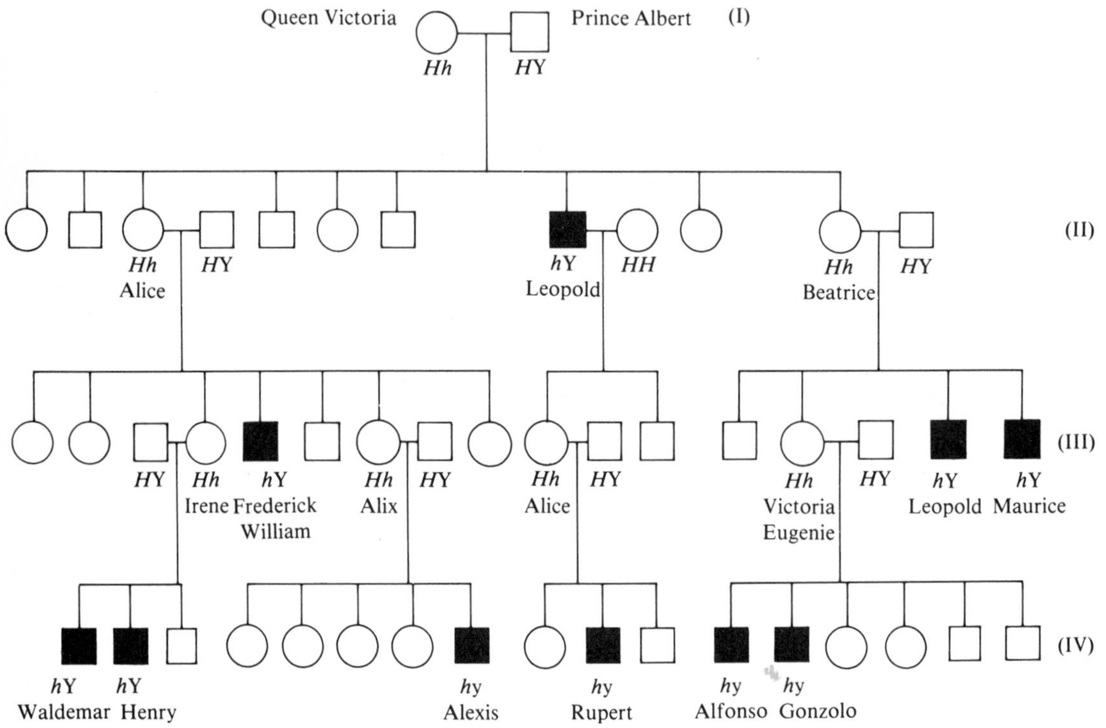

Figure 4-12. Hemophilia and the pedigree of Queen Victoria. Queen Victoria transmitted the gene for hemophilia to one son, Leopold, who expressed the disorder, and to at least two of her daughters, Alice and Beatrice, who in turn transmitted it, producing affected sons and carrier daughters. Alice produced a carrier daughter, Alix, who in turn bore the afflicted Tsarovitch, Alexis. Note that Queen Victoria's son Leopold produced a carrier daughter, but his son was not affected, since he gave the Y and not the X to his son. All males in the pedigree who do not express the hemophilia trait must be genotype *H*Y, as was Prince Albert. This pedigree does not show those lines in which the hemophilia disappeared. For example, one of Queen Victoria's daughters produced two daughters and three sons among whom no hemophilia appeared and from whom the disorder was not transmitted. This can be readily explained on the assumption that this daughter was homozygous for the dominant, normal allele.

Tsarovitch, who consequently was a victim of the clotting defect. The monk Rasputin in some way achieved a mysterious influence on the boy and seemed to be able to alleviate some of the bleeding episodes. As a result of the hold he exerted on both mother and son, Rasputin gained an undue amount of political power, and this may have been one factor which precipitated the Russian revolution. As the pedigree shows, the gene was passed to several other of Queen Victoria's descendants. Even more recently, it has figured in European politics. General Franco of Spain, before naming Prince Juan Carlos de Borbon as his possible successor, pointed out that the lineage of the Prince traces back to Queen Victoria. The General contemplated the risk of hemophilia among children of the Prince. However, since the Prince himself is not a hemophiliac, he cannot be carrying the sex-linked recessive, and thus the matter of the blood disorder was finally eliminated as a cause for concern.

Since hemophilia is an X-linked recessive trait, its pattern of inheritance is identical to that for colorblindness. Many more affected males than females should arise. Actually, the very occurrence of hemophiliac females was questioned for many years, and it was suspected that the homozygous condition *hh* might produce such serious upsets that the embryo would not survive. However, there have now been authenticated a few cases of female hemophiliacs. Their extreme rarity is most likely due to the highly improbable set of conditions necessary to produce a female with hemophilia. The normal mother of such a person must be a carrier, and the father must have hemophilia. Since the hemophilia gene is uncommon (only about 1 in 10,000 males is afflicted), such matings would be most infrequent. The unlikelihood is compounded, since most males with the disease do not survive long enough to become the fathers of large families. Those who do become parents would most likely marry women of the genotype *HH,* since the gene is rare in the population. Marriage within the family, between cousins for example, would increase the chance of the woman's being *Hh.* However, we would expect very few female hemophiliacs to be born because of the unlikely set of circumstances required for their origin.

Mention was made of the fact that many genes control factors which operate in the process of blood clotting. So it is to be expected that various different kinds of defects are found which can be called hemophilia. One milder form of the disorder is known as *Christmas disease* and accounts for about 20% of the cases diagnosed as hemophilia. This form of hemophilia is so named because it was first recognized in a family with the surname, Christmas, during Christmas week. The gene is also sex-linked and is therefore inherited precisely in the manner just described for the commoner form, hemophilia A. The gene controls a different clotting element, the Christmas factor, which is also needed in the chain of reactions which lead to normal blood clotting (Fig. 4-11).

Less than 1% of hemophilia victims are deficient in still another clotting element, PTA (plasma thromboplastin antecedent). The gene which controls the production of this substance, however, is not X-linked but is inherited as an autosomal gene. The recessive allele results in absence of the essential clotting factor. The inheritance of this kind of hemophilia, therefore, shows simple Mendelian inheritance, in the manner of oculocutaneous albinism and the other traits discussed in Chapter 1.

It should come as no surprise after our discussion of colorblindness that blood clotting is influenced by many different genes. Any characteristic involves the interaction of more than one pair of genes, and blood clotting is a characteristic which is no exception to this statement. Since many genes *do* influence any character, caution must always be exercised before a decision is reached on the nature of the gene which is responsible for a disorder in a particular case or family. A disorder inherited as a sex-linked recessive in one family can possibly be inherited in a different manner in another. At first, it may appear that the two cases are identical, but closer inspection may show two different forms of the disorder which involve different genes.

Sex-linked Recessives and Autosomal Recessives

Defective color vision, as noted, is usually a sex-linked recessive trait, but at times it is inherited as an autosomal recessive. Indeed, when we speak of albinism, we usually think of a form in which an autosomal recessive gene interferes with the formation of pigment in the skin and the eye. However, there may be different genes located at different loci which are involved. This may become evident when various pedigrees are examined. Two completely different genes, both inherited independently as autosomal recessives, can prevent the formation of melanin pigment in the eyes and the skin. These are known as the oculocutaneous forms of albinism. That two different autosomal genes are indeed involved in these conditions is revealed from the study of certain pedigrees which show only normally pigmented offspring from two albino parents. Figure 4-13 illustrates the basis of this. One parent expresses albinism due to the presence of one kind of recessive gene in double dose. The locus for this gene is on a particular autosome. The other albino parent carries recessives associated with a completely different gene location, probably on another autosome. These two albino-causing genes are therefore not alleles, since they are associated with very different loci. This means that each albino parent in this case carries a normal, dominant gene at the pigment locus for which the other parent carries only a recessive. Consequently, each albino parent produces gametes which carry a recessive on one of the autosomes and a dominant on the other. All the zygotes, and hence all the offspring, will be normally pigmented, since the dominant allele of each recessive is present. In still another form of albinism, pigment is missing from only the eye. The gene responsible for this ocular albinism, as mentioned above, is transmitted

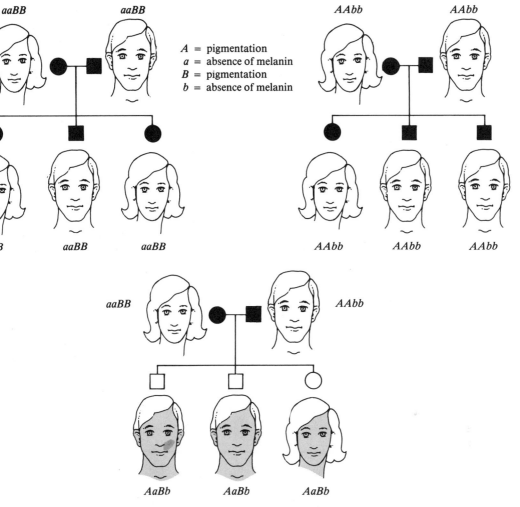

Figure 4-13. Two loci for oculocutaneous albinism. For normal pigmentation, *both* dominant autosomal genes, *A* and *B,* are required. One family (above left) in which both parents are albinos may produce all albino children, since both parents are homozygous for the same recessive gene. Another family (above right) may also show this same picture, but the recessive gene responsible for the lack of pigmentation may be associated with a very different locus. That two different loci and hence two different pairs of recessives are involved becomes apparent after the mating of certain albino persons (below). In these families, all the children will have normal pigmentation, since each offspring receives two required dominant genes, one from each parent. Each parent contributes a dominant which the other parent lacks.

as a sex-linked recessive and shows a very different inheritance pattern from the two just discussed for oculocutaneous albinism. The pattern would be the same as that for any sex-linked recessive, such as sex-linked colorblindness and hemophilia.

These examples emphasize that each family must be considered independently before generalizations are made on the specific gene and type of inheritance involved. In the course of general discussions, the commonest forms of inherited disorders may be presented as examples. This must not be taken to mean that other forms, caused by different genes, do not exist.

Sex Chromosomes and Mendel's Laws

When studying a pedigree, the geneticist may be following at the same time both a sex-linked gene and one which is carried on an autosome. The inheritance of two different pairs of genes located on separate chromosomes was considered at the end of Chapter 3. The genes in that example were found on autosomes and were found to undergo independent assortment as a result of the independent behavior of the chromosome pairs at meiosis. Just as Mendel's law of segregation applies to any pair of genes, autosomal or sex-linked, Mendel's law of independent assortment holds in relation to any sex-linked gene and any one on an autosome. Figure 4-14 shows that the sex chromosomes behave independently of the autosomes. This means that just as any two different pairs of autosomes will undergo independent assortment, so will the sex chromosomes in relation to any pair of autosomes. Therefore, if the sex-linked recessive for deutan colorblindness is followed along with the autosomal recessive for the inability to taste PTC, different combinations of the genes can arise in the eggs of a dihybrid female (*TtDd*) or in the sperms of a male heterozygous for tasting ability and hemizygous for colorblindness (*Ttd*Y). The mating shown in Fig. 4-14 will produce several combinations of the autosomal and sex-linked genes. This in turn results in the formation of an assortment of different genotypic and phenotypic combinations. Figure 4-15 shows the results to be expected following reciprocal matings between persons with oculocutaneous albinism (autosomal) and those with ocular albinism (sex-linked).

If the genotypes of two parents are known for two pairs of genes located on separate chromosome pairs, whether they are on different autosomes or on an autosome and the X chromosome, the expected ratios can be predicted, since the laws of segregation and independent assortment apply. In Chapter 6 we will discuss inheritance patterns resulting when two genes are followed whose locations are on the same chromosome, as is the case of the two X-linked loci for deutan and protan colorblindness.

A genetic counsellor is frequently concerned with ratios, particularly when the need arises to estimate the chance that a certain trait or combination of traits will appear in a family (Chapter 12). One pair of genes will be more frequently followed in human pedigrees than will two or three at the same time. However, in certain cases, it is desirable to follow more than one pair at a time if at all possible. Knowledge of whether two genes are on the same chromosome or on different ones can be a valuable guide in the detection of defective offspring before birth (Chapter 12). It is therefore important to grasp Mendel's first and second laws and to be able to follow two pairs of genes at the same time. This will become more evident in the discussion of gene linkage (Chapter

Taster, normal vision, $TtX^{-D}X^{-d}$

Taster, colorblind, $TtX^{-d}Y$

Gamete formation: $T\ t\ X^{-D}\ \ X^{-d}$ $T\ t\ X^{-d}\ \ Y$

Gametes: *TD* *Td* *tD* *td* *Td* *TY* *td* *tY*

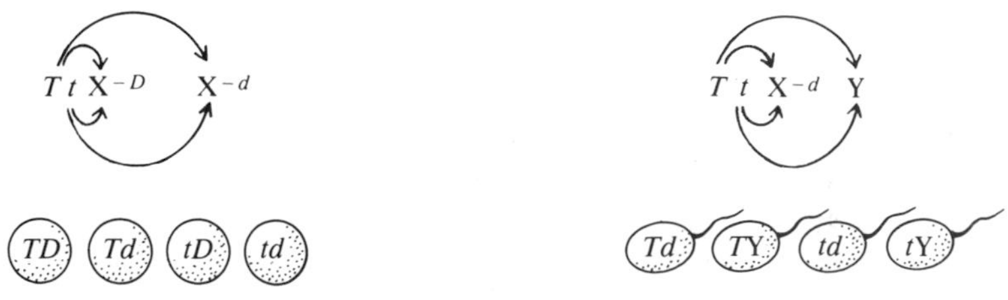

EGGS

SPERMS

	TD	*Td*	*tD*	*td*
Td	*TTDd*	*TTdd*	*TtDd*	*Ttdd*
*T*Y	*TTD*Y	*TTd*Y	*TtD*Y	*Ttd*Y
td	*TtDd*	*Ttdd*	*ttDd*	*ttdd*
*t*Y	*TtD*Y	*Ttd*Y	*ttD*Y	*ttd*Y

♀♀

Taster, normal vision	= 3
Taster, colorblind	= 3
Non-taster, normal vision	= 1
Non-taster, color-blind	= 1

♂♂

Taster, normal vision	= 3
Taster, colorblind	= 3
Non-taster, normal vision	= 1
Non-taster, color-blind	= 1

Figure 4-14. Independent assortment of sex-linked and autosomal genes. The sex chromosomes assort independently of the autosomes. A cross such as this one is almost identical to a dihybrid cross involving two loci on autosomes. It must be remembered, however, that for a sex-linked gene, the male is always hemizygous, and the Y chromosome *must* be represented in a diagram of the cross. Whether or not the X chromosome is designated, as shown here in both parents, is a matter of choice. The gametes show simply a *D* or *d,* the linkage to the X chromosome being understood.

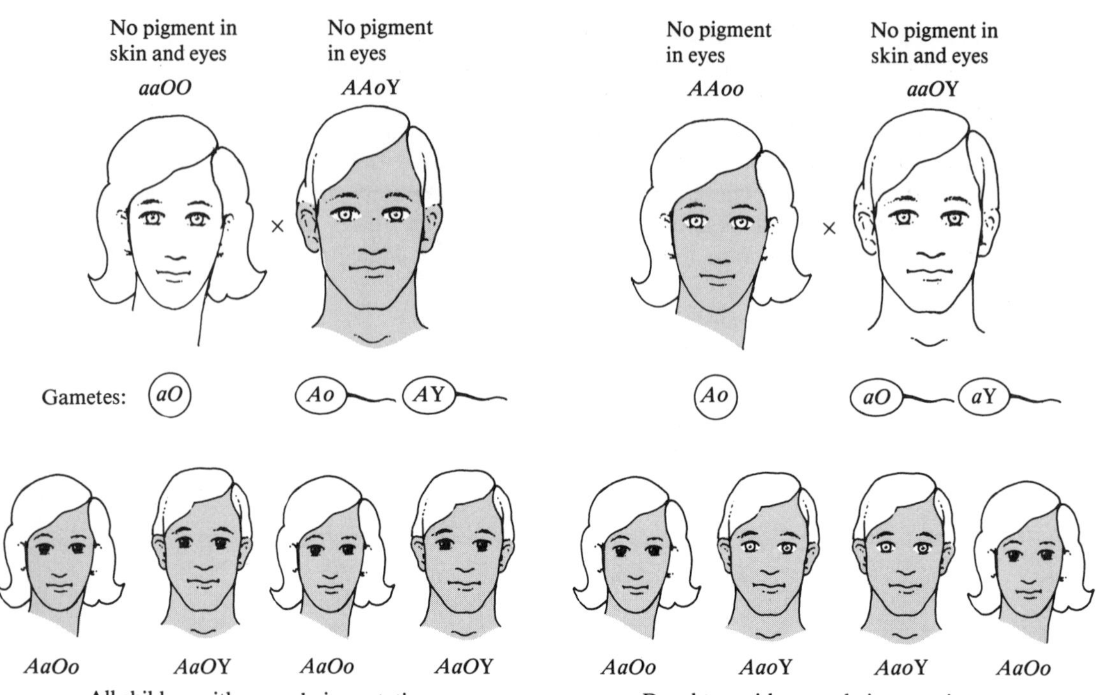

Figure 4-15. Autosomal and sex-linked albinisms. Since oculocutaneous albinism is autosomal and the ocular form is sex-linked, the genes will assort independently in a cross where both of them are involved. Note that, in both crosses above, all the children will have pigment in the skin and hair, since the dominant autosomal gene is present and was contributed by one parent. However, the reciprocal crosses are different. The one on the right shows a crisscross pattern, since the female is homozygous for a sex-linked recessive. Therefore, all her sons must show the trait, in this case ocular albinism. This is another example of more than one pair of alleles which interact and affect the expression of a characteristic. Compare this figure with Fig. 4-13, showing the crosses involving two different forms of oculocutaneous albinism.

6), the association of two or more genes on the same chromosome. Before proceeding to this and other topics, there remain other important aspects of inheritance and sex to discuss in the following chapter.

REFERENCES

De La Chapelle, A., Schroeder, J., and Pernu, M., "Isochromosome for the Short Arm of X, a Human 46, XX Pi Syndrome." *Ann. Human Genet.,* 36:79, 1972.

Ferguson-Smith, M., "Chromosomal Abnormalities II: Sex Chromosome Defects." In *Medical Genetics* (pp. 16–26), McKusick, V. A., and Claiborne, R. (Eds.). HP Publishing Co., N. Y., 1973.

Hamerton, J. L., *Human Cytogenics,* Vols. I and II. Academic Press, N.Y., 1971.

Hook, E. B., "Behavioral Implications of the Human XYY Genotype." *Science,* 179:139, 1973.

Kalmus, H., *Diagnosis and Genetics of Defective Color Vision.* Pergamon Press, 1965.

Lazerson, J., "The Prophylactic Approach to Hemophilia A." *Hospital Practice,* Feb: 99, 1971.

McKusick, V. A., "The Royal Hemophilia." *Sci. Amer.,* Aug: 88, 1965.

Mittwoch, U., *Genetics of Sex Differentiation.* Academic Press, N. Y., 1973.

Ohno, S., Tettenborn U., and Dofuku, R., "Molecular Biology of Sex Differentiation." *Hereditas,* 69:107, 1971.

Robinson, J. E., and Buckton, K. E., "Quinacrine Fluorescence of Variant and Abnormal Y Chromosomes." *Chromosoma,* 35:342, 1971.

Rushton, W. A. H., "Visual Pigments and Colorblindness." *Sci. Amer.,* March: 64, 1975.

Stern, C., Centerwall, W. R., and Sarkar, S. S., "New Data on the Problem of Y-Linkage of Hairy Pinnae." *Amer. Journ. Human Genet.,* 9:147, 1964.

REVIEW QUESTIONS

1. What is the number of autosomes in a human male and in a human female?

2. In chickens, the bar feather pattern is due to a sex-linked dominant gene, *B,* whereas the recessive allele, *b,* results in non-bar. In birds, the female is the heterogametic sex. Therefore, what would be the result of a cross between a bar female and a non-bar male?

3. *A.* In the human, the gene for cleft iris of the eye is inherited as a sex-linked recessive, *i.* Normal iris is the dominant trait, *I.* The trait for normal skin pigmentation is dominant (*A*) to that for albinism (*a*). The locus for this gene pair is autosomal. Write as much as possible of the genotypes of each of the following: (1) An albino woman and an albino man. (2) A woman with cleft iris and a man with cleft iris. (3) A woman with normal iris who is an albino. Her father has cleft iris and normal skin pigmentation. (4) A man with normal skin pigmentation whose albino mother has cleft iris. ***B.*** Show the different kinds of gametes which can be formed by (1) a woman who is heterozygous for both the skin and the iris characteristics; (2) a man heterozygous for the skin characteristic who has a cleft iris.

4. A certain type of migraine headache is believed to be dominant (*M*) to the normal condition (*m*). The allelic pair is autosomal. Deutan colorblindness (*d*) is an X-linked recessive, whereas normal color vision is dominant (*D*). A phenotypically normal woman consults a physician because her daughter has migraine. The doctor discovers that the girl is also colorblind. Write as much as possible of the genotypes of (1) the woman; (2) the daughter; (3) the husband of the woman.

5. *A*. Assume that in a woman who is heterozygous for deutan colorblindness, nondisjunction of the X chromosomes occurs at first meiotic division. What kinds of offspring would result if one or the other of the possible eggs are fertilized by sperms from a non-colorblind male? *B*. What could the result be if eggs from the above-mentioned woman have the normal number of X chromosomes but are fertilized by unusual sperms resulting from nondisjunction in the non-colorblind male at first meiosis?

6. A certain type of rickets is inherited as an X-linked dominant, *R*. The normal trait is recessive, *r*. A man with rickets marries a woman who is free of the disorder. What are the genotypes of these two persons and what kinds of offspring are to be expected?

chapter 5

SEX AS A COMPLEX CHARACTERISTIC

Sex-limited Genes

The expression of sex, as in the case of any characteristic such as eye color and blood clotting ability, involves the interaction of many genes with one another and with the environment. While the genes for the switch to maleness causing the onset of testis development are holandric (Y-linked), they are not the only ones which play a role in the development of the male reproductive system or of other masculine features. Other genes, including those for normal differentiation of testes and other parts of the male reproductive system, may be on autosomes or even on the X. Similarly, while the X chromosome carries genes necessary for the normal differentiation and expression of many female traits, other genes located on autosomes also exert an influence on female development. Moreover, as pointed out in the preceding chapter, the X is a large chromosome which carries the loci for many genes which influence characteristics unrelated to sex. All the observations are compatible with the concept that expression of sex and other human characteristics depends upon the actions of an assortment of genes found throughout all the chromosomes. And these activities may be influenced by the environment in which they take place.

Once these points are appreciated, it comes as no surprise that certain traits which are commonly considered male or female are determined by genes carried by both sexes. For example, beard growth is considered masculine, since it normally takes place only in males. However, genes determining beard features are not confined to the Y chromosome, even though the expression of the characteristic is normally confined to males. That genes on autosomes are involved is indicated by the fact that beard features found in a father and his son may be very different. Genes which a male receives from his mother can evidently affect his beard growth. Therefore, a female must carry these genes which typically do not express themselves in her. Such genes whose expression is normally limited to just one of the sexes are called *sex-limited genes.* Their limited expression results from the nature of the hormonal environment. In the internal environment of members of both sexes, both male and female hormones are present. It is the proper balance of these hormones which determines the normal expression of certain genes. Genes for beard growth will act only when male hormone is present above a certain level. In some pathological conditions, there may be an elevation of male hormone in a female. The adrenal cortex secretes many types of hormones, among them a class of androgens or male hormones. The presence of a tumor can cause the level of androgens to rise in a female; as a consequence, beard growth may occur.

The role of the internal environment is also clearly seen in the expression of breast development, which usually takes place only in the female. A male as well as a female, however, carries genes for this female characteristic. Therefore, a girl may show breast features which are due to genes transmitted to her by her father. She would receive a corresponding set from her mother as well, but some of these may not be expressed due to dominance or to other kinds of genic interactions. Genes for breast development depend upon a certain level of estrogen for their expression; so they are sex-limited, since the necessary level is not usually reached in the male. Klinefelter males (Fig. 4-3) often show some breast development, whereas the Turner female shows little. The atypical expression of sex-limited genes such as those for the breast and the beard characteristics undoubtedly ensues from upsets in the female: male hormonal balance caused by unbalance in the chromosomal constitution. Since hormones of both sexes are present in every person, any factor which increases the amount of male hormone in the female can bring about virilization; likewise, an abnormal increase in the proportion of female to male hormones in the male may cause feminization.

The role of hormones in triggering the expression of sex-limited genes is evident in certain cases of precocious puberty, where sexual maturation takes place at a rate faster than the normal. Development of a deep voice, which depends on male hormone, as well as other traits associated with masculine maturity, may be displayed by a boy who has not reached the age typically associated with the expression of certain secondary sex characteristics. The cause of such precocious expression may be a tumor of certain regions of the testis or of the adrenal cortex which brings about increased levels of male hormone. From cases such as these, it becomes clear that hormones may influence the time of expression of a gene. Any factor which alters the normal hormone balance at a given time can thus have a pronounced effect on the rate and development of a characteristic. Precocious puberty may also occur in a girl and may stem from pathological conditions in which estrogen levels become elevated at a premature age.

Sex-influenced Genes

Another class of genes can be recognized whose expression varies according to the sex of the individual. These are *sex-influenced genes,* genes which are expressed in both sexes but which behave as dominants in one sex and recessives in the other. Sex-influenced genes may be carried on any chromosome, but the majority of them is not sex-linked. The genes involved in pattern baldness appear to be sex-influenced. In the typical pattern baldness, onset of hair loss occurs about the age of 30 in the males. During the next 10 years or so, hair is lost in a typical fashion, usually starting from the top of the head and proceeding outward in all directions. Studies of human pedigrees are compatible with the concept that a pair of alleles is involved: *B* for baldness and *b* for non-bald. The genes are autosomal and are inherited in a typical Mendelian fashion. However, the gene *B* for baldness is expressed as a dominant in males and as a recessive in females. Its allele *b* for the non-bald trait acts as a dominant in females and a recessive in males. Therefore, for a woman to become bald, she must have the homozygous constitution *BB.* For a male to retain his hair, his genotype must be *bb.* The expression of the heterozygous state, *Bb,* is influenced by the sex of the individual, producing a bald phenotype in a male and a non-bald one in a female [Fig. 5-1(A)]. There is thus a decided effect on the dominance of each of these 2 alleles which is caused by the sex of the person. As in the case of sex-limited genes, the hormonal environment appears to be responsible for the difference. That this is the case is illustrated by those conditions where hormone therapy has been prescribed. A male with a sparse head of hair, when given female hormones for clinical reasons, may start to grow hair on the top of his head. Such a person may be genotype *Bb.* Under the usual circumstances, his male hormone level permits *B* to be expressed as a dominant. When given estrogens, the increased amount of female hormone permits the gene *b* to be expressed. However, this hair growth is often accompanied by breast enlargement, since the sex-limited genes for breast development, carried by the male but normally suppressed, will be called to expression. In a comparable manner, a woman taking male hormone may suddenly begin to lose her hair. This could happen to a woman of genotype *Bb* in whom the increased amount of male hormone will allow the *B* gene for baldness to mask the expression of its allele, *b* (Fig. 5-1(B)]. Moreover, a male who is castrated and loses his prime source of male hormone will express the genes for pattern baldness in the same way as would

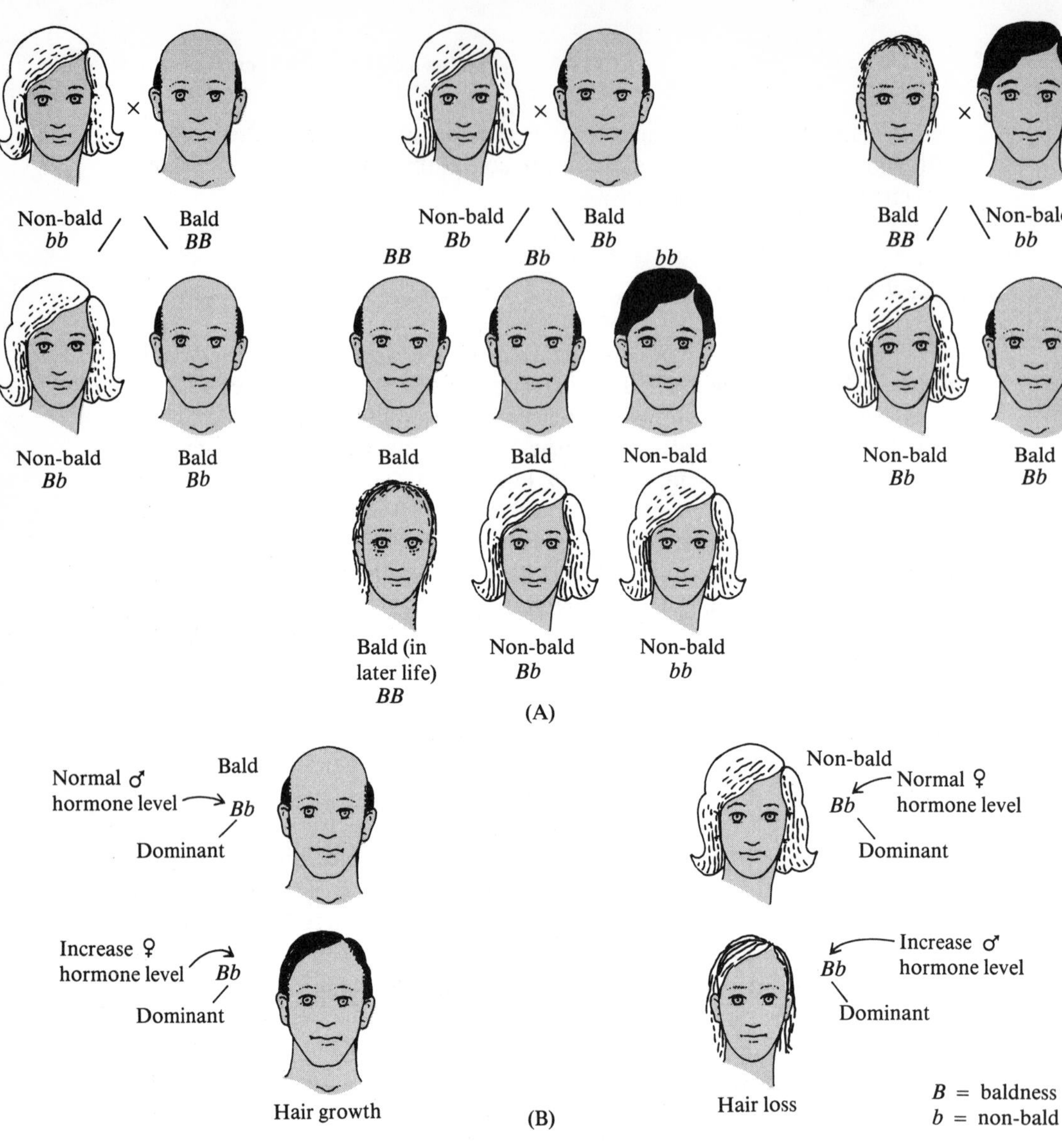

Figure 5-1. Sex-influenced genes. (A) The inheritance of pattern baldness is no different from that of any other autosomal trait. The expression of the genes involved in baldness, however, is influenced by the hormonal environment so that the gene *B* behaves as a dominant in males and as a recessive in females. Note that a woman (as in the cross on the right) will not show the extreme degree of baldness as will a male, even when she does experience hair loss. Moreover, the onset of hair loss will not occur until later in life. (B) The interaction between the genetic and the hormonal environment becomes very apparent when males must take female hormones or when females are prescribed male hormones. If two such persons are heterozygotes, the increased female hormone level will trigger gene *b* to expression in the male who is bald, and hair growth will occur. On the other hand, the increased male hormone in the woman under treatment will call gene *B* to expression, and hair loss will occur.

a female; and so a male castrate who is genotype *Bb* would tend to retain his hair.

Hair loss in a female of the genetic constitution *BB* will not be as pronounced as in a male of the same genotype. This is understandable in view of the fact that any normal woman would have a higher level of female hormone than of male hormone, and this would influence the degree of expression of the *B* gene, even when it is present in the homozygous condition. As might be expected, other genes play a role in baldness, and at least one pair is suspected to influence the time of onset of baldness. Young men may go bald if they carry genes for early hair loss, but these genes appear to be sex-limited. In women, any genes for early loss of hair cannot express themselves; consequently a female with normal hormonal balance will not lose her hair at any early age, even though she may have the appropriate genetic constitution (Fig. 5-2).

It must be pointed out that not all cases of baldness are genetic. Hair may be lost for a variety of strictly environmental reasons: high fever which can destroy hair follicles; certain chemicals and microorganisms which can cause

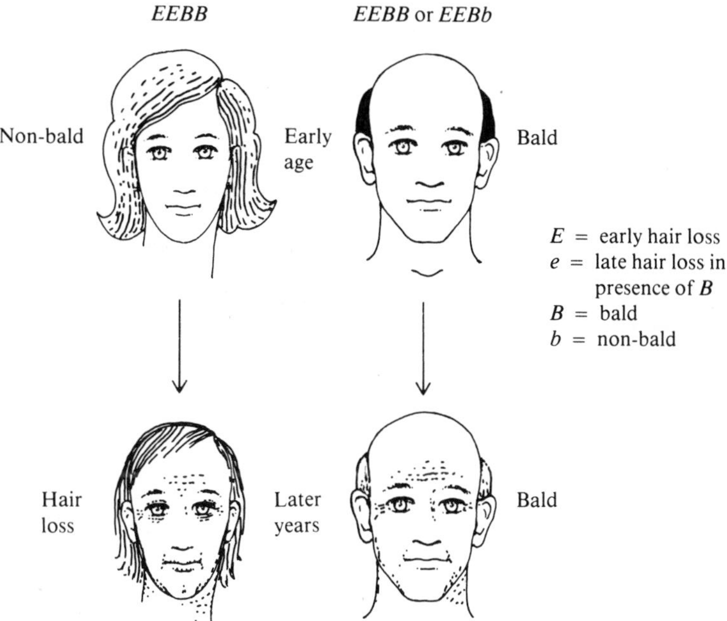

Figure 5-2. Sex-limited genes. Hair growth is under the influence of more than one pair of genes. At least one pair of them (*E, e*) appears to influence the *time* at which hair loss takes place. This pair of genes seems to be sex-limited. The gene for early loss of hair, *E*, cannot express itself in young women. It is limited in its expression to males. Therefore, even if a woman has the genetic constitution, *BB*, for hair loss, she will express these genes only in later life.

hair to fall. Some persons may thus be bald even though they have the genotype conducive to good hair growth. These persons would be phenocopies (p. 25) whose phenotype resembles that resulting from a specific genotype.

Though genes which can result in unusual or undesirable traits are often discussed, it must be remembered that the attention is often focused upon those genes whose effects are somewhat dramatic. As a result, the erroneous impression is sometimes given that genes are responsible only for undesirable effects. The absurdity of such an idea is evident when we remember that any undesirable gene can be recognized *only* because it has a normal form or allele. Moreover, normal human development obviously depends on a host of *normal* or *standard* genes (often referred to as *wild-type* genes). So any person must have a majority of normal or wild-type genes whose effects are typically human. An alternative form of a standard gene is usually referred to as the *mutant allele* and may produce an undesirable effect. There are also many gene pairs in which both forms seem to be about equivalent in effect and to have no associated disadvantages. The gene pair for the condition of the earlobes (p. 60) is an example. Another illustration of an apparently innocuous characteristic in humans is the length of the index finger. A pair of sex-influenced genes seems to play a role. In some persons, the index finger is longer than the ring finger. In others, it is shorter. A survey of a large group

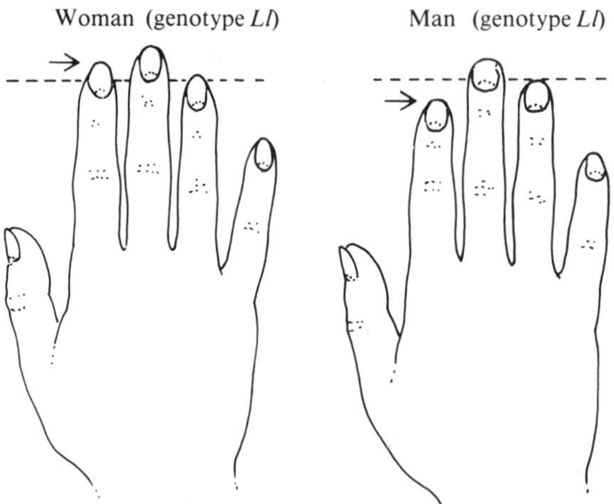

Figure 5-3. Length of index finger. This is an example of an insignificant characteristic which is based on the expression of a pair of sex-influenced genes, *L*, (long index finger) and *l* (short index finger). When the ring finger is placed on a line, it will be found that the index fingers of some persons extend above the line (left), whereas those of others fall below it (right). The gene *L* behaves as a dominant in females and as a recessive in males. Therefore, the genotype *Ll* is expressed as long finger in a woman and as short finger in a man.

usually shows an excess of long index fingers among females and short among males. This can be explained by assuming the gene L for long fingers to act as a dominant in females and as a recessive in males. The inheritance patterns would in all respects correspond to those discussed for pattern baldness (Fig. 5-3). This example is given to emphasize that genes with relatively inconspicuous effects are found in every human genotype and that genes whose expression may prove drastic for the individual are really not the commonest ones.

Aspects of Hermaphroditism

In discussions related to sex, the subject of hermaphroditism and related topics often arise. Biologically, a hermaphrodite is an individual capable of producing both sperms and eggs. The common earthworm is a truly hermaphroditic organism. In its strictest sense, true hermaphroditism does not occur in the human or any other mammal. The term, however, is applied to those cases in which both ovarian and testicular tissues are present, even within one gonad. The external sex organs typically exhibit various degrees of abnormality in which both male and female features can be recognized. The term *pseudohermaphrodite* is generally used for those persons having only one kind of gonad tissue (ovarian or testicular), but whose primary and secondary sexual characteristics are either intermediate or show a mixture of both male and female attributes. The direct cause of abnormalities of this kind is not always known. Some persons apparently have a normal chromosome constitution, such as an XY individual who is a pseudohermaphrodite with testes, small penis, some breast development, and certain other abnormalities of the reproductive system. Some cases of male pseudohermaphroditism have been attributed to X-linked as well as to autosomal genes which cause the testes to secrete female rather than male hormones. Both male and female hormones are related biochemically. It is conceivable that a particular defective gene could cause the building materials in cells of the testes of an otherwise normal XY male to be converted to the female rather than to the male sex hormone.

Some female pseudohermaphrodites with a normal XX chromosome constitution result from an overgrowth of the adrenal cortex during development, which produces excess male hormone. Virilization of the female can then take place. Some of the cases have been attributed to certain recessive genes which, when homozygous, result in adrenal overgrowth. This overgrowth can occur in a male as well as in a female, and it may produce severe symptoms leading to death in infancy.

Certain other hermaphrodite types have been found to possess two different classes of cells. For example, some body cells may have the chromosome constitution XY; others may be XO. Any individual possessing cells of more than one type of genetic constitution is said to be a *mosaic*. A mosaic always

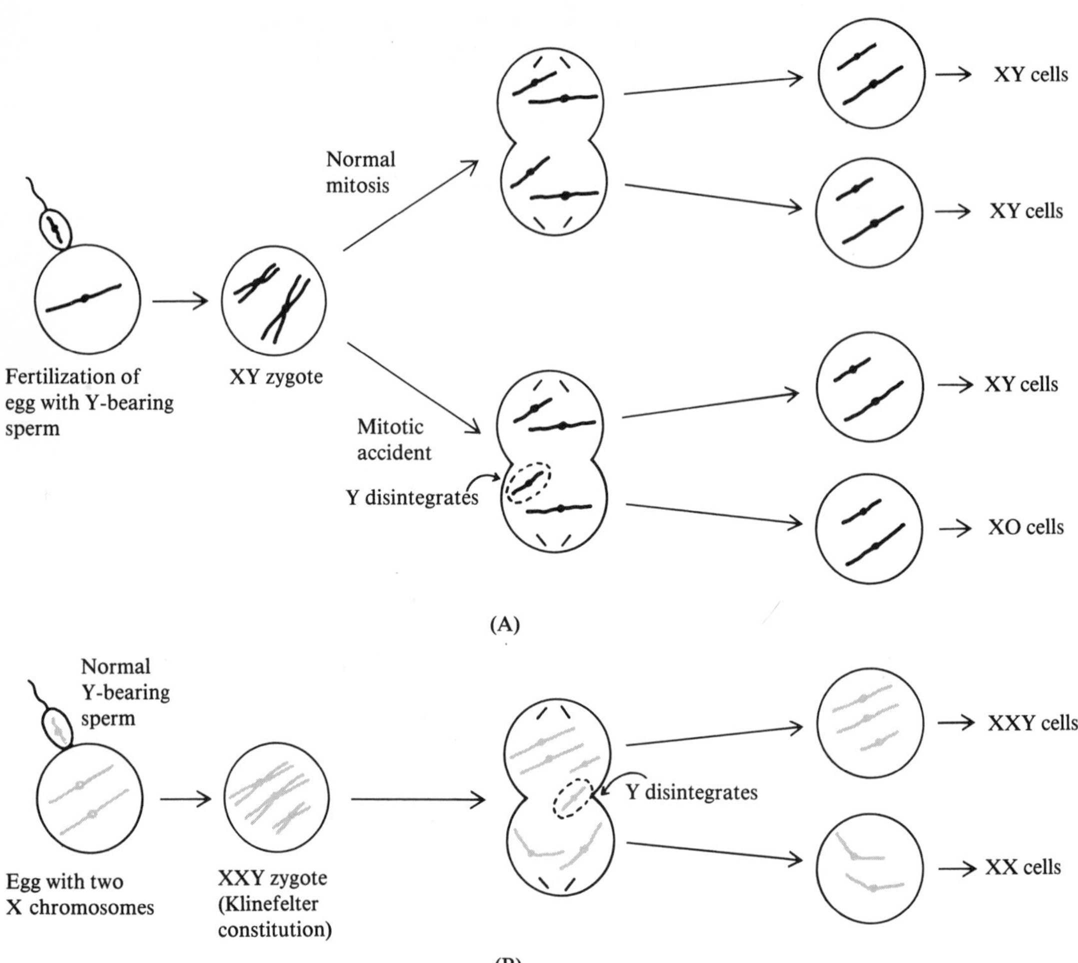

Figure 5-4. Origin of mosaicism. (A) An XY/XO mosaic can arise due to some failure in a mitotic division of the embryo. Normally each cell of an XY embryo will carry one X and one Y chromosome. If the Y chromosome fails to reach one of the poles during a mitotic division, it, like any chromosome not incorporated into a nucleus, will break down. Therefore, the product of the mitosis will be an XY cell and an XO cell. Further mitotic divisions will thus give patches of XO cells among those of the normal XY constitution. The extent of the mosaicism will depend on when the mitotic accident occurs during development of the embryo. (B) An XXY/XX mosaic can arise as a result of nondisjunction at meiosis (as in the production of an XX egg) followed by an accident at mitosis, in which a Y chromosome gets lost. The diagram indicates an error at first mitotic division of the zygote, but this could occur at any time. Therefore, different degrees of mosaicism occur.

has more than one cell line composing the body. The origin of two cell lines differing in chromosome constitution can be explained on the basis of an error during the mitotic divisions of the embryo [Fig. 5-4(A)]. Loss of a Y in an XY embryo will give patches of XY and XO cells, the extent of the latter depending on when and where during development the accident takes place. The presence of a Y in some cells but not in others could result in the abnormal development of tissues or organs associated with the reproductive system. Mosaic persons are known who are mixtures of XX and XO cell lines and who show symptoms of the Turner syndrome. Klinefelter males who are mosaic, such as XXY/XX have also been recognized. These unusual chromosome constitutions can be traced to accidents at meiosis in the formation of the egg or sperm, to accidents at mitotic divisions of the embryo, and even to a combination of both [Fig. 5-4(B)].

Sexual Preferences

Homosexuality, the performance by an adult of a sexual act with a member of the same sex, is a subject which is receiving wide attention today. About 2–4% of adult American males and 1% of the females are believed to practice homosexuality exclusively. Masculinity and femininity are often difficult concepts to define and frequently have a cultural basis (Chapter 15). A homosexual male, for example, may have all the behavior traits which his society attributes to masculinity except for his sexual preference. Just as a spectrum of masculine and feminine characteristics is evident within any heterosexual group, the same is true for a homosexual one. There is absolutely no evidence that the anatomy or the chromosome constitution of a homosexual are any different from those of a heterosexual. No sound basis exists for the idea that homosexuality is inherited, and there is no gene known which elicits homosexual preferences.

Experiments with rats suggest that alterations in the balance of the hormonal environment of the fetus and the newly born animal can cause them to exhibit certain behavior traits typical of the opposite sex. Comparable studies in the human concerning the influence of the maternal hormones on the fetus or concerning changes in the hormonal balance of the infant have not been made. There have been reports on work with the human that male homosexuals, when compared to heterosexual counterparts, show a reduced level of male hormone. However, many points still must be answered before such data can be properly evaluated. The reasons why one person develops heterosexual preferences and another homosexual ones still remain obscure. Little is known about the multitude of factors and the interactions which can determine human behavior of any kind. Certain genetic factors which can influence behavior have been recognized in a few lower species. Man's behavior certainly involves both a genetic component as well as an environmental one. However, the relative importance of each in the expression of human sexual behavior has by no means been established.

Publicity has been given to a number of persons who have undergone "sex change" operations. Such individuals are known as *transsexuals,* those who consider themselves to be members of the opposite sex. A transsexual does not classify himself or herself as a homosexual, and psychiatrists accept this viewpoint, since the homosexual person *does* identify with his or her own anatomical sex. The transsexual's identification with the opposite anatomical sex is often so intense that the person elects to undergo surgery. Most transsexuals are males, and their operation involves removal of the penis and testes, and the construction of an artificial vagina and clitoris from surrounding tissues in the pelvic area. Cosmetic surgery is required to remove the beard and to reduce the size of the Adam's apple. Beard growth, once initiated under the influence of the male hormone, will continue unless the beard follicles are removed or destroyed. Development of a conspicuous Adam's apple is a sex-limited trait, usually expressed only in males. Hormone therapy is required for breast enlargement and for the development of a degree of female musculature. The corresponding operation in a female transsexual entails removal of the breasts, construction of a penis, and the administration of male hormones. The procedure, however, is much more difficult and the results are not quite as satisfactory as in the male-to-female transformation. Regardless of either operation, the transformed person still remains genetically a member of his or her original sex. A male transformed to a female still retains his XY genetic constitution in all cells of the body. Likewise, the XX genetic constitution will persist in all the body cells of the person who was born a female. Neither surgery nor hormones will alter the kinds of chromosomes nor change the genes they carry. These transformed persons are in a sense phenocopies of the sex with which they identify. The reasons for a person's complete identification with the opposite sex are just as obscure as those which lead to homosexual preferences. The untransformed transsexual is in all detectable ways normal anatomically. Indeed, several of them have become parents, one male having sired five children before his transformation. (The transformed individual, of course, will never be able to sire or to bear children.) Moreover, there is no known gene or genes which predispose an individual to transsexuality. Any specific causative factors, environmental or genetic, which may be involved have yet to be identified.

The Female and Mosaicism

We have noted that the X chromosome carries many genes whose effects are on characteristics unrelated to sex: blood clotting, color vision, etc. Since a female carries two X chromosomes and a male only one of them, this implies that a female has a double dose of all sex-linked genes, whereas only a single dose is carried by the male. It is well known that an extra dose or a reduced dose of even a small amount of genetic material can cause pronounced defects in an individual. Normally, in the cells of any organism, there exists a balance of the different kinds of genes. Departure from that balance may produce serious consequences. Therefore, the question arises as to how a female can have a double dose of the many genes on the X which definitely affect the body, whereas the male possesses only a single dose. Great differences might be expected between males and females in relation to certain characteristics which are not associated with sex; yet this is not so. The reason is that a device exists which compensates for this dosage difference between the sexes, so that in the cells of the female only those genes on one of the X chromosomes actually express themselves. The explanation of this seeming paradox begins with the discovery in 1949 by Dr. Murray Barr of a body which is characteristic of the nuclei of cells from mammalian females. First discovered in the neurons of the cat, this body appears as condensed chromatin material which may be found in various locations within the nucleus (Fig. 5-5). Since the body is absent from cells of males, it was designated *sex chromatin.* It is also commonly referred to as the *Barr body.* The Barr body, or sex chromatin, can be recognized very easily in certain kinds of cells, such as the epithelial cells of the buccal mucosa which lines the mouth cavity. A simple scraping with a toothpick is all that is needed to obtain these cells which are then stained with any one of several common dyes. The body is typically found in these cell types at the edge of the nuclear membrane [Fig. 5-6(A)]. In certain of the white blood cells, the polymorphonuclear leukocytes, sex chromatin assumes an interesting appearance. In a simple blood smear of a female which has been stained with Wright's stain, some of the polymorphs can be seen to bear a

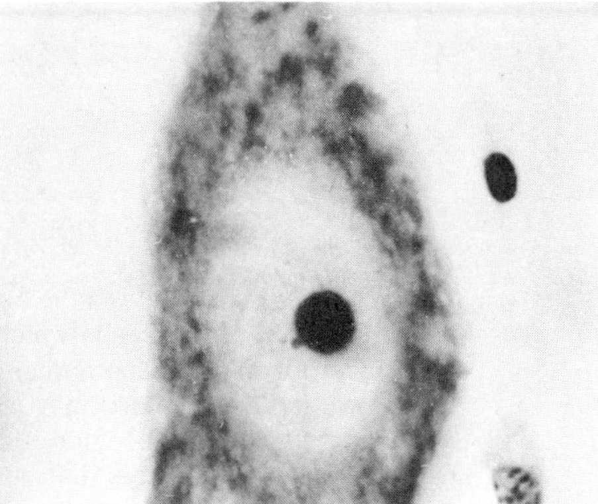

Figure 5-5. Barr body in neuron of cat. The sex chromatin may be positioned anywhere within the nucleus. It is typical (as shown here) for it to be adjacent to the nucleolus, the dark-staining body. The rest of the nuclear region appears as a clear, lightly stained area.

small appendage attached to one of the lobes of the convoluted nucleus [Fig. 5-6(B)]. Such an appendage is absent from the white blood cells of males. From its appearance, the body has been called a "drumstick." It is now known that the drumstick is also sex chromatin. Although sex chromatin may not be evident in every cell of a female, it is believed that it is indeed present but that its location in the nuclei of some cells makes resolution difficult.

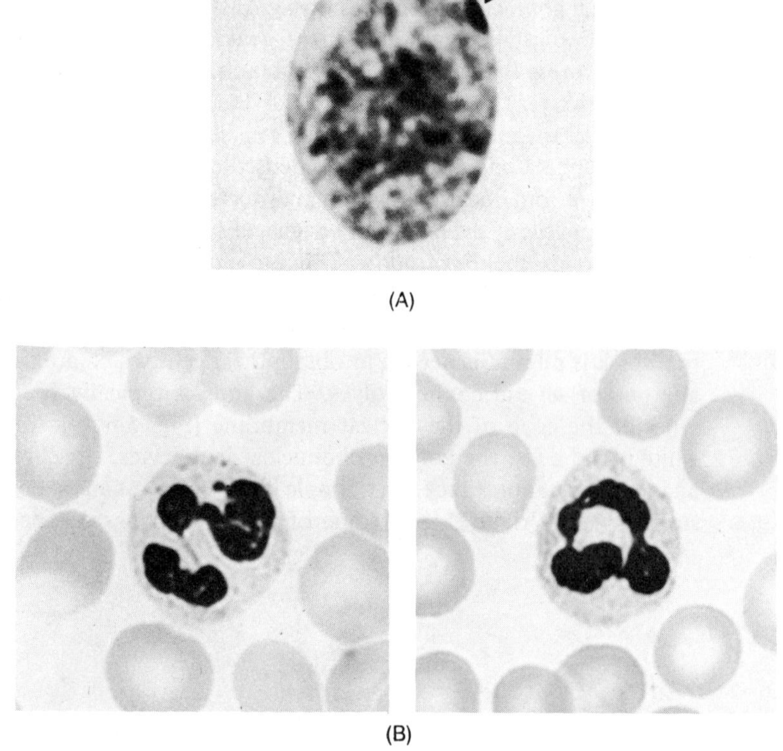

Figure 5-6. Sex chromatin in the human female. (A) A simple scraping with a toothpick will provide cells which can be placed on a slide and stained, using acetocarmine or acetoorcein. The majority of the cells in a normal female will show one Barr body (arrow). (B) In a small percentage of the polymorphonuclear leucocytes, those white blood cells with very lobed nuclei, a small appendage, the drumstick, may be observed in the female (left). This is equivalent to the sex chromatin or Barr body seen in other cell types. The position of the sex chromatin in a cell may make its detection difficult, but it is believed to occur in the majority of all kinds of cells from a normal female. The white blood cell of a normal male shows no drumstick (right). Figure 5-6(B) courtesy of Carolina Biological Supply Company.

Following its discovery, a correlation was made between the amount of sex chromatin and the number of X chromosomes in a cell. A Turner female, XO in chromosome constitution, shows no Barr body or drumstick. In contrast, a Klinefelter male who is XXY shows one Barr body and one drumstick. When persons of various unusual chromosome types are studied (such as XXXX, XXX, XXXY, and XXYY), the following correlation is seen: the number of Barr bodies or drumsticks equals one less than the number of X chromosomes. This relationship suggested that the sex chromatin actually represents an X chromosome, or a portion of an X, which is condensed in the nondividing nucleus. A series of cytological studies has supported this idea. It is also known that when any chromosome material is in a condensed state, it is relatively inactive as a participant in the metabolic activities of the cell. Actually, when we see the condensed chromosomes at mitosis or meiosis, they are quite inactive. Their movements at the time of nuclear divisions are concerned only with their distribution, not with the chemical activities of the cell.

All the observations were assembled by Dr. Mary Lyon, who has presented a very convincing argument on the nature and significance of the sex chromatin. All of the propositions of the Lyon hypothesis have now been supported by observation or experimental analysis. The process of condensation of an X chromosome in a female is called *X-inactivation.* Some geneticists refer to the condensed X as a *lyonized X* and call the condensation *lyonization.* According to the Lyon hypothesis, all mammalian females are mosaics in the sense that two cell lines are found in their bodies. In some of the cells, the X chromosome the female received from her mother is condensed and thus inactive while the X from the father remains active. The contrary situation is true in other cells where the active X is the one from the mother and the paternal X is lyonized and inactive. Which X chromosome becomes condensed in any one cell is a matter of chance. Some mechanism becomes operative to inactivate one of the X's early in the development of the embryo. This statement is supported by the fact that both X's in the female appear the same during oogenesis. In the very early female embryo, both X's continue to look and to behave alike in all the cells. However, when the embryo is about 16 days old, sex chromatin appears in the cells which will become body cells. The device which is responsible for the sudden condensation of one of the X's is still unknown, although several ideas have been offered.

Evidence for the Lyon Hypothesis

If one of the X chromosomes of a female is indeed inactive, this means that she is actually hemizygous for the X in any one cell. Therefore, the dosage of her X-linked genes in any cell is just the same as that for any normal male. Proof of this dosage compensation is obtained from the following type of observation. Persons can be compared who are of different genetic constitutions for the pair of genes *H* and *h,* which influence blood clotting. As already discussed, the normal gene *H* controls the protein AHF which is required for normal blood clotting; its allele, *h,* results in the absence of this essential clotting factor. The AHF protein can be detected in the blood serum.

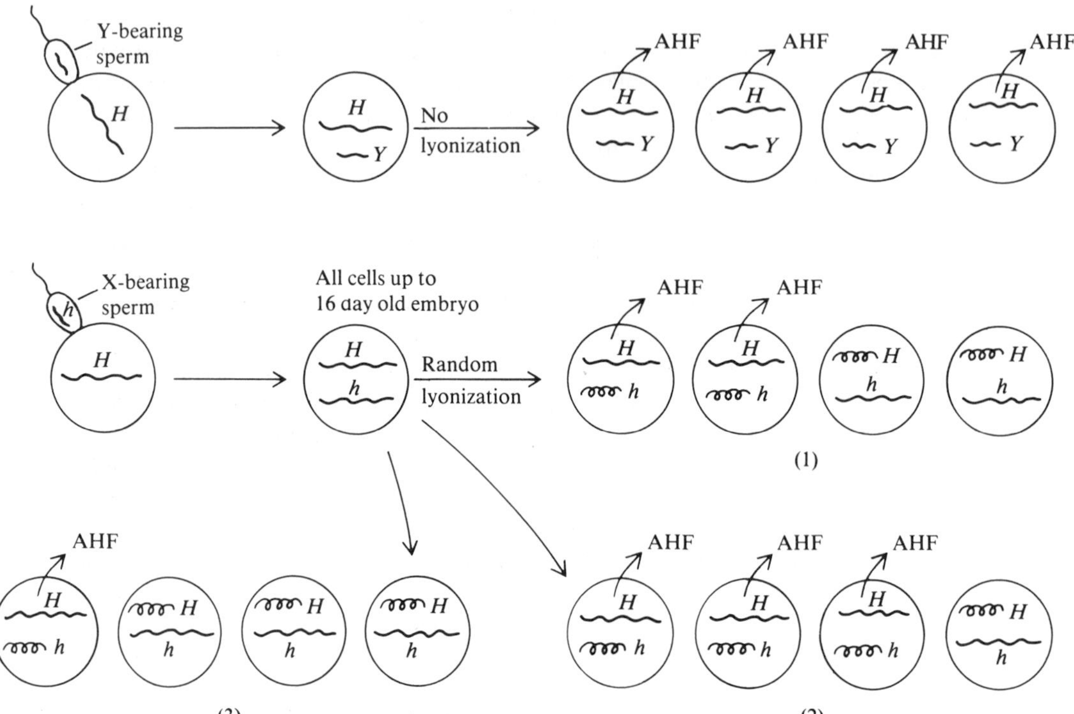

Figure 5-7. Random inactivation of the X chromosome. In the male, no lyonization occurs. Therefore, a male carrying one dose of the normal allele for AHF production (above) will have a potentially active gene for production of the clotting substance in all his cells. A female who is a heterozygote will also carry a single dose of the normal gene for AHF production, but she will show a lower AHF level than that of the hemizygous male. There will also be variation in AHF level from one carrier female to the next, as would be expected on the basis of the Lyon hypothesis (below). On the average, half of the cells of a carrier female will have the X with the recessive allele (*h*) condensed, while the X with the normal allele remains active (1). In a smaller number of carrier females, the X with the recessive allele has been inactivated in most cells (2). Therefore, these females will have AHF levels which approach those of the *HY* male and the *HH* female. As expected, a small number of carrier females will have very low AHF levels (3), since by chance the X with the normal allele has been rendered inactive in the majority of the cells.

When the amount of AHF from a homozygous woman (*HH*) is compared to that from a normal man (*H*Y), the levels of AHF in both are found to be the same. This means that the genotype, *HH,* does not result in the production of a double amount of the clotting factor. However, when a heterozygous female, *Hh,* is so examined, it is found that the amount of AHF is lower than that of either the *HH* female or the *H*Y male. Even though both the heterozygote, *Hh,* and the male, *H*Y, carry one dominant gene for AHF production, they nevertheless produce different levels of the substance which is controlled by that gene. This is actually what is expected on the basis of the Lyon hypothesis (Fig. 5-7). In some cells of the female, the X chromosome with the normal gene is operating, since that X is not condensed; in other cells, the gene is inactive, since by chance that chromosome with the *H* gene is condensed in that cell. Therefore, not all the cells will be producing AHF protein. Since the condensation of the X is at random, it is to be expected that carrier females, *Hh,* will vary from one to the other when they are compared for their AHF levels. This prediction has been borne out. A few heterozygous females have levels of AHF almost as high as that found in the *HH* female. This indicates that the X chromosome with the normal allele *H* is active in most of the cells. A few other carrier females show very low levels of AHF (but enough to prevent hemophilia). In this case, the X with the normal gene is condensed in a majority of the cells. Most heterozygous females have been found to have levels of AHF between these two extremes, again as would be expected according to the Lyon hypothesis.

One additional line of evidence will be given, although several others could be presented. There is a rare sex-linked recessive gene which prevents the red blood cells from making a certain enzyme named glucose-6-phosphate dehydrogenase (G6PD). This enzyme is normally present and plays a minor role in the utilization of glucose by the cell. Its absence is generally not significant and typically remains unknown in the average situation. However, the absence of this enzyme somehow renders the membrane of the red blood cell fragile when certain drugs are taken, especially primaquine, used in the treatment of malaria, and sulfanilomide. Red cell destruction and anemia ensue. (We see here another good example of the role of the environment in the expression of the genotype.) Most persons who lack G6PD are males, since a sex-linked recessive is involved. A comparison can be made among males and females whose genotypes are known for the pair of alleles, *G* and *g,* carried on the X chromosome (Fig. 5-8). As in the case of AHF, the *GG* female and the *G*Y male show the same level of G6PD. However, in the case of this enzyme, it is actually possible to detect whether or not G6PD is present or absent in any *one* cell. It is seen to be present in every cell of those persons of the two genotypes, *GG* and *G*Y. Again, the heterozygous female, *Gg,* has less enzyme than persons of the other two genotypes. But more important, some *single* cells can be shown to produce the enzyme, whereas others can be shown to lack it entirely. This means that the enzyme production in a cell is all or none, exactly what is to be expected if one of the X chromosomes is condensed and inactive in any particular cell. The X with the gene for the production of enzyme, *G,* operates in some cells; in others, it is inactive, since the X chromosome to which it is linked is condensed. There is little doubt that all mammalian females, including the human, are somatic mosaics; two lines of cells compose their bodies, some with an active maternally derived X chromosome, others with an active X which was contributed by the paternal parent.

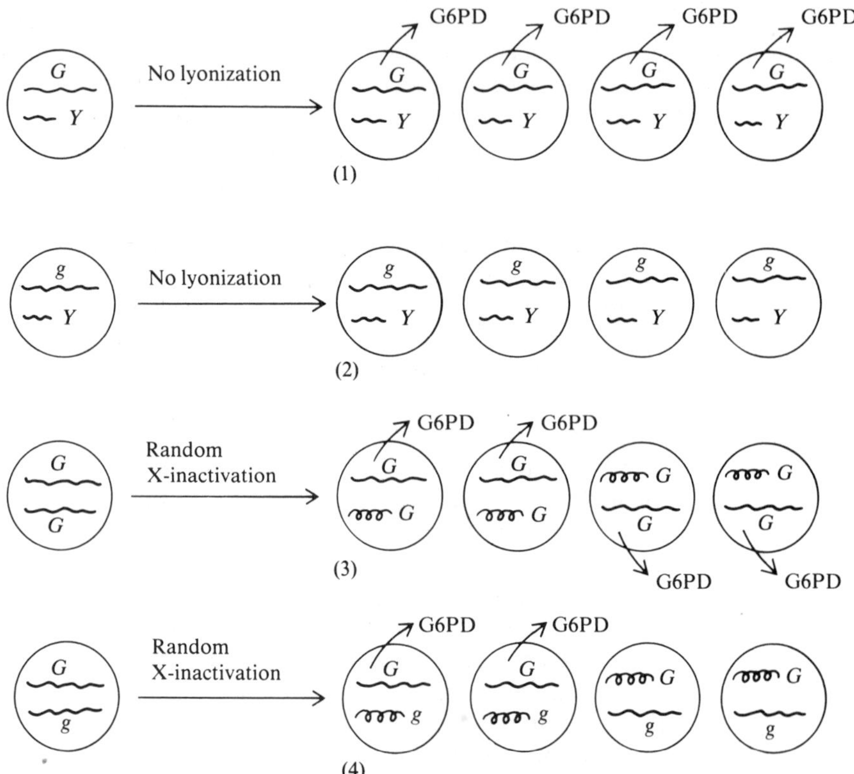

Figure 5-8. X inactivation and G6PD production. Single red blood cells can be analyzed for the presence or absence of G6PD. In every cell of a male with the dominant gene for enzyme production, G6PD is present, since his single X is active in all the cells (1). In a male with the recessive allele, no enzyme is present in any cells (2). In a female homozygous for the dominant gene, every cell has the enzyme, since there will be one X chromosome carrying a dominant gene no matter which X is lyonized (3). In the heterozygous female (4), however, a given cell will either contain the enzyme or will lack it entirely, since some cells will have a condensed X which carries the recessive allele, whereas others will have a condensed, inactive X with the gene for enzyme production.

It should be noted that X inactivation or lyonization in a female is probably partial, not affecting the entire chromosome. A normal XX female is quite distinct from the XO Turner female. If any X chromosome in addition to a basic one were completely inactivated, this would mean that the XX and the XO females would both possess the same number of active X chromosomes in their body cells. It is much more likely that some small portion of the X does *not* become inactivated by the mechanism of lyonization which represses most of the X-linked genes. Certain regions of the X chromosome may be necessary in double dose for normal female development. Indeed, two complete X chromosomes are known to be necessary for female fertility. The loss of just a portion of an X produces infertility. Studies of deletions and other types of aberrations in the X chromosome show that the loss of just the short arm of the X will result in a Turner female, similar to a person who has just one X. The absence of the short arm of the X chromosome would seem to be responsible for the body features associated with the Turner syndrome. If a deletion is present in the long arm of the X chromosome, infertility again results, but none of the physical features of the Turner syndrome are present. It would seem likely that it is most of the long arm which becomes inactive as a result of lyonization, most or all of the short arm of the X remaining active. Arms of the sex chromosome will be discussed later (Chapter 8) in reference to genes they may carry which affect height.

Practical Implications of a Knowledge of Sex Chromosome Constitution

The presence or absence of a Barr body or sex chromatin is more than just a curiosity. A simple cell scraping may aid in the diagnosis of suspected cases of the Turner or Klinefelter syndromes before the more tedious karyotype analysis is made. In cases of hermaphroditism or pseudohermaphroditism where anatomical features may appear intermediate, the presence or absence of sex chromatin in the cells can help in the recognition of the true genetic sex of the person, either XX or XY. Surgical procedures may then be undertaken to develop those anatomical features which are appropriate to the genetic sex. It should be noted that an XY transsexual who has become a female will not show sex chromatin. And the XX person who identifies as a male will continue to show the Barr body and the drumstick, no matter how extensive the surgery or the hormone therapy.

Often the knowledge of the sex of an unborn child is of paramount importance in counseling for a genetic disorder. This information can be obtained through the technique of *amniocentesis,* which is assuming increasing importance in prenatal diagnosis. In the procedure, the physician inserts a hypodermic syringe into the abdomen of the pregnant woman (Fig. 5-9). The

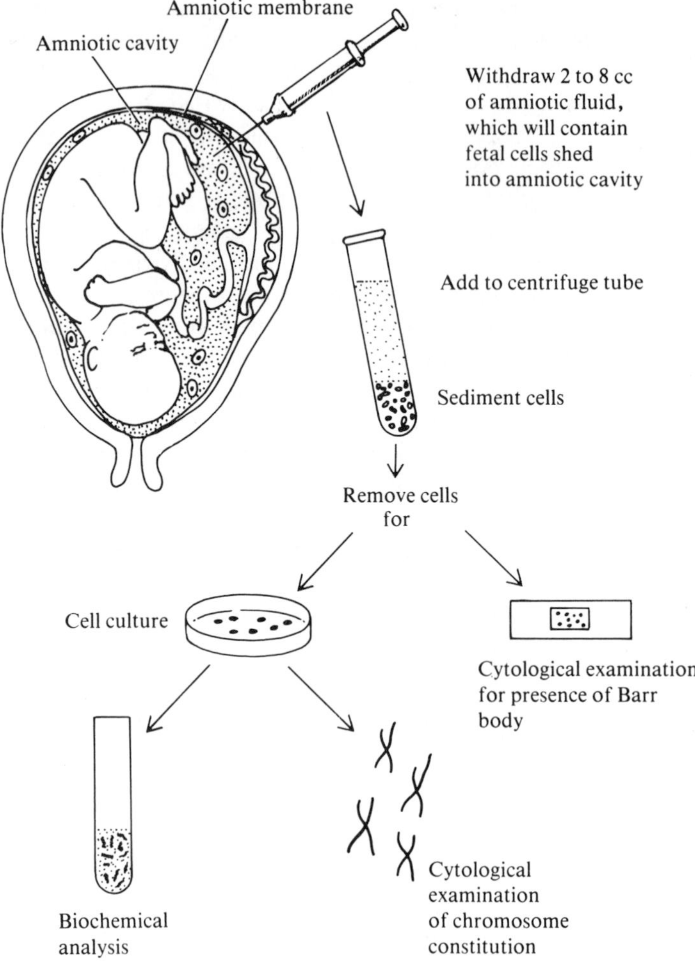

Figure 5-9. Amniocentesis. The amniotic fluid contains cells which are all fetal, derived either from the body of the fetus or from the amniotic membrane which is genetically identical to the fetus proper. When fluid with cells is withdrawn, it may be processed in a variety of ways and is thus extremely valuable in prenatal diagnosis. Sex determination, by noting the presence or absence of sex chromatin, is one of the chief uses of the procedure. Cells may be raised in tissue culture and kept for weeks. They can be subjected to biochemical analysis and tested for the presence or absence of one or more chemical substances. This is of great aid in the detection of certain genetic disorders *in utero.* Another important application is the analysis of cells growing in culture for their chromosome content. Down's syndrome, for example, can be readily detected by the presence of an extra chromosome #21. Amniocentesis is being more routinely used in analyses of this kind on women over the age of 40 (see Chapter 10 for more details).

needle penetrates into the amniotic cavity, the fluid-filled space in which the child develops. As growth takes place, cells are shed from the embryo or fetus into the fluid, and a sample of these can be obtained by amniocentesis. The cells are then concentrated by centrifugation and can finally be grown under test tube conditions. This makes it possible to analyze them in a variety of ways. If the embryo or fetus is XX in chromosome constitution, a Barr body will be present. A complete karyotype analysis would be performed to establish precisely the number and appearance of sex chromosomes and autosomes. The identification of the sex *in utero* can have very important applications. For example, suppose a pregnant woman is found to be a carrier of the gene for hemophilia. She has a 50:50 chance of passing the gene to any male offspring and producing a hemophiliac son. If amniocentesis followed by karyotype analysis indicates that such a woman is carrying a male child, special precautions may be instituted for the safety of the baby at birth. In addition, the knowledge that a particular unborn child is a male can, in other kinds of cases, be evaluated along with genetic information known about the parents. This often permits the calculation of the chance for the birth of an afflicted child with a high degree of accuracy. In Chapter 12, the application of amniocentesis to medical genetics and genetic counseling will be pursued further, not only in relation to sex-linked disorders but also to those caused by genes carried on any of the autosomes.

REFERENCES

Avers, C. J., *Biology of Sex.* John Wiley & Sons, N. Y., 1974.

Beutler, E., "Glucose-6-Phosphate Dehydrogenase Deficiency." In *The Metabolic Basis of Inherited Disease,* Third edition, Stanbury, J. B., Wyngaarden, J. B., and Fredrickson, D. S. (Eds.), p. 1358. McGraw-Hill, N. Y., 1972.

Lyon, M. F., "Sex Chromatin and Gene Action in the Mammalian X-Chromosome." *Amer. Journ. Human Genet.,* 14:135, 1962. (Reprinted in S. H. Boyer, *Papers in Human Genetics.* Prentice-Hall, Englewood Cliffs, N. J., 1963.)

Lyon, M. F., "Possible Mechanism of X-Chromosome Inactivation." *Nature New Biol.,* 232:229, 1971.

Money, J., and Ehrhardt, A. A., *Man and Woman, Boy and Girl.* The Johns Hopkins Univ. Press, Baltimore, 1972.

Moore, K. L. (Ed.), *The Sex Chromatin.* W. B. Saunders Co., Philadelphia, 1966.

Mittwoch, U., "Sex Differences in Cells." *Sci. Amer.,* July: 54, 1963.

Ohno, S., "Evolution of Sex Chromosomes in Mammals." *Annu. Rev. Genet.,* 3:495, 1969.

REVIEW QUESTIONS

1. Select the proper answer(s). A sex-influenced gene is one which (1) is found only on the regions of the X and the Y which are homologous; (2) is found only on the Y chromosome; (3) will be expressed only in one sex; (4) may be found on autosomes as well as on sex chromosomes; (5) can be expressed as a dominant in one sex and a recessive in the other.

2. Select the proper answer(s). Analysis of a buccal smear from a female reveals the presence of three Barr bodies. The following would seem likely: (1) The woman is a Turner female. (2) A Klinefelter condition probably exists. (3) The individual is probably a transsexual. (4) Three drumsticks will probably be present in white cells of this person. (5) Four X chromosomes can be expected in a karyotype analysis.

3. In cats, the gene for black fur (*B*) acts as a codominant to its allele for yellow fur (*b*). The heterozygote is a tortoise cat. The pair of alleles is X-linked. In the clover butterfly, the autosomal gene associated with white color (*W*) is dominant to yellow (*w*). However, the dominant allele is limited in expression to females. You notice a tortoise kitten and a yellow kitten playing with a white butterfly. What can be said about the sex and genotypes of (1) the tortoise kitten; (2) the yellow kitten; (3) the butterfly?

4. Dairy farmers consider the pedigrees of both the cow and the bull for high milk yield. Why should the bull be considered?

5. *A*. In the human, the autosomal gene pair for long index finger (*L*) and short index finger (*l*) is sex-influenced. *L* acts as a dominant in the female, whereas *l* acts as a dominant in the male. Write as much as possible of the genotypes of the following: (1) a male with long index finger; (2) a female with short finger. *B*. What kinds of children would these two persons produce with respect to index finger length?

6. The enzyme G6PD is present in a red blood cell only if the X-linked dominant (*G*) is being expressed. The recessive allele (*g*) cannot elicit formation of the enzyme. Suppose that 100 red cells are tested from persons of the genotypes given below. How many cells on an average would be expected to contain the enzyme? (1) *GG;* (2) *G*Y; (3) *Gg;* (4) *g*Y; (5) *gg*.

chapter 6

LINKAGE, CROSSING OVER, AND HUMAN VARIABILITY

Linked Genes Tend to Stay Together

Independent assortment always occurs among pairs of genes whose locations (loci) are on separate chromosomes (Fig. 3-14). A moment's thought, however, tells us that the number of genes in any creature must exceed by far the number of chromosomes. The human with only 23 pairs of chromosomes certainly must carry thousands of different kinds of genes. There must therefore be a large number of gene loci on any one kind of human chromosome, even on the very small chromosomes such as #21 and #22 (Fig. 2-4). In the course

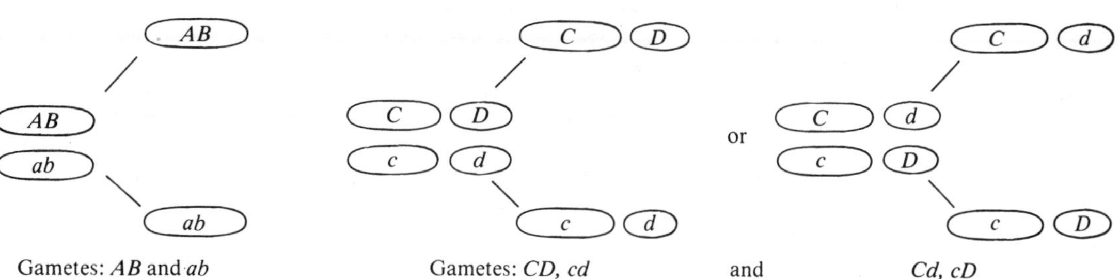

Figure 6-1. Linkage versus independent assortment. If two genes are associated with two different loci located on the same chromosome (left), any gene combination on one chromosome will tend to stay together and thus be inherited together. This is to be contrasted with the situation in which two different loci on two different chromosomes are considered. In the latter, four different kinds of gametes arise in equal proportions from a dihybrid (right). If two genes are completely linked, a dihybrid will form only two classes of gametes, since the old gene combination will remain intact.

of studying the genetics of any living organism, it is certain that one would encounter 2 or more genes which are carried on the same chromosome. Indeed, it is a common procedure in genetic studies of most species to follow simultaneously the inheritance of 2 or more genes associated with the same chromosome. In certain microorganisms, this is all that can be done, since many

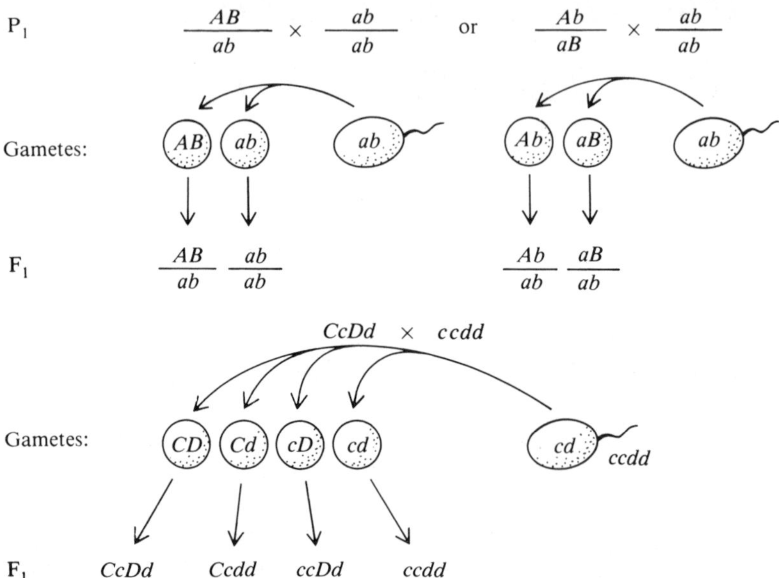

Figure 6-2. Linkage and testcross results. Note the manner in which linked genes are represented in a diagram. They are written as a fraction to distinguish them from genes which are found on different chromosomes. The two genes above the bar are found in that association on one chromosome. The association of genes below the line indicates the arrangement of those two genes on the homologue. The bar or line can be considered to represent two chromosomes (one pair of homologues), so that the genes *A* and *B* (above left) are found on one chromosome and the alleles *a* and *b* are associated together on the homologue. This arrangement in a dihybrid of two dominants linked together on one chromosome and two recessives on the other is known as the *cis* arrangement. Two linked genes need not always occur in the same arrangement, such as this. Note (above right) that on one chromosome a dominant gene may be associated with a recessive (*Ab*), and on the homologue the corresponding recessive and dominant may be associated (*aB*). This arrangement in a dihybrid is known as the *trans* arrangement. From a dihybrid with either the *cis* or the *trans* arrangement, only two classes of gametes can be formed when linkage is complete. When crossed to a double recessive, the testcross in each case produces only two classes of offspring. In contrast to this is the behavior of two pairs of genes which are not linked (below) and which can assort independently. Four classes of gametes are formed. When crossed to a double recessive, four classes of offspring arise.

lower cell types (some bacteria) contain only 1 chromosome per nuclear region. Consequently, all of the loci and the genes which can be found at them are tied together. The genes of viruses are found together on 1 DNA molecule or virus "chromosome." Genes are said to be *linked* when their loci are found on the same chromosome.

Two linked genes do not tend to undergo independent assortment. The reason for this is easy to grasp. If the two genes which are being followed in a cross happen to be on the same chromosome, this simply means that they will tend to travel together at meiosis and to be transmitted to the next generation in exactly the same arrangement as the one found in the original parent (Fig. 6-1). The fact that linked genes tend to stay together rather than to behave independently is no contradiction to Mendel's second law, since the law of independent assortment was formulated from the patterns of inheritance of two or more *unlinked* genes, those on separate chromosomes. As it happens, all of the genes which Mendel chanced to follow in his pioneer work assorted independently. He did not encounter complications of linkage in his studies with the pea plant. This is perhaps fortunate, since Mendel knew nothing about chromosomes, mitosis, or meiosis. The complication of linkage in these early genetic experiments could have led to confusion and slowed progress in the development of the science of genetics.

The recognition of linkage between two genes is usually established through genetic analysis by performing different types of crosses and following the behavior of two or more genes together. Figure 6-2 gives some of the conventions used in representing linked genes, and it also illustrates that the results of dihybrid testcrosses are very different when two genes are completely linked in one kind of cross and when two other genes are unlinked in another kind. In the former case, only old combinations of genes are transmitted, whereas independent assortment yields new combinations as well as the old ones. Detailed linkage studies have been made on a number of experimental plants and animals. In these well-studied species, this has permitted the assignment to specific chromosomes of specific genes with known phenotypic effects.

Linked Genes May Separate

Linkage analyses have revealed that, although two linked genes tend to stay together and to be inherited together from one generation to the next, the linkage is by no means 100% as shown in Figs. 6-1 and 6-2. The genetic data obtained from crosses in species like the mouse, the fruit fly, and corn clearly demonstrate that linked genes may come apart, that is, they may separate and enter into the formation of new combinations. A further consideration of chromosome behavior during meiosis supports this concept. In early meiotic prophase, it will be recalled, the homologous chromosomes engage in specific pairing. At anaphase I, they separate and move to opposite poles of the spindle. Before their separation, however, the chromatids in a bivalent appear to cross at one or more points, as revealed by observations with the microscope (Fig. 3-3). Each point where the chromatids seem to cross is called a *chiasma.* Two or more chiasmata per bivalent can be seen in the chromosomes of a primary spermatocyte or primary oocyte. According to the most widely accepted interpretation, these chiasmata are associated with the breakage and the reunion of chromatids (Fig. 6-3). The breakage takes place at precisely the same site on each member of a chromosome pair. Following the breakage, a reciprocal exchange may take place between the maternally derived and paternally derived chromosomes. A segment from the chromosome contributed by the female parent joins the remaining segment of the chromosome from the male parent. In an identical manner, the corresponding segment from the chromosome donated by the male parent joins the remaining segment of the maternally derived chromosome. The outcome of this reciprocal exchange is a new association of genes contributed by the parents. Figure 6-3 shows that a new combination of the genetic material arises. This separation of linked genes is called *crossing over.* The number of new combinations which arise between linked genes depends on the number of times such a breakage and

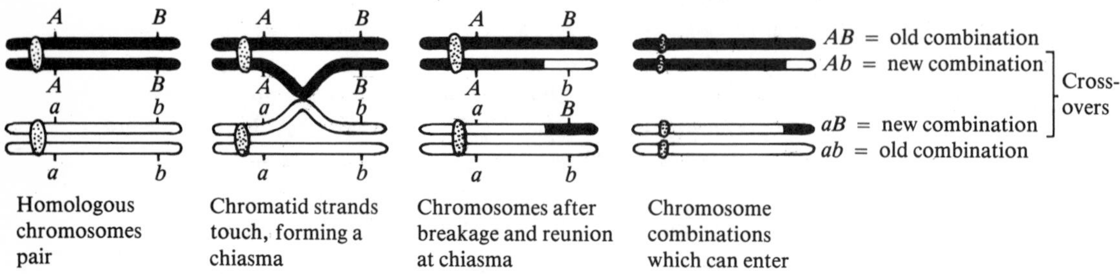

Figure 6-3. Chiasmata and crossing over. According to this concept, the homologous chromosomes pair at meiotic prophase and form an initimate association in which the corresponding loci on the homolgues pair (such as the *a* locus and the *b* locus). Therefore, any alleles at the loci will be paired. Each chromosome is composed of two chromatids. Therefore, each gene locus would be present four times. Thus at the time of crossing over, there would be four *a* and four *b* loci. An *A* or a *B* gene or an *a* or a *b* could be present at one of the corresponding loci on a chromosome. Any two of the chromatids may cross at one or more points, each called a *chiasma.* If the threads break, an exchange of chromatid segments can occur so that portions from each homolgue are switched reciprocally. At the end of meiosis, after the centromere of each chromosome has divided, four combinations are present, the old ones (the parentals) and the new ones (the crossovers or recombinants). If no crossing over had taken place, all the gametes would carry the old combinations.

Chromosomes	Crossover	*a* forms new combinations with:	New combinations		
ABCDE F / *abcde × f*	Between *e* and *f*	*f*	*ABCDEf*		*abcdeF*
ABCD EF / *abcd × ef*	Between *d* and *e*	*e* and f	*ABCDef*	and	*abcdEF*
ABC DEF / *abc × def*	Between *c* and *d*	*d, e,* and *f*	*ABCdef*	and	*abcDEF*
AB CDEF / *ab × cdef*	Between *b* and *c*	*c, d, e,* and *f*	*ABcdef*	and	*abCDEF*
A BCDEF / *a × bcdef*	Between *a* and *b*	*b, c, d, e,* and *f*	*Abcdef*	and	*aBCDEF*

Figure 6-4. Frequency of crossing over. The probability of crossing over between any two loci on a chromosome is constant. However, the frequency varies when different loci on a chromosome are compared, some showing very little crossing over, others undergoing a high frequency of recombination (formation of new gene combinations). This can be explained simply on the basis of the fact that genes occur in a linear order on a chromosome and that two genes close to each other will tend to remain together if a crossover occurs anywhere along the length of the chromosome. The farther apart two genes, the greater the chance that a chiasma and a break can occur between them and separate them. The simplified figure above shows only one chromatid per chromosome for the sake of clarity, and it considers only new combinations between locus *a* and the other loci. Note that *any* cross to the right of *a* will bring about new combinations of any genes to the right of the crossover. It would be less likely for any single crossover to occur between the *a* and *b* loci than it would be for any one to occur elsewhere along the length of the chromosome. Note also that when *a* is separated from *b,* it forms new combinations with all genes to the right of *b.* So the farther apart two linked genes are, the greater is the chance of crossing over between them.

mutual exchange takes place between paired homologous chromosomes. It is known from linkage studies in many species that the frequency of crossing over between two linked genes depends largely on the distance between their locations on the chromosome. As Fig. 6-4 shows, the closer together two linked genes are, the less chance there is that they will become separated, assuming that breakage and crossing over can take place at random along the chromosome. The greater the distance between two genes, the greater the chance that a crossover will occur between them and thus the greater the chance that these two linked genes will form new combinations.

Gene Locations Can Be Mapped

If the amount of crossing over is a reflection of gene distance, it should be possible to map gene locations on the chromosome. This possibility was realized over 50 years ago during the course of genetic studies with the fruit fly, the first species for which chromosome maps were constructed. The details of mapping are not essential to an understanding of human genetics, but some points must be appreciated, since rapid progress is being made in the construction of human chromosome maps. As we will learn in Chapter 12, these have practical application for the genetic counselor. The logic behind chromosome

Figure 6-5. Chromosome mapping. The basic logic behind chromosome mapping is summarized here. With experimental plants and animals, crosses can be easily manipulated by the investigator. When 2 linked genes are being studied, the number of new combinations between them can be determined by making a testcross, a cross to a complete recessive for the genes under consideration. The number of new combinations is a reflection of the amount of crossing over (see Fig. 6-6 for qualification). This in turn reflects their distance apart on the chromosome (see Fig. 6-4). Genes can thus be mapped by noting the percentage of new combinations which arises from a testcross (many testcrosses would be made). The percentage of crossing over or new combinations is simply translated as map units. In cross #1 (above), genes at the *c* and *e* loci give new combinations among the offspring in the amount of 10%. We can thus say that loci *c* and *e* (and hence the genes which can occur at those locations) are 10 map units apart. In cross #2 (middle), a cross follows *c* and *b* and gives 20% new combinations, meaning *c* and *b* are 20 map units apart. However, we do not know which map arrangement is correct. All we know is that *c* and *b* are 20 map units apart and that *c* and *e* are only 10. The order could be *ceb* or *bce*. We cannot tell which locus is between the other 2 loci. To settle this, a cross must be made which follows *b* and *e*. If a dihybrid testcross gives about 30% new combinations between *b* and *e* then we know that the second arrangement is the correct one. If it gives about 10%, then the first is the correct order. Therefore, to build up a map showing the proper sequence of genes, crosses must be made in all possible combinations when 3 genes are being considered.

mapping is summarized in Fig. 6-5. A testcross is commonly performed with experimental organisms. The number of new gene combinations can be readily detected by observing the phenotypes among the offspring of a testcross. If 2 linked genes, genes 1 and 2, are found to undergo new combinations among 10% of the total offspring, they are considered to be 10 map units apart. These 2 genes are then followed in further crosses with other genes, genes known to be linked to them. New combinations between genes 1 and 3, 2 and 3,

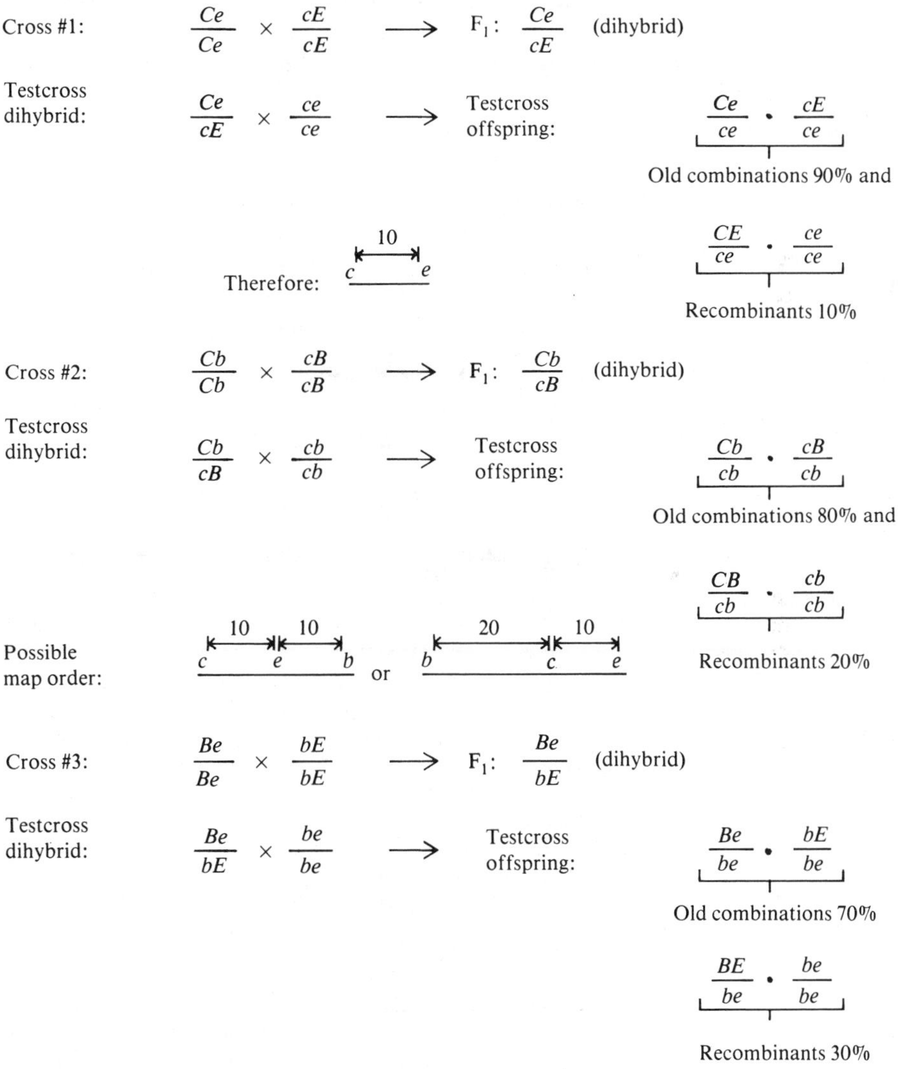

Cross #1: $\frac{Ce}{Ce} \times \frac{cE}{cE} \longrightarrow$ F_1: $\frac{Ce}{cE}$ (dihybrid)

Testcross dihybrid: $\frac{Ce}{cE} \times \frac{ce}{ce} \longrightarrow$ Testcross offspring: $\frac{Ce}{ce} \cdot \frac{cE}{ce}$

Old combinations 90% and

$\frac{CE}{ce} \cdot \frac{ce}{ce}$

Recombinants 10%

Therefore: c —10— e

Cross #2: $\frac{Cb}{Cb} \times \frac{cB}{cB} \longrightarrow$ F_1: $\frac{Cb}{cB}$ (dihybrid)

Testcross dihybrid: $\frac{Cb}{cB} \times \frac{cb}{cb} \longrightarrow$ Testcross offspring: $\frac{Cb}{cb} \cdot \frac{cB}{cb}$

Old combinations 80% and

$\frac{CB}{cb} \cdot \frac{cb}{cb}$

Recombinants 20%

Possible map order: c —10— e —10— b or b —20— c —10— e

Cross #3: $\frac{Be}{Be} \times \frac{bE}{bE} \longrightarrow$ F_1: $\frac{Be}{bE}$ (dihybrid)

Testcross dihybrid: $\frac{Be}{bE} \times \frac{be}{be} \longrightarrow$ Testcross offspring: $\frac{Be}{be} \cdot \frac{bE}{be}$

Old combinations 70%

$\frac{BE}{be} \cdot \frac{be}{be}$

Recombinants 30%

Figure 6-5

1 and 4, etc. are counted among testcross offspring and translated into map units. In this way a map is built up.

In actuality, complications enter the picture which is not as simple as presented here and in Fig. 6-5. For example, 2 crossover events (a double crossover) may occur between any 2 genes. This will cut down on the amount of recombination, since it will place the 2 genes back in their original arrangements (Fig. 6-6). This will obscure the true amount of crossing over, since that amount is calculated by recognition of the new combinations between the genes. In experimental crosses with plants and animals, procedures can be followed which permit the detection of those events such as multiple crossovers which can decrease the number of new combinations between 2 linked genes. So strictly speaking, the number of new combinations is *not* always

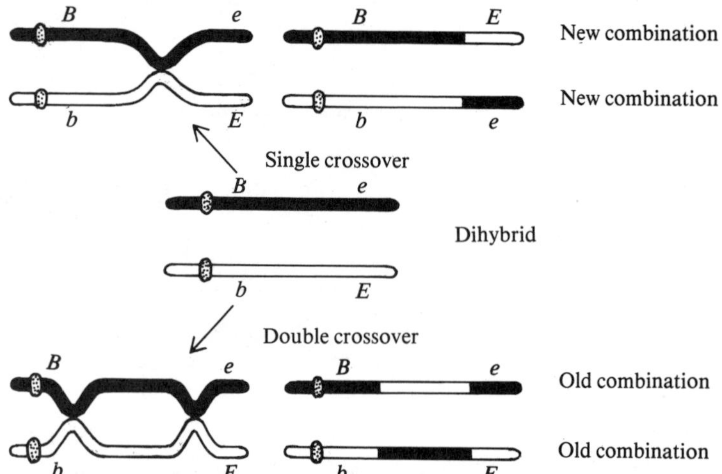

Figure 6-6. Consequences of a single and of a double crossover between two genes. A dihybrid is represented here. The gene combination *Be* is shown on one chromosome strand and the combination *bE* on the homologous strand (middle). If a single crossover occurs between the two strands (above), two new combinations result (*BE* and *be*). These would be detected by observation and the number of these new combinations would then be used to calculate map distance. However, if two crossovers occur between the loci (below), the genes remain in the original arrangements. A double crossover places genes on either side back in the old combinations. Therefore, although two crossovers took place between the genes at the *b* and the *e* loci, we would lose sight of the fact, since no new combinations would be detected. The diagram here shows only two strands, one for each chromosome. (In actuality, each chromosome is composed of two chromatids, so that four strands are present at the time of crossing over. However, for any one crossover event, only two strands participate. For the sake of simplicity and diagrammatic clarity, only two chromatids or strands are shown in the figure.)

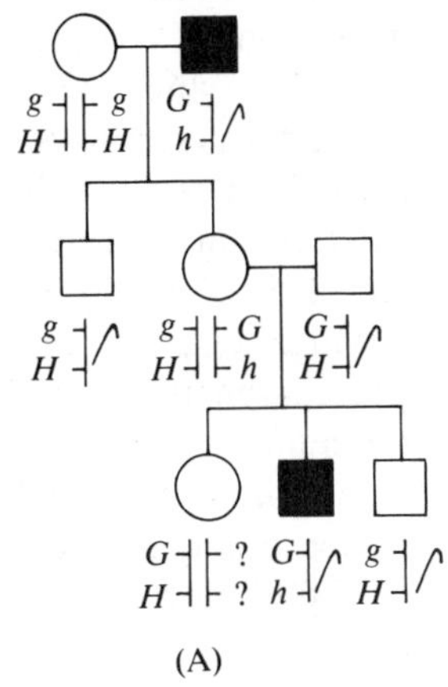

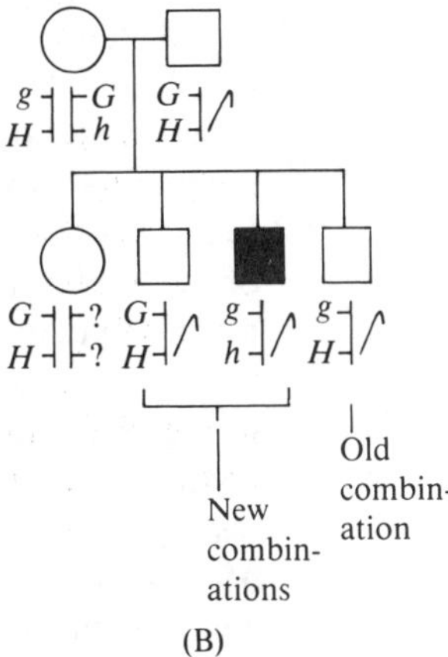

G = G6PD production
g = absence of G6PD
H = normal blood clotting
h = hemophilia
/\ = Y chromosome

Figure 6-7. Linked genes and pedigree analysis. Since sex-linked inheritance patterns are easy to identify, it has been possible to associate many genes with the X chromosome. All X-linked genes are obviously linked to one another, and two of these may be followed together at times in some pedigrees. (A) The exact arrangement of genes may at times be readily determined. The P_1 female must have the arrangement shown here if she does not produce G6PD, is not a hemophiliac, and comes from a family with no history of the rare gene. if the P_1 male has hemophilia and is an enzyme producer, then we know his gene arrangement for his single X chromosome. His other sex chromosome is a Y. Studies of many pedigrees similar to this for certain 2 sex-linked genes show that the old combinations tend to be transmitted together. (B) New combinations, however, would be detected when a large enough number of families is followed. If it is seen that males with the new combinations of the linked genes arise in about 2% of all the male offspring, then the amount of crossing over between the loci is 2% and the map distance is estimated as 2 map units. Note that the female offspring are disregarded, since they receive an X chromosome from both parents. If a female receives an X from her father which carries dominant genes, we may not be able to tell which X she received from her mother, since a recessive on that X from the mother would not be expressed if the dominant allele were on the X from the father. This complication does not arise in the case of sex-linked genes in a male, since he receives a Y from his father. Therefore, both dominant and recessives on the maternal X can come to expression.

directly proportional to the true amount of crossing over and hence the map unit distance between 2 genes. The amount of recombination is accurate as a reflection of map distance only for genes which are a few map units apart. When longer distances are involved, multiple crossovers will tend to yield fewer recombinants and cause 2 genes to appear closer together than they really are. Let us say that 2 genes are actually 100 map units apart. The maximum amount of recombination possible for them is 50%. This is so because multiple crossovers will take place in the distance which separates them. The overall effect is to place the genes back in their old combinations as shown for the doubles in Fig. 6-6 and thus to decrease the number of new combinations which can be observed. Maps, therefore, are built up accurately by studying small map distances. The general order of the genes on the chromosome and some estimate of map distance can, however, be established by working with genes more widely separated, as shown in Fig. 6-5.

Regardless of the complications, the basic logic used in mapping is that which is summarized in Figs. 6-5 and 6-6. The testcross in one form or another is used widely in linkage studies and mapping, since it is the most direct way to detect new combinations. Even in the human, the same reasoning presented

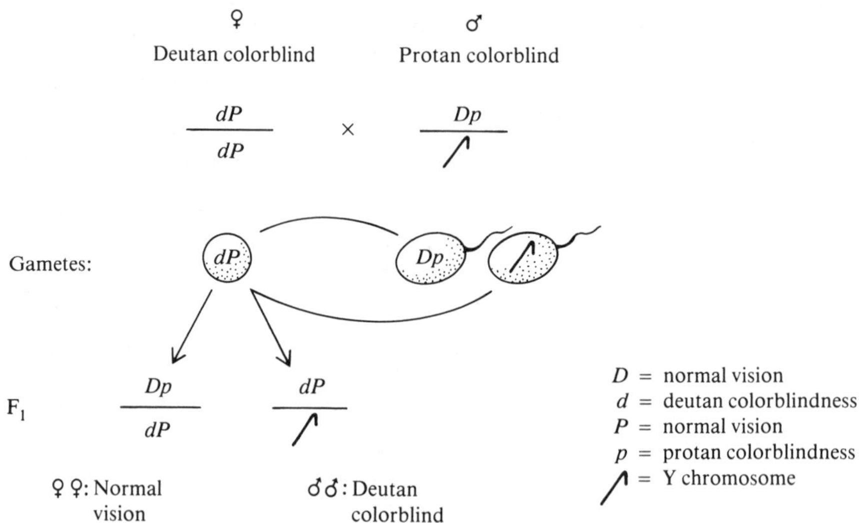

Figure 6-8. Two X-linked loci. Two separate X-linked loci are known which affect the same characteristic, color vision. A cross between a woman with one form of color vision defect (the deutan form) and a man with another form (the protan) will produce sons who show the same vision defect as the mother. All the daughters, however, will be dihybrids who are normal in color vision, since each parent contributes to the daughters an X chromosome with a normal gene which the other parent lacks.

here is the basis of mapping, although special approaches to the problem are utilized. Crosses cannot be performed at will with the human, but pedigrees from a large number of families can be studied. Evidence for linkage can be obtained when an analysis of pedigrees consistently shows that two genes tend to be inherited together in a given combination in a family history. Crossover values may even be estimated by comparing many pedigrees and noting the percentage of new combinations that appear between two genes which seem to travel together. This is a slow and laborious process, but it has yielded estimates of map distance for some human genes.

Without certain special techniques, the assignment of genes to specific chromosomes would be an almost impossibly slow task in the human. An exception is the X chromosome to which well over 100 loci have been assigned. Map locations have also been determined for several of these. The reason for the relative ease with which genes can be assigned to the X chromosome should now be obvious. Typical sex-linked inheritance patterns, such as those shown in cases of hemophilia and colorblindness, are easy to identify from analyses of family histories. Typically, these would show a recessive condition afflicting more males and would indicate the transmission by carrier daughters from grandfather to grandson (Fig. 4-8). Two genes known to be sex-linked can at times be followed together by compiling and analyzing a large number of family histories. The original gene combinations in the parents can be compared with the number of new combinations that appear among the male offspring. Map values may be estimated from the percentage of new combinations originating between the genes [Fig. 6-7(A)–(B)].

Figure 6-8 is a diagram of a cross between persons who are colorblind due to different sex-linked recessives, the deutan recessive and the protan. It should be obvious from the figure how two separate loci on the same chromosome may be detected, each affecting the same characteristic. Note that the F_1 daughters in this cross will all be normal. Therefore they must have received two dominant genes from each parent. The daughters are dihybrids in whom the recessives cannot be expressed since each parent contributed a dominant which the other lacked.

Assignment of Human Genes to Chromosomes

Except for genes on the X, not many genes have been specifically assigned to chromosomes. However, a great deal of progress is now being made in this direction, and increasing numbers of reports continue to announce the assignment of human genes to chromosomes. A major reason for this surge of activity is the refinement of a very interesting procedure and its application to studies of human linkage. Several years ago, it was found that fusion could be achieved between two cells originating from very different species, such as the fusion of mouse and chicken cells or mouse and human. Fusions between cells of diverse origin were made possible by the discovery that certain viruses which are harmless to man (the Sendai viruses), after being inactivated by exposure to ultraviolet light, are able to alter the properties of cell membranes and render them susceptible to fusion with other cells. A human cell type often used in fusion studies is the fibroblast, a cell which can develop into a connective tissue cell. Fibroblasts can be obtained from pieces of skin which have been kept in tissue culture. Human blood lymphocytes are also commonly employed. Human fibroblasts can be fused with mouse cells after the cells are exposed to ultraviolet-treated Sendai virus particles. Following cell fusion, the two nuclei of the original cells fuse to produce a hybrid cell with a hybrid nucleus carrying chromosomes from two distinctly different species. These hybrid cells can be maintained in cell culture where they may continue to divide. However, for unknown reasons, human chromosomes tend to be lost at divisions of the hybrid nucleus. The specific human chromosomes which become eliminated vary from one cell to the next. Eventually, hybrid cells can be derived which tend to be stable and which carry a small number of human chromosomes. From these cells, several diverse cell lines can be established, each line having its unique combination of human chromosomes.

Cell fusions can be manipulated as follows to identify a specific human gene with its chromosome. The gene that is being followed must be associated with some activity or product which can be detected biochemically. For example, suppose it is known that enzyme A is required by a cell to convert substance A to substance B and that the ability to produce enzyme A depends on the presence of the normal gene *A*. If an accumulation of substance A in the cellular environment is toxic or harmful, a cell may die if it carries a defective allele in place of the normal *A* gene needed for enzyme A formation and the conversion of substance A. Knowing this, we can proceed to identify enzyme A with a gene on a specific chromosome. We select human cells which carry the normal *A* gene and which thus produce enzyme A. These cells are then fused with mouse cells which cannot produce the enzyme (Fig. 6-9). The hybrid cells are grown on a medium which contains the necessary growth substances but which contains the toxic substance A. If A is not removed under the action of enzyme A, the cells will die. Since the mouse cells cannot make the enzyme, the hybrid cells' survival depends on the presence of the human chromosome which carries the gene associated with enzyme A. If that chromosome is eliminated from a hybrid cell, the cell will die. Stable lines of hybrid cells can be isolated following loss of human chromosomes from different hybrid cells. Any growth medium which screens out certain kinds of cells and which permits growth only of those cells which possess certain traits is called a *selective medium.* In this case, the selective medium permits growth of only those cell lines with the ability to make enzyme A and which must therefore carry Gene *A*.

The chromosomes of these cell lines can be observed with the microscope, and a comparison of them can be made with regard to the chromosomes they carry. As Fig. 6-9 shows, the chromosome associated with the specific gene can be surmised by the fact that it is always present in a surviving line. In this example, it is chromosome #10. This chromosome appears in each line that can grow on the selective medium. When it is absent, the line does not survive.

This fusion procedure can be adapted in a variety of ways to identify a gene with a chromosome. The presence of an enzyme or some other product

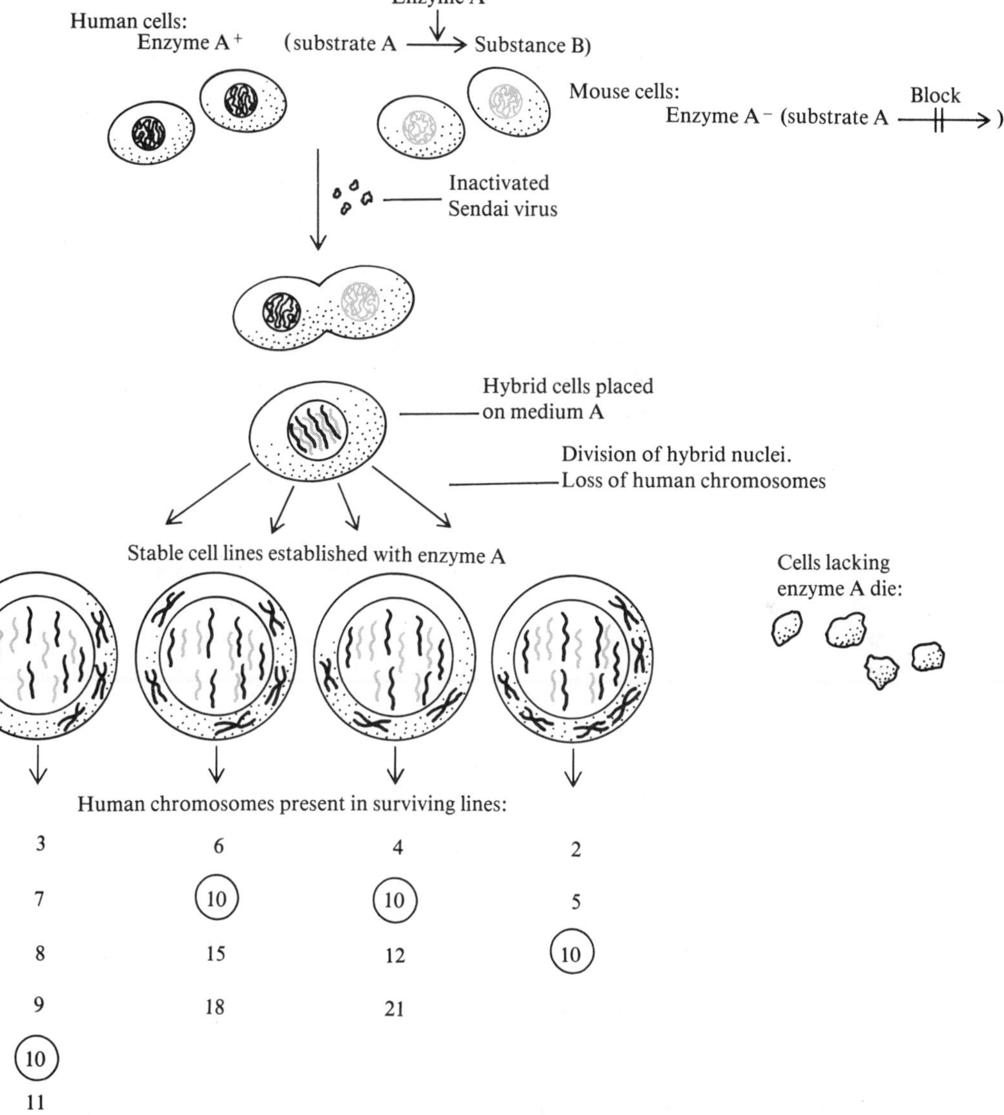

Figure 6-9. Summary of cell hybridization and its use in associating a human gene with a specific chromosome (see text for details).

may be detected by chemical analysis. If the product is present only in those cell lines which carry a certain chromosome, and if it is always absent when that chromosome is absent, then the gene controlling that substance can be assigned to that specific chromosome. Using cell fusion studies, more than one human gene may be found to be associated with a single chromosome. In this way, a human linkage group can be established and identified with a certain chromosome.

Map distance between linked genes cannot be obtained by these cell fusion studies. For this kind of information, pedigree analysis is still essential. The use of computers in conjunction with linkage studies is now facilitating the processing and the interpretation of the assembled data. While two or more genes may be assigned to a specific chromosome as a result of cell fusion studies (and assigned to the X chromosome on the additional basis of X-linked inheritance patterns), there may be no evidence for genetic linkage from pedigree analysis. If two genes, for example, are sufficiently far apart on the same chromosome, they may appear to assort independently, as if they were on separate chromosomes. This would be a consequence of the high incidence of crossing over which would take place between them. The term *synteny* is frequently used in human genetics to refer to genes known to be located on the same chromosome, although genetic linkage may not have been demonstrated. Some geneticists prefer to reserve the term "linkage" for only those genes found on the same chromosome which do not undergo independent assortment, as seen by their inheritance patterns. In those cases where there is no evidence for genetic linkage even though the genes are on the same chromosome, some would say that the genes are *syntenic* (associated on the same chromosome) but are not linked (they appear to assort independently). As used by most geneticists, however, linkage implies association of genes on the same chromosome even though a high incidence of crossing over may give the impression that independent assortment is taking place.

Linkage is a most important concept in the genetic analysis of all species and must be kept in mind in human genetic studies. The knowledge of linkage and map distance in the human, obtained largely from cell fusion studies and pedigree analysis, is important in many contexts. Not only does it improve our insight into the nature of the human genetic material, but it is also being applied to the prenatal detection of certain specific human genetic disorders. Specific examples will be discussed at some length in Chapter 12.

Sexual Reproduction and the Generation of Variation

The crossing over which occurs between linked genes is a normal accompaniment of meiosis and the sexual process. The full biological significance of crossing over can be appreciated best by a reappraisal of sexual reproduction and the independent assortment which it makes possible. As stressed in Chapter 3, meiosis generates new combinations of genes, and one way this comes about is through independent assortment. New gene combinations are responsible for producing variation, and variation is essential for evolutionary progress. Sexual reproduction is a process which speeds up the appearance of new types, since it is responsible for bringing together in new arrangements the genetic material from separate lines or stocks. Figure 6-10 should clarify this point. If a species is completely asexual, the same combination of genes will be transmitted by mitosis generation after generation. If the environment should suddenly change, one or more of the lines will be wiped out if the combination of genes within them is unable to provide the information which could enable them to survive under the changed conditions. The only way a completely asexual line can obtain new information is through spontaneous mutation, a very rare event. In contrast to this, sexual lines are at an advantage. Each line, as in the asexual forms, may carry combinations which by themselves permit survival under set environmental conditions. However, the sexual process permits pooling of genetic information. The information from two lines can be shuffled through meiosis into a variety of new combinations. Some of these may confer a distinct advantage on the ability of a line to reproduce or to cope with the environment. In the case of a sudden environmental change, there is a greater chance that one or more of the new types will survive or even have a decided advantage in the new situation. Sexual reproduction itself is actually a characteristic which gives advantage to a species. The sexual process became established early in the evolution of life, and we find that only those species with sexual reproduction went on to produce the forms which are responsible for the diversity of life types we see today. Sexual reproduction is the primary factor which has generated the variation we see among all living things, within members of a species as well as among different species.

Within any major plant or animal group, the more diverse the members, the larger the number of environments that can be exploited. One need only consider the insect group, an assemblage of diverse forms which includes more species than any other group of living creatures. There are few environments which fail to be exploited by some kind of insect. It is the extreme variation found from one insect species to another which has made this possible. Within any one species of plant or animal, the more highly variable the members of the populations, the greater the chance that that kind of organism will be able to survive as new environments continue to arise with time. Also greater is the ability of a variable species to spread over wide areas. The human is an example of such a variable creature, even though there is only one species of human. Since his origin, *Homo sapiens* has spread over most regions of the planet Earth, living under numerous sorts of conditions and utilizing the features of the many environments to the utmost. The human is more independent of the environment than is any other species, and he can manipulate

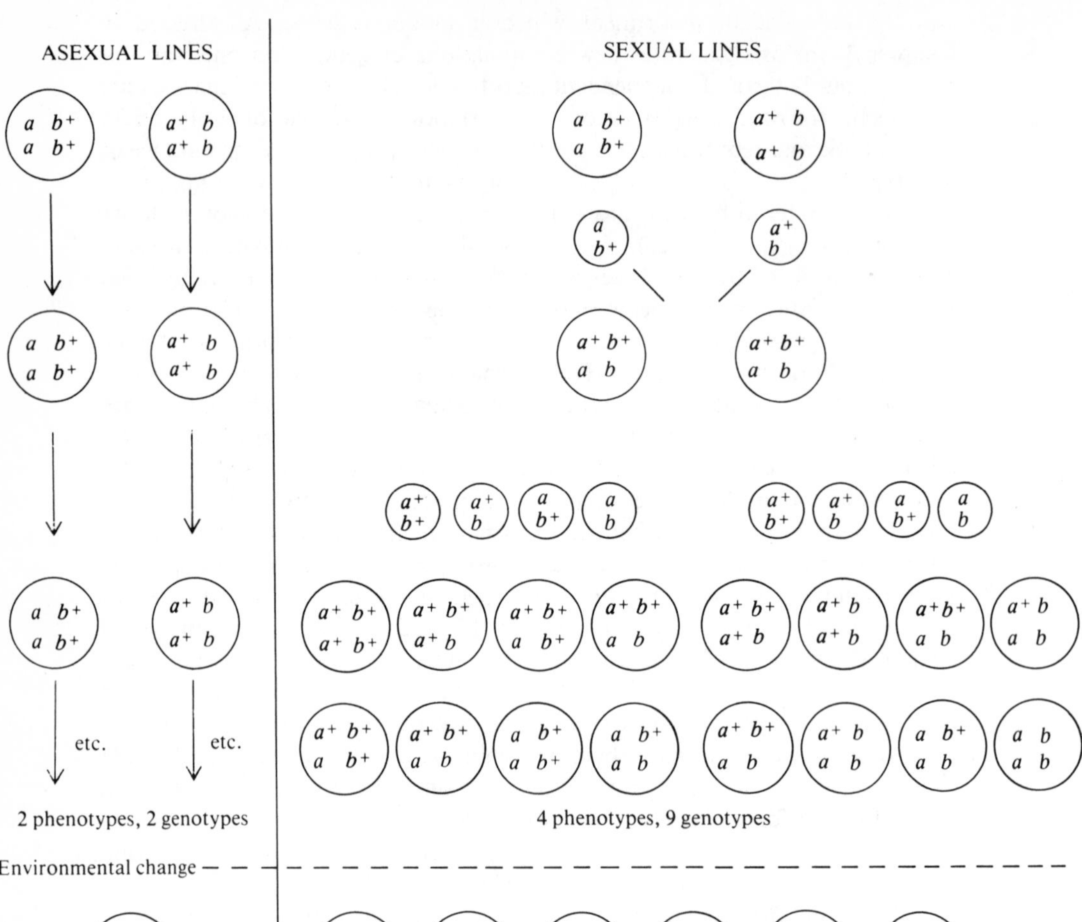

Figure 6-10. Advantage of sexual reproduction. Individuals in an asexual species continue to produce the same genetic types generation after generation. Only a rare mutation can bring about a new combination. In this example, which assumes dominance of a^+ and b^+, we see that the asexual form produces only two phenotypes and two genotypes. A sexual species, however, can pool genetic information from all the individuals. Independent assortment makes possible a large number of phenotypes and genotypes. If the environment suddenly changes, there is a greater chance that the sexual species will have genetic types which can cope with the new conditions. In the asexual form, there is a greater chance of extinction, since one or even all of the lines may not carry favorable gene combinations.

it to the best advantage. Without the flexibility which accompanies variability, it is certain that the human would not have been so successful.

To appreciate more fully the role of the sexual process in the generation of variability, consider first the number of different kinds of gametes which independent assortment makes possible. Table 6-1 shows that if we are following just 1 pair of genes, each heterozygous parent will form 2 different types of gametes. Assuming that dominance is operating, 2 different phenotypes occur among the offspring of such heterozygotes. The number of genotypes, however, will be 3. This will also be the number of phenotypes if dominance is not operating, and as we have noted, dominance is usually not complete. Two dihybrid parents will each produce 4 classes of gametes. Assuming dominance in 1 member of each gene pair, 4 kinds of offspring can arise. The number of genotypes, however, is 9, and this also represents the number of phenotypes if dominance is incomplete. Table 6-1 shows that the number of different kinds of gametes, phenotypes, and genotypes increases rapidly with the number of heterozygous gene pairs that are present. The handy expressions 2^n and 3^n summarize the situation. Substituting for n the number of heterozygous gene pairs being followed in a particular case, 2^n represents the number of different classes of gametes and also the number of phenotypes if dominance exists in 1 member of every gene pair. Substituting in like fashion for 3^n, the number of genotypes and phenotypes (if there is no dominance) is easily calculated. The table shows that just by independent assortment alone, the number of possible new combinations becomes tremendous when only a small number of heterozygous gene pairs is involved. Now consider the following fact. Each human is a highly heterozygous individual. While each of us is homozygous for genes at some loci, the amount of heterozygosity is greater by far. Current research is indicating that a normal gene may come in many

Table 6-1. Summary of degrees of hybridity and the generation of genotypes and phenotypes

Number of allelic pairs in hybrid	Number of kinds of gametes which hybrid can form	Number of phenotypic classes when such hybrids are crossed (assuming dominance)	Number of phenotypes in testcross of the hybrid	Number of genotypes when two such hybrids are crossed (also the number of phenotypes when no dominance pertains)
1	2	2	2	3
2	4	4	4	9
3	8	8	8	27
4	16	16	16	81
5	32	32	32	243
6	64	64	64	729
n	2^n	2^n	2^n	3^n

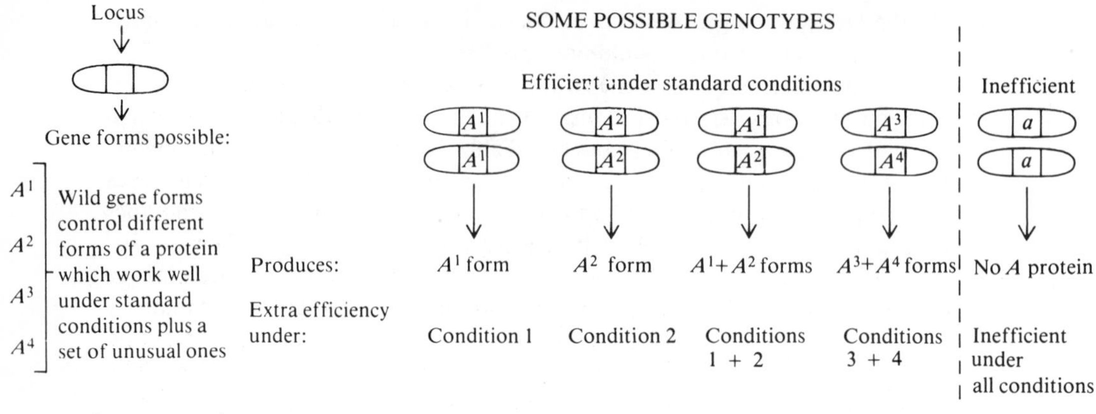

Figure 6-11. Forms of a wild-type gene. At a particular locus on a chromosome (left), one or more allelic forms of a gene may occur. Several of these may be called wild-type genes. Each guides the formation of a slightly different form of some protein which has a specific task to perform in the cell, such as the control of a chemical step. Each variant form of the protein may be able to do this efficiently under ordinary or standard conditions. If an unusual set of conditions arises, however, one of these protein forms may operate better than another one under the changed environment. An individual heterozygous for any pair of wild-type alleles would have an advantage over any homozygote, since the heterozygote will be efficient over a wider range of environments. The individual homozygous for the defective allele is inefficient under all conditions. Many allelic forms of wild-type genes apparently exist in the human species. Humans are evidently very heterozygous at most loci. Any person is apt to contain variant forms of the same protein as a result and to differ from another person in the combination of alleles he carries.

forms. This means that even for standard or wild-type genes, the chance is great that each human is heterozygous, carrying at each particular locus 2 forms of the normal gene (Fig. 6-11).

To keep the matter at its simplest for the moment, let us consider just 1 pair of alleles on each of the human chromosome pairs. Since a human carries 23 pairs of chromosomes, each person can form 2^{23} different kinds of gametes, *if only 1 pair of genes is being followed on each chromosome.* The minimum number of phenotypes possible among the offspring of any 2 persons would thus be 2^{23}. This immense figure is still an absurd underestimation. The number of genotypes possible would be 3^{23}. These numbers are so large that they defy the imagination. They should, however, enable us to understand why each of us (except for identical twins) is a unique entity. The chance of 2 human offspring being identical, even though they have the same parents, is so improbable that it would be ludicrous to expect it to occur.

The Biological Significance of Crossing Over

The amount of variation generated by meiosis becomes even more staggering when we consider crossing over in addition to independent assortment. It is obvious that the human or any other sexually reproducing organism is heterozygous, not for just 1 pair of alleles per chromosome but for large numbers of them per chromosome pair. As Fig. 6-12 shows, if an individual is dihybrid,

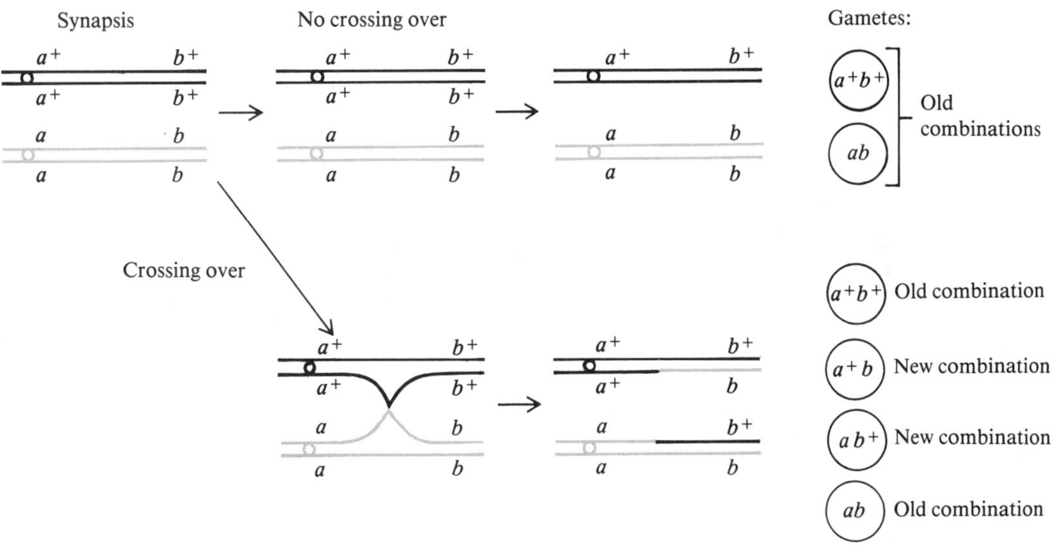

Figure 6-12. If no crossing over occurs between 2 loci, the genes at these positions will continue to remain in the original combination. When a crossover occurs in a cell at meiosis, 4 kinds of gametes result. Note that crossing over takes place when each chromosome is composed of 2 chromatids. Only 2 of the 4 present engage in any single crossover event. This means that if a crossover occurred in every meiotic cell between 2 given loci, *a* and *b,* 4 different kinds of gametes would be produced in equal numbers, the half with the old combinations and the half with the new. It would appear as if the 2 gene pairs were on separate chromosomes and were assorting independently. Genes that are far apart on a chromosome may become separated in the majority of the cells by a crossover and approach 50% recombination. They may appear as if they are not linked. For the sake of simplicity and diagrammatic clarity, the figure shows the two middle chromatid strands involved in a crossover. In actuality, any of the strands may engage in crossing over, not just the same two for every event. Moreover, multiple exchanges occur; not just 1 but 2 crossovers may take place between any 2 genes (Fig. 6-6). These multiple crossovers may involve all 4 strands in some meiotic cells. It is the proportion of the different kinds of crossovers (single events, double, triple) which lead to the approximate 50% recombination. It is not just a single crossover between the same 2 strands which approaches the value of 50%.

heterozygous at 2 loci associated with the *same* chromosome pair, only 2 classes of gametes are possible if no crossing over occurs between them. These are the old combinations, those present to begin with. Assuming no crossing over, all the genes on 1 chromosome would remain forever together in that parental combination until a rare gene mutation arose. And so, in the complete absence of crossing over, only 2 types of gametes are possible and only 2 phenotypic classes can be produced if dominance exists. Three genotypes would occur. It is just as if we were following 1 pair of genes instead of 2. Two genes on separate chromosomes undergo independent assortment, and the comparable figures would be 4 phenotypes and 9 genotypes (Table 6-1).

Now assume that a crossover occurs in the dihybrid shown in Fig. 6-12 so that the 2 linked genes may form new combinations. The outcome is a significant increase in the number of possible kinds of gametes, phenotypes, and genotypes. If a crossover between the 2 genes were to take place in every meiotic cell, it would be just as if the 2 genes were located on separate chromosomes, since they would then become separated in each cell. While such separation does not actually occur during the maturation of each primary spermatocyte and primary oocyte, those genes located far enough apart on the same chromosome may become separated by a very high incidence of crossing over and may appear to be assorting independently, as if they were on separate chromosomes. For genes closer together, the number of new combinations, as noted earlier in this chapter, will be demonstrably smaller than for the old ones. This reduction will vary with distance between the genes; the closer the 2 genes, the smaller the number of new combinations. Therefore, while independent assortment does not occur between genes which are closely linked on a chromosome, they can nevertheless still form new combinations. This is possible because of crossing over which generates new combinations between linked genes. Those genes far enough apart will actually form new combinations in amounts which approach those formed by independent assortment. The importance of crossing over to the sexual process cannot be overestimated. Without it, evolution would not have progressed at the rate capable of producing the wealth of variation seen today among living things. Only those lines with independent assortment *and* crossing over formed the stocks from which the main groups of plants and animals arose.

It can now be seen why the figure 2^{23} is an oversimplification, since any individual can produce new combinations of the genetic material far in excess of that number. To appreciate even more the individuality of each person, a simple exercise can be performed with a group of approximately a dozen persons. Table 6-2 lists several common and easily recognized human traits along with their modes of inheritance. Each person in the group should classify himself on the basis of these. Everyone will then stand up, and one person will proceed to describe his or her phenotype by calling out one trait at a time. As a trait is called, only those persons will remain standing who also have the trait; those without it will be seated. By the time about six have been called, the caller will very likely be standing alone. This simple exercise shows that, for just a small number of inherited traits, the number of combinations is so overwhelming that the chance of any two persons in a group having the identical combination for all of them is improbable.

Table 6-2. Some common human traits and their modes of inheritance

Characteristic or feature	Phenotype
Earlobes	Hanging free: Dominant Attached: Recessive
Pigmentation of iris	Brown pigment in front (eyes brown, green, hazel): Dominant No brown pigment in front (eyes blue, gray): Recessive
Length of index finger	Longer than third finger: Dominant in females, recessive in males Shorter than third finger: Dominant in males, recessive in females
Finger hair (one or more fingers with hair on segment between first and second joints)	Some hair present on at least one middle segment: Dominant Hair completely absent from all middle segments: Recessive
Little finger position	Bent inward when hands at rest: Dominant Straight when hands at rest: Recessive
ABO blood type	A: Dominant to O type B: Dominant to O type AB: *A* and *B* alleles present O: Recessive to A and to B
Rh blood type	Rh positive (D antigen present): Dominant Rh negative (D antigen absent): Recessive
Ability to taste phenylthiocarbamide (treated papers available from supply houses)	Ability to taste at all (usually bitter): Dominant Inability to detect taste: Recessive
Ability to taste sodium benzoate (treated papers available from supply houses)	Ability to taste at all (salty, sour, sweet, bitter): Dominant Inability to taste at all: Recessive
Tongue rolling	Ability to roll or curl tongue between lips: Dominant Inability to curl or roll tongue: Recessive
Widow's peak	Hairline coming to a point on forehead: Dominant Hairline straight: Recessive

Types of Twins and the Variation in a Twin Pair

The topic of twinning is relevant to the subject of genetic variation. Twins arise in approximately 1 pregnancy in 83. It is well known that human twins fall into 2 main categories, commonly designated *identical* and *fraternal.* Identical twins are monozygotic, meaning that they develop from the same zygote or fertilized egg. Since only 1 sperm and 1 egg are involved, the genetic material carried by monozygotic twins is identical. Such twins must therefore always be of the same sex. Not only will the physical resemblance between them be striking (Fig. 6-13), but they will also be identical for their blood groups and all other inherited characteristics. The variations which do occur between them are due to environmental differences which begin to exert their effects well before birth. Identical twins arise if the 2 cells forming from the very first division of the zygote become sufficiently separated to continue their development and later divisions more or less independently. Separation of groups of cells can take place a bit later. This separation can occur in more than 1 way, but the outcome is the same. Each group proceeds to produce an embryo. Separation of cells at an early stage can give rise to 2 complete individuals, because these cells all contain the same set of information needed to direct the formation of a human. Moreover, the cells at the earliest stages of embryology have not yet begun to specialize to any extent. If, however, a division into 2 cell groups occurs later, or if the separation of the groups at any stage is incomplete, the twins may be born conjoined. Such twins are commonly called *Siamese twins,* a reference to the fact that the first pair of conjoined twins to receive wide publicity (approximately 100 years ago) were of Siamese birth. There is no genetic or biological significance to the term whatsoever.

On the other hand, dizygotic or fraternal twins represent 2 genetically different zygotes. Two different eggs and 2 sperms are involved. Such twins may be of either sex. It is just a matter of chance. Half the time they are brother and sister. In one-fourth of the cases, they will both be boys, and in the other one-fourth both will be girls (Fig. 6-14). If the twins are of opposite sex, they obviously must be dizygotic. Fraternal twins experience a more similar intrauterine environment than that experienced by the average brothers or sisters; but otherwise these twins are no more alike than any 2 children born separately within the family. The number of fraternal twin births is at least twice that of monozygotic or identical twin births. There is evidence for the influence of both genetic and environmental factors in dizygotic twinning. Dizygotic twinning occurs with different frequencies in different ethnic groups. It is slightly higher among American blacks than whites (13.4 pregnancies/1000 vs. 10/1000). Dizygotic twins have a high incidence in some families, and women from such families have a greater chance of bearing twins than women from families with no history of twinning. These observations suggest a genetic

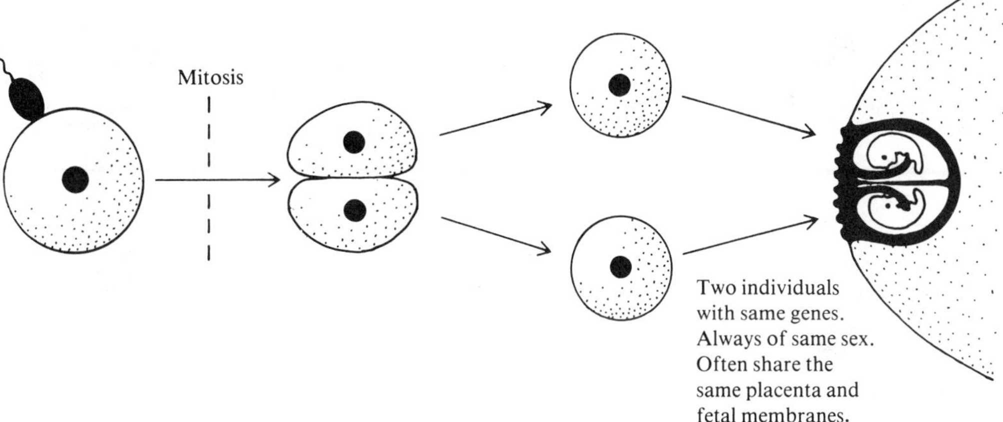
Mitosis
Two individuals
with same genes.
Always of same sex.
Often share the
same placenta and
fetal membranes.

Figure 6-13. Identical twins and their origin. As the photo shows, there is a striking resemblance between identical twins. This is understandable, since they are derived from the fusion of one sperm and one egg. The embryo may separate after the first mitotic division into two separate cells, or it may occur at a slightly later stage, producing two separate masses of embryonic cells. In either case, the separate entities contain exactly the same genetic material and go on to produce two separate individuals. Identical twins must therefore always be of the same sex.

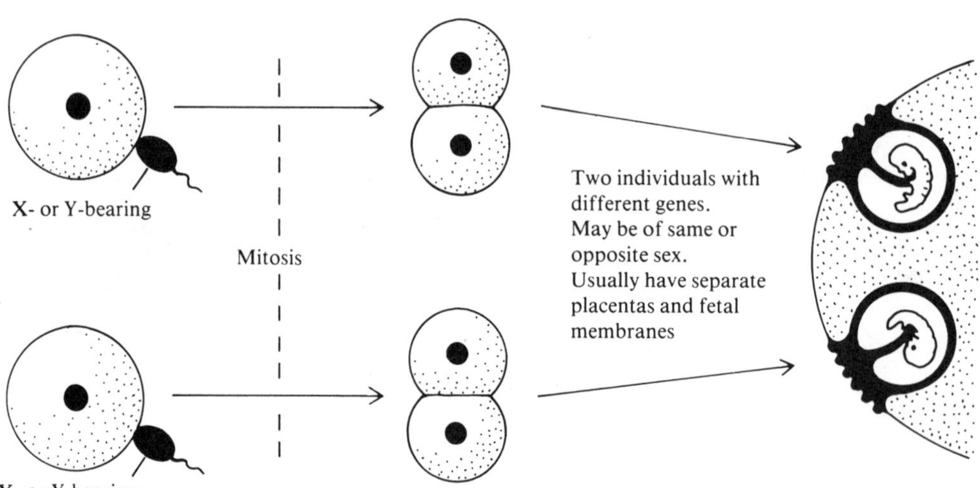

Figure 6-14. Fraternal twins and their origin. Fraternal twins are derived from two separate eggs which have been released from two follicles that matured at the same time. Each of these eggs is fertilized by a different sperm. Fraternal twins, except for having almost identical prenatal environments, have no more in common than any two brothers or sisters. Two members can therefore be of different sexes, since one of the sperms involved may have been X-bearing, the other one Y-bearing.

component in the origin of dizygotic twins, but an environmental influence is strongly indicated by the observation that the frequency of twins in the U.S. and in other countries seems to be declining! Moreover, populations of persons of the same ethnic background may show different rates of twinning when they are living in different localities. The age of the mother also influences the frequency of dizygotic twinning; older mothers have the higher chance. However, the frequency of identical twins does not seem to be influenced by maternal age.

The multiple births publicized recently result from the taking of fertility drugs by the woman and are examples of fraternal twins. If more than two eggs are released from ovarian follicles, as also occurs spontaneously at times, they may be fertilized by separate sperms. Triplets or other types of multiple births are then possible. But again, these children are no more alike genetically than the average brothers and sisters. Rarest of all are the identical triplets, quadruplets, etc. These arise from the separation at any early stage of the embryo into more than two relatively independent masses of cells, each with the potential to produce a complete individual.

Identical twins are no real exception to the statement that no two genetically identical persons will arise from one set of parents, since these twins actually represent *one* genotype which was formed by the chance combination of the genes in one egg and one sperm. From the genetic viewpoint, these twins are one individual. Identical twins are the focus of study in many areas of research. They are particularly valuable for studies which attempt to evaluate the relative influences of environmental and genetic factors in the expression of a characteristic. Since these twins have the identical hereditary constitution, they would be expected to exhibit the same traits at the same time if these are under strict genetic control. Any variation would be due to environmental influences. Particularly valuable in this type of study are those identical twins who have been separated in infancy and reared in different settings. Certain traits (blood types, eye color, colorblindness) would always prove to be the same regardless of the environment, since the hereditary component almost completely controls their expression. Others, however, will show different degrees of variation. In these kinds of traits, the environmental effect may be a small one, so that little variation is seen for them when separated identical twins are compared. This would indicate some role of the environment, but a larger one for the genetic factors. At the other extreme would be traits or characteristics which vary greatly when the twins are compared. In these cases, the environmental component is exerting a highly significant influence. For certain complex characteristics, such as intelligence (Chapter 15), attempts have been made to evaluate the relative importance of environment and heredity using this kind of approach. As we shall see when we return to the subject of twin studies, the interpretation of the data is frequently controversial.

Effects of Inbreeding on Variation

The advantage provided by sexual reproduction through its generation of variation is so great that a study of living things shows that most sexual species have some mechanism which insures some degree of outbreeding or crossing with other lines or stocks. The desirability of crossing with other lines rather than within the same family line can be seen from a study of Fig. 6-15, which shows the effects of exclusive self-fertilization which is found in some plant species. With each generation, the number of heterozygotes decreases by 1/2. The number of homozygotes continues to increase as a result. For any single pair of genes, in just 7 generations, almost 100% of the individuals will be homozygous at that locus. Exclusive self-fertilization does not occur naturally in animals, but highly inbred lines can be maintained in the laboratory for studies of mice, etc. Any species in which exclusive self-fertilization occurs (as in peas) forms gametes by meiosis, but no benefits at all are being received from the sexual process. It is as if the same type were being perpetuated generation after generation, since eventually almost every individual will be homozygous at each locus. The species might as well be asexual. No genetic information is being introduced from any other line. The genetic information within the line will remain invariable, except for rare mutations. New combinations of this information with that from other lines cannot be formed by self-fertilization. It is also evident that continued inbreeding increases the chances of expression of detrimental recessives. All individuals carry some harmful genes which tend to remain unexpressed. The deleterious genes in unrelated stocks are apt to be at different genetic loci. The chances are that 2 unrelated stocks do not carry the same detrimental genes. Consequently, outbreeding decreases the risk of bringing harmful alleles, many of them recessives, to expression. Inbreeding has the opposite effect by increasing the chance that any given allele, dominant or recessive, will come through both the sperm and the egg so that a homozygote arises. Animal and plant breeders have known since historical times that crossing inbred lines with other highly inbred lines is often accompanied by the appearance of very vigorous offspring. Hybrid corn is just one example of this (Fig. 6-16). Part of the reason may be that the inbred lines have become homozygous for many recessives which affect their capacity for growth or reproduction in some adverse way. When the lines are crossed, heterozygotes are produced. One of the lines carries along with its particular deleterious alleles the "standard" genes which the second line lacks. The hybrid offspring in such a case would therefore carry more dominants than either parental stock, and the deleterious effects of the recessives could remain unexpressed. It must be kept in mind that homozygosity for certain *dominant* genes may also entail deleterious effects. It is not always a recessive allele which has deterimental consequences when homozygous. The increased vigor of the hybrid is often due to the heterozygosity, *not* just to the presence

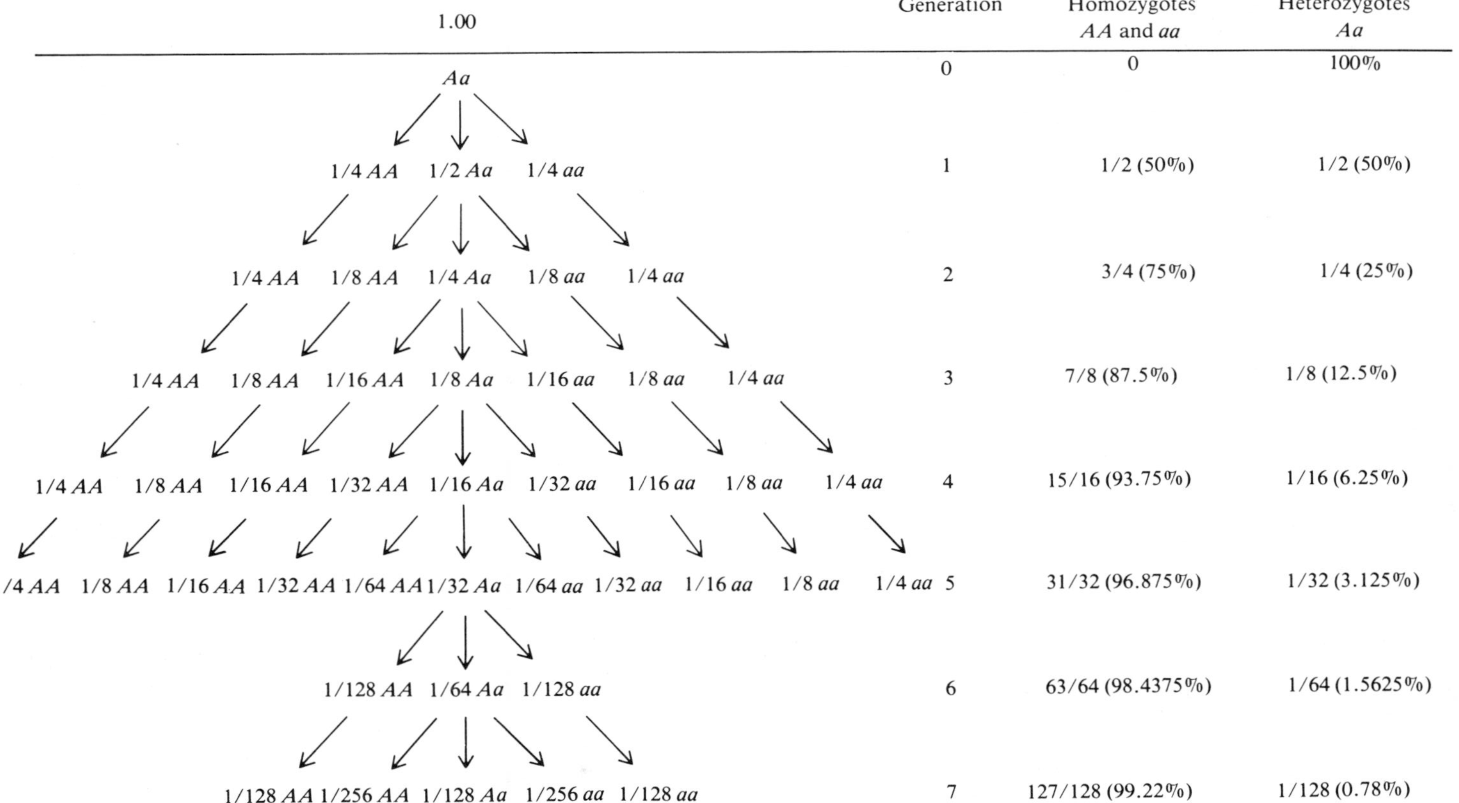

Figure 6-15. Effects of inbreeding. If self-fertilization, the most extreme form of inbreeding found in some plant species, occurs exclusively, the benefits of sexual reproduction are lost. With each generation, the number of heterozygotes decreases by one-half as the number of homozygotes increases. In just a few generations, most of the members of a population will be homozygous for any given locus.

of dominant alleles. The hybrid carries fewer dominants and recessives in the homozygous condition, and this feature decreases any deleterious effects of *either* dominants *or* recessives.

The human tends to be a highly outbreeding species. Persons tend to mate with others who are not from within the family. The benefits of this are precisely the same as those for other species. The probability that 2 persons taken at random have in common some rare deleterious allele is very unlikely. However, within a family line, there is a greater likelihood that 2 persons have any specific gene in common. This allele would have been derived from some common ancestor. The chance for 2 related persons to be carrying such a gene derived from a common ancestor increases with the degree of relationship. This likelihood can be expressed by the *coefficient of relationship,* which also

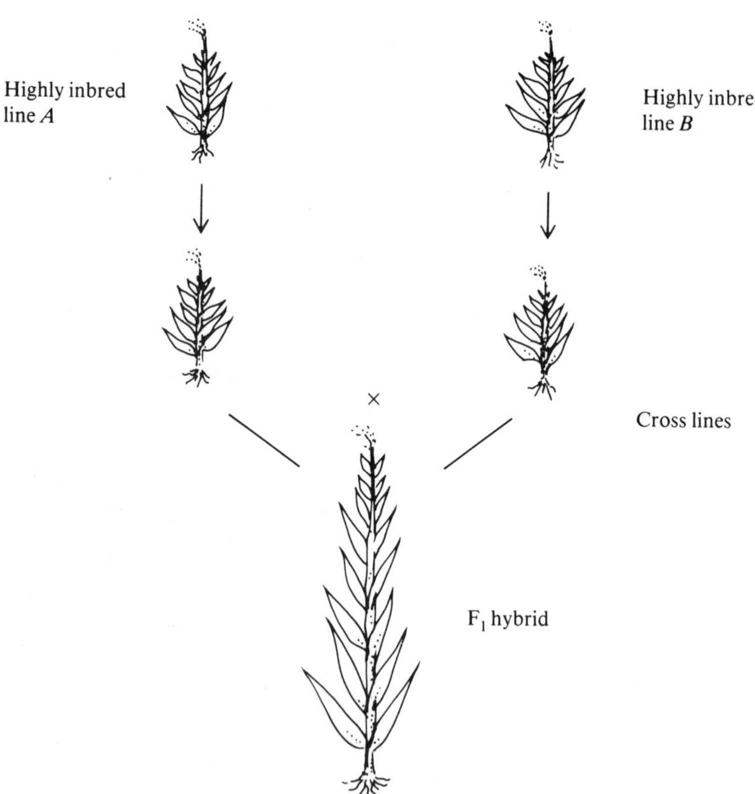

Figure 6-16. Hybrid vigor. When highly inbred lines of plants are crossed, the F_1 offspring are often more sturdy and robust than either parental line. Not only may this be reflected in plant height as suggested here, but grain yield and disease resistance are other characteristics which may be improved.

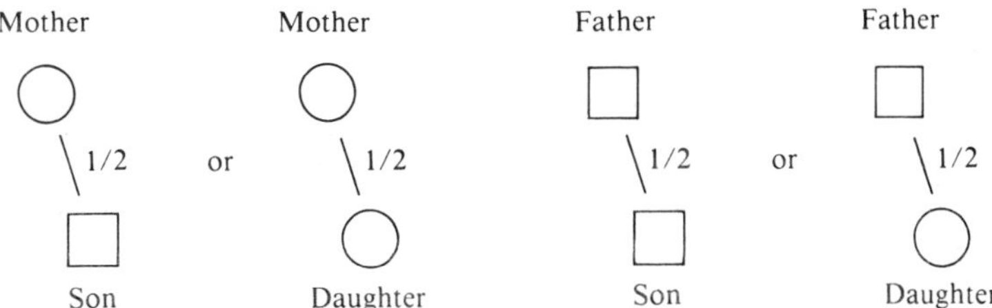

Figure 6-17. Coefficient of relationship between parent and any offspring. This value must be 1/2, since any individual receives one-half of his or her genes from the maternal parent and the other half from the paternal. Therefore, there is a chance of 1/2 that any given gene on a specific chromosome will be present in a parent and in his or her offspring. It is only for this relationship, parent and offspring, that the fraction of genes which is shared is exact (1/2). In all other relationships, the given value is an average. For any brothers and sisters (except for identical twins), the value is 1/2, but it is an average, not exact as in the case of parent and child. Therefore, *on the average,* one-half of the genes are identical for any two children with the same parents.

indicates the proportion of all the genes in 2 related persons which on the average are identical. Figure 6-17 shows that the coefficient of relationship between a mother and son (or any parent and an offspring) is exactly 1/2. This is quite obvious, since one-half of the genes in any person have been descended from 1 of the parents.

Two first cousins have a coefficient of relationship of 1/8, again meaning that, on the average, 1/8 of their genes have come from a common ancestor and therefore, on the average, 1/8 of their genes are identical. To appreciate more fully degrees of relationship among cousins and other relatives, a few elementary mathematical concepts are needed, and these will be presented in Chapter 11 where genetic relationships will be treated at more length with regard to inbreeding and genetic counseling. It is important at this point to appreciate the fact that the closer the relationship, the greater the chance that 2 persons will be carrying a specific allele from an ancestor they have in common. The chance that 2 related persons will each transmit a specific deleterious gene to an offspring is consequently increased. The advantage of sexual reproduction, which makes possible combinations of genes from many stocks within the species, is realized to its fullest extent only by continued crossing with members from outside a family line.

REFERENCES

Bulmer, M. G., *The Biology of Twinning in Man.* Clarendon Press, Oxford, 1970.

German, J., and Chaganti, R. S. K., "Mapping Human Autosomes: Assignment of the MN Locus to a Specific Segment in the Long Arm of Chromosome No. 2." *Science,* 182:1261, 1973.

McKusick, V. A., "The Mapping of Human Chromosomes." *Sci. Amer.,* April: 104, 1971.

Newman, H. H., Freeman, F. N., and Holzinger, K. J., *Twins, a Study of Heredity and Environment.* Univ. of Chicago Press, Chicago, 1937.

Renwick, J. H., "The Mapping of Human Chromosomes." *Annu. Rev. Genet.,* 5:81, 1971.

Rothwell, N. V., *Understanding Genetics.* Williams & Wilkins, Baltimore, 1976.

Ruddle, F. H., "Linkage Analysis in Man by Somatic Cell Genetics." *Nature,* 242:165, 1973.

Ruddle, F. H., and Kucherlapati, R. S., "Hybrid Cells and Human Genes." *Sci. Amer.,* July: 36, 1974.

Salthe, S. N., *Evolutionary Biology.* Holt, Rinehart and Winston, N. Y., 1972.

Volpe, E. P., *Understanding Evolution,* Second edition. William C. Brown, Dubuque, Iowa, 1970.

Whitehouse, H. L. K., "The Mechanism of Genetic Recombination." *Biol. Rev.,* 45:265, 1970.

Winchester, A. M., *Laboratory Manual. Genetics,* Second edition. William C. Brown, Dubuque, Iowa, 1972.

REVIEW QUESTIONS

1. *A*. How many different kinds of gametes can be formed by individuals of the following genotypes (just give the number)? (1) *AaBbCc*; (2) *AaBBCCDDEe*; (3) *AaBbCcDdEe*; (4) *AABBCCDDEEFF*. *B*. (1) What would be the number of phenotypes and genotypes possible among the offspring of two parents of genotype (1) above? (2) Answer the same question for genotype (2).

2. Assume that the two gene pairs *M, m* and *N, n* are so closely linked that crossing over can be ignored. What kinds of gametes can be formed by (1) a double heterozygote with alleles in the *trans* arrangement? (2) a double heterozygote with alleles in the *cis* arrangement?

3. Assume the two gene pairs *O, o* and *P, p* are linked. (1) What kinds of gametes can be formed by a dihybrid with the genes in the *trans* arrangement, and what would be the expected frequencies if the amount of crossing over is 20%? (2) Answer the same question for a dihybrid with the *cis* arrangement.

4. Assume that in the dihybrid with the *trans* arrangement for the above genes a double crossover occurs in a cell. What would be the combination of the genes resulting from the double crossover between the two loci?

5. The amount of crossing over between the linked loci *d* and *t* is found to be 14%. The locus *m* is found to be linked to *d* and *t*. Alleles at the *m* locus engage in crossing over with those at the *t* locus with a frequency of 10% and with those at the *d* locus with a frequency of 4%. What is the order of these genes on the chromosome and the distance between them in map units?

6. In chickens, the gene *I* results in white feathers and is dominant to the recessive *i* for colored feathers. The gene *F* causes the feathers to be frizzled. It is codominant to its allele *f* for normal feathers. The following cross is made:

$$\frac{iF}{If} \text{ (white, mildly frizzled)} \times \frac{if}{if} \text{ (colored normal)}$$

The results among the offspring are:

white, mildly frizzled	19
colored, mildly frizzled	60
white, normal	65
colored, normal	12

Calculate the amount of crossing over.

7. In the human, the gene pair *D, d* which is associated with red-green colorblindness is found at a locus on the X chromosome. The gene pair *P, p* is also X-linked and associated with the color vision characteristic. In each case, the recessive is responsible for defective color vision. The two loci are so closely linked that the amount of crossing over is very low. (1) Ignore crossing over and diagram a cross between a woman with deutan colorblindness (associated with the *d* locus) who is homozygous for the normal allele at the locus associated with protan colorblindness (the *p* locus) and a man who has protan colorblindness. Show the expected offspring. (2) Suppose a dihybrid woman carrying the alleles associated with color vision in the *trans* arrangement marries a man who carries neither recessive. Diagram the cross and show the offspring, assuming no crossing over. (3) In the last case, show the results if crossing over does occur.

8. The gene *O* is required for pigmentation of the iris. Its allele *o* is responsible for ocular albinism. The locus involved is X-linked. What would be the results of a cross between a woman with ocular albinism who is homozygous for the *D* allele for normal color vision and a man who has pigmented iris but has deutan colorblindness? (Assume no crossing over.)

9. Give the results of a cross between a female dihybrid for both ocular albinism and deutan colorblindness (*cis* arrangement) and a man with both normal traits. (Assume no crossing over.)

10. The locus for the production of the enzyme G6PD is X-linked and approximately 6 map units from the locus associated with protan colorblindness. Diagram a cross between a dihybrid woman (*trans* arrangement) and a man who produces the enzyme and has normal vision. Considering crossing over, show the expected offspring and the frequencies.

chapter 7

GENES AND THE IMMUNE SYSTEM

Antigens and Antibodies

In 1930, Landsteiner was awarded the Nobel Prize for his pioneer work of 1900 on the recognition of the ABO blood grouping, which includes types A, B, AB, and O. Most of us are aware of the particular type to which we belong because of the importance of these blood types in transfusions. Today a great deal of information has been assembled on the molecular basis of the ABO types as well as on their inheritance.

Besides the ABO grouping, humans may be assigned to a minimum of 15 other blood groups, some of which are also significant in transfusions and other sorts of blood incompatibilities. The recognition of any blood grouping or specific type within it depends on the ability to detect the presence or absence of specific *antigens* on the red blood cells. Such antigens are found in the membranes of the red cells (erythrocytes). Antigens are any large molecules which are able to bring about the formation of other large protein molecules, *antibodies,* with which they can react. The antigens themselves usually are large protein molecules or contain a protein component. Some of the antibodies which are found in humans comprise the gamma globulin fraction of the circulatory system. Familiar to most of us are the antigen-antibody reactions which constitute a major part of the natural defense system of the body against disease-causing organisms. In a normal person, any large molecule that is foreign to the system tends to act like an antigen and usually provokes a response by antibodies which react with it and destroy it. The antigen-antibody reaction is very specific; a certain antigen reacts only with a very definite kind of antibody. Therefore, introduction of a smallpox virus into the body causes the formation of a specific kind of antibody against the smallpox virus. The antibody will not react against a typhoid-causing organism which elicits its own kind of antibody. Vaccines are really preparations which contain weakened microorganisms incapable of producing a severe disease reaction but which do cause antibodies to form in quantities sufficient to confer protection against a later infection. But a person must be immunized against each kind of antigen or disease-causing entity if he is to have a reservoir or assorted kinds of antibodies.

When an antigen and an antibody react, there may be cellular clumping if the antigen is cellular in nature. Or, the cell may undergo *lysis,* in which case it is destroyed. A person may develop specific antibodies very soon after birth without ever having been in actual contact with the particular antigens, producing a *natural immunity.* Certain naturally occurring antibodies against specific blood antigens are found in everyone. The ABO blood grouping is an example of a system in which such natural antibodies develop in the blood serum. Classification of a person as blood type A, B, etc. (Table 7-1) is possible because of the presence of the detectable antigens. If a certain antigen is present in the red blood cells, these cells may be clumped by the corresponding specific antibody when it is present. In the human, if a specific antigen is *not* present in the red blood cells, the corresponding antibody, anti-A or anti-B, *is* present in the blood serum. A person of type AB blood, therefore, is born with both A and B antigens in the red blood cells, but *no* anti-A or anti-B antibodies are found in the serum. The type O person lacks both A and B antigens in the membranes of the red blood cells, but the serum contains *both* antibodies, anti-A and anti-B.

Table 7-1. Major ABO blood types

Blood type	A-B antigens on red blood cells	A-B antibodies in serum
A	A	Anti-B
B	B	Anti-A
AB	A and B	None
O	None	Anti-A and Anti-B

From Table 7-1, it can be appreciated why certain transfusions are safe whereas others are not. The blood cells of a donor, if they are carrying a specific antigen, will be clumped or destroyed if the recipient's serum contains the corresponding antibody. Thus, the cells of a person of type A will be clumped by anti-A in the serum of a type B person. The type B individual cannot safely receive from type A, since the clumping may clog small blood vessels and the destruction of cells may impair kidney function, resulting in death. Only persons of type O blood can donate safely to all the others insofar as the A and B antigens are concerned. While the serum from an O person contains anti-A and anti-B antibodies which will cause some clumping of the cells of the recipient who is type A, type B, or type AB, this is usually not severe enough to prove critical, since the donor's serum tends to become diluted by that of the recipient. So if the cells of the *donor* are *not* clumped, a transfusion may be reasonably safe.

Inheritance of ABO Blood Types

The inheritance of the specific ABO blood type is known to depend on a particular genetic locus, known as the *ABO* locus. The specific chromosome on which this locus occurs is still unknown, but it is certain that it is not sex-linked. At the *ABO* locus on a chromosome, one of at least three different allelic forms of a gene may be present (Fig. 7-1). A particular chromosome may carry the allele for A blood type (gene *A*), for B type (gene *B*), or for O (gene *O*). The inheritance of genes for blood type is no different in principle from that of other genes which have been discussed so far. However, it must be kept in mind that *more than two* gene forms or alleles can exist and that any one of them can occur at a locus. A series of three or more alleles is known as a *multiple allelic series.* Any individual may carry at most only two of the forms, one on each of the homologous chromosomes. In the case of the ABO system, the genes *A, B,* and *O* are alleles. Both genes *A* and *B* are dominant to *O,* but *A* and *B* act as codominants when both are present in a red blood cell. This means that persons of certain blood types must have certain specific genotypes. For example, an O-type person must be homozygous, genotype *OO,* since the presence of either gene *A* or gene *B* would result in the presence of the corresponding antigen in the red blood cells. Similarly, an AB type

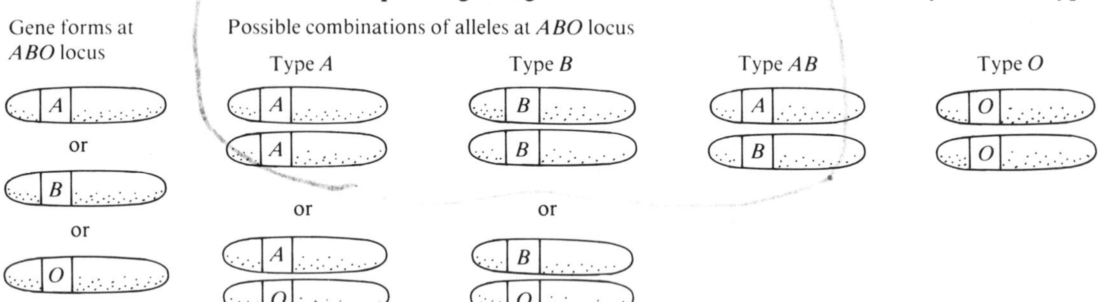

Figure 7-1. At least three different gene forms are possible at the *ABO* locus. Any one person may carry any combination of the three alleles at that site. Since the locus would be represented only twice in the body cells of any individual, any one person can have no more than two forms of the gene. The alleles for A and for B antigen are both dominant to the allele for *O* blood type.

must carry both the *A* gene and the *B,* since these alleles are codominant. Each gene will express itself and both A and B antigens will be present.

Actually, the number of alleles at the *ABO* locus is greater than three. The A type has been recognized to have at least two subgroups, A_1 and A_2. The same is indicated for type B as well. The A_1 allele seems to behave as a dominant to A_2. The existence of genes A_1 and A_2, rather than the single gene *A* as first believed, increases the number of different ABO types which can be detected by antigen-antibody reactions (Table 7-2). For the sake of simplicity in this discussion, we will ignore any subgroupings of types A and B. They exert no major influence on the question of transfusions, and their existence does not contradict any of the principles being presented.

Table 7-2. Subgrouping of the type A classification and the possible combinations of alleles

Genotype	Blood group
A_1A_1	A_1
A_1O	A_1
A_1A_2	A_1
A_2A_2	A_2
A_2O	A_2
BB	B
BO	B
A_1B	A_1B
A_2B	A_2B
OO	OO

In addition to their importance in transfusions, a knowledge of the inheritance of blood groups may prove valuable in certain legal matters. Table 7-3 shows that persons of certain ABO types cannot be the parents of children who fall into some other particular groups. A type AB individual, for example,

Table 7-3. ABO blood types and parental combinations

ABO blood type	Possible parental combinations	Impossible parental combinations
A	A × A; A × B; A × O; AB × O A × AB; B × AB; AB × AB	O × O; O × B; B × B
B	B × B; A × B; B × O; AB × O A × AB; B × AB; AB × AB	O × O; O × A; A × A
AB	AB × AB; A × B; A × AB; B × AB	O × O; A × A; B × B; O × A O × B; O × AB
O	O × O; A × A; B × B; A × B; O × A; O × B	AB × AB; B × AB; A × AB O × AB

cannot be the parent of a child who is type O, since any offspring of an AB type must receive either gene *A* or *B* and will consequently have antigen A or B in the red cells. Similar reasoning tells us that two type O parents can produce only O children. While the genotype of type AB must be *AB* and that of type O *OO,* the exact genetic constitution of a type A or type B person is not directly known just from a knowledge of his blood type. An A or a B person may be carrying the recessive gene *O,* and this possibility must be recognized in the absence of any other information. Therefore, if one parent is type A and the other B, all the ABO types could possibly occur among the offspring. Information on blood groups may disqualify a particular individual as the parent of a child in a paternity suit (a type O man, for example, cannot sire an AB child). However, the possession of a blood type which is compatible with the parenthood of a child does *not* by itself settle the matter. Any man who is type A *could* be the father of a child who is also type A, and so this knowledge of the blood type alone in such a case would have little if any significance. Besides its use in some matters of disputed paternity, blood group inheritance is at times used as evidence to establish the identity of heirs in those cases where appreciable sums of money or property are involved.

Blood grouping is a characteristic which, like any other, entails the interaction of genes. The expression of alleles at the *ABO* locus may be influenced by genes found elsewhere on the chromosomes. There is a locus on another chromosome which is known as the *secretor locus* and which interacts in an interesting way with the *ABO* locus. About 70% of the population carries the dominant gene *Se* at the secretor locus. If a person possesses A or B antigens in the red blood cells, as well as the dominant gene *Se* at the secretor locus, the saliva and other body fluids will also contain the A or B antigen. A person who is either genotype *A*__ *Se Se* or *A*__ *Se se* (the blank indicates that either *A* or *a* may be present) will be type A blood group, and A antigen may be detected in the saliva. On the other hand, an individual who is *A*__ *se se* will have no detectable amounts of A antigen in the body fluids. Since gene *Se* enables A and B antigens to be secreted, the O person, lacking antigens A and B, will secrete neither of the antigens, even if he carries the dominant *Se.* However, he can pass this dominant gene down to a child who may be a secretor if he received a gene for A or B antigen from the other parent. The secretor trait is more than just a curiosity, for at times it enables the genetic counselor to reach certain conclusions concerning the chances that an unborn child may be carrying an undesirable gene. More on this important point will be found in Chapter 12, where the knowledge of genetic linkage is applied to genetic counseling.

Still another locus on another chromosome is known which interacts with the *ABO* locus. This is the *h* locus at which the dominant gene *H* is usually present. Almost everyone has the genotype *HH.* The alternative form of the gene, *h,* is recessive, and when it is present in homozygous condition, *hh,* the person will be blood type O, regardless of the presence of genes *A, B,* or *O*! This unusual situation is due to the fact that the normal gene *H* is responsible

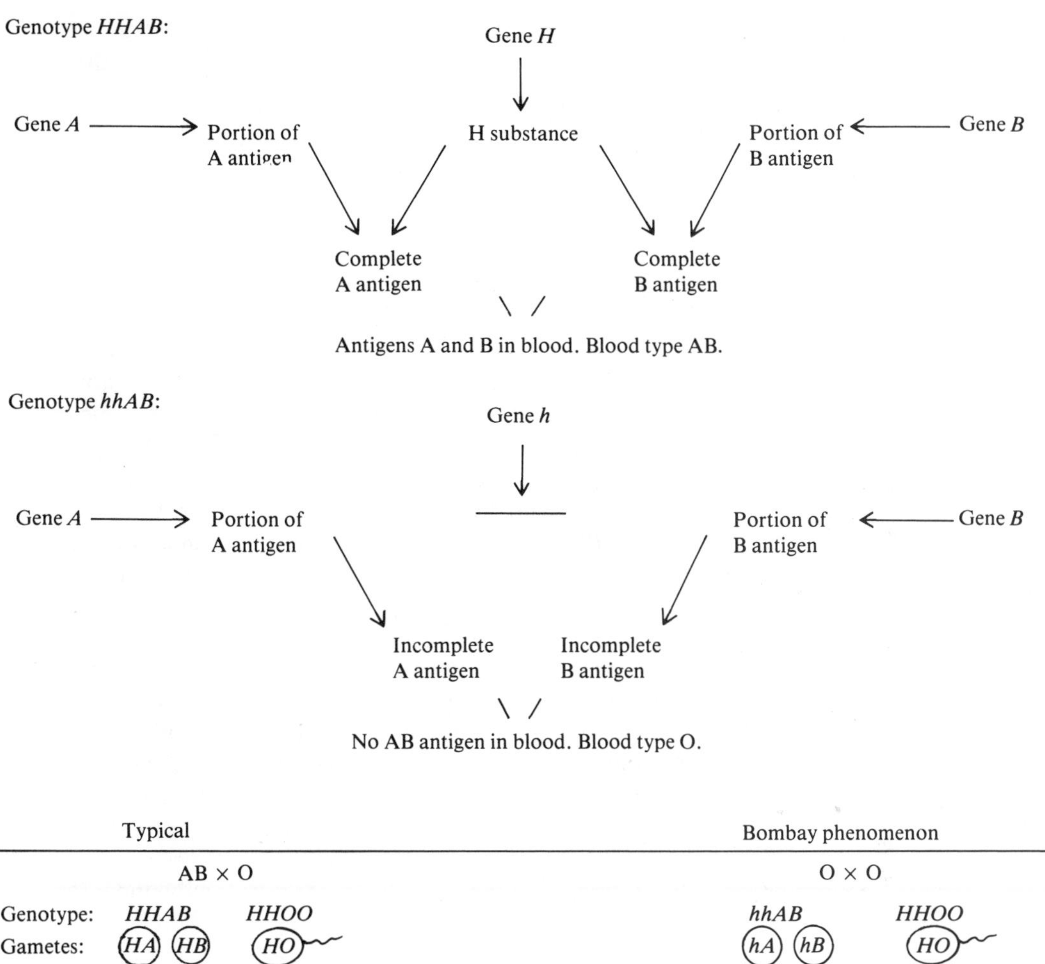

Figure 7-2. Genic interaction and the Bombay phenomenon. Most persons, regardless of A, B, O blood type, possess the dominant gene *H* required for completion of the A and B antigens. Normally a person of genotype *AB* (above) would have the A and B antigens in the red blood cells, since the dominant gene *H* will be present, and both antigens would be completed. In very rare cases (middle), a person may lack gene *H* and would be blood type O phenotypically, since no completed A or B antigens could arise, even if the rest of the genotype is *AB.* Typically, the cross of an AB and an O person (below) would produce offspring of type A and type B. Type O would be impossible among the offspring of an AB individual. However in the rare case of genotype *hhAB,* the very unexpected appearance of A and B offspring may occur from two O parents. Since one O parent contributes the dominant gene *H,* the *A* and *B* genes contributed by the O parent who is genotypically *AB* may now come to expression.

for the production of a precursor substance in the blood of almost all persons. This product is needed before antigens A or B can be constructed in the body (Fig. 7-2). For example, the person of genotype *AB* will almost always possess both the A and the B antigens in his red cells, since he is most probably genotype *HH* as well, and the A and the B antigens can be constructed. However, if the very rare double-recessive condition, *hh,* is present in a person who is genotype *AB,* neither the A nor the B antigen can be manufactured, since the necessary H substance required for their completion is not available. And so, while he is genotype *AB,* such a person would appear to be type O. This rare condition is known as the ***Bombay phenomenon,*** since it was first reported in that city. A few other cases have since been recognized elsewhere. The rarity of the gene *h* makes an encounter with the Bombay phenomenon highly unlikely, but it is presented here to emphasize once more genic interaction. In this case, expression of a gene (*A* or *B*) may be completely suppressed by one at another locus, or even one on another chromosome. We see here an excellent example of genic interaction. Even though the allele *A* or the allele *B* may be present at the *ABO* locus for the construction of A or B antigen, the presence of gene *h* in the double-recessive condition at a different gene locus prevents the expression of either the *A* or the *B* allele. Therefore, a person of genotype *hh* cannot develop A or B antigens, regardless of the genetic constitution at the *ABO* locus. This phenomenon is commonly called *epistasis,* the masking of the expression of a gene by another which is at a different locus and is thus not its allele. The expression of the *A* or *B* gene would be reduced in families in which the Bombay phenomenon (and thus the *H* gene) occurs. Moreover, the Bombay phenomenon illustrates that certain human families may carry genes which are extremely rare in the population with the consequence that atypical inheritance patterns present themselves in pedigrees. A genetic counselor must always be alert to the individuality of each family under study.

The MN Blood Groupings

In 1927, Landsteiner and Levine discovered another set of red blood cell antigens, those responsible for the MN blood grouping, a very different one from the ABO. Every person can be classified as blood type M, N, or MN. The reason we hear little about these types is that the M and the N antigens are rather weak and do not tend to elicit antibody formation. Consequently, they are relatively unimportant in transfusions. Moreover, no one is born with naturally occurring antibodies against M or N antigens in the blood serum. However, the existence of the M and N antigens can be demonstrated by injecting human blood into a rabbit or some other animal. Antibodies will form in the animal's serum. Such serum will then clump the red blood cells which contain the corresponding antigens in their membranes. In this way, the three different phenotypes, M, MN, and N, can be recognized. Experimental animals are commonly used to generate antibodies against specific human antigens.

The MN blood types are inherited in a simple Mendelian fashion. A pair of alleles is associated with the MN locus. These behave as codominants and are designated L^M and L^N. (The L designation for the locus is in recognition of Landsteiner.) Table 7-4A gives the possible phenotypes and the corresponding genotypes. Since there is a lack of dominance between the two alleles, knowledge of the phenotype directly indicates the genotype, since a gene, if present, will be expressed through the production of the corresponding antigen. More recently, it has been found that differences exist within each of the MN types. For example, type M (as well as the MN and N types) may have three subtypes. One suggestion is that a series of alleles exists at the MN locus (Table 7-4B). However other interpretations which need not concern us here have been presented to account for the subtypes. The MN grouping is mentioned since

Table 7-4. MN blood groupings. A: Major MN types. B: Subtypes of MN types on the basis of a multiple allelic series at the MN locus. L^{MS}, L^{Ms}, L^{NS}, L^{Ns}. Codominance would exist among the different alleles.

A

Phenotype	Genotype	Antigens on red blood cells
M	L^ML^M	M
MN	L^ML^N	M and N
N	L^NL^N	N

B

Phenotype	Genotype	Antigens on red blood cells
MS	$L^{MS}L^{MS}$	M and S
Ms	$L^{Ms}L^{Ms}$	M and s
MSs	$L^{MS}L^{Ms}$	M, S, and s
MNS	$L^{MS}L^{NS}$	M, N, and S
MNs	$L^{Ms}L^{Ns}$	M, N, and s
MNSs	$L^{MS}L^{Ns}$ or $L^{Ms}L^{NS}$	M, N, S, and s
NS	$L^{NS}L^{NS}$	N and S
Ns	$L^{Ns}L^{Ns}$	N and s
NSs	$L^{NS}L^{Ns}$	N, S, and s

it is a familiar one to the geneticist and to the hematologist. In addition, information on MN type may be used along with that on the ABO type to aid in cases involving disputed relationships. Table 7-5 shows how this may be used to eliminate the possibility of parenthood. Even if the ABO blood type is compatible between a child and a supposed parent, the MN grouping may show that parentage is nevertheless impossible in a particular case. For example, an N-type individual cannot have an M-type parent regardless of agreement in relation to the *ABO* locus. But again, possession of the right blood type is not conclusive in the final identification of a parent.

Table 7-5. MN blood types and parental combinations

MN blood type	Possible parental combinations	Impossible parental combinations
M	M × M; M × MN	M × N; N × N; MN × N
MN	M × MN; N × MN; MN × MN	M × M; N × N
N	N × N; N × MN	M × N; M × M; M × MN

The Rh Types

An important and well-publicized blood grouping is the Rh, which also involves a set of antigens in the membranes of the red blood cells. The so-called *Rh factor* was discovered in 1939 by Landsteiner and Wiener, who had injected the blood of the *Rhesus* monkey into rabbits and guinea pigs. The response of the animals was to produce antibodies in their blood sera against *Rhesus* monkey blood. Blood cells of the monkey would then be clumped when exposed to this antibody-containing serum. When Landsteiner and Wiener added the same antiserum to human blood, they found that the blood of about 85% of the persons tested was also clumped. The investigators had made certain that there were no antibodies present in the serum which were known at the time to be able to clump human blood (such as MN, A, B). This observation meant that 85% of the persons tested carried an antigen or factor (Rh) in their red cells which is also carried by the *Rhesus* monkey in its red cells. Such persons were designated Rh^+, and those lacking the antigen, Rh^-. Studies of families indicated that the Rh^+ condition depends on a dominant, autosomal gene and that the double-recessive condition results in a person who is Rh^-. The dominant and recessive alleles are often represented as *D* and *d*, respectively.

It was soon realized that the Rh factor plays an important part in blood transfusions and that Rh incompatibilities between donor and recipient can account for many cases in which a serious reaction occurs, even though the ABO relationship of donor and recipient is compatible. As is true in the MN grouping, no one is born with naturally occurring Rh antibodies of any kind.

An Rh^+ person does not carry antibodies against Rh negative blood in his serum. Likewise, the Rh^- person is not born with antibodies which can react with Rh positive antigens. Therefore an Rh^- can donate blood to an Rh^+, and an Rh^-, lacking antibodies to Rh^+, can in turn receive blood. However, while the Rh negative person may be able to tolerate one transfusion of Rh^+ blood, the Rh^- individual will become sensitized and will respond by prompt formation of antibodies against the Rh factor in a second transfusion. This can prove fatal, just as in the case of an ABO incompatibility. The Rh^+ person, however, does not react by forming antibodies against the negative blood type and can thus receive more than one transfusion from Rh negative donors.

Shortly following its discovery, the Rh factor was shown to be implicated in cases of *erythroblastosis fetalis,* a condition of the newborn which commonly manifests itself in the production of anemia and jaundice, but which may entail damage to vital organs such as the liver and brain. In the severest cases, babies may be stillborn or may die shortly after delivery. The cause of the condition is an antigen-antibody reaction in the bloodstream of the infant which is brought about by an Rh incompatibility between mother and child. There is more than one way in which the unfortunate antigen-antibody reaction can be triggered, but in all cases, the mother is Rh negative and the infant Rh positive. An Rh^- woman may carry an Rh^+ baby if the father is Rh^+, genotype *DD* or *Dd.* If the man is heterozygous, *Dd,* there is a 50:50 chance at each conception that the zygote will be *Dd* (Rh^+) or *dd* (Rh^-). The first Rh^+ child from such an Rh^- woman would most likely be born with no Rh complications, since a person is born with no preformed anti-Rh antibodies. Actually, a second, third, or even a later Rh^+ child may be perfectly healthy at birth. However, a complication may arise at times when an Rh^- woman is carrying an Rh^+ child. The circulatory systems of mother and child are quite separate. Any exchange of nutrients or other metabolic products between mother and child takes place by diffusion across the placenta, generally a very effective barrier which restricts the passage of most substances other than those directly involved in the normal development and growth of the offspring (Fig. 7-3). At times, however, it appears that certain capillaries of the placenta may become defective so that they break or allow the seepage of blood from the circulatory system of the fetus into that of the mother. If Rh^+ blood from a fetus enters the circulatory system of an Rh^- woman in this way, the situation is almost the same as if the Rh^- person were receiving a transfusion of Rh^+ blood. The response is the formation of antibody against Rh^+ blood. The antibodies from the mother may then pass back across the placenta. Usually, the antibody buildup on the first exposure is insufficient to produce serious consequences. The transfer of fetal antigens appears to occur about the time of birth, so that not enough time is available for a high antibody level to be reached. However, the woman is now sensitized. This means that, while increased Rh antibody level caused by the first pregnancy will fall, a second exposure to Rh^+ antigen will trigger an immediate and highly effective response which will result in the formation of an antibody level against Rh antigen that is even higher than the first. It is as if the antibody-forming system were prepared to respond quickly and efficiently to produce antibodies to destroy Rh^+ cells. However, in a second pregnancy with an Rh^+ child, nothing dictates that a problem *must* arise. If there is no placental defect, no blood seepage occurs in either direction. If blood from the second Rh^+ child *does* enter the circulatory system of the Rh^- woman, there can then be an immediate response followed by an appreciable production of Rh antibodies. The mother may in turn pass these to the offspring in sufficient numbers to produce *erythroblastosis.*

Any Rh negative woman already sensitized prior to pregnancy by an earlier blood transfusion from an Rh^+ donor will, of course, be primed to produce antibodies against Rh^+ blood. If there is placental seepage of blood, even in the first pregnancy with an Rh^+ child, a large number of antibodies could be passed to the offspring with serious consequences. Figure 7-4 illustrates some

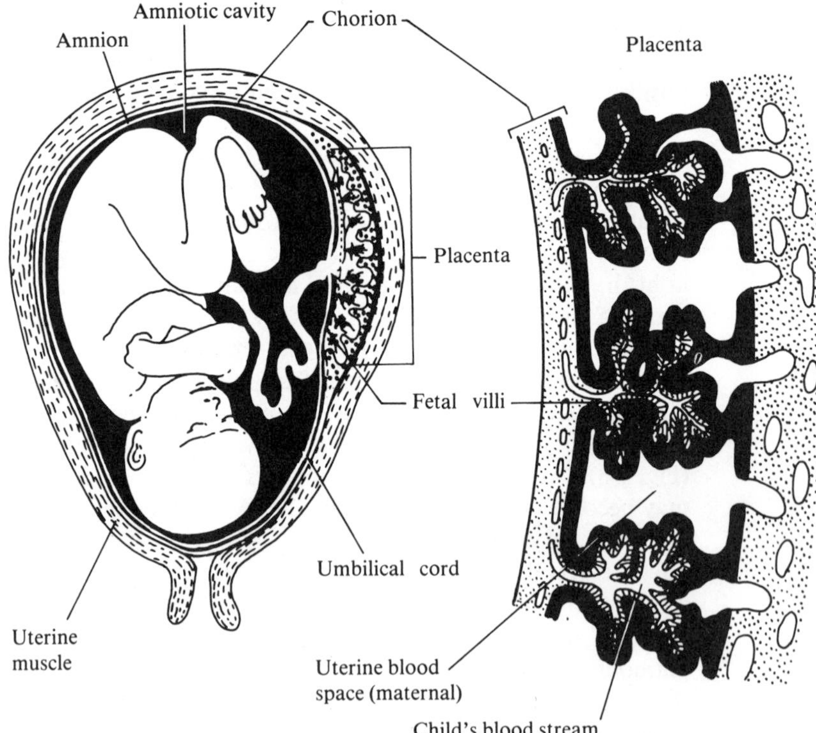

Figure 7-3. The placenta (right) is shown highly enlarged and greatly simplified. However, as indicated, there is normally no mixing of maternal and fetal blood systems. Exchange of nutrients and waste products occurs by diffusion across membranes in the placenta.

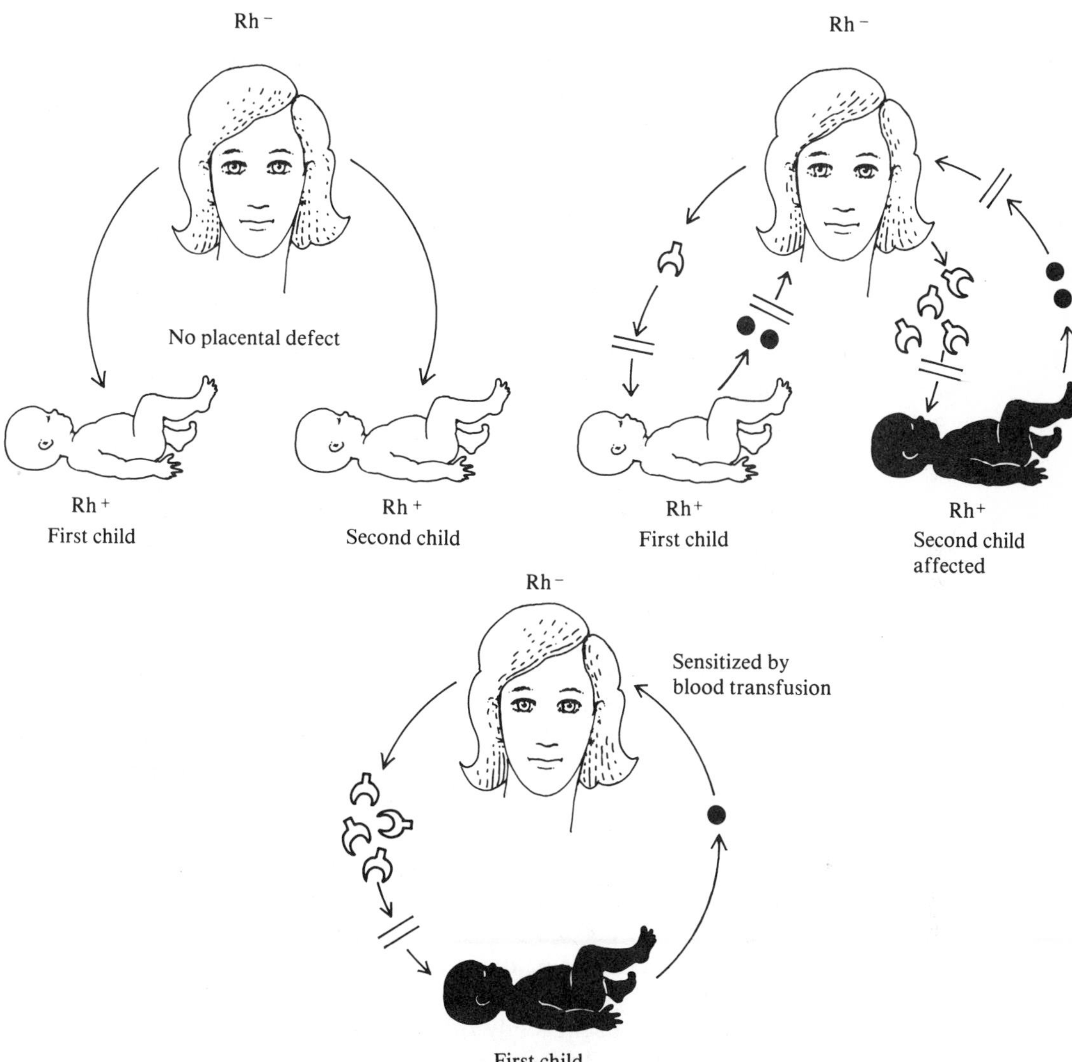

Figure 7-4. Rh negative mother and Rh positive offspring. An Rh$^-$ woman may bear several Rh$^+$ children with no complications if there is no exchange of fetal and maternal blood due to a placental defect (1). An Rh negative woman who has not been sensitized in any way to produce Rh antibodies will usually bear an unafflicted child at the first Rh$^+$ pregnancy, even if a placental defect is present, since there is little opportunity for a high level of Rh antibodies to arise (2). If a placental defect is present during a second Rh$^+$ pregnancy, the child may be affected since the mother has been sensitized and can respond to the child's antigens by producing a high antibody level. An Rh$^-$ woman already sensitized by a previous transfusion with Rh$^+$ blood (3) can bear an affected child at the first pregnancy, since her immune system has already been printed and will respond immediately to any Rh$^+$ antigens with the production of a high antibody level.

of the outcomes when an Rh^- mother bears an Rh^+ child. Although the Rh factor must be seriously considered in pregnancies, exaggeration by uninformed sources has led to an overstatement of the entire Rh story. In the first place, it must be appreciated that an Rh^- woman and an Rh^+ man may or may not have Rh^+ offspring, depending on the genotype of the man. But even if the woman does bear Rh^+ children, no problem necessarily must arise in a series of Rh^+ deliveries. There is nothing that demands that an exchange of blood must occur between mother and child. Typically, no such exchange takes place. So even if a woman *is* sensitized, there may still be no problem in a later Rh^+ delivery. Information on the Rh factor must be applied carefully and correctly to avoid unnecessary anxiety as well as to set the stage for precautionary measures.

Techniques are now available to prevent an Rh^- mother from producing damaging antibody levels in cases of a second or third Rh^+ pregnancy. Antibodies can be prepared against Rh^+ antigen (antigen D). These preparations can be given to Rh^- mothers in the form of gamma globulins after delivery of Rh^+ offspring. The antibodies will remove any Rh^+ cells that may have entered the maternal bloodstream and will thus prevent priming or immunization against a later Rh^+ pregnancy. The effectiveness of this treatment can be measured by comparing 2 groups of Rh^- women who have already experienced 2 Rh^+ pregnancies. Blood analysis of those in the group which received no gamma globulin preparation shows that almost 17% of the women carry antibodies against Rh^+ antigen. In contrast, of the Rh^- women in the group which received the gamma globulin treatment, less than 2% possess these Rh antibodies. This means there has been a 90% reduction in the risk of immunization of an Rh^- woman against Rh^+ antigen. The procedure is effective only if the woman has not been previously sensitized. However, other methods are also available to prevent loss of infants due to Rh incompatibilities.

While the inheritance of the positive antigen behaves as if it were based on a single pair of genes, *D* and *d*, the story is more complex by far. More refined procedures in the study of antigen-antibody reactions later revealed that several types of Rh^+ persons exist. Different kinds of Rh antigens are present in these various types of Rh positive people. Also, there is more than one type of Rh negative. We now know that Rh negative persons do not entirely lack antigens of the Rh genetic system. It is simply that these Rh-related antigens which they carry behave very weakly as antigens, as was true of the M and N. They do not readily elicit antibodies with which they react. Recognition of the several Rh^+ and Rh^- types has led to a great deal of confusion in their naming. It has also led to a continuing debate over the exact genetic basis for the Rh types. One school of thought holds that one gene locus is involved, while the other maintains that more than one plays a role. Again this is of no concern for our immediate purposes, since there is mainly one kind of Rh antigen which is implicated in the more serious cases of Rh incompatibilities. This is the D antigen, so named in one of the leading systems. It is responsible for 90% of the Rh difficulties. The D antigen, and hence the *D* gene, would be present in all persons defined as Rh^+. An Rh negative person would be genotype dd and would lack the D antigen, even though he carries other Rh-related antigens which do not normally induce antibodies.

Blood Groups and Associated Effects

The Rh system is not the only one involved in incompatible blood reactions between mother and child. There are several less common genes which are known in certain family pedigrees and which can cause problems (see below). However, the familiar ABO is not exempt. Placental defects may arise no matter what the blood types of the mother and offspring; they do not occur just because of the Rh factors. If seepage of blood should occur when an O mother is carrying an A, B, or AB child, the two kinds of antibodies, *a* and *b,* may enter the fetal blood stream and cause clumping of cells. It is suspected that incompatibilities involving differences in the common ABO system may prevent approximately 5% of human embryos from surviving to birth. The ABO types may also interact with the Rh type. Data suggest that when an Rh negative mother is carrying an Rh positive child there is less chance of *erythroblastosis* if there is also an incompatibility between the ABO blood types of mother and child. Though this sounds paradoxical, it can be explained in the following way. Suppose blood of an Rh^+ type A child reaches the circulatory system of an Rh^- type O mother. The mother's blood carries antibodies against A- and B-type blood, and so there is a chance that the fetal cells will be destroyed *before* Rh sensitization can occur. If the baby were type O, then the cells introduced to the mother would not be quickly destroyed by Anti-A or Anti-B anitbodies, and they could then cause Rh antibodies to build up. An Rh problem could arise at a later birth due to the priming of the mother's immune system.

The Rh genotypes which are possible can be added to those of the MN and the ABO groups to aid in legal matters of relationships among persons. Two Rh^- individuals (*dd*) cannot have an Rh^+ child, since the D antigen would express itself. So even if the other groups support the possibility of a relationship, the specific Rh grouping can negate it. To these major groups we have discussed may be added several others which are less familiar. Some of these are not involved in serious incompatibilities, and the variant gene form involved may be rare. For example, the Kell factor is another red blood cell antigen, and its presence is determined by the dominant gene *K*. Most persons are homozygotes, *KK,* but the alternative form of the gene, *k,* does occur in a low frequency in the population. Double-recessive persons, *kk,* lack the Kell antigen. A woman who is Kell negative and who carries a Kell positive baby may bear children who have *erythroblastosis.* The story is the same as that for the Rh^- mother and an Rh^+ child. The names given to many of the other known blood groupings (Kell, Lewis, Duffy, etc.) are those of the family in which the grouping was first recognized. Since an extensive number of blood groupings exists and since several different types can be distinguished within each, it is no understatement to say that each one of us is unique in the combination of blood antigens and hence our antigen-determining genes. To obtain by chance another combination of the many types and subtypes of the various blood groups as they are found in any one person is highly unlikely.

While any person can be easily classified as to his various blood types, the actual biological significance of the major groups themselves is uncertain. The blood antigens may perform certain roles in the actual structure of the red blood cell membrane. Their full significance, however, may be much more subtle and complex and may involve other systems of the body. The reason for this suspicion is the higher association of persons of certain blood types with certain disorders. Individuals of blood type O seem to be more predisposed to bleeding peptic ulcers; yet O-type persons seem to be less susceptible than those of the other ABO types to certain kinds of circulatory disturbances. Among victims of the disease *arteriosclerosis obliterans,* characterized by hardening of the arteries and blood clots, there is a significantly higher proportion of type A. There also appears to be a higher risk of gastric cancer for persons of type A than for the other types. The reason for these apparent associations is far from clear, but they raise the possibility that the genes which control the formation of the blood antigens also influence other proteins. Conceivably, they may do this by their control of a single step in the chemistry of the cell which can in turn influence not just one but two or more other metabolic pathways. Genes known to exert such an effect will be discussed in Chapter 10.

That the *ABO* locus may indeed have a significant function is also suggested by the fact that most humans are blood type O. This is somewhat unexpected, since the gene for type O results in a phenotype which is inactive antigenically. (Since most persons are genotypically *HH,* H substance is, of course, present in the blood of O persons and can be detected antigenically.) Cells of blood type O do not cause any antibodies of the AB type (anti-A or anti-B) to form. In other words, the genes *A* and *B* control *active* substances; yet the frequency of these genes is not as high as the recessive *O.* This suggests that the O type may be associated with some kind of advantage. The correct interpretation of the high frequency of gene *O,* however, is still unknown. The *ABO* locus, as well as the MN, Rh, and other blood antigen loci, may prove to have important pleiotropic effects which are still unrecognized.

Transplants and Grafting

Today a great deal of research is devoted to the problem of organ transplants and grafting of tissue. It is common knowledge that a person, unless he has an identical twin, cannot readily accept tissue from another individual. He can, however, have successful grafts made from one part of his own body to another part. The reason for the difficulty encountered in tissue transplants is not too different from the story of transfusions, which are actually exchanges of tissue. Any tissue (skin, heart, blood, kidney) possesses antigens in the membranes of the cells which compose it. Therefore, when an organ is taken from a donor and transplanted to a recipient, foreign antigens are being introduced into the recipient. The response is the production of antibodies and the priming of certain cell reactions which succeed in the rejection of the tissue carrying the antigens. Drugs which tend to suppress the body's immune response are usually given when either a graft or transplant is made so that attack on the introduced tissue will not be severe enough to cause rejection. Unfortunately, a suppressor of the immune system also renders the body vulnerable to infections from disease-causing agents.

The immune response system entails one of the most intricate series of interrelated mechanisms operating in the human body. It is by no means understood in all its details. The genetics of histocompatibility is perhaps the most complex of all the fields of genetic research. Since 1970, several painstaking and exceedingly intricate experimental studies, employing among other tools inbred strains of mice and human cells in tissue culture, have clarified many important aspects of the genetics of human tissue compatibility. Only the barest outline can be presented here in a simplified form.

It should be clarified at the outset of the discussion that the actual rejection of tissues or organs does not entail the operation of antibodies to a very great extent. Rather, the pronounced effects of rejection are based on *cellular* immune reactions. These involve lymphocytes, white blood cells which may be formed in the thymus gland and which are able to recognize tissues carrying foreign antigens. They then attack such tissues, causing rejection.

It has been recently demonstrated that two distinct types of tissue differences exist between foreign donated tissue and the tissue of the recipient or host. One of these involves a response in which antibodies are elicited in the body. These antibodies can be detected *in vitro* using routine antigen-antibody procedures. This category of responses can be referred to as *serologically defined responses,* or SD. The second difference causes another kind of response and involves lymphocytes which originate in the thymus gland. This produces the so-called "lymphocyte derived response," or LD. The LD response does not by itself entail the formation of circulating antibodies. The genetic information which controls these two types of responses, the SD and the LD, is known to reside at different genetic loci. One of the main regions involved has been known as the HL-A region or locus. Actually, it is not a single locus but is instead a chromosomal region or complex which includes two very closely linked loci, the SD_1 and the SD_2 loci. At each one of these two different SD loci, any one of a number of different allelic gene forms can reside. These two loci control antigens which bring about antigen-antibody reactions and which are thus referred to as *serologically defined.*

However, in addition to the 2 SD loci which form part of the HL-A complex, there is at least 1 other locus, and most probably 2 (and we will consider 2). These loci are very closely linked to the SD_1 and the SD_2 loci and control LD antigens, those involved in the lymphocyte response. These can be designated the LD_1 and the LD_2 loci. As Fig. 7-5(A) shows, 1 of the 2 LD loci may actually be located between the SD_2 and the SD_1 loci. However, at each 1 of the 4 loci, SD_1, SD_2, LD_1, and LD_2, a series of multiple alleles exists. The exact number of genes which can possibly occur at each locus is unknown, but at least 9 and 14 possibilities exist at the SD_2 and SD_1 loci, respectively. Since humans are highly heterozygous, this means that for each of these 4 loci, any person will probably inherit from each parent a different gene [Fig. 7-5(B)]. He would thus produce *4* different SD antigens and *4* different LD antigens. Any combination of the possible gene alternatives on any single chromosome would be conceivable. Since the genes are so very closely linked, a combination would tend to be inherited as a unit, although crossing over can separate them at times, break up the combination, and produce a new one. Evidence indicates that certain combinations are more common among humans than are others. The suggestion has been offered that, since more than 2 loci definitely exist at the so-called HL-A region, the term *major histocompatibility complex* (MHC) be used instead for the entire region. The MHC would include the SD and the LD loci and all the alleles which can reside at each one of them. The mouse and the dog have histocompatibility complexes similar to that in the human. Evidence suggests that in the animals even additional histocompatibility loci exist in this small chromosomal segment. It is quite possible that, as the ability to detect immunologic responses increases, still more loci may be identified in the human HL-A complex.

The foregoing summary should give some idea of the complexities of the histocompatibility story and should indicate why it is quite unlikely that any two unrelated persons would be genetically identical for histocompatibility genes at the MHC region. In transplants, attempts are made to match persons who are as similar as possible for all known antigens. Even the ABO system is considered, since differences in blood types may also exert some influence on transplant toleration. It has been reported that if two persons are identical for the SD antigens (and would thus have the same four genes at the SD_2 and SD_1 loci on the two homologous chromosomes), there is a less vigorous response of the lymphocytes whose activity is controlled by the LD genes. If a difference does exist between individuals with reference to genes at the SD loci, the response of the cells involved in rejection is greater.

As the term "*major* histocompatibility locus" implies, this genetic region is not the only one which plays a role in tissue compatibility, as the reference to the *ABO* locus also implied. There are probably many others with less

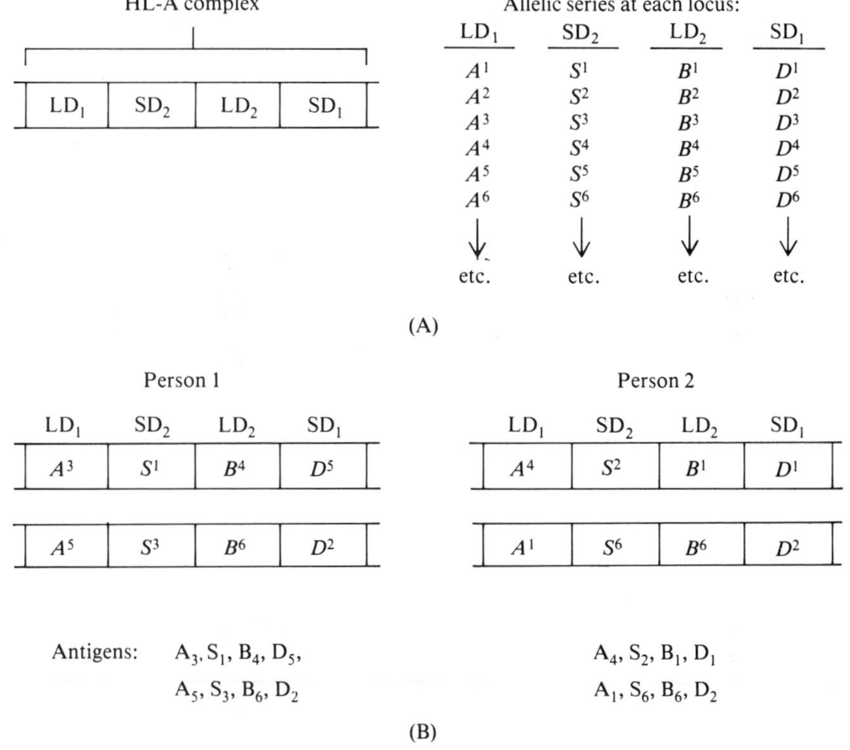

Figure 7-5. The HL-A complex. (A) Often called the HL-A region or locus, this segment of the genetic material (left) includes four loci, two involved with lymphocyte-derived responses, the LD loci, and two which control serologically derived responses, the SD loci. At each one of the four loci, a series of multiple alleles exists (right). Any one of the members of a series can occur at the corresponding locus on any one chromosome. (The designations given here are arbitrary and solely for illustrative purposes). (B) Any person would have each LD and each SD locus present twice, four on each of the two homologous chromosomes he or she carries. Since humans are highly heterozygous, eight different alleles are likely to occur in each person. A large number of different combinations is possible and would vary from one person to the next as indicated here. Two people taken at random would therefore differ in many of the histocompatibility antigens they possess.

pronounced effects which nevertheless may be additive and which in the long run may determine the success or failure of a transplant. So even in the rare case of complete compatibility at the SD and the LD loci, other genes controlling other immune responses add to the risk of rejection. Fortunately, matching of genes at the loci in the HL-A region seems to increase greatly the chance that a kidney will be accepted; thus it is possible that two children within a family will match in these genes about one-quarter of the times. This is so because of the very close linkage among the loci in this region. Only rarely would they become separated by crossing over. Figure 7-6 attempts to explain this, treating the two loci in the complex on one chromosome as one gene because of the

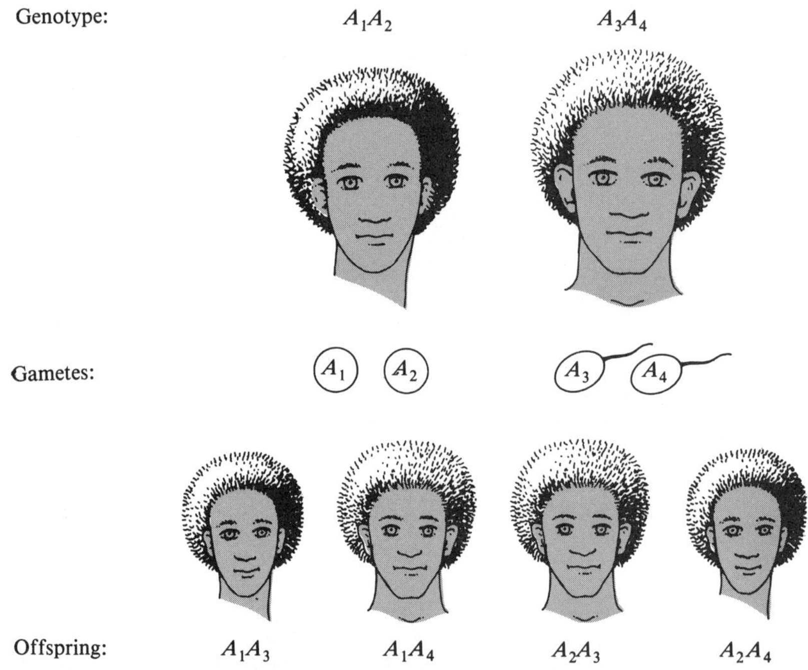

Figure 7-6. Compatibility at the HL-A region. As Fig. 7-5 shows, there are really four loci, two SD and two LD, at each of which any one of a number of alleles is possible. Since the loci are so closely linked, the chance of a crossover which could separate them is not very high. For the sake of simplicity, the four loci may be thought of here as one locus. According to the idea shown here, any one gene at the locus can control the formation of four antigens. In this diagram, gene A_1 would control the formation of four antigens and gene A_2 would control the formation of four different ones. Two parents would most probably differ in the alleles they possess at the HL-A region. Among their children, four combinations would occur. It would be possible for two children in the family to possess the same genotype with respect to the SD and LD loci. Hence they would have the same SD and LD antigens and would match antigenically in this respect.

very close linkage between them. The unlikely chance of a crossover which could separate them is ignored for the sake of simplicity. The LD loci in the region play the most important role in graft acceptance. Identity of genes at these loci gives greater ability to accept a graft than does identity at the SD loci alone. Successful transplants of organs other than the kidney apparently depend to a large degree on other histocompatibility loci in addition to the loci in the HL-A region. This would account for the lower incidence of success in transplantation of organs other than the kidney.

What is the reason for such a complex mechanism in the human body? Does it possess any value? Certainly, the response to invasion by disease-causing agents is an important function of the immune system, but such intricacies as exist must surely involve still other reactions. Certainly, the system has not evolved to reject tissue transplants which are performed by man! There are very good reasons to suspect that a primary role of the immune system is to guard against changes in the body's very own cells which can lead to cancerous growths. A malignant transformation (Chapter 14) entails changes in surface properties of a cell. The assortment of alterations which may affect a normal cell as it becomes transformed to a malignant one may necessitate a system intricate enough to recognize any subtle changes and to remove the affected cell before it can grow at the expense of other tissues, with the final destruction of the entire body. One observation which supports such a concept is the increased risk of cancer entailed by persons who must take drugs to suppress the immune system so that grafts may be tolerated.

The genes which control immune responses are, like any others, capable of undergoing mutation. There are several disorders known which have altered the efficiency of the immune mechanism, causing a decrease in the response of the body to foreign antigens. The general term *agammaglobulinemia* includes several different kinds of immune deficient disorders which are due to the presence of different defective genes. Some of these are inherited as sex-linked recessives, whereas others are autosomal in transmission. The inheritance of still others has not been clarified. As a result of the genetic defect, there may be a decrease in the antibody response or in the production of cells which are able to attack other cells carrying foreign antigens.

In one of the most extreme of the agammaglobulinemias, both major parts of the immune system are impaired so that no antibody or lymphocyte responses occur to any extent to defend the individual against infection. Death in early childhood results from this form of the disorder, since no treatment has been devised. A sex-linked recessive is responsible in some families, whereas in other families an autosomal recessive is the cause. Some of the less severe immune deficient disorders can be treated. For example, if the cell response is normal but no antibodies can be formed, injections of gamma globulins, the blood fraction in which antibodies arise, can alleviate the condition.

The genes which control the immune responses in the body make each of us a unique entity. They also provide an important line of defense against the introduction of foreign matter which could interfere with normal body functions. They most probably serve to protect the body against pathological changes arising within the body itself. Unfortunately, the uniqueness and the overall protection bestowed impart a certain disadvantage, one which is not usually apparent but which can be critical at a time of emergency when healthy tissue from a donor is needed for survival.

REFERENCES

Bach, F. H., "Transplantation: Pairing of Donor and Recipient." *Science,* 168:1170, 1970.

Bach, F. H., Widmer, M., Segell, M., Bach, M. L., and Klein, J., "Genetic and Immunological Complexity of Major Histocompatible Regions." *Science,* 176:1024, 1972.

Bodmer, W. F., "Evolutionary Significance of the HL-A System." *Nature,* 237:139, 1972.

Clarke, C. A., "The Prevention of "Rhesus" Babies." *Sci. Amer.,* Nov: 46, 1968.

Cooper, M. D., and Lawton, A. R., "The Development of the Immune System." *Sci. Amer.,* Nov: 58, 1974.

Edelman, G. M., "The Structure and Function of Antibodies." *Sci. Amer.,* Aug: 34, 1970.

Jerne, N. K., "The Immune System." *Sci. Amer.,* Jul: 52, 1973.

McDevitt, H. O., "Genetic Control of the Antibody Response." In *Medical Genetics,* McKusick, V. A., and Claiborne, R. (Eds.), pp. 169–182. HP Publishing Co., N. Y., 1973.

Notkins, A. L., and Kaprowski, H., "How the Immune Response to a Virus Can Cause Disease." *Sci. Amer.,* Jan: 22, 1973.

Race, R. R., and Sanger, R., *Blood Groups in Man,* Sixth edition. Blackwell Scientific, Oxford, 1973.

Reisfield, R. A., and Kahan, B. D., "Markers of Biological Individuality." *Sci. Amer.* June: 28, 1972.

Vogel, F., "ABO Blood Groups and Disease." *Amer. Journ. Human Genet.,* 22:464, 1970.

Watkins, W. M., "Blood Group Substances." *Science,* 152:172, 1966.

Wiener, A. S., "Blood Groups and Disease." *Amer. Journ. Human Genet.,* 22:476, 1970.

REVIEW QUESTIONS

1. Give the correct answer(s). A man who is blood type O marries a woman who is type AB. Which of the following would be very likely? (1) None of the children will be like the parents in blood type. (2) The children will be either blood type AB or type O. (3) The children will all be type AB. (4) The children will all be type O. (5) The children will all be type B. (6) None of the statements is likely.

2. A woman is blood type M, Rh negative. Her baby is type MN, Rh negative. Give the correct answer(s) from among the following. (1) The father could be type N. (2) The father cannot be Rh positive. (3) The mother must be heterozygous for the Rh blood type. (4) The father cannot be type M. (5) The father could be type MN.

3. A woman is blood type A, Rh positive. Her husband is type B, Rh positive. With just this information, determine what blood types are possible among the offspring.

4. A man who is blood type O and a woman who is type B bear one child who is type B and another who is AB. Explain this unexpected result.

5. In fowl, the genes C and O are both necessary for colored feathers. Duplicate recessive epistasis operates so that the absence of either dominant allele results in white birds. What will the feathers be like (colored or white) among the offspring of these two colored birds? $CcOo$ and $CcOo$.

6. Assume that animal 1 has the following genotype with regard to histocompatibility genes: $A^1A^1S^2S^2B^3B^3D^4D^4$. Animal 2 is $A^2A^2S^3S^3B^4B^4D^5D^5$. Select the correct statement(s): (1) Animal 1 can accept tissue from animal 2. (2) Animal 2 can accept from animal 1. (3) Neither animal can accept from the other. (4) An offspring between the animals will be able to accept tissue from neither parent. (5) An offspring will be able to accept from either parent. (6) The parents can accept from the offspring.

7. Suppose a woman and a man have the following genotypes, respectively, for certain closely linked histocompatibility genes:

$$\frac{A^1S^2B^1D^2}{A^2S^1B^2D^1} \quad \text{and} \quad \frac{A^3S^4B^3D^4}{A^4S^3B^4D^3}$$

Ignoring crossing over, what kinds of offspring would be possible? Can the children accept tissue from the parents or donate to them? Is it possible that two family members could be compatible in respect to these genes?

chapter 8

SINGLE GENES, POLYGENES, AND POPULATIONS

Types of Variations

As we observe any group of people, large or small, we become aware of the phenotypic differences among them. No two persons are exactly alike. Since environment interacts with the hereditary factors in the final production of a phenotype, even identical twins exhibit differences. The human species includes a countless variety of types. What is responsible for such diversity among members of the human species? To answer this, it is first necessary to define the variation which is seen among people. The evident differences in certain human characteristics are rather easy to describe. For example, without difficulty we can classify persons on the basis of several eye color traits—blue, brown, gray, hazel, etc. We have already encountered many characteristics in the preceding chapters in which two or more traits are readily identifiable: normal pigmentation vs. albino; attached earlobes vs. free lobes, etc. Besides these conspicuous phenotypic differences among people, there are also those which are more subtle and which entail chemical tests or microscopic examination: the various ABO, MN, and Rh blood types; the ability to produce the enzyme glucose-6-phosphate dehydrogenase. But even for characteristics such as these, the differences are clearcut ones. Variation of this kind is called *discontinuous,* meaning that the distinction among the traits or alternatives for that characteristic is sharp and well defined.

We would soon encounter difficulty in any attempt to describe all the variations in a group of people. One conspicuous difference among individuals is height. Some persons can be readily described as tall or short. However, most persons fall somewhere in between the obvious extremes. No sharp boundaries exist among the size classes. In contrast to a characteristic such as eye color, height varies *continuously,* without distinct traits to set one group apart from the next. Characteristics which exhibit *continuous variation* are therefore often difficult to describe. To do so requires measurements of many individuals and the use of statistical tools. The most familiar statistic is the *mean,* or average, which is commonly presented to give some idea of a central tendency such as an average height or average weight (Fig. 8-1).

When making any description of the overall appearance of a group of people, it soon becomes evident that most of their obvious characteristics show continuous variation. Length and breadth of arms and legs, shape of face and nose, weight, and skin color are among the more obvious. What is the basis for differences such as these? All of us are aware of the fact that weight and height may be influenced by environment. An abundance of data illustrates that growth may be stunted by dietary deprivation. Environmental factors play a decided role in contributing to the differences seen in those characteristics which show continuous variation. Since the environment can influence height, weight, or the length and breadth of a part of the body, it is no small wonder that attempts to determine the genetic factors involved are beset with difficulties. The nineteenth-century geneticists who chose to follow such human features failed to uncover the basic principles of heredity. On the other hand, Mendel was able to formulate the laws of segregation and independent assortment since his study of the easily classified discontinuous traits in characteristics of the garden pea was not clouded by pronounced effects of the environment. Moreover, the discontinuous traits which Mendel selected were in each case governed primarily by one pair of contrasting genes (tall versus short plants and the responsible gene pair *T* and *t*). The discontinuous variation shown by many characteristics typically involves one major locus or a few loci which interact and produce an effect which can be recognized as a distinct kind of trait. For example, the *ABO* locus, the Rh locus, and the several loci involved in blood clotting are associated with genes which have a readily discernible effect. While such genes interact with others (as seen in the case of the Bombay phenomenon) and while their expression may be influenced by the environment (a PKU child can escape mental retardation), they do not express themselves in the production of a range of variation in which one class or type gradates unto the next.

Deep-seated characteristics such as size or shape or weight of a body part involve a large number of genes whose expression can be influenced greatly by the environment. All of the data which have been assembled from a variety of plant and animal species indicate that continuous variation entails no real contradiction to the basic principles of genetics. Rather, the complexity results from the large number of gene pairs involved and the many subtle influences of the environment in which they act.

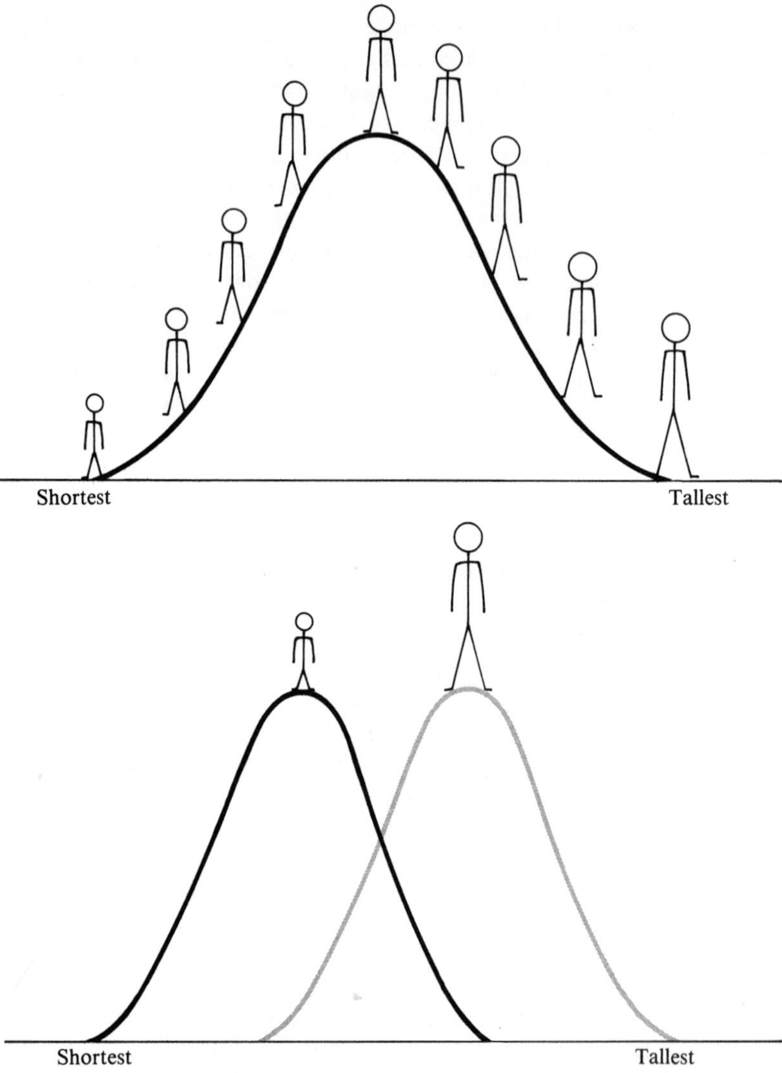

Figure 8-1. A characteristic such as height which exhibits continuous variation has no sharp delineations among the various classes (above). If many persons in a population are measured, some would be found at the extremes, very short and very tall. Most persons would fall in between. The central tendency around which the majority would cluster is the *mean.* This value is the commonest of the statistics used to describe a population. When two different populations are measured (below) with respect to a given characteristic, the mean value may be very different. A second population studied for height will exhibit continuous variation, but the mean may be a higher value, reflecting the fact that a person taken at random in population 2 would probably be taller than a person taken at random in population 1.

Inheritance of Skin Pigmentation

Let us examine the inheritance of a continuously varying characteristic in the human and then attempt to account for the differences seen among human populations with regard to certain obvious phenotypic features. Skin pigmentation is a characteristic which varies greatly in the human. An explanation for the difference in pigmentation between "white" and "black" populations was presented as early as 1913 by Davenport, whose concept was basically correct. However, more refined methods are available today to permit an estimate of the amount of pigment present in the skin.

The pigment which is largely responsible for skin color differences is *melanin,* a dark-brown substance which associates with protein to form granules of pigment in the pigment cells (melanocytes) found in the living layers of the epidermis. The recessive gene responsible for oculocutaneous albinism interferes with the formation of melanin so that little of this dark pigment, if any, can form. The responsible recessive gene is found among all populations of humans. Except for albino persons, everyone possesses pigment granules with melanin. People of dark-skinned populations have large amounts of melanin distributed throughout different skin layers. Light-skinned persons possess smaller amounts which tend to be found in one epidermal layer. The amount may be small enough to allow the blood pigment, hemoglobin, to show in certain parts of the body, as evidenced by pink cheeks.

Light-skinned people reflect as much as 40% of the light which falls on their skin, whereas dark-skinned persons may reflect only 1%. The techniques that are used to estimate the amount of pigment measure the amount of this reflected light. Crosses between humans cannot be dictated by an investigator, but persons in different family groups have been measured for their amount of skin pigment. In effect, crosses between white and black persons have been followed, as well as further matings of their offspring (F_1) and their grandchildren (F_2). All of the data indicate that skin pigmentation, while showing continuous variation, is actually a relatively simple characteristic. Only a few pairs of genes (3 to 5) seem to control almost all the pigment difference between an extreme "white" person and a very dark-skinned one. The white person and black person both possess genetic factors for melanin production.

In estimating the number of alleles responsible for pigmentation differences between black and white populations, we consider *only* those genes which contribute to an *effective* pigment difference *above* a certain basic level. Figure 8-2 depicts a cross between a dark-skinned person and a light-skinned person on the assumption that only 2 pairs of genes account for the effective difference. Although at least 3 pairs are more likely involved, the same principles can be very simply illustrated using the smaller number. We see from the figure that the dark person has 4 pigment-producing genes above the level *common* to both a dark- and a light-skinned person. The white

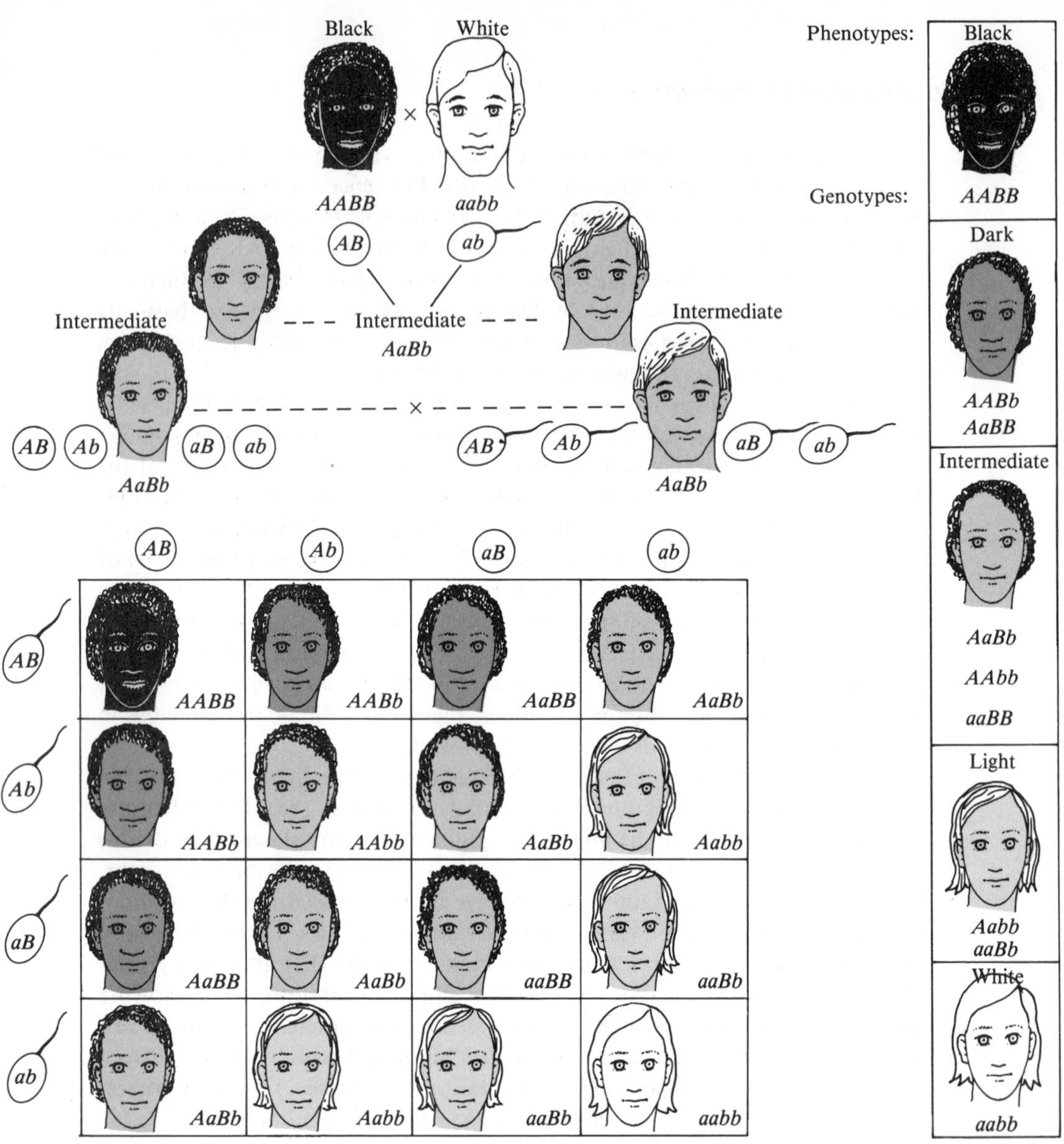

Figure 8-2. Skin pigmentation differences. Both black and white persons carry genes which control the formation of melanin pigment. The conspicuous differences seen between persons considered black or white result from those genes which control pigment dose above a certain level. On the concept of a difference involving two pairs of genes, *A* and *B* would control one pigment dose each. Their alleles *a* and *b* would be ineffective in pigment formation. The dark-skinned person would thus have four effective pigment-controlling genes (*AABB*), whereas the white person would have none (*aabb*). A mating between two such persons would produce offspring who are intermediate in shade. Two persons of genotype *AaBb* who are intermediate in color can give rise to offspring who may vary in shade from dark as the dark parent to light as the white one.

parent carries genes which contribute no effective pigment dosage above that common level. The offspring between them are intermediate, since 2 effective pigment genes are present. The effect of the genes responsible for the pigmentation is quantitative, as seen clearly from the mating of two individuals of genotype *AaBb* (Fig. 8-2). This means that we can add up the effects of the gene dosage: 4 effective doses of pigment, 3 doses, 2, 1, 0. As mentioned, these dosages can be estimated by tools which measure light reflection. Neither dominance nor recessiveness pertains here. The effect of each gene is the contribution of a certain amount of pigment. This type of inheritance involving several pairs of genes whose effects are cumulative or additive is called *quantitative* or *polygenic inheritance.* This means that several pairs of genes interact and exert an influence by somehow producing a measurable effect on a character. *Multiple-factor inheritance* is an earlier designation which is also used for quantitative or polygenic inheritance. The term *multiple factors* implies that many pairs of genetic factors (genes) can indeed influence a character and that factors in the environment, many largely unknown, also exert an effect.

One early concept in genetics was that of *unit characters.* According to this idea, there would be one gene for a particular characteristic or trait. The unit factor would have just one effect and would not be influenced by other factors. The multiple-factor concept indicated that this was not the case for those characteristics showing continuous variation. We now know, of course, that the unit factor idea is erroneous even for characteristics in which the variation is discontinuous, since all characters involve the interaction of many genes, and any gene may have an effect on more than one character (Chapter 1).

Quantitative or polygenic inheritance is the genetic basis of most of the more obvious features of an organism. The skin pigmentation characteristic is a rather superficial one which does not involve many pairs of genes. Many more allelic pairs are probably involved in the inheritance of such attributes as size and shape of limbs, weight, and height. The genetic details have yet to be worked out in these cases, although there is no question that the genetic component of such characteristics is polygenic. The environmental component is so intimately involved that it may prove impossible to identify the precise contribution of each component, the genetic and the environmental. Even in the case of skin pigmentation, effects of the environment are evident. All persons tan, since pigment formation is stimulated following exposure to ultraviolet light. Consequently, no sharp boundaries exist among pigment classes even in the oversimplified representation shown in Fig. 8–2.

Figure 8-3 illustrates a cross between 2 persons of intermediate shade on the more likely assumption that 3 pairs of genes determine pigmentation differences between 2 such individuals. Note from Fig. 8-2, which assumes that 2 gene pairs are involved, that 9 different genotypes can arise and that these may be recognized as 5 different degrees of pigmentation. If 3 pairs are

AaBbCc × *AaBbCc*

Gametes formed by each parent: *ABC, AbC, aBC, abC, ABc, Abc, aBc, abc*

	ABC	*AbC*	*aBC*	*abC*	*ABc*	*Abc*	*aBc*	*abc*
ABC	6	5	5	4	5	4	4	3
AbC	5	4	4	3	4	3	3	2
aBC	5	4	4	3	4	3	3	2
abC	4	3	3	2	3	2	2	1
ABc	5	4	4	3	4	3	3	2
Abc	4	3	3	2	3	2	2	1
aBc	4	3	3	2	3	2	2	1
abc	3	2	2	1	2	1	1	0

	6	5	4	3	2	1	0
Genotypes and their frequencies:	(1) *AABBCC*	(2) *AABBCc* (2) *AABbCC* (2) *AaBBCC*	(4) *AABbCc* (4) *AaBbCC* (4) *AaBBCc* (1) *AABBcc* (1) *AAbbCC* (1) *aaBBCC*	(8) *AaBbCc* (2) *AABbcc* (2) *AaBBcc* (2) *AabbCC* (2) *AAbbCc* (2) *aaBbCC* (2) *aaBBCc*	(4) *AaBbcc* (4) *AabbCc* (4) *aaBbCc* (1) *AAbbcc* (1) *aaBBcc* (1) *aabbCC*	(2) *Aabbcc* (2) *aaBbcc* (2) *aabbCc*	(1) *aabbcc*
Phenotype:	Very dark	Dark	Fairly dark	Medium	Fairly Light	Light	Very light (white)

Figure 8-3. The larger the number of effective gene pairs which contribute to continuous variation, the larger the number of possible genotypes and phenotypes. This in turn is responsible for the production of additional classes which intergrade even more. Compare this figure with Fig. 8-2. It is seen that 3 pairs of alleles, in contrast to 2, raises the number of genotypes from 9 to 27 when two intermediates mate. The number of phenotypic classes is increased from 5 to 7. The numbers shown here in the squares simply indicate the number of effective pigment-promoting alleles.

involved (Fig. 8-3), 27 distinct genotypes are possible and these fall into 7 phenotypic classes. This comparison between 2 pairs and 3 pairs of quantitative factors is given here to show that the larger the number of gene pairs involved in any case of polygenic inheritance, the wider the range of variation and the less pronounced the distinction among the classes. This is so even ignoring the complication of the environment. It is easy to see why we know a great deal more about the inheritance of characteristics which show discontinuous traits than we do about those with a polygenic basis.

A study of Fig. 8-2 and 8-3 will reveal several other points. It can be seen that persons having the same degree of pigmentation may actually possess different genetic constitutions. For example, assuming only two gene pairs, persons of intermediate color may be *AaBb, AAbb,* or *aaBB.* In each case, two effective genes are present to allow the formation of two doses of melanin pigment. This point can account for the very different results following a mating of two persons similar in shade. Mating between certain persons of the same intermediate phenotype can produce F_1's with a wide range of variation in the amount of pigmentation, all the way from very dark to white (Figs. 8-2 and 8-3). We see that in other cases [Fig. 8-4(A)] the offspring will have exactly the same number of effective pigment genes as the parents and thus will be similar to them in pigmentation. On the other hand, a crcss between the dihybrid, *AaBb,* and either of the other two intermediate types, *AAbb* or *aaBB,* can produce some children somewhat darker or somewhat lighter than either parent [Fig. 8-4(B)]. However, if one parent is white and thus carries no effective pigment genes, no children can be darker than the darker parent [Fig. 8-4(C)]. The interpretation of skin pigmentation in this way on the basis of a small number of polygenes explains why the offspring of certain parents resemble them in color, whereas in other families, the children are of various phenotypes, some markedly different from either parent. For the sake of simplicity, the discussion here has ignored the complications of environmental effects, which could quite possibly produce similar shades in two persons with different numbers of effective pigment genes. *Aabb,* for example, could conceivably be as darkly pigmented as a person of genotype *AaBb* as a result of increase in pigment in the former due to an environmental influence.

Problems in Analysis of Polygenic Inheritance

It should be appreciated from what has been said so far why quantitative inheritance is so difficult to analyze. When a study is undertaken, large numbers of individuals must be measured and the environmental conditions must be closely controlled. In order to gain any idea of the number of gene pairs in the genetic component, known hybrids and their offspring must be followed. Suppose that we are studying quantitative inheritance of pigmenta-

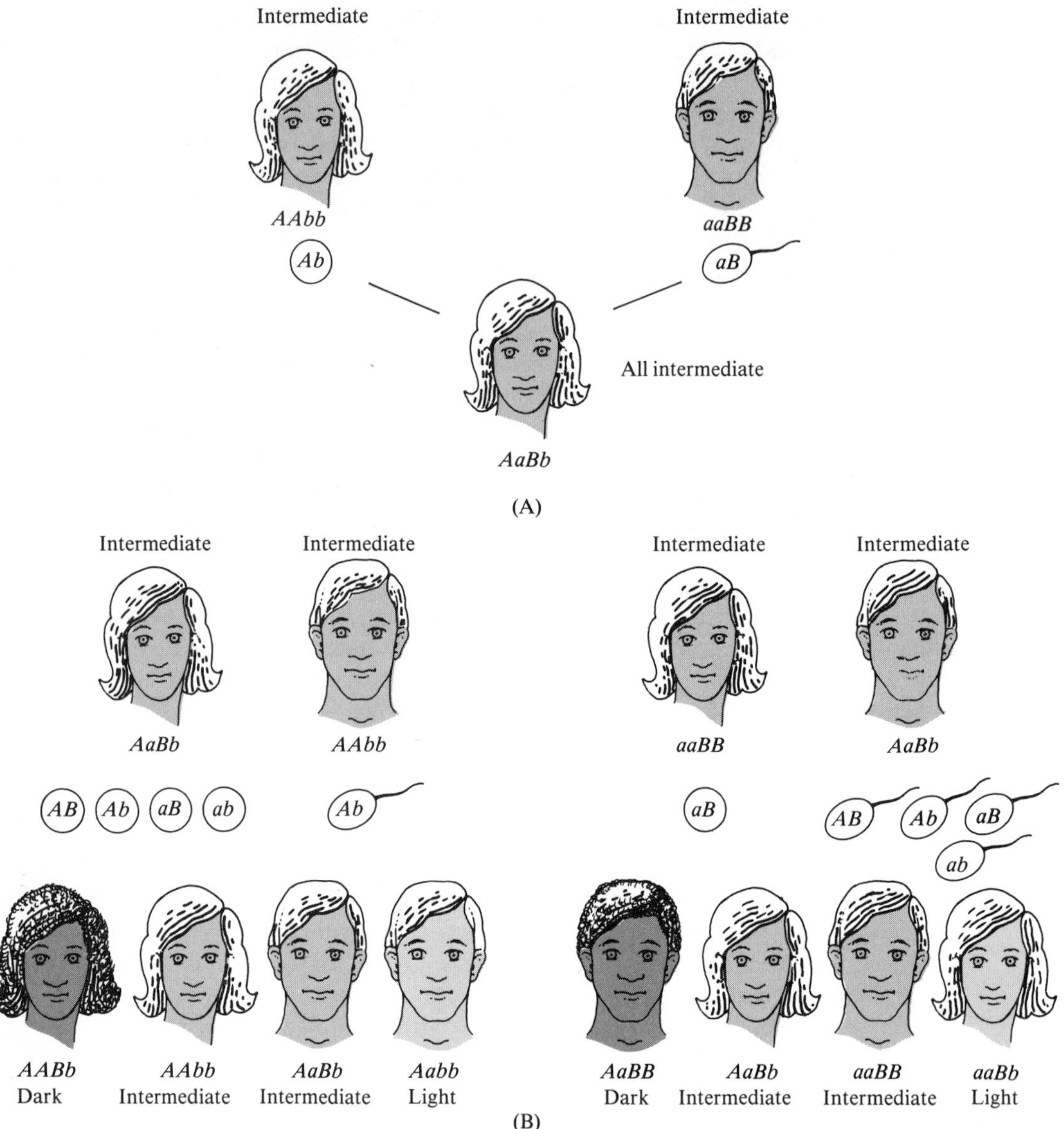

Figure 8-4. (A) A mating between some persons of intermediate shade produces offspring, all of whom are also intermediates. This is possible since two persons of the same phenotype may have different genotypes. In this case, each parent will always contribute one effective pigment-promoting allele to the offspring. (B) Two persons of intermediate shade can, however, produce offspring lighter or darker than either one of them, as seen here and in Figs. 8-2 and 8-3. (C) A mating between a white person and a person carrying any number of effective pigment-promoting alleles cannot yield offspring darker than the darker parent.

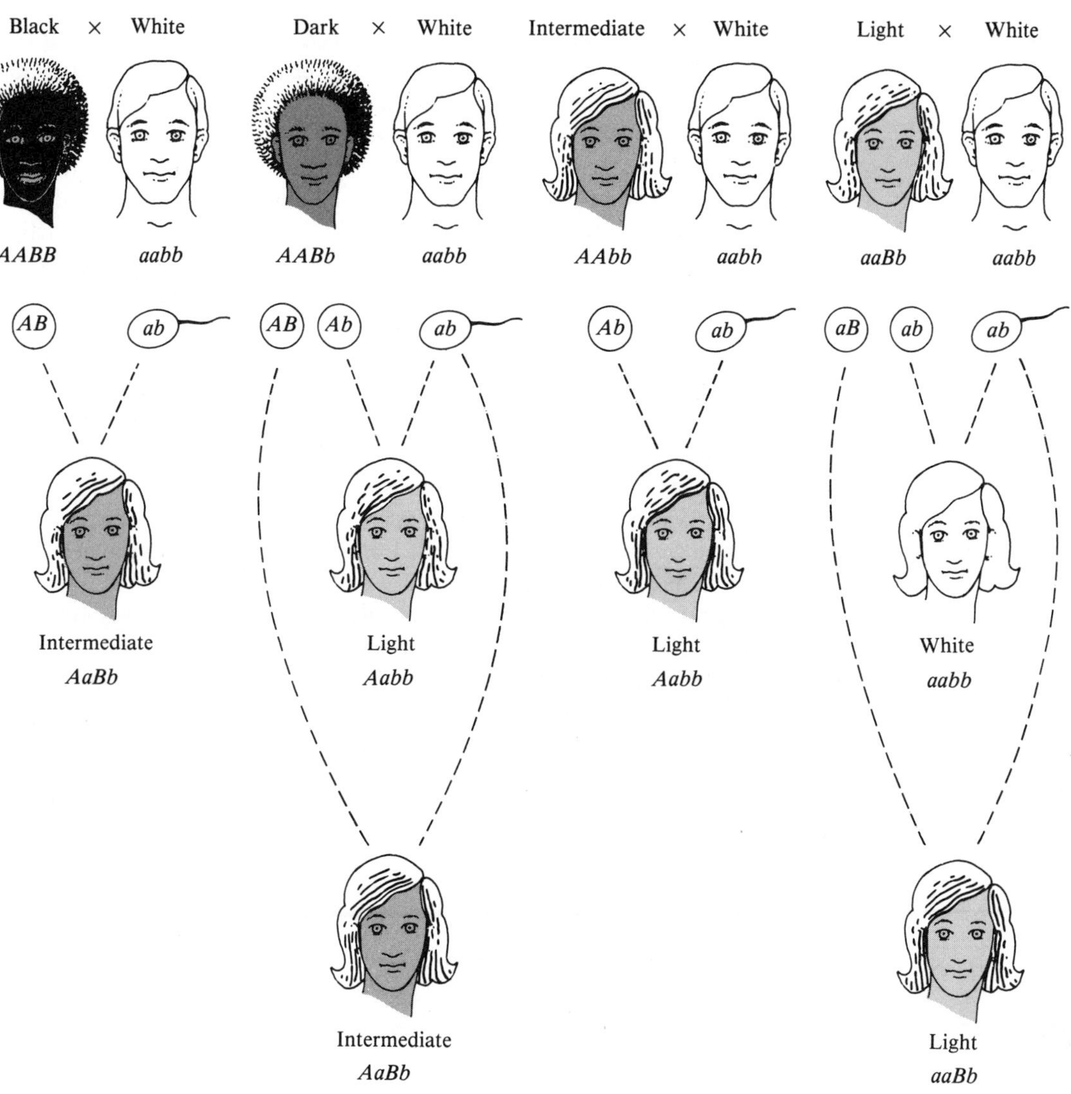

Figure 8–4. Concluded

tion in some animal and that 2 gene pairs are actually involved as diagrammed for the human in Fig. 8-2. Following a cross between 2 individuals (*AaBb*), only 1 animal in 16 will be as white as the white parent in the P_1, and only 1 out of 16 will be as dark as the dark one. If 3 gene pairs are involved (as can be seen from Fig. 8-3), only 1 out of 64 of the offspring of the trihybrids will show an extreme as great as one of the original parents in the pedigree. Using plants and animals (corn, flies, mice) to study polygenic inheritance, matings may be performed in a variety of ways. Large numbers of offspring from known hybrid parents can then be measured to determine what fraction of the total resembles one of the extreme types. For example, 1 out of 64 would indicate that 3 pairs of genes are interacting to produce the effect; 1 out of 256 would suggest 4 pairs; and so on (Table 8-1). To obtain this sort of information on humans demands the study of a large number of family pedigrees. A further complication which must be considered in any study of polygenes is that the contribution of each quantitative gene may *not* be exactly the same, as we have been assuming for the sake of simplicity. Work with various species indicates that such an assumption is not necessarily correct. Suppose that 3 effective gene pairs are involved in the weight of a particular organ and that the extreme types are as shown in Fig. 8-5. While the weight difference between the two may be 18 grams, the contribution of each effective gene, *A, B,* and *C,* may not be 3 grams. Perhaps $A = 4$, $B = 2$, and $C = 3$. Therefore, we can only estimate the *average* contribution of all when considering them together. Various kinds of interactions between 2 alleles as well as among several genetic loci could also very possibly cause suppression or enhancement of the quantitative effects of 1 or more of the effective genes.

Height is a characteristic which helps to illustrate why progress in our understanding of discontinuous variation, which typically involves one or relatively few major genetic loci, has advanced more rapidly than that of continuous variation and its polygenic basis. The role of the genetic component in stature is clearly seen when identical twins are compared. Even when they have been separated and raised in different environments, twins rarely differ from each other by more than one inch. The number of gene pairs involved in height is unknown. However, the loci of at least some of the genes which influence normal height would appear to be on the X and the Y chromosomes.

Table 8-1. Fraction of offspring from hybrid parents showing a measurement as extreme as one of the P_1 parents. The fraction becomes smaller as the number of gene pairs involved increases. This means that larger numbers of offspring must be examined to determine how many gene pairs are adding to the quantitative difference.

Allelic pairs	Different kinds of gametes	Fraction showing one parental extreme
1	$2^1 = 2$	1/4
2	$2^2 = 4$	1/16
3	$2^3 = 8$	1/64
4	$2^4 = 16$	1/256
5	$2^5 = 32$	1/1,024
6	$2^6 = 64$	1/4,096
7	$2^7 = 128$	1/16,384
8	$2^8 = 256$	1/65,536
9	$2^9 = 512$	1/262,144
10	$2^{10} = 1024$	1/1,048,576

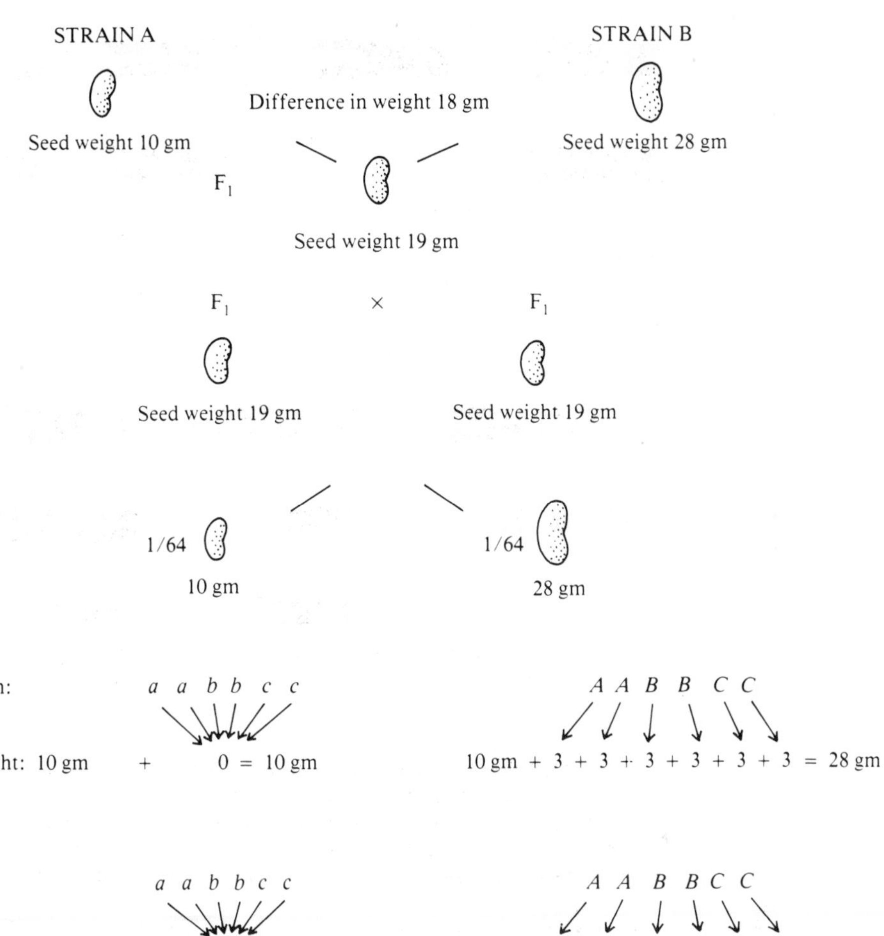

Figure 8-5. Assume that the basic seed weight in a species of plant is 10 grams and the heaviest seed weight is 28 grams. There would thus be a difference of 18 grams between a strain with the lightest seeds and a strain with the heaviest. F_1 hybrids between the strains would weigh about 19 grams, and a cross of 2 hybrids might show that among the F_2 plants 1/64 produces seeds which weigh 10 grams and 1/64 produces seeds which weigh 28 grams. This would indicate that 3 pairs of genes are involved (Table 8-1). It would appear that the heavier strain has a total of 6 effective alleles which add 3 grams each to the difference between the strains. In actuality, however, each allele might contribute very different amounts to weight. This would not be apparent. Only the average contribution of each could be estimated.

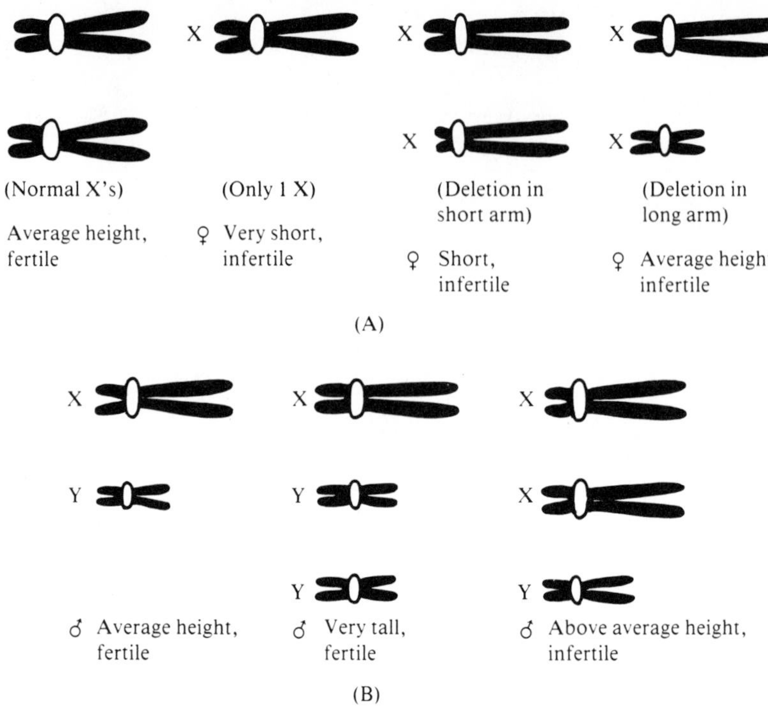

Figure 8-6. Certain genetic factors which affect normal stature must be distributed on the short arm of the X chromosome and on the Y. This is indicated by the fact that unusual chromosome situations may occur in females (A) and males (B), producing individuals who often depart from the average in stature as well as in other aspects of the phenotype.

Evidence for this lies in the fact that the sterile Turner female (Chapter 4) who possesses only one X chromosome is short in stature. Two complete X chromosomes seem to be involved in the normal development of both stature and female fertility in an XX embryo. Loss of just one arm of one of the X chromosomes (the shorter of the two arms) results in a Turner female [Fig. 8-6(A)]. If the long arm of one X chromosome is lost, infertility results but height is not affected.

Expression of normal height in a male must involve both the short arm of the X plus genes for stature on the Y chromosome [Fig. 8-6(B)]. This is supported by the fact that the XYY male (Chapter 5) is typically very tall. The XXY Klinefelter male is also typically above average in height. Such males would, if the idea is correct, have three instead of two doses of the genes on the sex chromosomes which influence height (one on the short arm of the X and one each on the two Y chromosomes). The number of genes we are referring to which compose a dose is not known. Other genes which influence height are probably located on one or more of the autosomes. The oversimplification in Fig. 8-7 enables us to understand how two persons of medium height may produce offspring who are much taller or much shorter than they. While the concept of quantitative inheritance as the genetic basis of height is undoubtedly correct, Fig. 8-7 by necessity assumes that all effective genes are adding the same number of inches to height and does not consider the complexities noted above which can result from genic interactions. There are

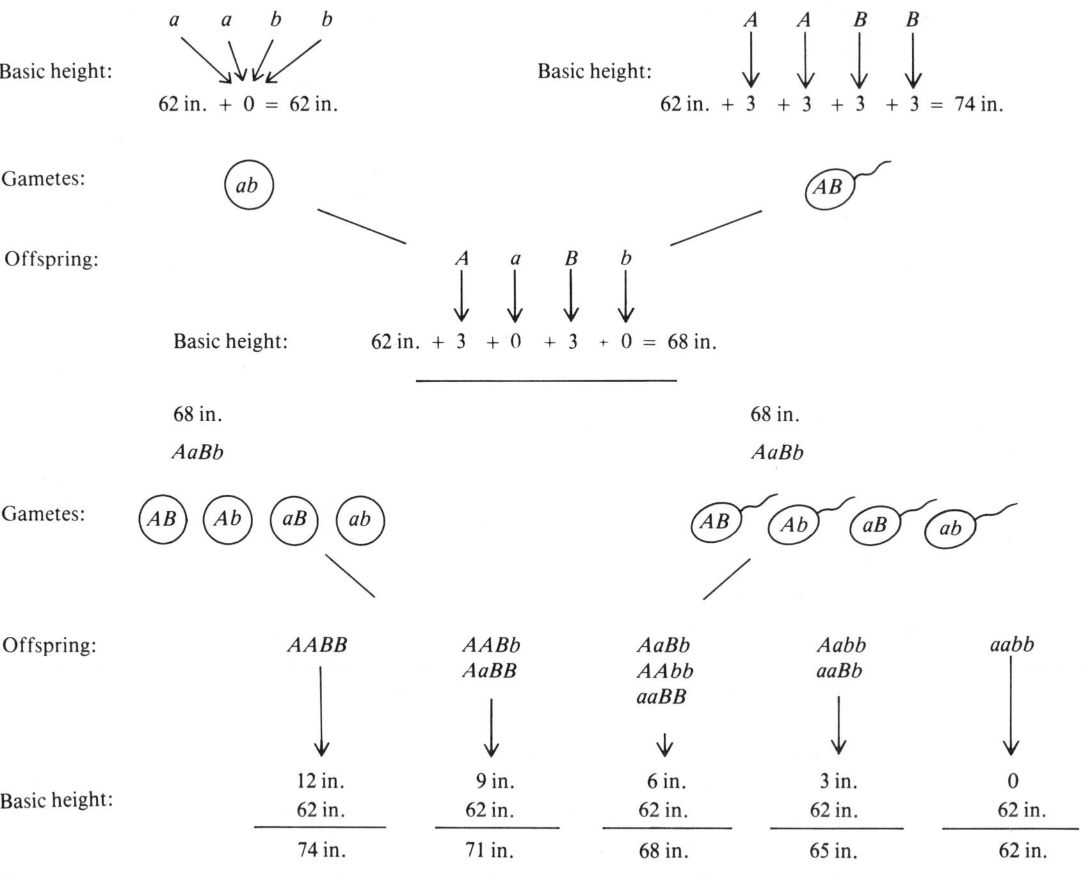

Figure 8-7. The average height for a male is approximately 68 inches. Assume that 2 pairs of quantitative genes are involved in height and that the tallest persons are genotype *AABB* and the shortest are *aabb.* Each effective gene would add 3 inches to the basic height of 62 inches which is common to both of them (above). A mating between the extremes will produce dihybrids who are intermediate in height since they carry 2 effective genes. A cross of 2 such individuals (below) can produce 9 different genotypes (compare with Fig. 8-2). Some of these will be individuals taller and shorter than either intermediate parent. (The figure is greatly oversimplified and does not consider influences such as hormones which may cause women to be shorter than men of the same genotype.)

many unknown environmental factors which can influence height and complicate the picture further. Exactly what all of them are and their relative importance in relation to the genetic component are matters of debate. There is no doubt, however, that height in the human has been increasing during the past 100 years or so in most modernized populations. The consensus is that the better living conditions which accompany industrial development have played a major role. Better nutrition is cited as the main single factor responsible for increase in stature. Exactly how nutrition and any other environmental factors interact with the genes that contribute to height has yet to be unraveled.

Certain human afflictions (hypertension, diabetes, atherosclerosis) may involve a polygenic genetic component which entails many gene pairs (Chapter 10). The variation in the severity of these disorders within a family and between any two individuals who show symptoms of the particular disorder rests on a complex genetic-environmental interplay. Consequently, afflictions such as these have been more difficult to analyze genetically than those in which only one or very few gene pairs play a major role (PKU, defective blood clotting, etc.).

Populations, Species, and Gene Flow

All of us are aware of the obvious fact that human populations native to different regions of the world may be distinguished from one another on the basis of certain distinctive combinations of phenotypic features. Many of the obvious differences among human groups are seen in characteristics which exhibit continuous variation. The generally tall, long-limbed, light-skinned peoples of Scandinavia can be distinguished from the generally shorter, darker peoples in various Asiatic countries such as Thailand. Within any one population, however, height, weight, limb length, pigmentation, etc. will vary from one individual to the next. So not only is there a difference among populations of the human with respect to those characteristics which show continuous variation, but within each population these same characteristics also vary. The reason we can recognize attributes which seem to describe this group of people or that race is that in any population most of the persons will tend to have measurements which fall within a certain range. For example, the height of most persons in Sweden tends to be over 5 feet 8 inches, whereas that of most Thai persons tends to be well under that measurement. This does not mean that every Swede is taller than every Thai. There would prove to be some overlapping. However, the chance or probability is higher to find a tall person in the Swedish group and the probability for a short person is greater in the Thai population. Therefore, height, while a highly variable characteristic, can be recognized as one feature in combination with others to describe differences between any two populations of humans. A question which may be posed is, "Why should any combination of characteristics exist at all to distinguish one group or population of humans from another?" To answer this demands a good appreciation of the meaning of certain common terms.

Most of us have heard the word "species" and realize that it refers to a unit of classification. Biologists recognize the species as the basic unit of classification, even though subgroups or varieties may be recognized within one species. While boundaries between many species are often difficult to define precisely, almost all would agree that any definition of the term "species" would take into consideration the ability of members from separate groups to mate and to produce fertile offspring. No one questions that the lion and the tiger are members of different biological species. This seems obvious on the basis of phenotypic differences. However, it is still apparent that they are cats and that they have many cat-like characteristics in common which distinguish them from non-cat groups. Still, in nature, the tiger and the lion do not mate. Not only does their geographical distribution prevent this, but their mating preferences or behavior would make this highly unlikely and would discourage pair formation between members of these two groups. In the artificial environment of the zoo, mating has been achieved and an offspring produced, the tiglon. The tiglon however, is a sterile animal. This means that exchange of genes cannot take place between lion and tiger populations, since any rare hybrids which do arise cannot pass their genetic factors to either group. They are thus effectively isolated and represent "genetic dead ends," so to speak. The lion and tiger populations would retain their distinctive combinations of features, even if lions and tigers did become associated in the same area. We say that the two groups are *reproductively isolated.* It is thus clear that the two are separate species on all criteria, including both phenotypic distinctions and the inability to exchange genes. It is the latter criterion, however, on which the greatest emphasis is placed in the recognition of a species.

Now let us consider the assemblage of very different varieties which we recognize as the dog. While poodles, spaniels, and hounds look very different, it is common knowledge that such types are not reproductively isolated. Left to their own devices, very fertile offspring will arise which can then mate not only with other hybrids but with members of each parental group. Eventually, the distinctions between the original parental forms would become eradicated. If all dogs were allowed to interbreed freely, it is believed that eventually a generalized wolf-like creature would result. We see that there is a very profound difference between groups such as the tiger and the lion and between others such as poodles and hounds. This difference is based, not on the conspicuous visible distinctions, but rather on the ability of genetic material to flow from one group to another. Therefore, individuals are members of the same species if they are not reproductively isolated and can freely mate and exchange genetic information. On the other hand, reproductive isolation

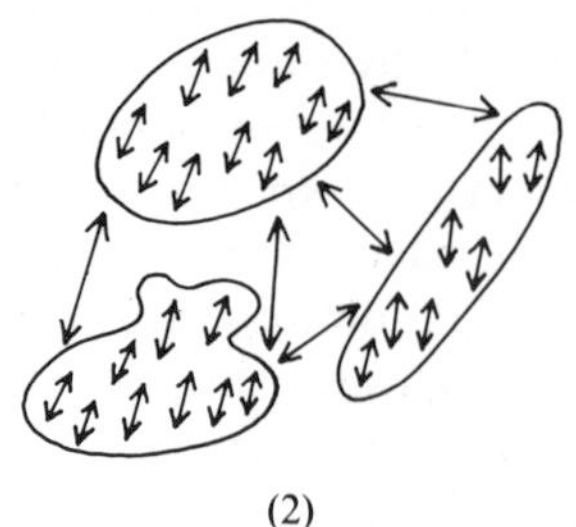

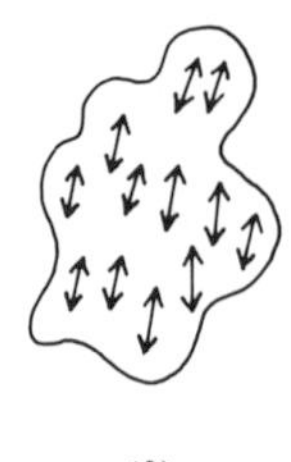

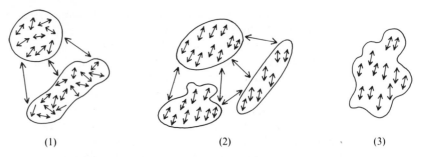

Figure 8-8. Populations and species. Members of a population tend to mate with other members within the population (as indicated by small arrows). Some species are composed of two or more populations, each of which may have a distinct combination of phenotypic features. Nevertheless, members of two populations within a species can freely exchange genes and produce viable offspring when they come in contact and mate (large arrows). Two groups are considered separate species if no genetic exchange occurs between them (absence of arrows). The figure illustrates three species. Species 1 is composed of two populations, species 2 is composed of three, and species 3 of only one.

between individuals in different groups can prevent or restrict gene flow. In this case, the particular groups concerned are separate species. The members of each share the pool of genes within that group without contributions of genetic information from any other group from which it is reproductively isolated (Fig. 8-8). In a very real sense, then, all members of one species share the same collection of genes; they have a *gene pool* in common. As a result of random mating within the species, the different genes in the pool are found in an assortment of combinations carried by the individuals composing the species. The genes are reassembled into still new combinations as a result of meiosis. The genes in the new combinations are in a sense returned to the gene pool as individuals mate and pass them to other individuals, who in turn can pass genes back again to the gene pool in new arrangements.

A detailed study of living creatures shows that distinctions among species are not always as sharp as presented here, since hybrids between some groups may prove to be only partially sterile and only a limited amount of gene flow can occur between parental groups. This often causes debate among biologists concerning the boundaries of a particular species. Nevertheless, the concept of gene flow still remains basic to the recognition of the species category.

Factors in Variation Among Populations

When the concept of genetic isolation is applied to the human, there is no difficulty in recognizing the human as one species. Despite the obvious phenotypic features which may distinguish geographical groups of humans from one another, no known genetic isolation exists. Isolation may be geographic, so that members of a group in one part of the world do not usually exchange genes with those on the opposite side. However, such distantly separated human populations have not been separated sufficiently long for the establishment of reproductive isolation. Moreover, exchange of genes between neighboring populations which are only partially isolated by distance takes place. This means that two very widely separated groups may actually be connected genetically by a series of other populations which are found between them. The fair-skinned Scandinavians and the darker Mediterranean peoples have exchanged and shared genes, since they have been connected by a continuum of populations. The human species is a highly diversified one composed of groups (or populations) whose collection of phenotypic features makes possible the recognition of a variety of types. But no matter how different the many subdivisions of the human may appear, no reproductive barrier is found. All the groups of mankind share a pool of genes.

We must now examine some of the factors responsible for the origin of differences among populations of humans. The human species has been composed of many partially isolated populations whose sizes vary greatly. These populations are spread over a large area of the earth. The mere fact of geographic separation dictates that individuals living within a more-or-less defined location will mate and exchange genes more frequently with one another than they will with individuals in another group which is somewhat distant from it. Consequently, members of a geographically defined population come to share more genes in common than they do with members of another population. Therefore, we can recognize population gene pools within the human species. Within any one of these, the frequencies of certain genes may be much higher than they are in the gene pools of other populations. Still, the many population gene pools, while differing in the frequencies of various alleles, are actually subdivisions of the *single* gene pool of the entire species. This is so because the population gene pools are not completely isolated. While members of one population tend to share genes from their immediate population gene pool, they still share genes with members of the entire species, since all the population gene pools are connected (Fig. 8-9).

A population in one sense is a compartment, since it tends to be somewhat isolated from any other. When any group of living creatures becomes distributed into compartments, the stage is set for the production of diversification among them. This comes about through the operation of the forces of evolution on the separate populations or compartments. Gene mutation is one of these forces. Gene mutation will be discussed at greater length in Chapter 13, but it must be recognized at this point that gene mutation occurs spontaneously in all populations without induction by any specific known factor in the environment. A gene at any locus on a chromosome may undergo sudden mutation at any time. The rate at which spontaneous mutation takes place varies from one kind of gene to another. Some genes mutate at a significantly

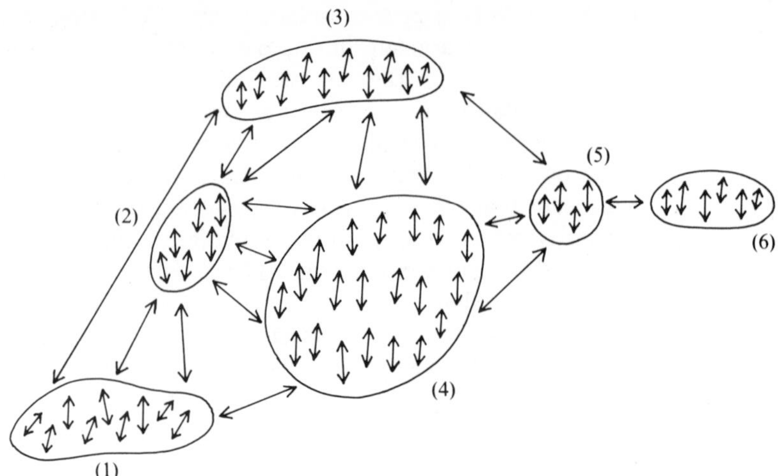

Figure 8-9. Populations and species gene pools. Members of a population share a population gene pool, since they tend to mate with each other (small arrows). Geography may isolate populations within a species. Still, genetic exchange is possible when members of two different populations contact and mate. Genetic exchange may be more frequent between some populations than between others (as indicated by the number of large arrows). Some populations may not mate directly with members of other populations within the species. Population 6 exchanges genes only with population 5; 5 only with members of 6, 3, and 4. Nevertheless, all the populations here are tied together, since genes are being passed from populations at the extremes of the species range through intermediate populations. While gene frequencies may differ from one population to the next as a result of the tendency of population members to mate with each other, the different populations still share a species gene pool, since gene flow is occurring from one population to the next.

higher rate than others. However, no environment is known which favors any one specific kind of gene mutation, nor does any natural environment direct the origin of mutant forms which will be better adapted to it. When a certain kind of gene mutates, it may not always change in exactly the same way; therefore several forms of a gene may exist in a population. We have seen examples of this from our consideration of multiple allelic series (Chapter 7) in which more than two alternative forms of a gene (alleles) may exist at a chromosome locus. Mutation, therefore, occurs spontaneously in a population without environmental direction. Studies of mutation rate in microorganisms and in experimental plants and animals have shown that genes at different loci have different spontaneous mutation rates, some loci having a mutation rate much higher than others. We refer to this constant spontaneous rate of gene mutation in a population as the *mutation pressure.*

The consequence of gene mutation to a population depends on several

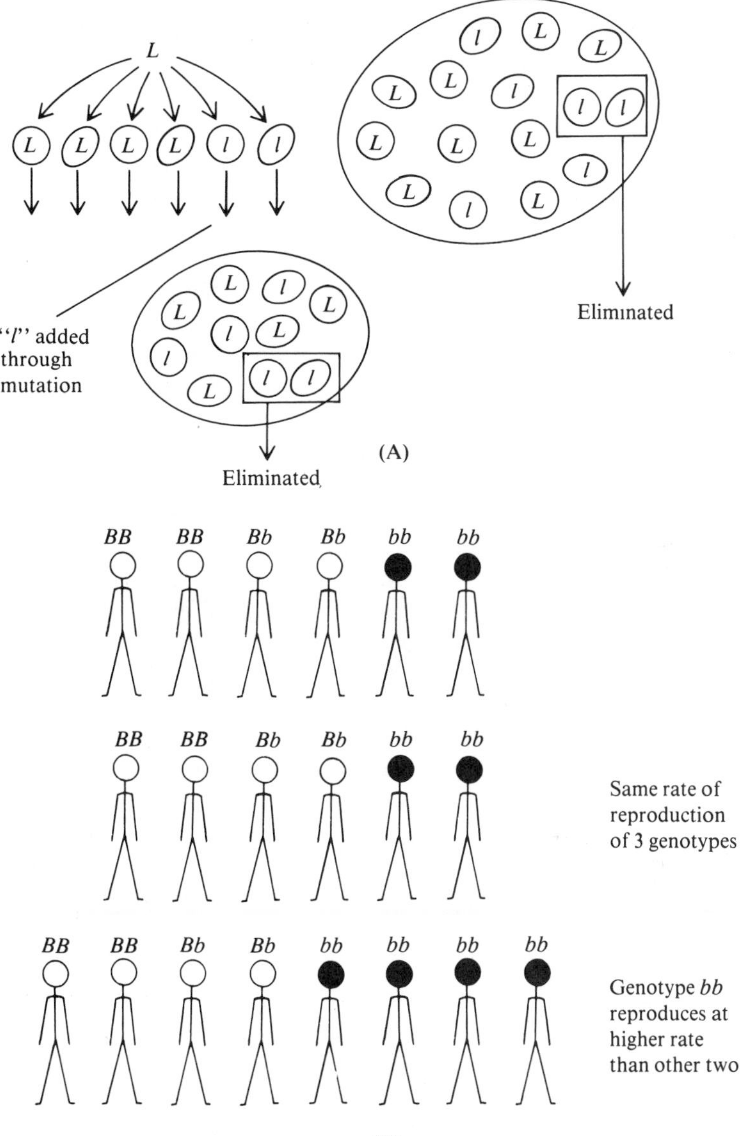

Figure 8-10. Natural selection and gene frequency. (A) If a gene is fully lethal in the homozygous condition, it will tend to be eliminated (above, right), since each individual homozygous for the gene will be eventually eliminated without leaving offspring. The only way a fully lethal gene can persist is through recurrent mutation from the wild form of the gene to the mutant form (left). (B) If a gene confers an advantage to those carrying it (in this case to the homozygote), it will tend to increase in frequency in the population, since those carrying it will leave more offspring.

factors. Most gene mutations prove to be harmful. A gene mutation is an unplanned change which causes a random alteration in the genetic information. Any such change is much more apt to be detrimental in some way rather than beneficial. This is so, since any population of living things is today the product of millions of years of evolution, and the population as a whole has become adapted to its set of living conditions. A change which occurs in the genetic information without any planning or guidance is not very likely to confer advantages, since it arises without any kind of forethought for the individual or the population. Since most spontaneous mutations are indeed harmful, they tend to be eliminated once they arise and, in general, are thus kept at a low frequency in the population. This comes about through the operation of another evolutionary force, *natural selection.* While evolution itself cannot continue indefinitely in the absence of mutation pressure, it is natural selection which is primarily responsible for the rate of evolutionary progress. Basically, natural selection means *differential reproduction.* If a gene mutation imparts a disadvantage to the individual, that individual will tend to leave fewer offspring than one lacking the mutant gene. An extreme case would be a fully lethal gene which completely prevents the reproduction of the individual carrying it. The gene would be eliminated by the death of that individual, and the only way it could occur in a population would be through recurrent mutation, mutation pressure which replenishes it [Fig. 8-10(A)]. If, however, a gene imparts an advantage to an individual, the carrier will tend to leave more offspring than those who lack it [Fig. 8-10(B)]. If this continues, the frequency of the gene will increase in the population. The relative benefit which any gene confers depends largely on the complex of environmental conditions under which the population members live. A gene which carries an advantage under one set of environmental factors may be neutral or even detrimental under another. Therefore, a particular allele or even a certain combination of genes may accumulate in one population but be kept at a low level in another. This results from the *relative* advantage which the gene or genes give in the particular environment in which a population exists. Therefore, the force of natural selection will tend to favor different genes in separate populations of the very same species, since the environmental conditions are never the same in two or more geographical compartments. There is thus a different selection for a particular gene or gene combination which depends on the environment. This operation of natural selection on the frequency of genes in a population is called *selection pressure.* Selection pressure, therefore, differs from one kind of environment to the next. However, *even if* two environments are very similar, very different gene mutations can occur which can produce the same kind of effect. This means that two populations may solve an environmental challenge in what appears to be the same way, let us say resistance to drought. However, the genes responsible for the beneficial trait may be very different ones at different loci on the chromosomes.

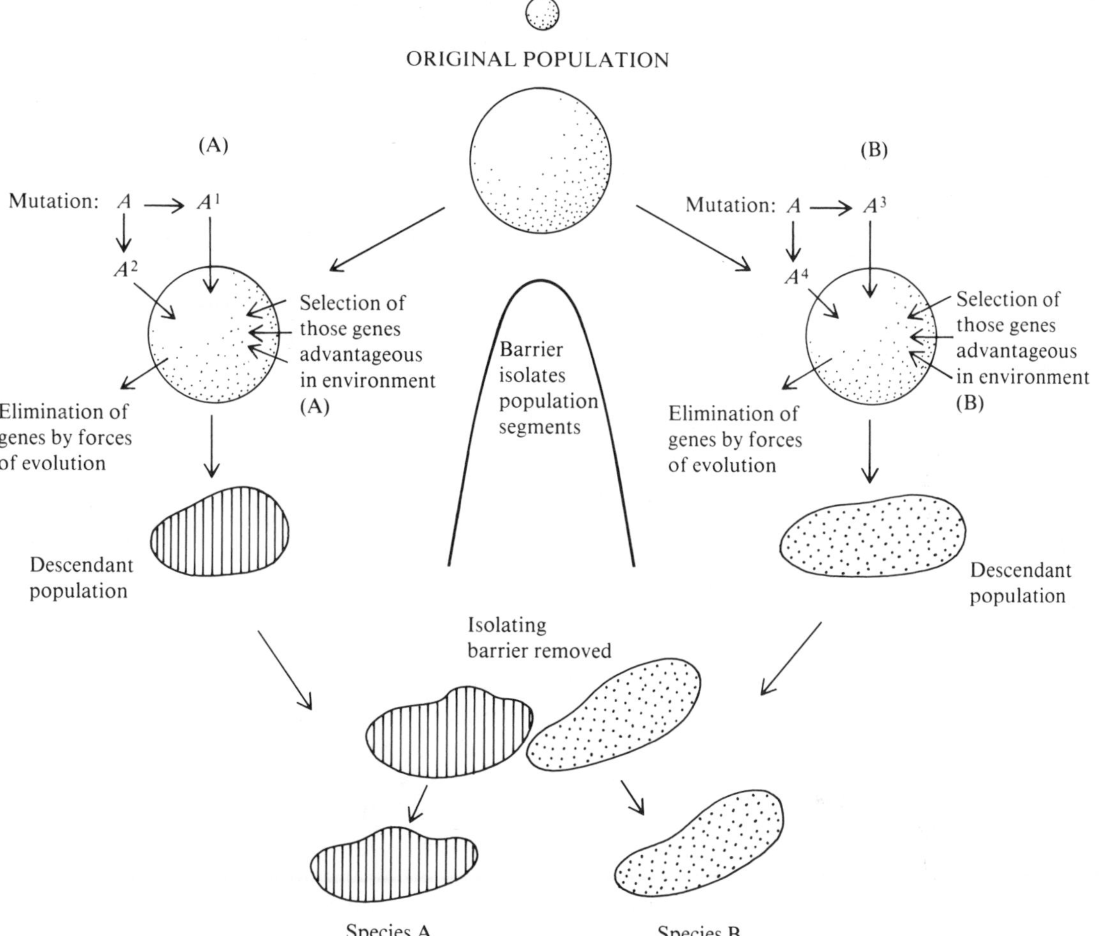

Figure 8-11. Isolation and the origin of species. A population may become divided into two or more segments. If these segments of the original population are prevented from exchanging genes due to distance or some other barrier, each segment will be subjected independently to the forces of evolution. Genes which are selected in one segment may be of no value or may carry a disadvantage in another. If separated long enough, the descendant populations may accumulate sufficient genetic differences to make them incapable of gene exchange when they are eventually brought together. Since genetic isolation would prevail, two species in this example would be recognized.

As gene mutation and natural selection continue to operate on two populations of the same species, the tendency is for the frequency of many alleles of a given kind to become different in the two groups. The less the gene flow between the two partially isolated groups, the greater the differences in gene frequencies which can become established. If two such populations should become so spatially isolated that no further gene exchange takes place between them, sufficient genetic differences may accumulate throughout time which can bring about genetic or reproductive isolation. This would mean that the two groups, if brought together again, could no longer cross successfully to produce fertile offspring. Instead of being two populations of one species, the two groups would now represent two distinct species (Fig. 8-11).

Differences Among Human Populations

In the human, no population has become reproductively isolated in this way from any other human population. However, human populations have accumulated sufficient genetic differences to cause obvious phenotypic distinctions. The term *race* is often used to designate a population or even a group of populations showing a collection of traits which distinguish them from other populations. The differences we observe among human populations have resulted mainly from the interaction of mutation and selection pressures operating over a period of time sufficient to result in a significant difference in the frequency of alleles among the various groups. Most of these differences (skin color, facial structure, height) are polygenic. In the course of thousands of years, different gene mutations have accumulated in the separate groups. Each genetic change has influenced some polygenic characteristics in a slight but measurable way. The outcome has been the establishment of distinctive phenotypic differences among human populations which enables us to recognize certain combinations of features which we consider typical of specific groups.

Fingerprint pattern is a clear case of a characteristic based on polygenes and of one which shows variation among populations. The precise number of genes responsible for fingerprint patterns is not known, but the type of pattern—whorls, arches, loops—has a decided polygenic basis (Fig. 8-12). While all three types are found among the major racial groups, there is a definite difference in their frequencies. In most white and black populations, loops are most common. Whorls are commoner among Mongoloid populations, native Australians, and the Melanesian populations of the Pacific. The arch pattern is least frequent among all populations, where it is found usually in less than 10% of the members. However, in Bushman populations, more than 10% have this pattern.

The influence of polygenes is clearly seen in the expression of the number of fingerprint ridges. Figure 8-12 explains how ridge counts are made. The number of ridges a person possesses is established at the fourth month of fetal development. While environmental influences do operate, their effect is generally small. The total ridge count has been shown to be highly correlated with the degree of relationship among persons. While identical twins usually show some difference in total count (and sometimes in pattern), the mean difference is much lower than that for fraternal twins of the same sex or for ordinary siblings. The number of ridges between a parent and a child or between siblings shows a higher correlation than does the ridge count between any two unrelated persons. The genes for ridge count appear to be distributed at random among persons in the population. All mathematical relationships show that a number of genes, each one exerting some additive effect on ridge number, is the basis of the fingerprint characteristic. The variation seen in ridge count among persons is due largely to genetic factors. The importance of the genetic component is clearly demonstrated by the fact that a person with Down's syndrome has a greatly reduced ridge count, indicating that chromosome #21 carries genes which affect ridge number.

Differences among populations are by no means the sole result of frequency differences in those alleles associated with polygenic inheritance or continuous variation. Differences between populations may also be found in relation to discontinuous variation. This is well illustrated by the distribution of alleles for blood groups among humans. While the allele for type O blood occurs with the highest frequency in all human populations, its incidence is by no

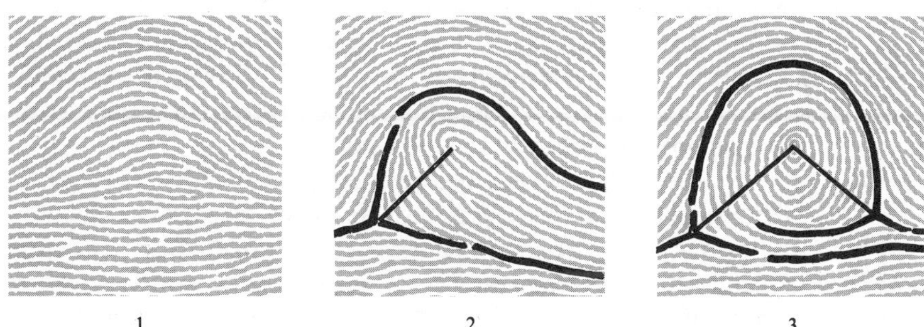

Figure 8-12. Fingerprint patterns. Example of an arch (1), a loop (2), and a whorl (3). Counting of ridges is based on a line drawn from the center of a loop or from the center of a whorl to a marking known as the *triradius* [indicated by heavy lines in (2) and (3)]. This is a formation formed by the forking of a ridge so that a Y-shape can be recognized. The ridge count is simply the number of ridges which the line from the center of the arch or loop crosses. Note that an arch (1) has no triradius; therefore the count is 0. There is 1 triradius for a loop (2) and 2 for a whorl (3). In the case of a whorl, the higher count is the one used. Counts are totaled for all 10 fingers and is called the *total finger ridge count.*

means the same in all. In certain American Indian populations, type O blood alone occurs. The allele for B blood type is completely absent from all American Indian populations, whereas populations in Eastern Asia show a higher incidence of the *B* allele than do other populations. Among Caucasian and Negroid populations, the *A* allele is commoner than the *B*. The Rh factor and therefore the responsible allele occurs in almost all the persons who compose American Indian and East Asian populations. Rh negative blood and the associated allele occur in black and Caucasian populations but are much more frequent among the latter. The Basque population of Spain has the highest incidence in the world of Rh negative blood.

Human Races and Their Distinction

Although we can recognize genetic differences among populations of the human, the question arises as to how many races exist. It should come as no surprise that there is no agreement on the point. Always keep in mind that the investigator himself erects the criteria to distinguish one group from the next. While the human is certainly just a single species, the criteria used to define race vary from one authority to the next. Using the broadest differences, only three racial groups are recognized: Caucasoid, Negroid, and Mongoloid. However, it is apparent that each of these is an assemblage of many diverse populations and that these in turn can be further subdivided (Table 8-2). Using more and more detailed measurements to distinguish groups, more and more races may be recognized. Obviously, the number will depend on the criteria used as the basis of *racial distinction.* These distinctions set up artificial boundaries which often fail to reveal the similarities that may exist among the various humans assigned to different racial groupings. Recent data on populations emphasize this point.

Techniques have been developed which enable the geneticist to detect gene differences between any 2 persons. As we will learn in some detail (Chapter 9), the primary role of many genes is the control of protein formation. The structure of all proteins in any living thing is determined by genes. Protein forms a most important part of the structural framework of a cell. Many of the most important chemical reactions taking place in a cell and necessary for life depend on molecules which are at least in part protein. Among these are enzymes, the organic catalysts which are required for most of the chemical steps to proceed in a cell. Many unfortunate genetic disorders (phenylketonuria, Tay-Sachs disorder, and others discussed in Chapter 10) result from the lack of a critical enzyme needed for an important chemical step. Deficiency in an enzyme in such cases is due to the presence of a mutant form of the normal gene. Any 2 normal individuals, of course, differ in many of their body proteins. This has been clearly demonstrated by detection of antigenic differences as discussed in Chapter 7. So differences in the number of proteins may be taken as 1 measure of the number of gene differences among persons. If any 2 unrelated white persons are compared, the procedures used for detecting protein (and hence gene) differences reveal that approximately 200,000 of them exist. This is true if 2 black persons are compared or if 2 Orientals are

Table 8-2. Classification of the human into 34 races. (From Dobzhansky, T., *Mankind Evolving.* Yale University Press. New Haven, Conn. p. 263, Fig. 10, 1962.)

1. *Northwest European*—Scandinavia, northern Germany, northern France, the Low Countries, United Kingdom, and Ireland
2. *Northeast European*—Poland, Russia, most of the present population of Siberia
3. *Alpine*—from central France, south Germany, Switzerland, northern Italy, eastward to the shores of the Black Sea
4. *Mediterranean*—peoples on both sides of the Mediterranean, from Tangier to the Dardanelles, Arabia, Turkey, Iran, and Turkomania
5. *Hindu*—India, Pakistan
6. *Turkic*—Turkestan, western China
7. *Tibetan*—Tibet
8. *North Chinese*—northern and central China and Manchuria
9. *Classic Mongoloid*—Siberia, Mongolia, Korea, Japan
10. *Eskimo*—arctic America
11. *Southeast Asiatic*—South China to Thailand, Burma, Malaya, and Indonesia
12. *Ainu*—aboriginal population of northern Japan
13. *Lapp*—arctic Scandinavia and Finland
14. *North American Indian*—indigenous populations of Canada and the United States
15. *Central American Indian*—from southwestern United States, through Central America, to Bolivia
16. *South American Indian*—primarily the agricultural peoples of Peru, Bolivia, and Chile
17. *Fuegian*—nonagricultural inhabitants of southern South America
18. *East African*—East Africa, Ethiopia, a part of Sudan
19. *Sudanese*—most of the Sudan
20. *Forest Negro*—West Africa and much of the Congo
21. *Bantu*—South Africa and part of East Africa
22. *Bushman and Hottentot*—the aboriginal inhabitants of South Africa
23. *African Pygmy*—a small-statured population living in the rain forests of equatorial Africa
24. *Dravidian*—aboriginal populations of southern India and Ceylon
25. *Negrito*—small-statured and frizzly-haired populations scattered from the Philippines to the Andamans, Malaya, and New Guinea
26. *Melanesian—Papuan*—New Guinea to Fiji
27. *Murrayian*—aboriginal population of southeastern Australia
28. *Carpentarian*—aboriginal population of northern and central Australia
29. *Micronesian*—islands of the western Pacific
30. *Polynesian*—islands of the central and eastern Pacific
31. *Neo Hawaiian*—an emerging population of Hawaii
32. *Ladino*—an emerging population of Central and South America
33. *North American Colored*—the so-called Negro population of North America
34. *South African Colored*—the analogous population of South Africa

compared to each other. So *within* a race, any 2 persons taken at random have about 200,000 protein differences between them. However, when the matter is studied closely, it becomes quite apparent that the protein (and gene) differences *within* a race can vary. Two persons of the same race from the same village or tribe tend to show fewer gene differences than 2 persons of the same race who are more widely separated geographically, such as a white person from northern Europe compared with one from southern Europe. This is exactly what we would expect on the concept of the human species distributed as partially isolated populations over the earth. Those persons belonging to the same population would tend to mate more within the group. They tend to share the same gene pool.

Genetic changes which arise by mutation pressure and which become established by selection pressure in one population may be different from those which arise and become established in another. Some genetic exchange takes place among the populations, but the farther away any two of them are geographically, the less the chance that genetic exchange will occur. And so, within a population, genetic differences *on the average* between any two individuals would tend to be less than that between two persons from neighboring populations. The farther away the two populations, the greater the genetic differences between any two individuals from the distant groups. And so we would expect to find a range of genetic variation *within* any so-called "race." It follows, therefore, that a comparison between two persons from two conspicuously different populations or races, say any white and any black, or any black and any Oriental, etc., should show differences even greater than that which is found between the extremes within a race. And it is true that genetic differences, as reflected in protein differences, *are* somewhat greater when persons from distinctly different races are compared, but surprisingly, the differences are *not* as great as might be expected. The protein differences which accompany obvious racial distinctions are really quite few and add little to the level of protein differences found between any two persons of the same race. In other words, the genetic differences between persons who appear so unalike that we designate them Caucasian, African, or Oriental are *not* that much greater than the genetic differences existing between any two Caucasians, any two Africans, etc. This means that those genetic differences responsible for the obvious physical differences which we often rely upon to distinguish one race from another are actually superficial. The genes which are responsible for most of the enzymes and other proteins in the human are very similar, regardless of the origins of the individual. It is these proteins which are fundamental to human physiology and structure. Those genes responsible for the conspicuous differences which we see affect such characters as skin color, type of hair, eye folds, and shape of face. These superficial physical differences have resulted to a large degree from the operation of mutation pressure and natural selection. They have a decided genetic basis which undoubtedly became established in relation to the relative advantages they conferred in the specific environments in which the various groups of the human have been evolving.

The precise value imparted by many of the visible differences among the races is not known. Nonetheless, these "racial" distinctions do *not* mean that the many populations of the human are separated by large genetic differences. As noted above, these interracial distinctions apparently do not add much more to those differences which may actually exist between two members considered to be of the same race. All populations of the human still share the common human gene pool. No local human population has been so geographically isolated from all others long enough to become genetically isolated or separated by large gene distinctions. The genetic similarities among all kinds of humans, as reflected in the similarities of their proteins, exceed those superficial phenotypic differences which tend to impress us.

Skin Pigment and Natural Selection

Skin pigment differences are undoubtedly related to different selection pressures in different environments of the earth. Let us examine this relationship a bit more closely. Melanin pigment screens out the ultraviolet of the sun which can cause severe damage to the skin, such as the induction of malignancies. In an environment where the population is exposed to intense sunlight, an individual would benefit from a deeply pigmented skin. Indeed, skin cancer is rare among black persons but not infrequent among whites who are exposed to excessive sunlight. However, the benefit conferred by the melanin in this way may be fortuitous and not the true reason for the selection of genes which increase the amount of pigment. Natural selection is effective only as it relates to the production of viable offspring. The occurrence of malignancies usually affects older persons, so that light-skinned individuals doomed to develop skin cancers would do so *after* having borne children. The development of the cancers would not be affecting the number of offspring produced. The reason for the accumulation of many pigment-producing genes in certain populations must be elsewhere.

A popular idea is that the amount of pigmentation is related to the amount of light required for the formation of the proper level of vitamin D. This vitamin is not found in sufficient quantities in the ordinary diet. However, it can be formed by the interaction of ultraviolet with certain substances normally present in the skin. A deficiency of vitamin D results in rickets, a skeletal defect, since it controls the absorption of calcium from the liver and the deposition of that element in the bones. But too much vitamin D can result in the deposition of calcium in the soft tissues of the body and the production of very harmful effects. Therefore, dark skin would benefit those in tropical regions where exposure to ultraviolet is intense. The pigment would afford protection against the formation of excessive amounts of vitamin D. The amount of pigment present would still permit the proper amount of vitamin D to form and hence normal bone calcification. This effect of vitamin D would be expressed at an early age in the life of an individual, so that genes which control pigment level would have a definite influence on the reproductive capacity.

As human populations spread northward, the ultraviolet exposure would be less intense. *Less* pigment would be more beneficial, since high melanin level (advantageous in equatorial regions) would screen out the smaller amount of ultraviolet reaching the skin. And so, lighter and lighter skin would be favored as populations spread northward. Seasonal differences in more northern climes may be very great, and so exposure to ultraviolet light can vary greatly from dark winter days to long summer ones. This problem may have been solved by the fact that some melanin formation is reversible. Sun tanning would thus be an advantage to fair-skinned groups during long summer exposures. Decrease in the amount of melanin would permit more ultraviolet to penetrate the skin during the shorter days of the year.

As mentioned earlier in this chapter, two populations may solve a similar environmental challenge in very different ways. Persons of the Mongoloid races develop a layer of skin which contains keratin, a horn-like material which imparts a somewhat yellowish color. This would screen out excessive amounts of ultraviolet light.

Detrimental Genes in Human Populations

Though a population tends to accumulate only those genes which adapt it to the highest efficiency level in its environment, there are unfortunately examples of human groups in which certain detrimental genes occur with a high frequency. The best understood of these is the gene responsible for the production of sickle cell hemoglobin. The individual homozygous for this allele, Hb^SHb^S, produces abnormal hemoglobin which results in distortion of the red blood cells upon their exposure to low oxygen levels, as occurs in some of the vessels of the body. A gene with an effect so harmful that it usually kills the homozygote before reproductive age is typically kept at a very low level in any population by the force of natural selection. Only by recurrent mutation does such a harmful gene continue to exist in the population. However, the frequency of the sickle cell gene is anything but low among certain African populations. As many as 30% of the persons in some regions may be carrying the harmful gene in the heterozygous condition along with the normal allele, Hb^A. It was suspected, therefore, that some environmental factor was somehow responsible for the high incidence of the detrimental Hb^S gene. A clue was obtained when it was demonstrated that the frequency of this harmful gene parallels the incidence of malaria. The greater the incidence of the mosquito-borne disease, the higher the incidence of the sickle cell gene and hence the incidence of the fatal genetic disease. Further investigation revealed an interesting fact. The infant who is heterozygous, Hb^AHb^S, and who thus produces *both* normal and sickle cell hemoglobin in the red blood cells, has a much better chance of resisting malaria than does the child of normal genotype, Hb^AHb^A, who produces only normal hemoglobin. For some reason still not fully understood, the malarial parasite has greater difficulty in invading a red blood cell which contains both kinds of hemoglobin.

The sickle cell trait (genotype Hb^AHb^S) does not confer any protection against malaria on an adult. However, some protection is afforded a small child who is given an increased chance of surviving the first malaria attack and who will then develop antibodies against the parasite which will prepare against a later invasion. The advantage given to the heterozygote also explains why certain black populations were able to live in malaria-laden regions, environments which proved fatal to many a European.

The result of this combination of circumstances is that in areas heavily infested with the malarial parasite, more heterozygotes will survive than will those who are homozygous for the normal gene (Fig. 8-13). This comes about since the heterozygote, Hb^AHb^S, has a greater malarial resistance and will thus tend on the average to leave more offspring than the homozygote, Hb^AHb^A, who will succumb more frequently to malaria. Consequently, the detrimental gene accumulates through the increased survival rate of the heterozygote in malaria-infested regions. This accumulation means an increased incidence of homozygotes who suffer and die from the anemia (Hb^SHb^S). As many as 2% of the population in some areas may be victims. This story illustrates one important way in which the force of natural selection operates: selection in favor of the heterozygote. In the more typical situation where the heterozygote who carries some particular gene is favored, the homozygote does not suffer such dire effects as seen in the case of the sickle cell anemia. The sickle cell gene is an example of an unfortunate solution in nature where an environmental challenge (malaria) has been met at the expense of the lives of some population members. The detrimental gene is carried by persons who no longer live in malarial regions but whose ancestry traces back to them. And so, we are aware of sickle cell victims in today's modern cities. Since the detrimental gene confers no reproductive advantage to a heterozygote who lives in a malaria-free region, we would expect the frequency of the gene to decrease, since those homozygous for the normal gene would survive as long as the heterozygotes. Indeed, the heterozygote may even suffer certain disadvantages at high altitudes and in other situations in modern society where oxygen levels may decrease at times. The evidence suggests that a decrease in the frequency of the sickle cell gene is actually taking place in certain areas of Africa as the incidence of malaria decreases with eradication of the malaria-carrying mosquito. The sickle cell story clearly illustrates the interaction of environmental factors with the human's genetic constitution.

No group of humans is without its share of detrimental genes. The reason for the relatively high incidence of specific ones in certain populations is not always as clear-cut as in the case of the sickle cell gene, but similar selection of the heterozygote is suspected in many cases. The high incidence in the Mediterranean region of the gene responsible for Cooley's anemia (*thalassemia major*), as well as of those which produce several other types of blood abnormalities, almost certainly stems again from an advantage imparted to the heterozygote in malaria-infested regions. Less clear are the exact reasons for

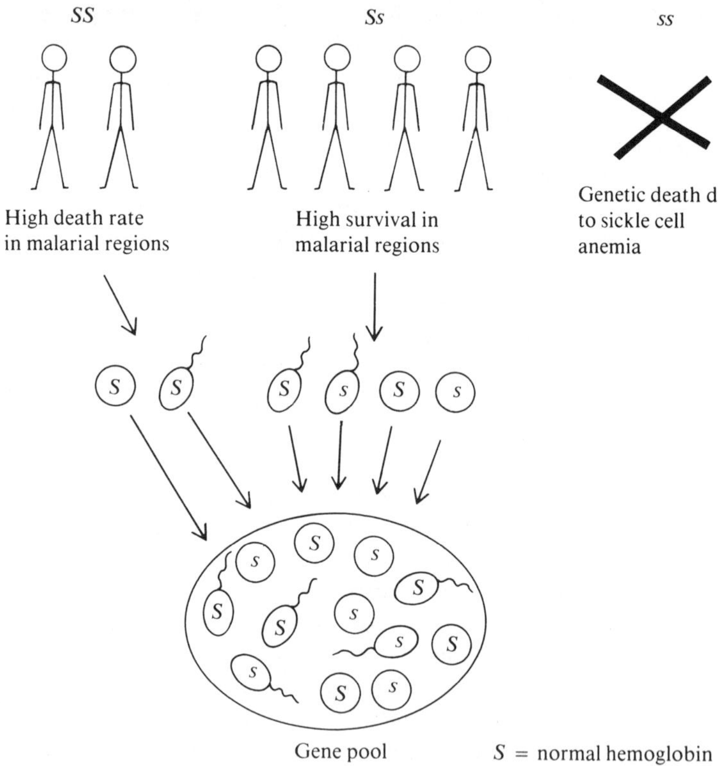

Figure 8-13. Selection of sickle cell heterozygote. Victims of sickle cell anemia usually die before reproductive age and cannot pass the sickle cell gene to the next generation. In malaria-infected regions, however, a higher percentage of individuals who carry the allele for sickle cell hemoglobin along with the one for normal hemoglobin tends to survive than of those who are homozygous for the standard allele for normal hemoglobin. The heterozygotes therefore continue to pass the defective allele into the gene pool with the result that the allele will accumulate up to a point and victims of sickle cell anemia will continue to arise with a high incidence.

the frequencies of still other detrimental genes. For example, the lethal gene responsible for Tay-Sachs disorder (Chapter 10) is much higher among Jewish persons of eastern European origins than among persons of other ancestry. There is reason to suspect that the high incidence of this gene in eastern European populations is again due to an advantage possessed by the heterozygote. Exactly what this advantage has been is unknown. Perhaps it may be the ability of the heterozygote to resist some type of childhood disorder related directly to some environmental factor. There are reports that in families where the gene is known to occur, the survival rate of the children is greater than in those families where the gene is absent.

A survival advantage to the heterozygote may apply to the gene responsible for cystic fibrosis. This disease is of great medical concern in the United States, where it occurs in approximately 4 births per 10,000. In this disorder, mucous-secreting tissues, such as those of the pancreas, bronchi, salivary glands, and small intestine, become overloaded with a mucoid substance produced in excessive amounts by cells of particular organs. The pancreas is so affected that its tubules often become so distorted with mucous that they appear as cysts. An impairment to the secretion of pancreatic enzymes results from this. The upsets in the various organs of the body as a result of the abnormal mucous secretions usually result in death before reproductive age. The recessive gene responsible for this severe disorder is unusually high among Caucasian populations. The frequency of heterozygotes in the U.S. has been estimated to be in the vicinity of 1 in 25. The gene is almost completely absent from Negroid and Mongoloid populations. To explain the high incidence in some groups, various suggestions have been offered. For example, it is possible that many generations ago, some Caucasian populations were subjected to a severe environmental stress of some type (lack of food, disease epidemics). The persons heterozygous for the fibrosis gene may have experienced some sort of survival advantage under the conditions of stress so that their survival was favored. Now that the stress has been removed, the incidence of the gene has decreased, but not as yet to the lowest possible level. Recently, evidence has been presented which suggests that families in which the fibrosis gene occurs tend to be larger than those in which it is absent, an indication that carriers of the gene may for some reason tend to leave more offspring. Why this may be so is not known, but it appears likely that this gene which has fatal consequences in the homozygous condition has persisted at a fairly high frequency in some populations due to an advantage it gives, or once gave, to the carrier.

The Role of Chance in Gene Frequency

A final example illustrates still another reason for the high incidence of some detrimental genes. Suppose, for example, that the normal gene, *A,* undergoes a rare mutation to a recessive form, *a,* which causes a severe disorder when homozygous. The *a* allele will be found in only a few members of the population, since natural selection tends to weed it out. Only by recurrent mutation will it be added back to the population gene pool. Now let us imagine that a small number of persons from this population migrate and form the basis for the establishment of a separate population. Among these few founders, one may be the carrier of the undesirable recessive. If the population that arises from the original few founders remains isolated, the detrimental gene will be passed to more and more members of the population, and eventually it will come to expression. More homozygotes for the rare disorder will be found in this population than in the larger original population from which the founders migrated. The high incidence of the gene in the second group is due just to the chance that it was present in one of the original settlers. This kind of chance effect which can result in a high frequency of a particular allele has been called the *founder effect.* There is no doubt that it is responsible for the higher incidence of some undesirable genes in certain human groups which tend to remain isolated. For example, a very rare form of dwarfism accompanied by an extra finger occurs with a significantly high frequency within a particular Amish population. The high incidence of the responsible gene can be traced back to the few founders of the population among whom at least one must have been a carrier.

The role of chance in the establishment of gene frequencies must be considered in small populations such as those which tend to isolate themselves for various religious or philosophical reasons and those which have become separated by chance due to geography. In a small population, simply by chance, one form of a gene may increase in frequency over its allele without any relation to the advantage or disadvantage it imparts. Indeed, in small human populations, the frequency of certain genes may be quite different from the frequency of these same genes in the large ancestral population from which the smaller ones have been derived. An example is seen in the case of the Dunkers of Pennsylvania, who represent an isolate derived from West Germany. The frequencies of the genes for the ABO blood types and the MN types have been found to be decidedly different from the frequencies of these genes in the area of West Germany from which the Dunkers migrated. The frequencies of these same blood grouping genes in the surrounding American population is similar to that in the German population. The difference seen in the Dunker group is readily explained as a result of chance. These chance fluctuations in gene frequencies are known to occur in small populations and have been very thoroughly studied in animal groups. The phenomenon has been named *drift,* the variation which occurs in the frequency of genes from one generation to the next due to chance factors. The role of drift (and the founder effect is an example of it) decreases as population size increases, and it plays no significant role where the population remains large and mobile. However, a present-day population may at one time in the past have been reduced to

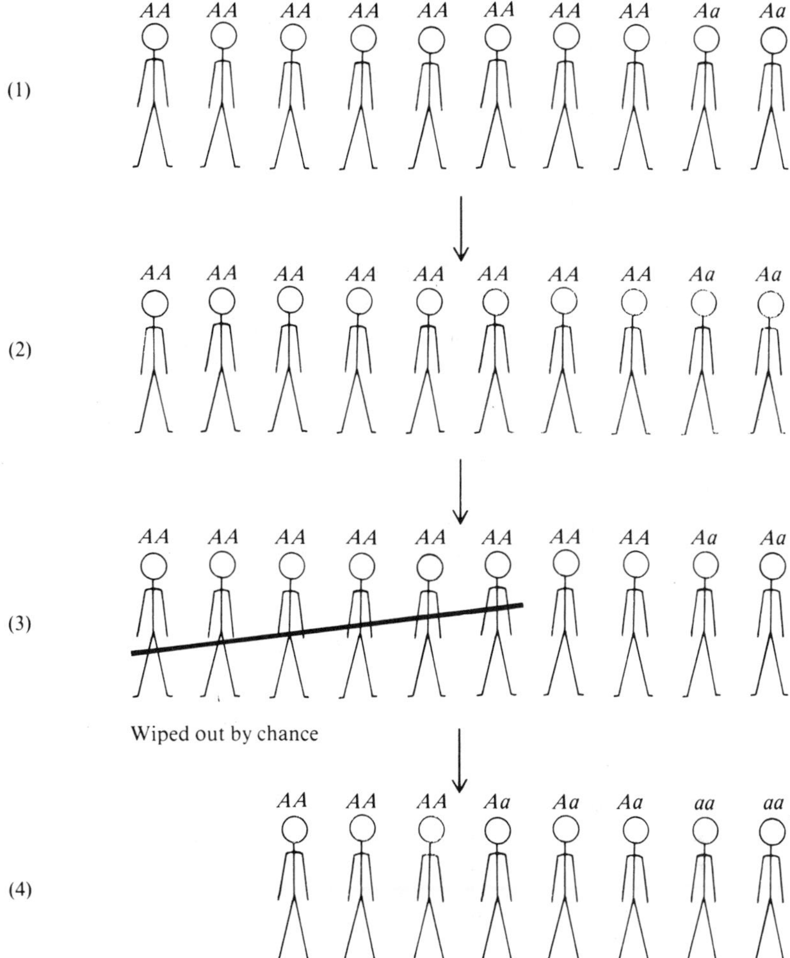

Figure 8-14. A gene of neutral or even deleterious significance may become established in a population by chance factors. Assume that a small number of individuals in a population carries a recessive which imparts little or no advantage to the carrier (1). As members of the population reproduce, the number of persons heterozygous for the gene remains low, since the allele imparts no advantage (2). By chance, a large segment of the population may not reproduce due to chance factors. Let us assume here that the segment was in an area where some cataclysm occurred (disease, earthquake, war). By chance most of those killed were homozygous for the wild gene (3). Among survivors, the incidence of carriers of the recessive is high. The effective breeding population after the event therefore contains a high percentage of carriers. In the ensuing generation (4) the mutant allele will have increased in frequency, and a significant number of individuals homozygous for it may arise.

small numbers for some chance reason, so that the breeding population at one time was quite small. As a result of drift, certain gene frequencies may have become established, even though no value was associated with the particular genes at the time (Fig. 8-14). Drift, then, is an evolutionary force which must be considered along with mutation and natural selection. Unlike natural selection, it may operate in small populations so that disadvantageous genes become established. As population size increases at a later date, the effects of drift may decrease, as natural selection becomes effective.

From what has been said about the sickle cell story, Tay-Sachs disorder, and cystic fibrosis, we must keep in mind that natural selection operates entirely without guidance or foresight. Overall, it *tends* to select those genes with the greatest advantage to the population, but some deleterious genes which happen to give the heterozygote an advantage at a given time may be selected. This can be at the expense of some population members who will suffer due to the serious effects of the gene when it is present in a homozygote. No one group of humans is completely free of detrimental genes which may have accumulated for one or more reasons during the history of the population. The accumulation of mutant genes will be discussed further in Chapter 13.

REFERENCES

Avers, C. J., *Evolution.* Harper & Row, N. Y., 1974.

Cavalli-Sforza, L. L., "Genetic Drift in an Italian Population." *Sci. Amer.,* Aug: 30, 1969.

Cavalli-Sforza, L. L., "Some Current Problems of Human Population Genetics." *Amer. Journ. Human Genet.,* 25:82, 1973.

Cavalli-Sforza, L. L., "The Genetics of Human Populations." *Sci. Amer.,* Sept: 80, 1974.

Cavalli-Sforza, L. L., and Bodmer, W. F., *The Genetics of Human Populations.* W. H. Freeman, 1971.

Crow, J. F., and Kimura, M., *An Introduction to Population Genetics Theory.* Harper & Row, N. Y., 1970.

Daniels F., Jr., van der Leun, J. C., and Johnson, B. E., "Sunburn." *Sci. Amer.,* July: 38, 1968.

de Waal Malefijt, A., "*Homo monstrosus.*" *Sci. Amer.,* Oct: 112, 1968.

Eckhardt, R. B., "Population Genetics and Human Origins." *Sci., Amer.,* June: 94, 1972.

Falconer, D. D., *Introduction to Quantitative Genetics.* Ronald Press, N. Y., 1960.

Harrison, G. A., and Owen, J. J. T., "Studies on the Inheritance of Human Skin Color." *Ann. Human Genet.,* 28:27, 1964.

Holt, S. B., *The Genetics of Dermal Ridges.* Charles C. Thomas, Springfield, Ill., 1968.

King, J. C., *The Biology of Race.* Harcourt, Brace, Jovanovich, N. Y., 1971.

Lewontin, R. C., "Population Genetics." *Annu. Rev. Genet.,* 1:37, 1967.

Mettler, L. E., and Gregg, T. C., *Population Genetics and Evolution.* Foundations of Modern Genetics Series. Prentice-Hall, Englewood Cliffs, N. J., 1969.

Penrose, L. S., "*Dermatoglyphics.*" *Sci. Amer.,* Dec.: 72, 1969.

Population Genetics: The Nature and Causes of Genetic Variability in Populations. Cold Spring Harb. Symp. Quant. Biol., Vol. 20, 1955.

Stern, C., "Model Estimates of the Number of Gene Pairs Involved in Pigmentation Variability of the Negro-American." *Human Hered.,* 20:165, 1970.

REVIEW QUESTIONS

1. Height measurements were taken on a group of men. Determine the mean height for this sample using the measurement data below:

Height in inches:	65	66	67	68	69	70	71	72	73
Number of persons:	1	1	2	1	3	6	3	2	1

2. Recognizing five degrees of skin pigmentation—black, dark, medium, light, and white—represent the possible genotypes of the persons described. Assume only two pairs of alleles to be the basis of the skin pigmentation differences. (1) A dark-skinned person; (2) a person intermediate in shade; (3) a light-skinned person.

3. A dark-skinned person and one who is light-skinned have several offspring: a dark-skinned, an intermediate, and a light-skinned. What are possible genotypes of the parents involved assuming two gene pairs?

4. A white person and one who is intermediate in shade produce eight children, all of whom are light-skinned. What would be the probable genotypes of the parents?

5. Another white person and one who is intermediate produce three children: an intermediate, a light-skinned, and a white-skinned. What are the probable parental genotypes?

6. One variety of pumpkin has fruits which weigh 6 pounds on the average. In a second variety, the average weight is 3 pounds. When the two varieties are crossed, the F_1 yield fruit with an average weight of 4 1/2 pounds. When 2 of these are crossed, a range of fruit weights is found. Of 200 offspring, 3 produce fruits weighing about 6 pounds and 3 produce fruits about 3 pounds in weight. How many gene pairs are involved in the weight difference between the varieties, and how much does each effective gene contribute to the difference?

7. Assume that in rabbit variety 1, the length of the ear averages 4 inches. In variety 2, it is 2 inches. The hybrids between the varieties average 3 inches in ear length. When these hybrids are crossed to each other, there is a much greater variation in ear

length. Of 490 F_2 animals, 2 have ears about 4 inches in length and 2 have ears about 2 inches long. How many gene pairs are involved and how much does each effective gene seem to contribute to the length of the ear?

8. In corn, two varieties each average 64 inches in height. When the two varieties are crossed, the F_1 also average 64 inches. Crosses of the F_1's to each other produce plants which range greatly in size, from 32 to 96 inches. About 4 in 1000 reach 96 inches, and the same number are only 32 inches tall. Offer an explanation for these results.

9. Assume that height in the human depends on four pairs of alleles. Give an explanation to account for two persons of moderate height producing children who are much taller than they. Assume that environment is exerting a negligible effect.

10. Assume that two human populations, one in Canada and one in Peru, both originated from migrants who formed part of the same European population. In the Canadian population, a high incidence of a hereditary disease occurs which causes death of young adults. In the Peruvian population, the disorder is unknown. In the ancestral European population, only a small incidence of the disease continues to be reported. Offer an explanation.

11. Suppose that the members of three separated populations of plants have many similar phenotypic features but also have several traits which make them distinct. When members of population 1 are crossed with those of population 2, the F_1's are intermediate, but they are completely sterile. When members of population 2 are crossed with those of 3, highly fertile intermediate F_1's result. Crosses of members of 1 with 3 produce intermediate F_1's which are sterile. Offer an explanation which includes recognition of species.

chapter 9

CHEMICAL ASPECTS OF GENETICS

Since the early 1950s, a wealth of knowledge has been acquired on the chemical nature of the gene and its molecular interactions. In just two decades, more information was assembled than in all the years since the discovery of the cell in the seventeenth century! The advances which continue to be made on the chemistry of inheritance are of the utmost significance to human genetics. Once the scientist understands how the genetic material is constructed, he is in a position to devise experiments which can contribute information on the ways in which the gene exerts its effects on the cell. This information in turn enables him to ascertain what has gone wrong with the chemistry of the cell when a mutant gene replaces the normal one and produces an abnormal phenotypic effect. Knowing the precise cellular alterations which accompany a genetic disorder, the biologist can undertake a search for ways to correct or alleviate the condition. The strides which are being made in this direction in the case of several genetic defects offer hope that some human suffering will be prevented as a result of an understanding of genetics at the chemical and molecular levels. This progress has been due to contributions from many outstanding investigators. Only a bare outline of some of their efforts will be presented here to demonstrate the nature of certain hereditary disorders and the problems involved in their prevention.

Composition of the Hereditary Material

Before 1940, the chemical nature of the gene had not been identified. It was known for years that the chromosomes of most forms of life contain large amounts of protein joined in some way to other large molecules, the *nucleic*

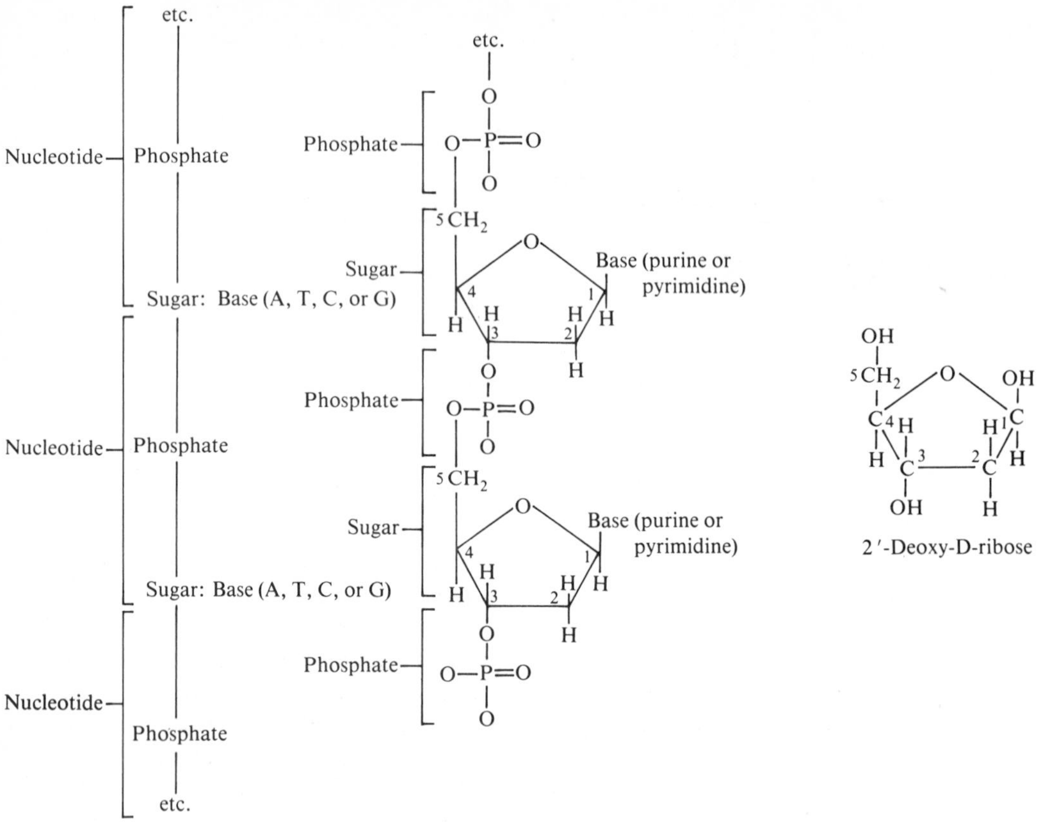

Figure 9-1. Nucleotide structure of DNA. A nucleotide is a unit formed by the union of a phosphoric acid molecule with a sugar and a nitrogen base (left). In DNA, the sugar is deoxyribose (right). The nitrogen bases are of four kinds. (See Fig. 9-2 for more details.)

acids. These nucleic acids were known to be of two main kinds, DNA (for deoxyribonucleic acid) and RNA (for ribonucleic acid). Of the substances found in the chromosome, the protein seemed most likely to be the substance of the gene. The idea was popular, since protein molecules are the most complex of all the molecules in the cell. However, experiments conducted with certain microorganisms (bacteria and viruses) in the 1940s and early 1950s clearly indicated that DNA alone was the genetic material of these lower forms. Later experimental work has left no doubt that DNA is the genetic material in all cellular species, whether bacteria, trees, dogs, cats, or the human.

Once DNA was implicated as the substance of heredity, research was focused on the way the DNA is constructed. It was demonstrated that each building block which can join with other similar building units to form DNA is composed of three kinds of substances (Fig. 9-1). Each of the single building units is called a *nucleotide* and is formed by the union of a phosphoric acid molecule, a certain kind of sugar, and a molecule of a type called a *nitrogen base* because of the presence of nitrogen atoms within it. All the nucleotides making up a DNA molecule are identical in their phosphoric acid and sugar components. The specific sugar is known as *deoxyribose.* One of the main distinctions between this sugar and common glucose is that it contains only five carbon atoms instead of six. Nucleotides, however, do not all contain the same nitrogen base. It is the nitrogen base, *not* the sugar or the phosphoric acid, which is responsible for the distinction among nucleotides. Nevertheless, the number of different varieties of nucleotides found in DNA is small, only four. In regard to their structure, the nitrogen bases are classified into two main categories, *purine bases* and *pyrimidine bases.* Both of these types have the form of rings [Fig. 9-2(A)]. The purines, however, are composed of two rings condensed together, whereas the pyrimidines contain only one. The two kinds of purine bases which are found in DNA are *adenine* and *guanine*; the pyrimidine bases are *thymine* and *cytosine.* Early in the 1950s, it was established that DNA is composed of just these four different kinds of nucleotides [Fig. 9-2(B)]: a kind with adenine (A), a kind with thymine (T), another with cytosine (C) and a fourth with guanine (G). It was still not known, however, exactly how these four kinds of building units are joined to form DNA.

The Watson-Crick Model and Some of Its Implications

In the 1950s, some of the most outstanding landmarks in biology were achieved. These resulted in the birth of a new science, molecular biology. Several of the most important and fundamental questions relating to the chemistry of heredity were finally answered. Paramount among the contributions is the work of Watson, Crick, and Wilkins, who were awarded the Nobel prize in 1961 for their achievements. Watson and Crick assembled an immense amount of data from studies on DNA performed in many laboratories. They interpreted these in relation to extremely detailed X-ray analyses made on DNA by Wilkins and his associates. The result of the combined efforts was the presentation by Watson and Crick of a model of the structure of DNA which bears their names. The announcement of the Watson-Crick model marked the beginning of molecular biology as a separate discipline. The model made possible all kinds of experimental approaches to the study of gene action on the molecular level. Many experiments conducted throughout the years have shown that the Watson-Crick model of the DNA molecule is essentially correct as proposed by Watson and Crick in their original paper. The model has enabled the biologist to gain an insight into the very nature of living activities, and its implications are profound for all species of life. The basic features of the model must therefore be understood to provide a fuller appreciation of human heredity.

Watson and Crick knew that the DNA is composed of nucleotides, and they were familiar with the molecular details of the four different varieties. Their interpretation of all the data from biochemistry and physical chemistry led them to certain novel conclusions about the architecture of the DNA. According to the Watson-Crick model, the DNA is composed not of one but

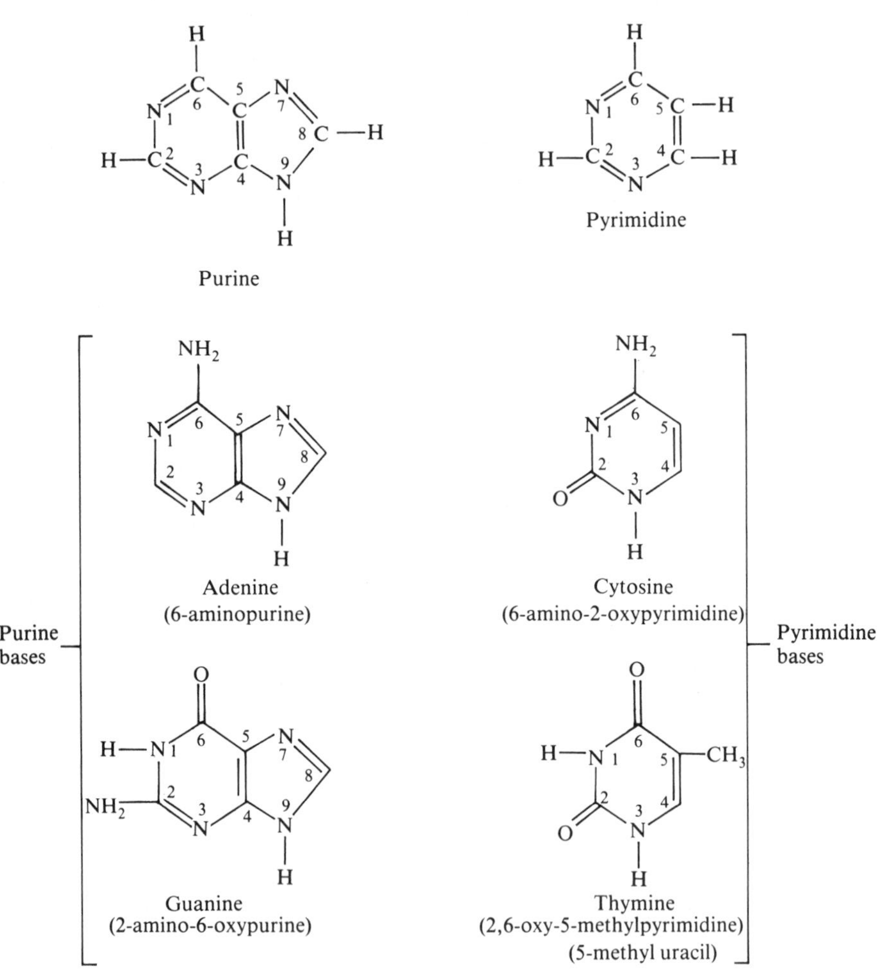

Figure 9-2. Substances composing DNA nucleotides. (A) The purine and pyrimidine bases of DNA. The parent compounds are shown above. The two derivatives of each, which occur in DNA, are shown below. (B) Structure of the four kinds of nucleotides which occur in DNA.

of two long chains. Each chain in turn is composed of nucleotides linked together. In each chain [Fig. 9-3(A)], the phosphoric acid component of one nucleotide is attached to the sugar portion of another one. This type of linkage provides each chain with a backbone composed of alternating phosphate and sugar. The nitrogen base of each nucleotide (a purine or a pyrimidine) is attached to the sugar, *not* to the phosphate, just as if the base were suspended on the backbone. Either one of the chains of the DNA can be called a *poly-*

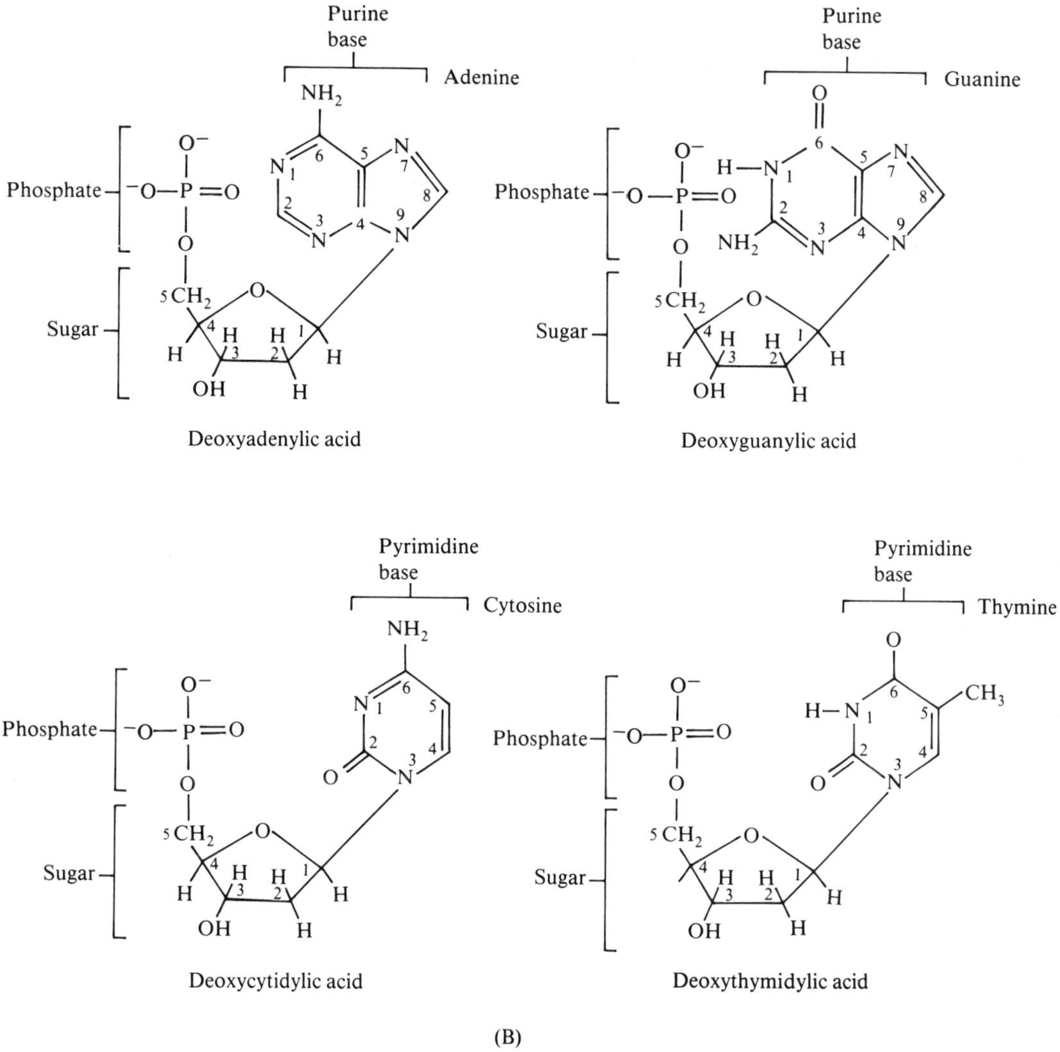

Purine base — Adenine
Phosphate
Sugar
Deoxyadenylic acid

Purine base — Guanine
Phosphate
Sugar
Deoxyguanylic acid

Pyrimidine base — Cytosine
Phosphate
Sugar
Deoxycytidylic acid

Pyrimidine base — Thymine
Phosphate
Sugar
Deoxythymidylic acid

(B)

Figure 9-2. Concluded

nucleotide, since it is composed of one nucleotide after another joined through the sugar and phosphate portions. The two polynucleotide chains which compose every DNA molecule do not just lie side by side. Instead, they are twisted around each other in a spiral or helical fashion [Fig. 9-3(B)]. The association of the two chains is by no means haphazard, as emphasized in one of the important features of the Watson-Crick model. The two chains are bound together in a rather fixed association because the purine base adenine (A) on one chain pairs only with the pyrimidine base thymine (T) on the other chain. This pairing is highly specific. Adenine will normally pair only with thymine (T), *not* with cytosine (C) or guanine (G), or another adenine (A). Similarly, the other purine base, guanine (G), pairs with the other pyrimidine base, cytosine (C), and this pairing, G with C, is also specific. This very important feature of the model means that the two chains of each DNA

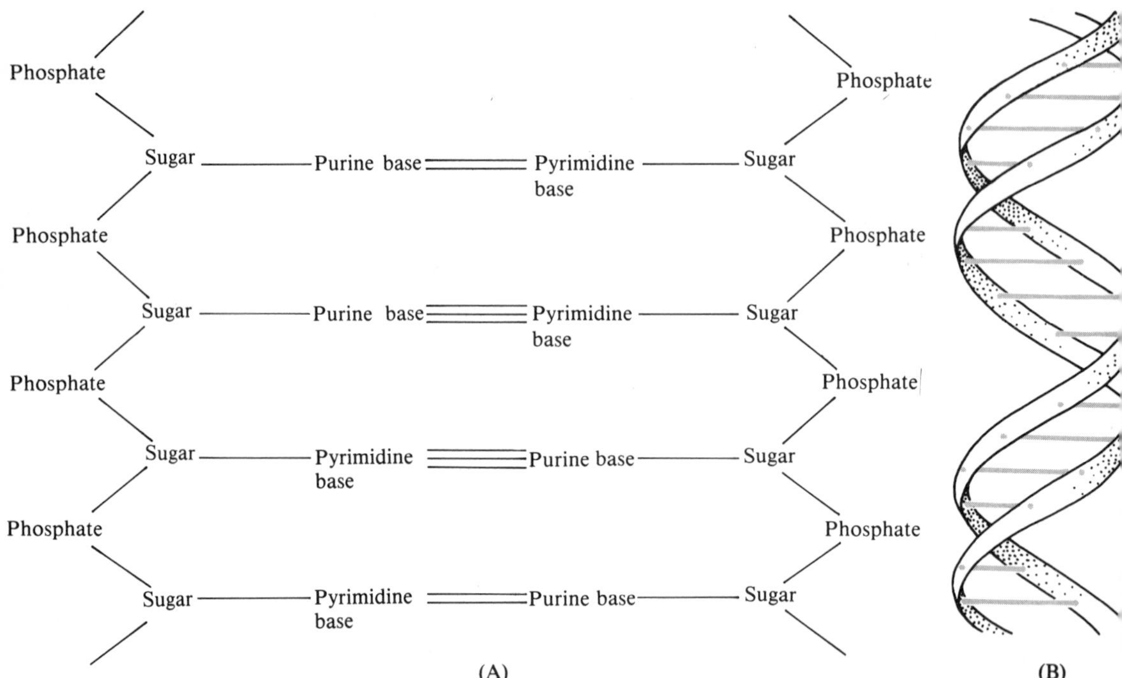

Figure 9-3. DNA structure. (A) The DNA molecule is composed of two long polynucleotide chains. In each chain, the phosphate and sugar forms a backbone. The purine or pyrimidine base in each nucleotide is attached to the sugar component. The two chains are held together by chemical bonds which form between a purine on one chain and a pyrimidine on the adjacent chain. (B) The two chains, as first proposed by Watson and Crick, are wrapped in a double helix and can be compared to a spiral staircase in which the base pairs form the "steps" and the sugar-phosphate backbone forms the "rails."

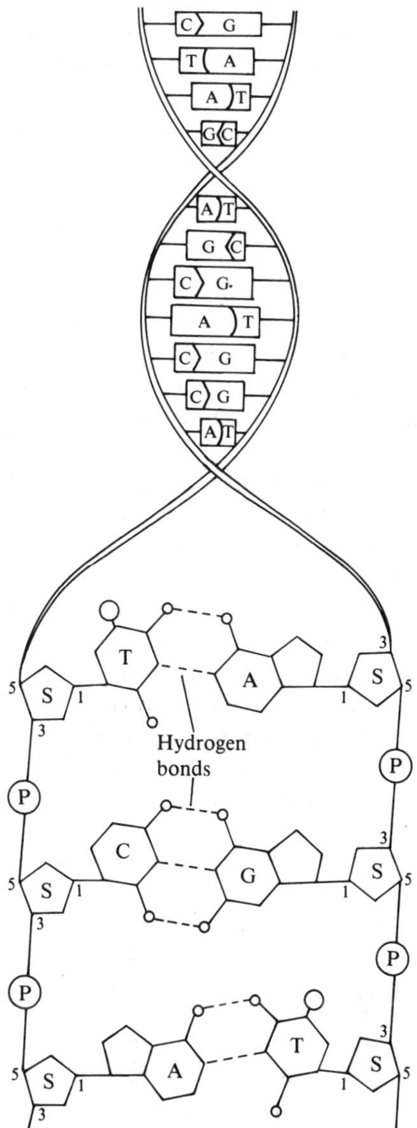

Figure 9-4 Specific base pairing. The pairing of the bases in DNA is very specific. Thymine (T) on one chain pairs only with adenine (A) on the adjacent chain. These two bases are held together by two chemical bonds (hydrogen bonds). Cytosine (C) on one chain normally pairs only with guanine (G) on the adjacent chain. The two are held together by three chemical bonds.

molecule must be complementary, as shown in Fig. 9-4. For each adenine (A) on any one chain, there is a thymine (T) on the other chain at the same position; for each guanine (G), there is a corresponding cytosine (C). So if one knows which nucleotides follow one another along one chain (for example, A-T-T-A-G-C), one automatically can determine the sequence of the nucleotides at the corresponding positions on the chain which is paired with it (T-A-A-C-G).

The molecule is often compared to a spiral staircase with the rails representing the sugar–phosphate backbone and the bases forming the stairs. To summarize these points: The DNA molecule is made up of two long, complementary polynucleotide chains wrapped around each other in the form of a double helix. The backbone of each chain is composed of sugar and phosphate. The bases are attached to the sugars, and those bases on one chain form specific pairs with those on the complementary chain: A and T always form one kind of pair, and G and C form the other specific pair. The complete model is very detailed in its picture of the DNA molecule: how wide it is; the length of each turn of the spiral, etc. Such fine points as these are not necessary, however, to understand how the DNA controls activities in the cell.

The genetic material is composed of DNA, and the Watson-Crick model implies that each gene is a segment or stretch of DNA along the molecule. A gene would thus contain many nucleotides or more exactly, nucleotide pairs, since the DNA is a double helix. Genes, which may be of different lengths, follow one another along the DNA. Each gene differs from the next, because the sequence or order of nucleotide pairs in one gene is not identical to that in the following one (Fig. 9-5). There is no restriction on which base in one chain must necessarily follow another. This means there is an almost infinite number of possible combinations. This is not so difficult to understand when it is realized that simply a dot and a dash form the entire basis of the Morse code, which can represent all the words in a language. In the case of DNA, there is a four-letter alphabet, so to speak, provided by the four different kinds of nucleotides, each with its specific kind of base: A or T or G or C. Each gene has an individuality, since the order in which its base pairs follow one another is different from that of any other kind of gene. Moreover,

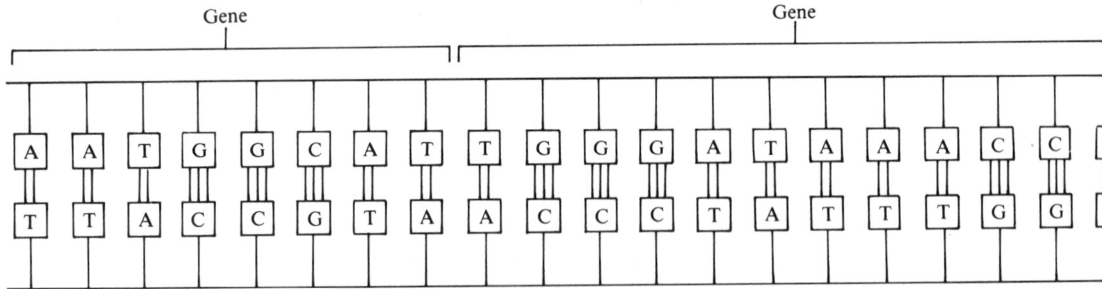

Figure 9-5. While the DNA exists in the form of a double helix, it is basically a linear molecule. When viewed as "untwisted," one nucleotide pair follows another in a linear order. A gene is a segment of a DNA molecule, and the difference between any two genes resides in the sequence of the base pairs contained in each. There is no set way in which the base pairs must occur in a gene. This fact makes possible an endless variety of genes which may be of different lengths. The figure is greatly simplified; in actuality, no gene would be composed of so few nucleotide pairs.

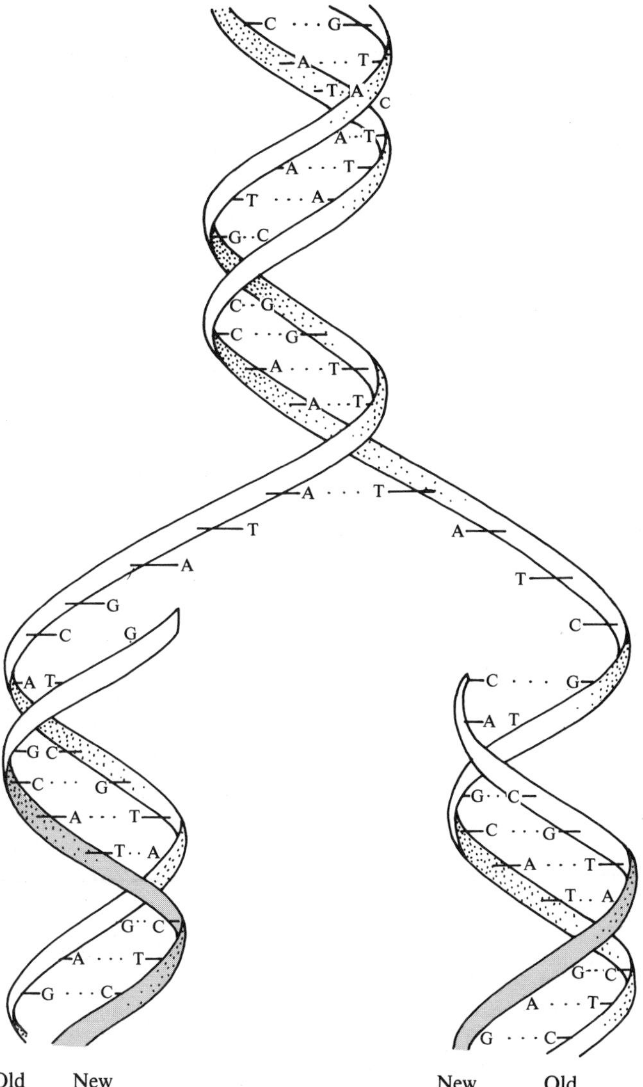

Figure 9-6. Replication of DNA. Note that each old strand acts as a blueprint for the formation of a complete new strand. Adenine always attracts thymine, and cytosine attracts guanine. The end result of replication is two molecules identical to the original. Each of the two is composed of one old and one new strand.

the lengths of genes are by no means the same, the number of nucleotide pairs varying from one specific gene to another.

Watson and Crick pointed out that their model of DNA could provide some insight into the way the genetic material replicates or duplicates itself. According to the model (Fig. 9-6), when the DNA is about to undergo replication, the chains in the region of duplication separate. This separation comes about when the attractive forces holding the adenine to the thymine and the guanine to the cytosine are enzymatically overcome. As the chains separate, however, each one of them remains completely intact. Each chain acts as a template (blueprint) for the assembly of a complementary chain from nucleotide building blocks present in the cell. This process requires energy and many other factors, but in essence, each chain guides the formation of a complementary one, A pairing with T and G pairing with C. As Fig. 9-6 shows, the result is two new DNA molecules formed from the original one. In each of the two new ones, there is actually one old chain which served as a blueprint and one new chain which is complementary to the template that guided its formation. This method of replication is known as *semiconservative replication.*

Genes and Protein Formation

Extremely important is the implication which the model holds for gene action, the manner in which each gene is able to achieve an effect on the molecular level. According to the model, each gene is unique because of the sequence of its base pairs. This implies that the sequence in some way establishes a sort of code which carries information or instructions. This information must somehow be transferred from the gene to other parts of the cell. Years before the presentation of the Watson-Crick model, it was known that many genes apparently control enzymes in the cell. This had been clearly established in species of fungi and bacteria, where a gene mutation was often found to result in a changed enzyme. Enzymes are the catalysts which are critical to the normal metabolism of all living things. Without them, chemical reactions cannot proceed in a cell. Therefore, if a gene mutation takes place which results in the lack of a particular enzyme, a chemical step may be stopped or blocked (Fig. 9-7). Numerous examples were assembled in lower forms of life which clearly demonstrated that a specific gene mutation produced an altered enzyme or a block in a chemical step. The idea was therefore proposed that a gene controls an enzyme and in this way can exert a cellular effect, since it consequently controls a specific chemical step in metabolism. While microorganisms provided the greatest number of examples relating genes to enzymes, it was actually the human (as noted in Chapter 1) that was the first species in which the chemical aspect of genetics was studied. Before 1910, Garrod in London, studying human hereditary disorders such as alkaptonuria (Chapter 10), focused attention on genetic blocks which can arise in the human and interrupt the normal chain of chemical reactions in a cell. Garrod's emphasis on the biochemical defects in some human hereditary diseases influenced later geneticists who proceeded to accumulate more and more concrete evidence for the association of specific genes with the control of specific enzymes.

All enzymes are protein in nature, and so the concept arose that the gene exerts its control over cell activities by governing the formation of all cellular proteins. Added support for this idea came from various lines of research, especially with microorganisms such as the viruses which are parasitic on bacterial cells (the *bacteriophages* or *phages*). Elegant experiments demonstrated that when a bacterial virus (or phage) attacks a cell, only the DNA of the phage actually enters the infected bacterium (more details are given in Chapter 14). Among the first products which the viral DNA causes the host cell to manufacture is protein. This protein is the protein of those enzymes that are needed for the construction of new virus particles. The phage is able to reproduce inside the host bacterial cell, because its DNA can cause the cell to manufacture certain proteins, specific viral enzymes. Once formed, these enzymes can initiate the building of viral DNA and of other kinds of viral protein. The end result is the production of many new viruses inside each infected cell (Fig. 9-8).

As more evidence accumulated to associate genes with the control of proteins, attention was turned to the mechanism by which the gene is able to accomplish this. A consideration of protein structure and its formation in the

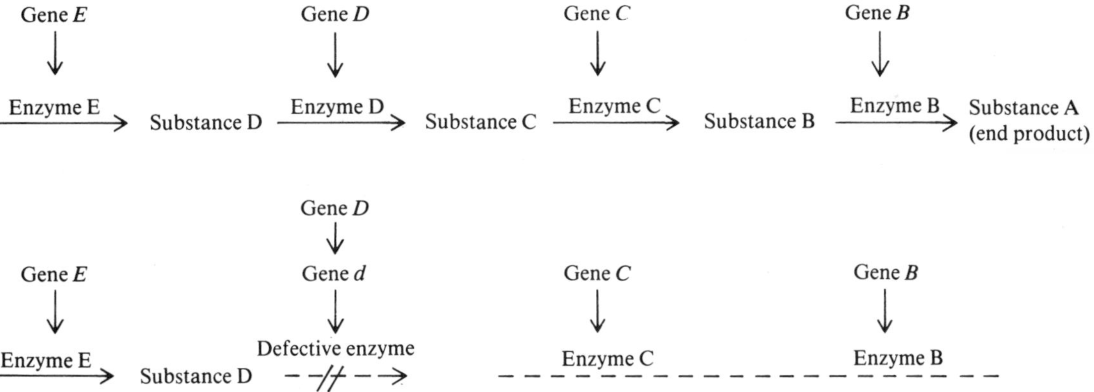

Figure 9-7. Gene control of enzymes. Most products in the cell are the result of a series of chemical steps, each one controlled by a specific enzyme (above). Each enzyme in turn is under the control of a specific gene. Therefore, a gene controls a step in a pathway leading to the formation of an essential product (substance A). If a gene mutation occurs which affects one of the enzymes in the chain of steps (below), a block results, and the steps following it cannot proceed even though normal enzymes for those steps may be produced. Consequently, one genetic block in a metabolic pathway may lead to the absence of a critical product.

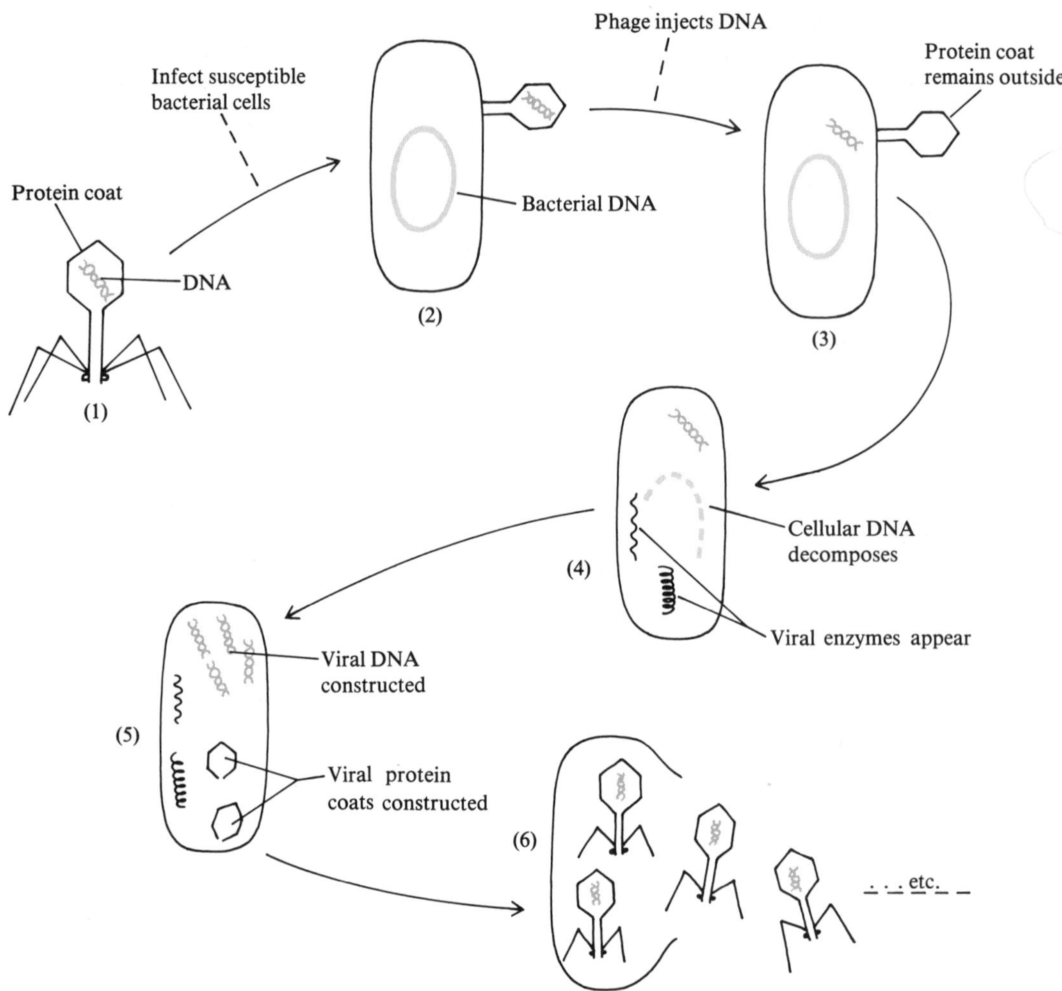

Figure 9-8. Infection of bacterial cell by a virus (phage). One of the common phages of the bacterium, *E. coli,* consists of a characteristic protein envelope in which the viral DNA is confined (1). When a phage enters a bacterial cell (2), it injects its DNA, and the protein coat remains outside (3). New protein soon appears in the infected cell. These are enzymes which are required to build more viruses (4). The cellular DNA soon decomposes. More viral DNA appears inside the cell and then the protein coats of new viruses (5). The DNA becomes inserted into the coats, and the bacterial cell eventually bursts (6) with the release of new, mature virus particles. The DNA of the virus, once inside the cell (3), can direct the formation of virus proteins which are enzymes (4) needed for the synthesis of DNA and other proteins of the mature virus.

cell raises several perplexing points. First of all, proteins are formed in the cytoplasm of the cell as a result of the activities of the *ribosomes,* bodies which can be visualized with the electron microscope and which are composed of RNA and protein. The ribosomes are able to construct protein molecules from their building units, the *amino acids.* However DNA, which supposedly guides the formation of protein in some way, is confined to the nucleus. Assuming that the DNA carries the information or instructions needed to direct the formation of specific proteins, how does the required information get from the nucleus to the cytoplasm? Since chromosomes do not pass through the nuclear membrane to deliver the information to the ribosomes, there must be some sort of intermediary or messenger which can accomplish this.

The Role of Messenger RNA

The answer to the enigma came through work with bacteria and their viruses. The findings were later supported by experiments with higher species. A type of molecule was recognized which apparently serves the function of the messenger in the cell. It is composed of ribonucleic acid. RNA is a substance that resembles DNA in many respects and is very widely distributed in both the nucleus and the cytoplasm of a cell. As a matter of fact, the ribosomes themselves are composed of both protein and RNA components, as noted above. However, this other kind of RNA molecule which was discovered is *not* the RNA which makes up part of the ribosomes. It was designated *messenger RNA,* or *mRNA.* Work with radioactive substances which permit the investigator to follow molecules or products as they move within the cell showed that the mRNA is formed in the nucleus and then travels out to the cytoplasm. Techniques finally permitted a closer chemical analysis of messenger RNA. It was clearly shown that mRNA is formed under the guidance of DNA and that the mRNA closely resembles specific segments of the DNA. The DNA actually acts as a template for the formation of the mRNA.

The general picture that has emerged is as follows. The DNA of the nucleus is able to undergo replication, as just discussed, in the manner proposed in the Watson-Crick model. Replication is, of course, necessary at the time of cell division to supply the nuclei of the new cells with their proper set of genetic instructions. Obviously, the genetic material cannot be concerned solely with its own replication. It must engage in some other kind of activity if it is to exert any sort of control or effect on the cell. One such activity is to send out messages. The DNA achieves this by way of the formation of mRNA. This process is called *transcription,* the assembly of RNA next to a DNA template. Transcription entails the unwinding of the two DNA chains and their partial separation so that single-stranded portions of the DNA can serve as blueprints [Fig. 9-9(A)]. Not all of the DNA of a nucleus undergoes transcription in a

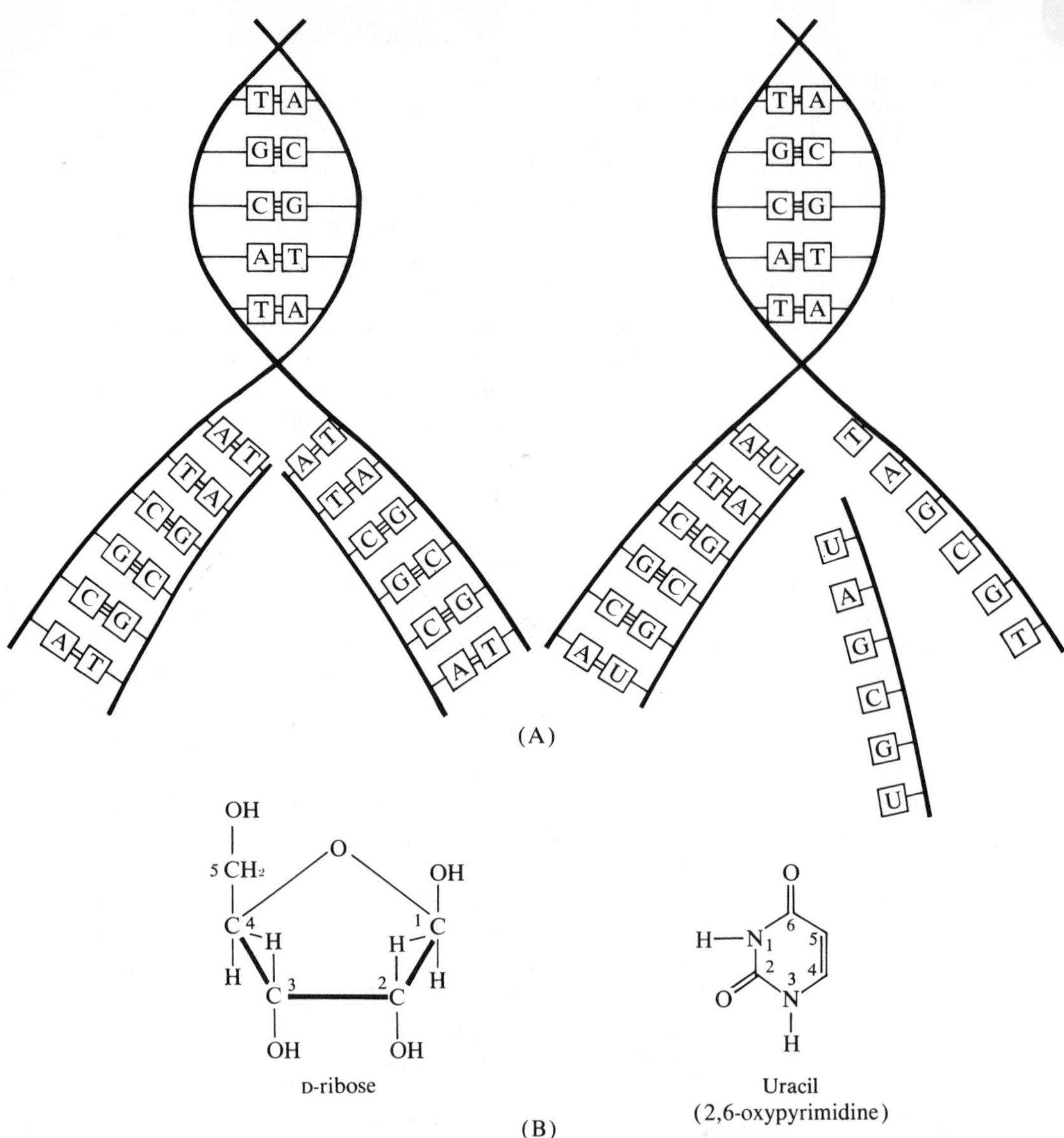

Figure 9-9. Replication and transcription. (A) When replication occurs, the double helix unwinds at the point of replication, and each old strand forms a complementary new strand until two new molecules are formed, each identical to the original (left). When transcription takes place (right), an unwinding occurs at the point of transcription, but only one strand is involved in the formation of mRNA. The messenger RNA then leaves the strand on which it was formed, and the two strands of the DNA again pair and form a double helix. (B) The RNA which is formed on a DNA template differs from a DNA strand which is formed on a DNA template. In RNA, the sugar is ribose instead of deoxyribose, and the pyrimidine base uracil (U) is found in place of thymine. Uracil occurs only in RNA; thymine is found only in DNA (refer to Figs. 9-1 and 9-2).

given type of cell (Chapter 14). Moreover, only certain DNA segments or genes become active and engage in transcription at specific times. The information from an active gene, encoded in the form of mRNA, is then carried to the cytoplasm. Therefore, during transcription only certain regions of the DNA unwind, just those parts or genes which are participating in the formation of mRNA. A perusal of Fig. 9-9(A) shows this and certain other important facts about the mRNA. Note that messenger RNA is similar, but not identical, to the single DNA strand which served as its template. First of all, the sugar in all types of RNA is slightly different from that in DNA. It is ribose sugar and contains one more oxygen atom than does the deoxyribose sugar in DNA. RNA, like DNA, is also composed of nucleotides, but RNA never contains nucleotides with the base thymine. Instead, another base, uracil (U), is present. The base uracil, on the other hand, is never found in DNA [Fig. 9-9(B)]. Moreover, RNA is never composed of two separate chains which form a double helix. Instead, it is one-stranded, although this single strand may loop back upon itself in various ways. Any one messenger RNA is formed on a DNA template from only one of the two DNA strands. This means that a particular messenger RNA carries the information from only one of the two strands within a gene. Note [Fig. 9-9(A)] that the mRNA, whose length will vary depending on the size of the gene, is *complementary,* not identical, to the DNA strand on which it was formed. The base T of the DNA template attracts an RNA nucleotide with the base A; G attracts C; C attracts G, but A on the DNA attracts U, since uracil is always found in place of thymine in RNA. Once the nucleotides of RNA are assembled on the template DNA, specific enzymes present in the cell join them together so that a strand of mRNA forms. This messenger strand then "peels" away from the DNA blueprint and can now enter the cytoplasm. The unwound region of the DNA then resumes the double-helical form that was present before transcription took place.

It is important to note (Fig. 9-9) that the mRNA carries within it a specific code in the form of its nucleotide sequence. This sequence reflects information coded in the DNA. The sequence of nucleotides within the mRNA strand is exactly complementary to the DNA template strand, the main difference being that uracil replaces thymine. Whatever the code may mean, it is obvious that it must be in the nucleotide sequence, since that provides the only bit of distinction among genes or among kinds of mRNA. The sugars and phosphates of DNA are the same, and so are those of RNA.

Amino Acids and Steps in Protein Construction

Once the messenger RNA is in the cytoplasm, it must be *translated* into protein, a specific protein for each kind of mRNA. A burst of research activity from many laboratories filled in the details of a remarkable set of interactions which occur on a molecular level in the cell. One problem that had to be settled concerned the genetic code and its exact relationship to proteins, whose structure it somehow directs. The code, as we have seen, is based on nucleotide sequence, but there are only 4 different kinds of nucleotides in DNA (A, T, G, and C). Proteins, the most complex of the molecules in the cell, are constructed from smaller building blocks, the amino acids. There are, however, 20 different kinds of amino acids which can enter into the formation of cellular protein. How can a code, based on a DNA alphabet of only 4 letters, direct the precise assembly of 20 different entities?

Before we can appreciate part of the explanation, it is necessary to review a few points about proteins. Figure 9-10 shows the 20 different amino acids that occur in proteins. It is completely unnecessary for our purposes to memorize the differences among them. What *is* essential is to note that all of them have certain things in common about their structure. Each amino acid has a so-called *amino group* which is the $—NH_2$ part of the molecule. Each also carries a —COOH group which is the acid portion and is called a *carboxyl group.* Any protein arises from the linking together of amino acids through their $—NH_2$ and —COOH groups (Fig. 9-11). The joining of these groups produces a linkage called a *peptide link.* One kind of protein differs from the next in the sequence of its amino acids, but all proteins are in essence long chains constructed of amino acids held together by peptide linkages. A protein chain, after its formation, may later associate with other chains to form a still more complex molecule. Proteins may become folded in a variety of three-dimensional configurations, but no matter how complex the architecture of a protein, it is the peptide linkage which is responsible for holding together the amino acid units of the chain. A protein chain is also referred to as a *polypeptide,* a term which indicates that the chain is composed of many amino acids joined by the peptide linkage. Any small group or fragment of linked amino acids from a polypeptide chain is often referred to as a *peptide.*

One major problem to be solved was how the 20 different kinds of amino acids, distributed in the cytoplasm of the cell, are recognized so that each kind receives an identity. In the 1950s, it was well established that before any amino acid reacts with a ribosome to enter into protein formation, it first reacts with another kind of molecule which is abundant in the cytoplasm. This molecule is also composed of RNA, but it is quite distinct from the mRNA or that RNA which forms part of the ribosomes. It is a small molecule compared to the other RNA types, but it is also composed of a single strand of nucleotides. The strand is believed to be twisted in the form of a clover leaf [Fig. 9-12(A)]. This small RNA molecule has been named *transfer RNA* (*tRNA*), since it is able to pick up an amino acid and transfer it to a ribosome [Fig. 9-12(B)]. Transfer RNA molecules, while all very similar, have slight differences among them. These differences are sufficient to permit each kind of tRNA to recognize only one specific kind of amino acid. For example, there is a certain tRNA

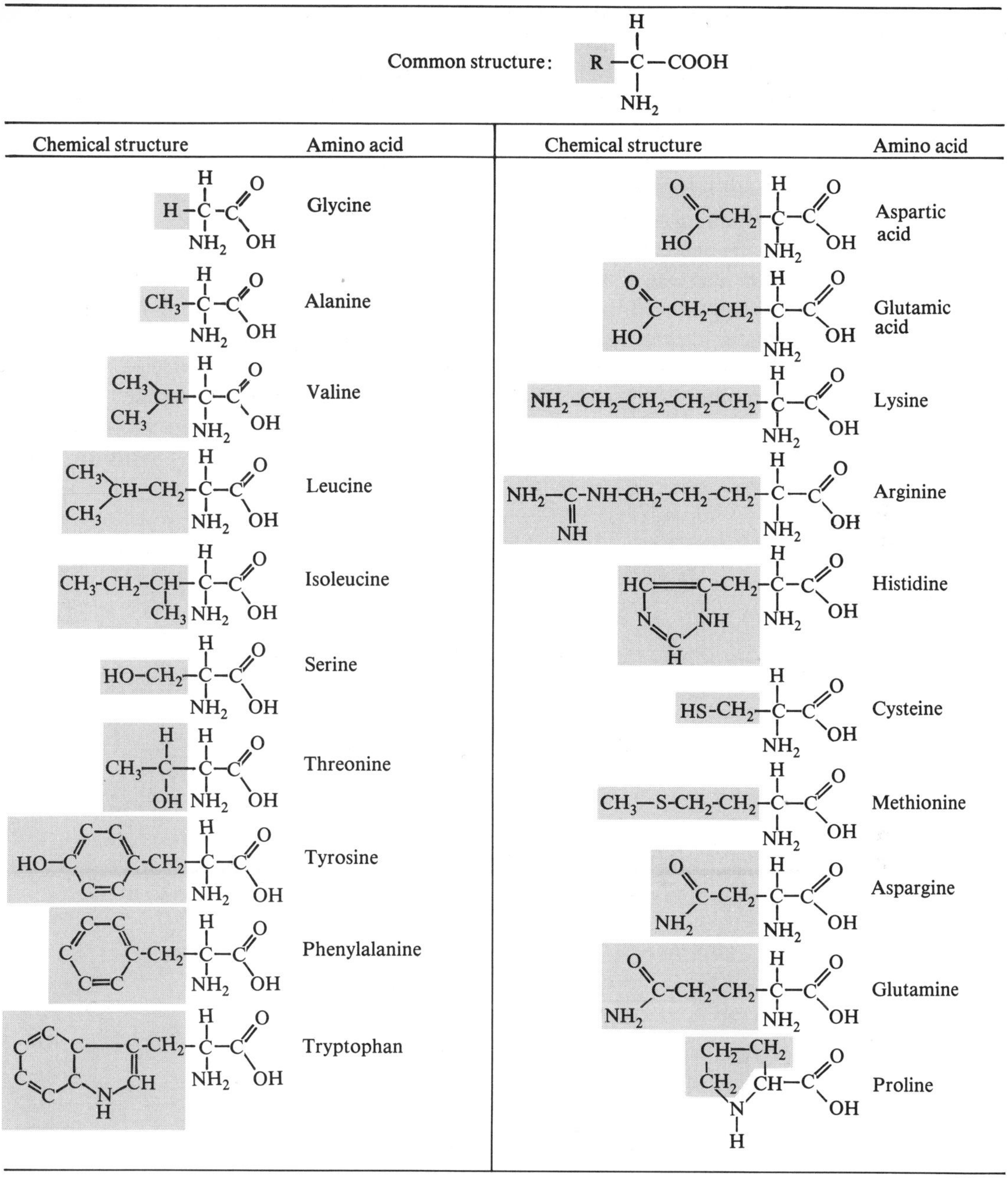

Common structure: $R-C(H)(NH_2)-COOH$

Chemical structure	Amino acid	Chemical structure	Amino acid
$H-CH(NH_2)-COOH$	Glycine	$HOOC-CH_2-CH(NH_2)-COOH$	Aspartic acid
$CH_3-CH(NH_2)-COOH$	Alanine	$HOOC-CH_2-CH_2-CH(NH_2)-COOH$	Glutamic acid
$(CH_3)_2CH-CH(NH_2)-COOH$	Valine	$NH_2-CH_2-CH_2-CH_2-CH_2-CH(NH_2)-COOH$	Lysine
$(CH_3)_2CH-CH_2-CH(NH_2)-COOH$	Leucine	$NH_2-C(=NH)-NH-CH_2-CH_2-CH_2-CH(NH_2)-COOH$	Arginine
$CH_3-CH_2-CH(CH_3)-CH(NH_2)-COOH$	Isoleucine	$HC{=}C(-NH-CH{=}N-)-CH_2-CH(NH_2)-COOH$ (imidazole ring)	Histidine
$HO-CH_2-CH(NH_2)-COOH$	Serine	$HS-CH_2-CH(NH_2)-COOH$	Cysteine
$CH_3-CH(OH)-CH(NH_2)-COOH$	Threonine	$CH_3-S-CH_2-CH_2-CH(NH_2)-COOH$	Methionine
$HO-C_6H_4-CH_2-CH(NH_2)-COOH$	Tyrosine	$NH_2-C(=O)-CH_2-CH(NH_2)-COOH$	Aspargine
$C_6H_5-CH_2-CH(NH_2)-COOH$	Phenylalanine	$NH_2-C(=O)-CH_2-CH_2-CH(NH_2)-COOH$	Glutamine
(indole ring, N–H) $-CH_2-CH(NH_2)-COOH$	Tryptophan	$CH_2-CH_2-CH_2-NH-CH-COOH$ (ring)	Proline

Figure 9-10. The 20 amino acids which commonly occur in cellular proteins. All the amino acids have a common structure (above). An amino group (NH_2) and an acidic group (COOH) are present in addition to an R group (indicated by the portion within the square). It is this R group which is responsible for the differences among the amino acids. In the case of glycine, the R group is simply one hydrogen atom, making glycine the simplest of the amino acids. The R group may be a complex grouping of atoms as seen in the aromatic amino acids.

type which reacts with the amino acid, alanine, and which can carry it to the ribosome. This alanine-tRNA cannot react with any other kind of amino acid, say tryptophan. The reaction between tRNA and its amino acid is not a simple spontaneous one; it involves several other factors including special enzymes and energy sources. Again, the details are not essential for a basic understanding of the relationship between the gene and a protein.

In the cell, therefore, the tRNA molecules identify specific amino acids. Until an amino acid is identified in this way, it cannot link up with other amino acids. Once the different tRNA's transport their specific amino acids to a ribosome, a series of reactions takes place which results in the joining of one specific kind of amino acid to another. It is essential to realize that the specific order or sequence of the amino acids is determined *not* by the ribosome or the tRNA, but by the instructions coded in the mRNA. The mRNA, it will be recalled, is a transcript of a DNA segment or a gene. The sequence of

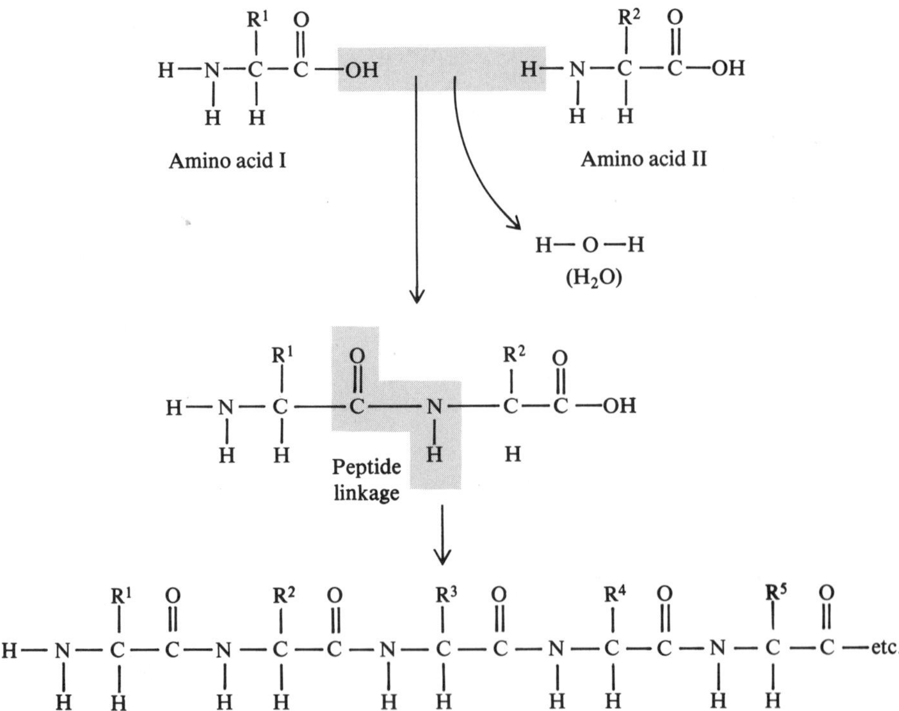

Figure 9-11. The peptide linkage. A protein is basically a chain of amino acids joined to each other through the union of the acid group (COOH) of one amino acid with the amino group (NH_2) of the following amino acid. When the linkage is formed between any two amino acids, a molecule of water is liberated. Note (below) that a protein will have an unattached amino group at its beginning and a free carboxyl group (COOH) at its end.

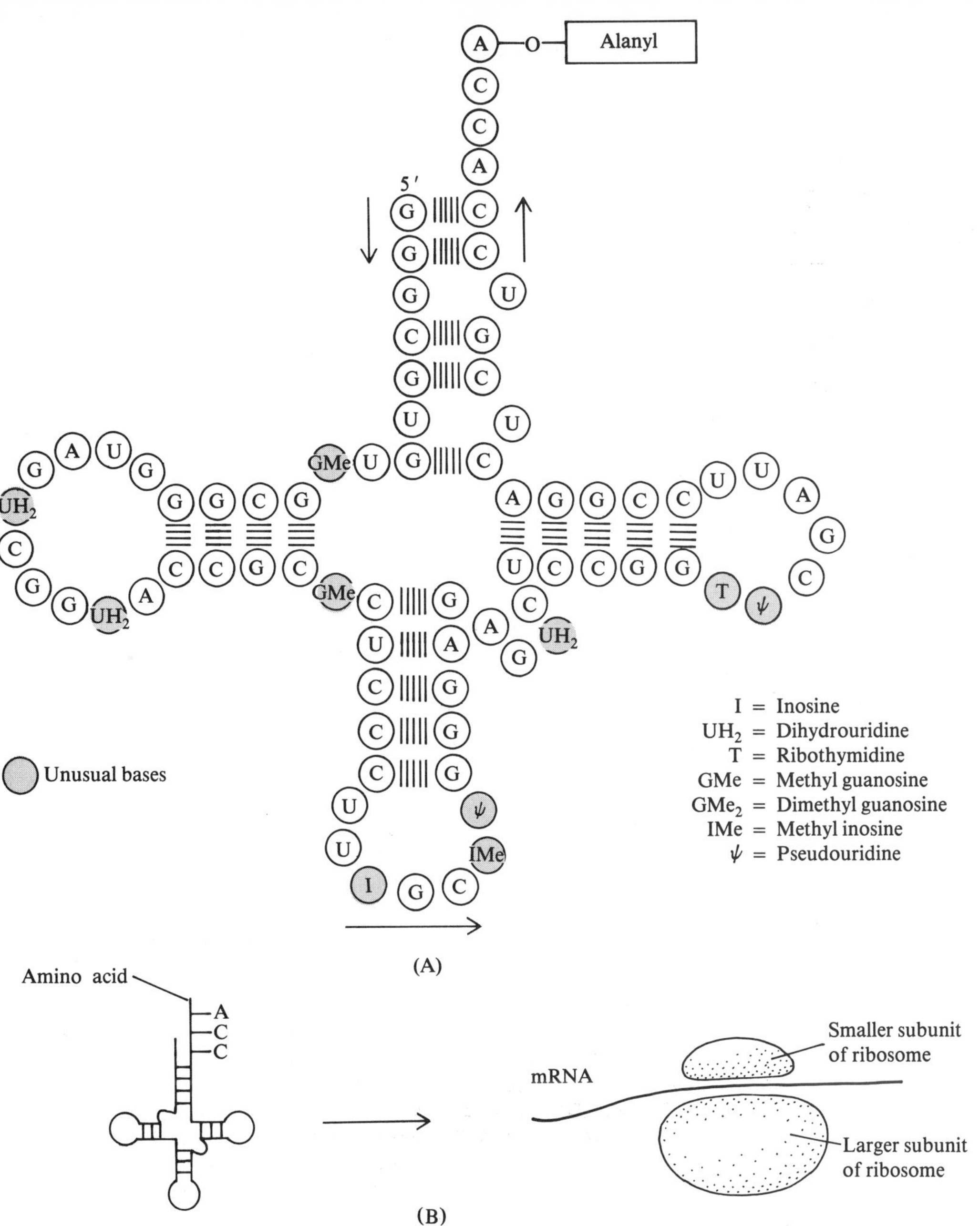

Figure 9-12. Transfer RNA. (A) Shown here is the complete structure of the tRNA which is specific for the amino acid alanine. Transfer RNA is a single polynucleotide chain. It contains some unusual nitrogen bases which are derived from the more familiar ones found in nucleic acids. (B) Any tRNA joins with its specific amino acid at the end of the tRNA molecule which ends in the nucleotide sequence CCA. It then carries the amino acid to a ribosome which is complexed with messenger RNA. It is here that the amino acid will be inserted into a protein by forming a peptide linkage with another amino acid that has been transported to the ribosome by another tRNA.

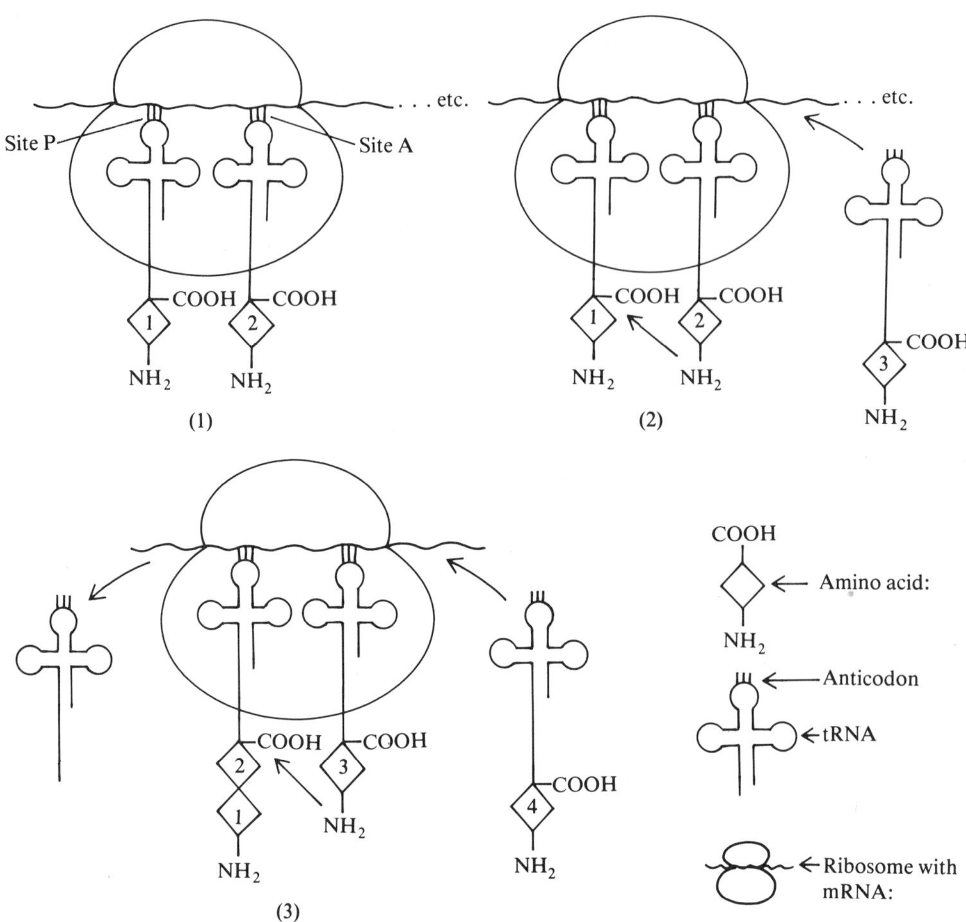

Figure 9-13. Some molecular interactions in protein synthesis. A ribosome has two sites, A and P, to which tRNA's may attach with their specific amino acids. A ribosome is associated with one mRNA. (However, as not indicated here, many ribosomes may be bound to the same mRNA at one time.) The first tRNA attaches to the P site of the ribosome, followed by the second tRNA which binds to the A site adjacent to the P site (1). The binding is governed by the mRNA, which contains a coded message specifying the sequence in which the amino acids are to be placed. Each tRNA has an anticodon, a coded nucleotide sequence that is complementary to a sequence in the mRNA. It is the code in the mRNA which determines which tRNA will attach to a ribosome site at a given time. A peptide linkage is formed between the COOH of amino acid 1 and the NH_2 group of amino acid 2 (2). A third tRNA is ready to bind to the ribosome-mRNA complex at the A site. After the peptide bond is formed (3), the tRNA which carried the first amino acid is ejected from the ribosome. The second tRNA moves to the P site previously occupied by the first one. The third tRNA now moves to the A site, and the process repeats, over and over, until the polypeptide chain is completed. Note that, except for the first tRNA, every tRNA moves from the A to the P site. Also note that the growing polypeptide has a free NH_2 at its beginning and a free COOH at its end.

nucleotides in the mRNA corresponds to a complementary sequence of nucleotides in a specific gene. This mRNA, therefore, is somewhat like a tape which is "read" by the ribosome and the tRNA. Although a tRNA molecule carries no complete message within it for amino acid sequence, each tRNA does have a small region, a sequence of three nucleotides known as the *anticodon,* which enables it to recognize a complementary region in the mRNA. The mRNA can therefore be pictured as a tape made up of segments (actually three adjacent nucleotides as we will see shortly), each segment designating a particular amino acid. Under the guidance of the ordered instructions in the mRNA, each tRNA gives up its specific amino acid (Fig. 9-13). As the amino acids, one at a time, leave their tRNA carriers, they are joined together through peptide links by the ribosome. Any one ribosome will continue to link together the amino acids into a polypeptide chain whose length will depend on the length of the information in the mRNA. This in turn is determined by the length of the DNA segment which engaged in transcription. Note (Fig. 9-13) that the first amino acid in the chain has an unattached or free amino (NH_2) group; the last one has a free carboxyl (COOH) group. This is so because the amino acids are joined by the reaction of the amino group of an amino acid with the carboxyl group of the proceeding one on the ribosome.

The ribosome, by its ability to form peptide linkages between amino acids, is a critical entity in protein synthesis. It is now known that these essential structures depend on the nucleolus of the cell for their formation. The RNA portion of a ribosome is transcribed from specific segments of DNA which are located on certain chromosomes in the cell, the nucleolar organizing chromosomes. These segments, the nucleolar organizing regions, are responsible for the reformation of the nucleolus at cell divisions and continue to lie in contact with the nucleolus in the nondividing nucleus. The RNA which arises as a transcript of the nucleolar organizing regions moves from the chromosomes into the nucleolus where it is processed into shorter pieces of RNA. These finally become complexed with protein to form the completed ribosomes of the cytoplasm. The nucleolus is therefore essential for the life of the cell since, lacking a nucleolus, no finished ribosomes can arise, and protein formation in the cytoplasm would cease.

The Genetic Code

The preceding summary of interactions between DNA, the types of RNA, the ribosomes, and the amino acids still leaves unanswered the question of the precise nature of the genetic code. It has been stressed in our discussion that the code resides in the nucleotide sequence. But what kind of sequence is involved and what does it mean? It is obvious that there could not be a 1-to-1 relationship between 1 kind of nucleotide and 1 kind of amino acid. There are only 4 different kinds of nucleotides in DNA (those with A, T, G, or C). The same is true for RNA, but T is replaced by U. And so any code based on single nucleotides allows the identification of only 4 of the 20 kinds of amino acids. If 2 nucleotides are taken at a time, they can be arranged in 16 possible ways, but this number is still insufficient for 20 amino acids. The hypothesis was proposed that 3 nucleotides taken at a time correspond to 1 kind of amino acid. An alphabet of 4 letters (nucleotides) taken 3 at a time permits 64 unique combinations. This is 44 more than is needed to identify 20 amino acids, but this idea of a *triplet code* seemed more likely than a code based on doublets.

A number of experimental works, some performed by Crick, indicated in the early 1960s that the code is indeed based on triplets. At this time, work was also in progress to decipher the code, that is, to determine what a particular triplet means or designates. Marshall Nirenberg was the pioneer in the perfection of test tube (*in vitro*) cell-free systems which he and his associates utilized to unravel the code. Nirenberg manufactured artificial mRNA whose exact nucleotide sequence he had dictated. This mRNA was then supplied to test tubes containing the ribosomes and all the other factors required for protein synthesis. To give just 1 example, a mRNA was prepared of nucleotides containing only uracil. This artificial messenger, polyuracil (U-U-U-U-U-etc.), was recognized by the ribosomes and by 1 kind of tRNA (Fig. 9-14). The final result was the production of a polypeptide containing only the amino acid, phenylalanine. This meant that *if* the code is indeed a triplet code, then the triplet UUU designates the amino acid, phenylalanine. Nirenberg refined his procedures to the point where he was able to manufacture all 64 kinds of triplets by themselves. He then used each individual species of triplet as a tiny piece of mRNA. A simple triplet composed of just the nucleotide sequence UUU would attract and bind the amino acid, phenylalanine, to a ribosome. In this way, all the triplets could be deciphered and related to a particular amino acid. In 1968, Nirenberg shared the Nobel Prize with Khorana and Holley, 2 other outstanding biochemical geneticists whose work facilitated deciphering the genetic code and supplied answers to the manner in which the gene determines the amino acid sequence of a protein.

At first it was debatable whether only 20 of the 64 triplets designate amino acids. As it turned out, 61, all but 3 of them, represent amino acids. The 3 which do not correspond to amino acids act as a sort of punctuation by providing "stop signals." Since genes follow one another along the length of the DNA, punctuation between adjacent genes is obviously required to set off or distinguish 1 gene from the following gene. When transcription occurs, the punctuation helps to insure that only those genes will be transcribed whose activities are needed at a given time. With no punctuation, all genes could be transcribed at 1 time, the messages all running on together. If, on the other hand, a stop signal were to occur somewhere *within* a gene, this could become reflected in the mRNA formed from it. When the mRNA is being read by the ribosome, the formation of peptide linkages would come to a halt, and an incomplete polypeptide chain would result. Mutations are known to arise in lower species which produce stop signals within a gene by causing the origin of a chain-terminating or punctuation triplet. These mutations are commonly referred to as *nonsense mutations,* and the punctuation or chain-terminating triplets are called *nonsense triplets,* meaning that they do not stand for amino acids. Nonsense mutations can account for some human disorders, as will be noted in later chapters.

The fact that 61 triplets designate amino acids means that 1 amino acid can be recognized by more than 1 kind of triplet. Table 9-1 lists the triplets and the amino acids they represent. A triplet or specific sequence of 3 nucleotides which corresponds to an amino acid or to a punctuation signal is referred to as a *codon.* Table 9-1 indicates that there are both DNA and RNA codons or code words. This is to be expected, since a codon in the mRNA has been formed under the direction of a complementary codon in the DNA acting as the template. The RNA codon, UUU, for example, was formed on the template

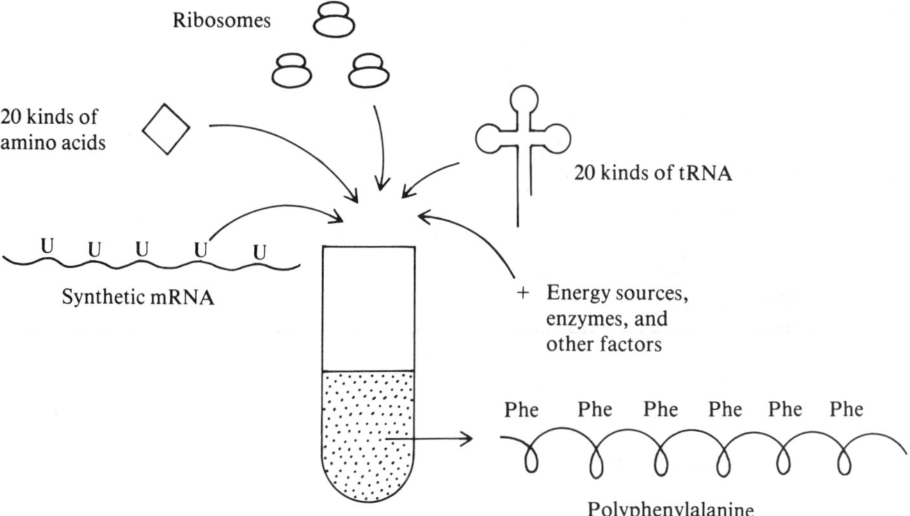

Figure 9-14. Deciphering the genetic code. Test tube systems are employed which are free of cells but which contain all the factors required for protein synthesis. Critical to the process is the preparation of a synthetic mRNA whose nucleotide content is known. This is achieved using an enzyme which links together separate nucleotides that are supplied to it. When a synthetic mRNA composed of nucleotides containing only uracil is added to an *in vitro* system as indicated here, a simple polypeptide is formed which contains only units of the amino acid phenylalanine linked together. Since none of the other amino acids in the test tube were incorporated, this indicates that UUU in RNA language prescribes phenylalanine if the code is a triplet code.

sequence AAA, and so on for all the others. The fact that 61 code words stand for amino acids does not mean that the genetic dictionary is imprecise or inexact. It simply means that 2 or more triplets or codons may designate the very same amino acid. The RNA codon, UUU, for example, always represents phenylalanine, never any other amino acid. Another specific RNA triplet, UUC, also designates phenylalanine, but like UUU, it never represents any other amino acid. In the chapter on mutation (Chapter 13), we shall see that there is an advantage to such a system where 1 or more specific codons designate 1 specific amino acid.

Table 9-1. Assignment of DNA and RNA codons to specific amino acids

RNA	DNA		RNA	DNA		RNA	DNA		RNA	DNA	
UUU	AAA	Phe	UCU	AGA	Ser	UAU	ATA	Tyr	UGU	ACA	Cys
UUC	AAG		UCC	AGG		UAC	ATG		UGC	ACG	
UUA	AAT	Leu	UCA	AGT		UAA	ATT	"Nonsense" (chain terminating)	UGA	ACT	"Nonsense"
UUG	AAC		UCG	AGC		UAG	ATC		UGG	ACC	Try
CUU	GAA	Leu	CCU	GGA	Pro	CAU	GTA	His	CGU	GCA	Arg
CUC	GAG		CCC	GGG		CAC	GTG		CGC	GCG	
CUA	GAT		CCA	GGT		CAA	GTT	Gln	CGA	GCT	
CUG	GAC		CCG	GGC		CAG	GTC		CGG	GCC	
AUU	TAA	Ileu	ACU	TGA	Thr	AAU	TTA	Asn	AGU	TCA	Ser
AUC	TAG		ACC	TGG		AAC	TTG		AGC	TCG	
AUA	TAT		ACA	TGT		AAA	TTT	Lys	AGA	TCT	Arg
AUG	TAC	Met	ACG	TGC		AAG	TTC		AGG	TCC	
GUU	CAA	Val	GCU	CGA	Ala	GAU	CTA	Asp	GGU	CCA	Gly
GUC	CAG		GCC	CGG		GAC	CTG		GGC	CCG	
GUA	CAT		GCA	CGT		GAA	CTT	Glu	GGA	CCT	
GUG	CAC		GCG	CGC		GAG	CTC		GGG	CCC	

Colinear Molecules and Their Significance for the Human

Another important piece of work was announced which showed a precise relationship between the DNA, the mRNA transcribed from it, and the specific protein whose structure they determine. As can be noted from the discussion of the structures of RNA, DNA, and protein, the three classes of molecules are essentially linear or in the form of chains. The two chains of the DNA, composed of DNA nucleotides, are twisted to form a double helix, but they are nevertheless linear in nature. When a stretch of the DNA unwinds to undergo transcription, a linear, straight chain of mRNA is formed composed of RNA nucleotides linked together. The mRNA is straight and unfolded as it is read (translated) by the ribosome. This is essential, for the codons must be exposed in the mRNA if they are to interact in any way with the ribosome and the tRNA. The mRNA undergoes translation into protein or a polypeptide, a molecule which can be very large and which can assume complex configurations as it folds upon itself and interacts with other chains or molecules. Regardless of such complexities, any protein is still essentially composed of polypeptides and is linear in nature. It can be "unwrapped" and shown to consist of linked amino acids. The relationship among the three kinds of linear molecules, DNA, mRNA, and protein, is now clear (Fig. 9-15), for it has been demonstrated that the position of a codon in the DNA corresponds to the position of a specific codon in the mRNA. This triplet in turn corresponds to the specific location of an amino acid in a polypeptide or protein. Another way to express the same thing is to say that the DNA, mRNA, and polypeptide molecules are *colinear.* A position in one of them can be related to the same relative position in the other two.

Most of the information on transcription, translation of mRNA into protein, and the relationships among the nucleic acids and proteins in the cell has been known only since the latter part of the 1950s. The wealth of knowledge is responsible for the progress being made in an understanding of the ways genes

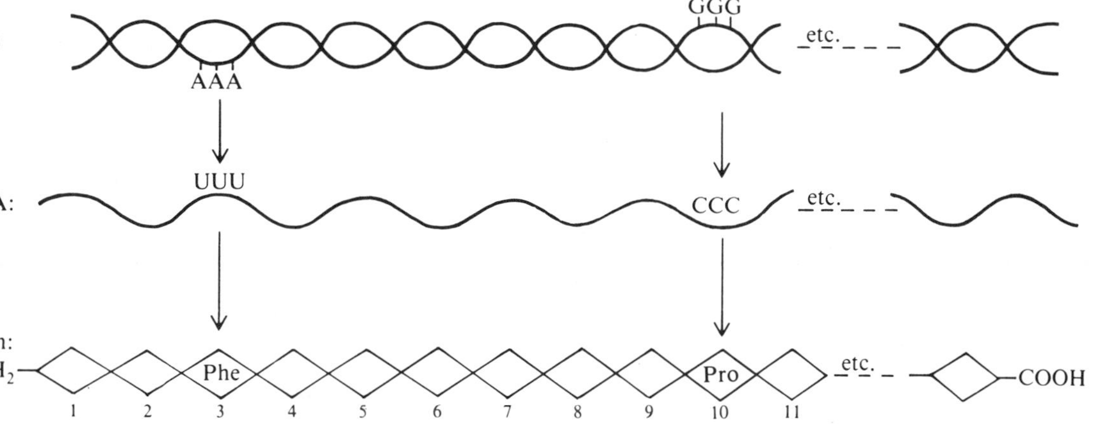

Figure 9-15. Colinearity. The position of a triplet in a strand of DNA corresponds to the position of a complementary triplet in the mRNA which is transcribed from it. This RNA triplet in turn corresponds to a related position of an amino acid in the polypeptide which is translated from that mRNA. We see here that a triplet near the beginning of the gene, and hence of the mRNA, corresponds to an amino acid (phenylalanine in this example) near the beginning of the polypeptide. A triplet farther away from the beginning of the gene and of the mRNA corresponds to an amino acid (proline here) farther from the NH_2 end of the polypeptide. Consequently, a gene mutation, a change in a DNA triplet, will be reflected as a change in a triplet in the mRNA and an amino acid substitution at a corresponding position in the polypeptide or protein translated from the mRNA. The three macromolecules, DNA, RNA, and protein, are colinear.

act in all forms of life. The importance to human genetics cannot be overemphasized. Now that we appreciate what constitutes the normal activities of a gene, we are in a position to attack a problem involving the appearance of a derangement or abnormality. In the human, the molecular basis of several disorders is well understood. With this knowledge, steps have been undertaken to alleviate and cure some of these afflictions. We will learn in Chapter 14 that some genes exert their control on the cell, not by governing the formation of structural protein in the cell, but by controlling the time and rate of activity of other genes. Before proceeding to this type of genetic control, which is important to an understanding of cancer and the aging process, we will first become familiar with the molecular nature of some hereditary disorders which involve changes in proteins. In a few cases, even the possible codon changes causing a disturbance have been narrowed down. If the foregoing summary of the extensive number of facts about molecular interactions has been grasped, the precise nature of some human afflictions can be better appreciated, as well as the problems and approaches related to their prevention and cure.

REFERENCES

Avery, O. T., MacCleod, C. M., and McCarty, M., "Studies on the Chemical Nature of the Substance Inducing Transformation of Pneumococcal Types." *J. Expt. Med.,* 70:137, 1944. Reprinted in *The Biological Perspective, Introductory Readings.,* Laetsch, W.M. (Ed.), pp. 105–125. Little, Brown & Co., Boston, 1969.

Clark, B. F. C., and Marcker, K. A., "How Proteins Start." *Sci. Amer.,* Jan: 36, 1968.

Crick, F. H. C., "The Genetic Code." *Sci. Amer.,* Oct: 66, 1962.

Crick, F. H. C., "The Genetic Code III." *Sci. Amer.,* Oct: 55, 1966.

Garen, A., "Sense and Nonsense in the Genetic Code." *Science,* 160:149, 1968.

Hartman, P. E., and Suskind, S. R., *Gene Action,* Second edition. Foundations of Modern Genetics Series. Prentice-Hall, Englewood Cliffs, N. J., 1969.

Kornberg, A., "The Synthesis of DNA." *Sci. Amer.,* Oct: 64, 1968.

Loewy, A. G., and Siekewitz, P., *Cell Structure and Function,* Second edition. Holt, Rinehart and Winston, N. Y., 1970.

Mechanism of Protein Synthesis. Cold Spring Harbor Symp. Quant. Biol., Vol. 34, 1969.

Miller, O. L., Jr., "The Visualization of Genes in Action." *Sci. Amer.,* March: 34, 1973.

Mirsky, A. E., "The Discovery of DNA." *Sci. Amer.,* June: 78, 1968.

Nirenberg, M., "The Genetic Code II." *Sci. Amer.,* March: 80, 1963.

Nomura, M. N., "Ribosomes." *Sci. Amer.,* Oct: 28, 1969.

Spiegelman, S., "Hybrid Nucleic Acids." *Sci. Amer.,* May: 48, 1964.

Tomasz, A. T., "Cellular Factors in Genetic Transformation." *Sci. Amer.,* Jan: 38, 1969.

Transcription of Genetic Material. Cold Spring Harbor Symp. Quant. Biol., Vol. 35, 1970.

Watson, J. D., *The Double Helix.* Atheneum Publishers, N. Y., 1968.

Watson, J. D., *Molecular Biology of the Gene.* W. A. Benjamin, Menlo Park, Calif., 1976.

Watson, J. D., and Crick, F. H. C., "Genetical Implications of the Structure of Deoxyribonucleic Acid." *Nature,* Lond., 171, 1964. Reprinted in *Papers on Bacterial Genetics,* Second edition, Adelberg, E. A. (Ed.)., pp. 127–131. Little, Brown, Boston, 1966.

Yanofsky, C., "Gene Structure and Protein Structure." *Sci. Amer.,* May: 80, 1967.

REVIEW QUESTIONS

1. The DNA molecule has the form of a double helix. Assume that one of the DNA strands contains a segment with nucleotides arranged as follows:

T-A-G-T-A-T-C-A-T-A-T-G

A. Give the sequence of bases in the corresponding region of the DNA strand which is paired with the above strand. *B.* Assume that the first-mentioned strand undergoes transcription. What would be the base sequence in the segment of the mRNA transcribed from this strand? *C.* How many amino acids would this segment code for? *D.* How many separate tRNA's are needed to translate this segment? *E.* What possible anticodons could designate the amino acids whose sequence in a polypeptide is determined by this segment? *F.* Assume that the second codon in this segment becomes changed to a nonsense codon following gene mutation. Briefly state the effect on translation.

2. Select the correct answer(s). In DNA: (1) The sugar component contains 1 more oxygen than the sugar found in RNA. (2) The sugar component exists as a 5-membered ring. (3) The purine and pyrimidine bases are attached to the sugar and not directly to the phosphates. (4) Carbon 3 and carbon 5 of the sugar are involved in internucleotide linkages. (5) The 2 polynucleotide chains are held together by peptide linkages.

3. Select the correct answer(s). There is evidence to support the following about ribosomes: (1) They carry within them the coded instructions for the amino acid sequence of a polypeptide chain. (2) There are at least 20 different kinds of ribosomes, 1 for each type of amino acid. (3) They are composed of RNA and protein. (4) Complete ribosomes cannot be formed in the absence of a nucleolus in a higher cell. (5) One ribosome normally reads more than one mRNA strand at a time.

4. Select the correct answer(s) about transfer RNA: (1) It is composed of two separate strands wrapped around each other in a double helix. (2) The tRNA molecules are able to form peptide linkages. (3) One kind of tRNA usually recognizes several kinds of amino acids. (4) Only one tRNA can be associated with a ribosome at any time during translation. (5) All tRNA's contain an anticodon.

5. The following is (are) true about the genetic code: (1) One triplet normally codes for more than one kind of amino acid. (2) Some triplets do not code for amino acids. (3) More than one triplet may code for a given kind of amino acid. (4) A change in a codon at the beginning of a gene will be reflected as an amino acid change at the beginning of a polypeptide. (5) There are DNA codons and RNA codons.

6. Give correct short answers for each of the following: *A*. Give an RNA codon which represents phenylalanine. *B*. Name the kinds of purine bases found in DNA and in RNA. *C*. Name the kinds of pyrimidine bases in DNA and in RNA. *D*. Name three major components of any nucleotide. *E*. Name two types of RNA which are not translated. *F*. Give three features which the amino acids have in common. *G*. In a polypeptide, how can the last amino acid in the chain be recognized?

chapter 10

GENES, MOLECULES, AND DISEASE

Genetic Blocks and Human Disorders

The recognition of the genetic control of enzymes stimulated research into the chemistry of several human disorders. As mentioned in Chapter 9, it was Garrod in 1908 who was the first to offer the suggestion that defective enzymes in the human might be responsible for certain metabolic upsets associated with various inherited conditions. However, it was not until 1952 that a specific enzyme defect was actually connected with a human genetic affliction. The particular disorder is one of a group which has been given the designation *glycogen storage disease.* As the name implies, victims of these disorders are unable to utilize glycogen, the form in which sugars are stored in the body. G. T. and C. F. Cori demonstrated that the activity of a specific enzyme, glucose-6-phosphatase, is almost completely absent from the liver and kidney cells of persons afflicted with 1 form of glycogen storage disease. This form (Gierke's) is inherited as an autosomal recessive and occurs in about 1 in 20,000 children. Figure 10-1 shows steps in the breakdown of glycogen. This specific enzyme is required for the last step in the reaction series. When this step is interrupted as a result of enzyme deficiency, glycogen breakdown is suppressed. As a consequence, deposits of glycogen accumulate in the liver and kidneys, the only two organs of the body where this enzyme is normally found. The accumulation leads to liver and kidney enlargement, as well as to an assortment of other effects. Among these is hypoglycemia, a reduction in level of the blood sugar, glucose. Figure 10-1 indicates that this could be expected, since normal amounts of glucose cannot form due to a blocking of the preceding step. A child's growth may become retarded as a result of this deficiency which also leads to upsets in other chemical pathways. Death may ensue from one or more of these.

In the very next year after the identification of the enzyme associated with this form of the glycogen storage disease, Jervis showed that the inherited disorder, phenylketonuria (PKU), is associated with a deficiency of the enzyme, phenylalanine hydroxylase, which normally is found in the liver where it develops shortly after birth. PKU, as noted in Chapter 1, is inherited as an autosomal recessive condition. Victims of the disorder are usually so seriously mentally retarded that they must be institutionalized. In the U.S., the frequency of babies born with the defect is approximately 1 in 10,000. Human populations differ in the incidence of the disease, the highest occurring among the northern Europeans and lower frequencies among blacks and Orientals (see Chapter 8 for possible reasons).

The biochemical basis of PKU is among the best understood of inherited human disorders. Figure 10-2(A) shows some of the steps in the metabolism of phenylalanine, a common amino acid. Phenylalanine is one of the so-called "essential" amino acids, meaning that the human body cannot form it from any other substances provided in the diet. It must be supplied ready-made as in the proteins of many common foods such as fish, eggs, and cheese. The ingested proteins are broken down in the body to their constituent amino acids. Some of the released phenylalanine is used in the construction of certain other proteins needed in various parts of the body. Most of it, however, is excess

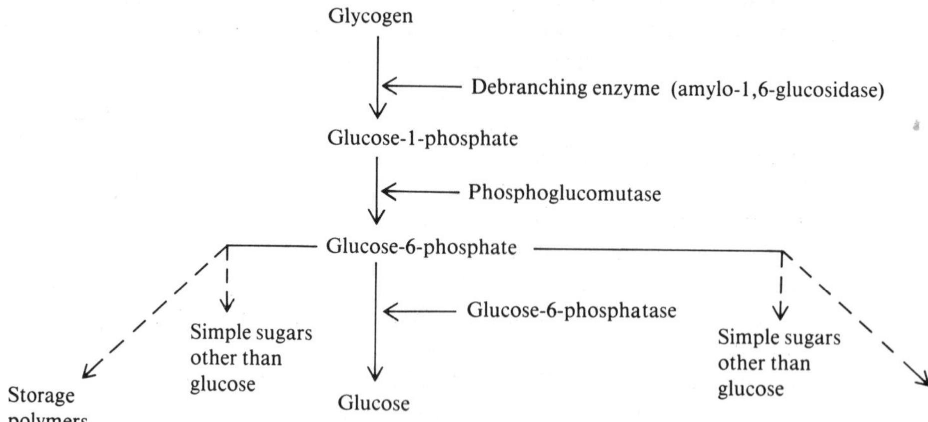

Figure 10-1. Summary of steps in glycogen breakdown. The figure shows enzymes only in the pathway leading to glucose formation. The enzyme, glucose-6-phosphatase, is normally present in the liver and kidney. In its absence due to a genetic defect, the final step leading to the formation of glucose cannot proceed. As a result, excessive glycogen deposits accumulate in these organs. Since liver glycogen is the body's main reserve of glucose, this defect also leads to hypoglycemia (decreased glucose level in the blood).

and is converted immediately to another amino acid, tyrosine. This step depends on the action of an enzyme, phenylalanine hydroxylase, found in the liver. Tyrosine, therefore, is not an essential amino acid, since it can be formed from another amino acid. It is also an amino acid constituent of many dietary proteins. Whether it is derived from phenylalanine or from dietary protein, tyrosine may also be used as an amino acid unit in the construction of body protein. Note, however, from Fig. 10-2(A) that tyrosine is a substance which is utilized in more than one pathway; among them are one involved in the formation of melanin pigment, one leading to the formation of the hormone thyroxine, and one leading through various intermediate steps which end in the complete breakdown to carbon dioxide and water. The bulk of the tyrosine actually enters this latter path.

Interruptions of the biochemical steps in pathways leading from phenylalanine and tyrosine may occur at various points, and many of these are well known. In a person with the double-recessive genotype, *pp,* there is a deficiency of the enzyme phenylalanine hydroxylase, and this lack interrupts the first step in the pathway leading from phenylalanine to its conversion product, tyrosine [Fig. 10-2(B)]. As the phenylalanine accumulates, the body diverts it into additional pathways, resulting in the formation of substances which normally do not reach high levels in the blood. Among these is phenylpyruvic acid. Excesses of both phenylpyruvic acid and phenylalanine can be detected in the blood and urine of PKU persons, and these provide the first clinical symptoms for detection of the condition.

The exact mechanism whereby the biochemical defect in PKU causes its serious effects is not completely understood. There is certainly sufficient tyrosine, despite the blocking of the step by which it is formed from phenylalanine. This is due to the fact that tyrosine enters the body in dietary proteins in amounts which are sufficient to accommodate other pathways. It is known, however, that some of the products to which the excess phenylalanine is converted can affect *other* pathways. The expression of phenylketonuria is pleiotropic; more than one phenotypic effect is usually detectable. Among these is a decrease in melanin pigment so that a PKU person is generally fairer in hair and skin color than other members of the family. It appears that excess phenylalanine and some of the products to which this excess is converted can inhibit one or more of the enzymes which convert tyrosine along the melanin pathway. We see here one way in which a single genetic defect can produce several metabolic disturbances. A primary block (such as the inability to form phenylalanine hydroxylase due to a defective gene) can result in the accumulation of substances or by-products which in turn can inhibit other chemical steps, even though the enzymes controlling the latter are normal. Interference with enzymes in still other pathways, such as those involved in the production of various hormones (thyroxine, epinephrine) may account for some of the nervous and behavioral disturbances often associated with PKU.

Accumulation of the phenylalanine and its by-products may operate in still

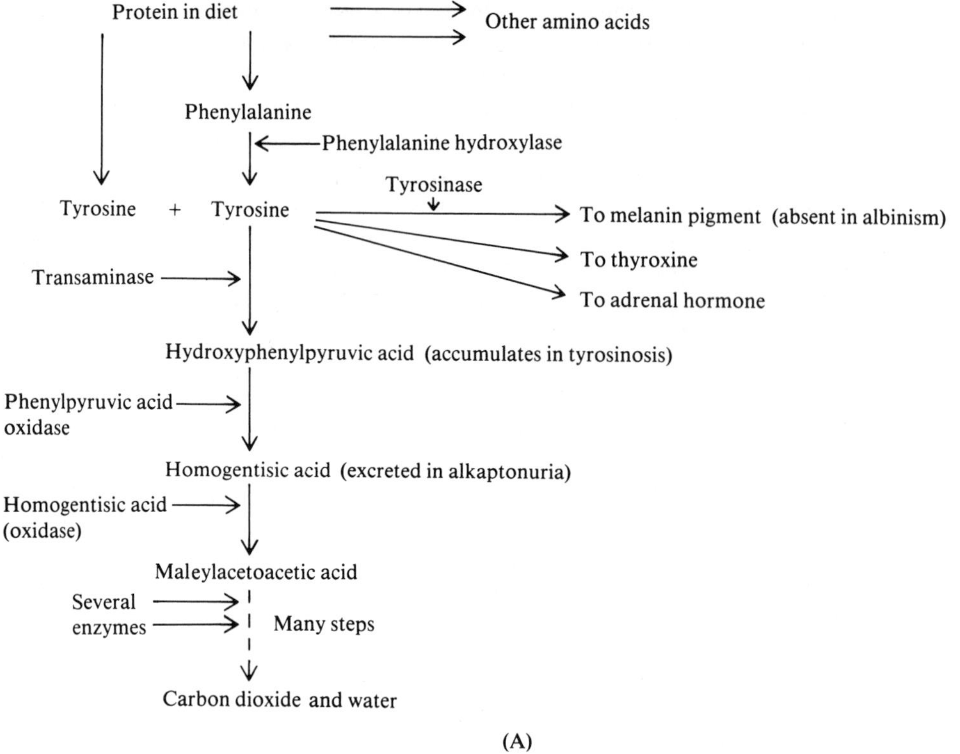

(A)

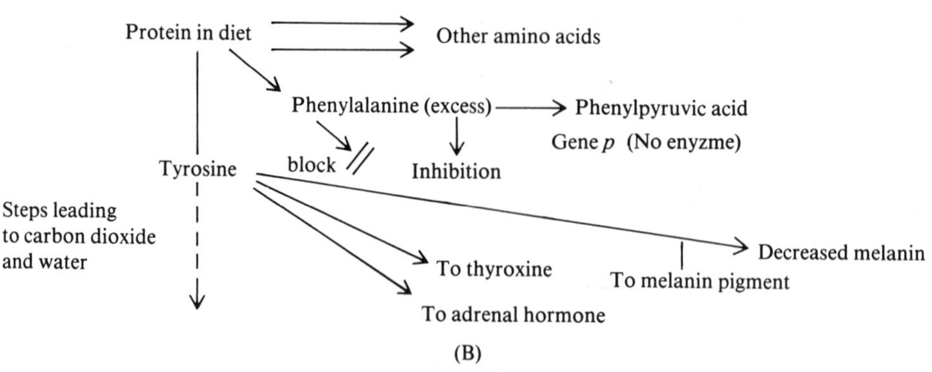

(B)

Figure 10-2.

other ways to produce deleterious effects, especially severe damage to brain cells. It has been shown that the *total* amount of all kinds of amino acids in the blood of PKU persons is about the same as that in a normal person, although in the former the amount of phenylalanine is about 30 times as high. This means that in a PKU victim there is more phenylalanine and less of the other kinds of amino acids, many of which may be essential to developing organs, in particular the very sensitive brain. In effect, the excess phenylalanine and its products may cause a decrease in the availability of vital substances essential to normal development of certain body parts. The assorted disturbances associated with PKU cause a decrease in the life expectancy; about 75% of the victims do not reach the age of 30.

Fortunately, certain steps can be initiated to offset many of the dire effects of PKU. As noted in Chapter 1, it is critical that the condition be detected shortly after birth. A delay can cause irreversible damage that no known treatment can correct. A newly born PKU child is normal, since excess phenylalanine is eliminated by the mother during intrauterine development. After birth, however, the baby must accomplish this elimination independently. Normally the enzyme phenylalanine hydroxylase is formed shortly after birth. In its absence, levels of phenylalanine and phenylpyruvic acid rise in the infant, and these can be detected by simple procedures which are now required by law in most states. Dietary regimens can then be instituted in which the amount of phenylalanine intake is restricted. A certain amount of this amino acid must be included, however, since small quantities of it are needed for the construction of some body proteins. Complete elimination of phenylalanine can retard growth and produce serious nervous disorders, an unfortunate result seen in some of the first cases that were treated.

Dietary restrictions on phenylalanine, when properly instituted, can permit children who otherwise may have become pathetically retarded to develop into persons of at least normal intelligence. There is a great deal of variation among treated PKU children in this respect, some having above average

---

Figure 10-2. Pathways in the metabolism of phenylalanine and tyrosine. (A) This simplified version of the pathways in tyrosine and phenylalanine metabolism indicates that many interrelated pathways are involved in normal utilization of substances within a cell. Each step in any pathway is controlled by an enzyme. Absence of a particular enzyme leads to failure of the step and results in an accumulation of a substance just behind the block. (B) Absence of gene *P* leads to absence of phenylalanine hydroxylase and a genetic block which results in a decreased amount of tyrosine and an accumulation of phenylalanine. Some of this excess is converted to other substances such as phenylpyruvic acid. The excess phenylalanine may also inhibit other pathways, as seen in its effect on an enzyme leading to melanin pigment formation. Still other pathways may be disrupted by an accumulation of the phenylalanine as well as by some of the substances to which it may be converted.

intelligence, others below. This might be expected even if the treatment were completely successful due to the complex of factors which may influence intelligence independently of PKU. Some of the treated children of normal intelligence, however, exhibit certain abnormalities in behavior (see Chapter 15). Still to be completely explained is the fact that a few PKU persons, who have *never* been treated at all, seem to incur little or no mental damage! The reasons for such exceptions could be due to the presence of certain specific genetic factors, peculiar to particular families, which can interact along with the PKU recessive. Very possibly, several kinds of PKU alleles occur. Not all may represent the same mutation. Some of the alleles may control the formation of enzymes which retain some function, enough to keep phenylalanine below the toxic levels, even though this level is above normal. Various unknown environmental variables may also be able to offset many of the metabolic disturbances.

There has been debate as to when dietary restrictions may be lifted—an important matter, since the special diet is expensive and inconvenient. Certainly it is required until normal brain development has been completed to avoid brain cell damage from the excess phenylalanine. This is approximately until the age of 6 years. However, there is also evidence that high levels of phenylalanine can cause various behavioral problems, even in later years. This may stem from a continued interference with one or more essential biochemical pathways, as noted earlier in our discussion.

The story of phenylketonuria has many interesting facets which are very informative. It illustrates well how knowledge of a genetic defect at the molecular level can spare many who might otherwise have spent their lives as helpless mental defectives. We also see the many ways in which just one gene defect can cause many seemingly unrelated phenotypic effects as a result of branches in metabolic pathways and the influence of excess metabolic byproducts on other basically normal pathways. The role of the environment in gene expression and the importance of the time element are clearly illustrated. In addition, many facts appear puzzling, but their final understanding may open up valuable insights into improved treatment of PKU and even other kinds of genetic disturbances.

Interruptions can occur at other points in the pathways leading from phenylalanine, and when this is the case, all enzymes may be normal except for a single one which controls one specific step. [Refer to Fig. 10-2(A) in the following discussions.] Oculocutaneous albinism occurs when the individual is homozygous for a certain recessive gene (genotype *aa*). This genotype results in a deficiency of the enzyme required to change DOPA (3, 4, dihydroxyphenylalanine) to a product essential for melanin formation. With a lack of the enzyme, no significant amounts of the black pigment can form. A block at a very different point may bring about a deficiency of the thyroid hormone and cause a form of genetic cretinism, a disorder encompassing retardation of mental and physical development.

Genetic blocks may occur at various points along the pathway leading to carbon dioxide and water. One of these interrupts a step which converts hydroxyphenylpyruvic acid. The accumulation of this substance, causing detectable amounts to appear in the urine, results in a rare disorder, tyrosinosis, which requires no treatment. The hydroxyphenylpyruvic acid is normally changed by an enzyme to another substance, homogentisic acid. In turn, this is converted by enzyme action to still another substance, maleylacetoacetic acid. Failure of this step leads to accumulation of homogentisic acid and the development of the rare clinical condition, *alkaptonuria.* The historical significance of this disorder has been referred to several times in our discussions. Garrod, in the early part of this century, analyzed biochemical findings on alkaptonuria and proposed that the genetic material controls steps in metabolic reactions and is thus responsible for certain biochemical defects. His conclusion that alkaptonuria is a recessively inherited biochemical disorder made it the first of this kind to be associated with a Mendelian pattern of inheritance.

Only about one person in a million has alkaptonuria. Infants homozygous for the responsible recessive gene can be easily identified, since the urine, upon exposure to air, turns dark brown or black. This striking effect results from the excretion of accumulated homogentisic acid in the urine. This acid oxidizes upon standing, producing a product similar to melanin. Alkali intensifies the darkening effect; consequently, washing diapers with soap makes stains deeper rather than removing them. The accumulation of the homogentisic acid in the body leads to formation of pigment which becomes deposited in cartilage and connective tissue, discoloring them. The eye whites and portions of the ear may become gray or blue-black. In later years, a distinct form of arthritis develops in association with the connective tissues. The joints of the spine, hip, and shoulders are typically affected. Alkaptonuria may have very serious consequences, since pigment may become deposited in the arteries and even in the valves of the heart. Fatality may result in such cases. The pleiotropic effects of the homozygous recessive genotype for alkaptonuria stem from just one genetic block. Garrod's prediction of such a genetic control was upheld by the discovery in 1958 of the enzyme, homogentisic acid oxidase, which carries out this step in normal persons but which is absent in those with alkaptonuria.

Our discussion so far has focused mainly on the metabolism of the amino acids phenylalanine and tyrosine. The steps shown in Fig. 10-2(A) are but a summary, and blockages at still other points are known. However, the examples presented above illustrate several important points which are relevant to other chains of reactions in the body involving the metabolism of very different substances. The story of the conversions in the phenylalanine-tyrosine pathways clearly demonstrates that most metabolic steps are genetically controlled and that this is achieved by the relationship between a gene and a specific enzyme. In Chapter 9, it was noted that a change at just one point in the DNA can bring about a change in an amino acid at a specific location in a protein. So the genetic change (a mutation) can produce a defective protein or enzyme which is unable to carry out a specific metabolic step. It may even result in the complete absence of the enzyme if the mutation changes a DNA codon to a stop signal. Only a fragment of an enzyme may be produced, but still a step is interrupted, and this blockage may exert influences on other steps in one or more ways. We have just seen that an accumulation of some intermediary substance may be responsible for an observed effect. In alkaptonuria, the accumulation of the excess homogentisic acid due to a blocking of the step in its conversion, leads to pigment deposition. An accumulation of some substance, such as phenylalanine, may have a direct toxic effect on developing cells or tissues. The accumulation (also seen in the example of phenylalanine) may cause the excess production of some other substance (phenylpyruvic acid, for example) which is usually produced in very small amounts, if at all. The second substance may also exert a toxic effect. Moreover, the accumulation of a product may in turn inhibit reactions in other pathways (phenylalanine excess inhibits an enzyme converting tyrosine). Pleiotropy can result from the accumulation of products, since they can influence chemical steps in one or more pathways.

A genetic block may cause its most pronounced effect due to the absence of a product rather than the accumulation of some substance. Albinism is a good example of this. Lacking the product needed for melanin formation, the black pigment cannot form. The metabolism of phenylalanine and tyrosine demonstrates clearly that very different disorders (albinism, PKU, alkaptonuria, cretinism) may be associated with closely related biochemical steps linked together in branched and unbranched chains of reaction.

Another inherited disorder known to result from the deficiency of an enzyme is Tay-Sachs disease (ganglioside lipidosis). This serious defect is receiving more publicity today as methods have become perfected for the detection of both carriers and of afflicted offspring *in utero.* Tay-Sachs is a fatal disease caused by the lack of an enzyme, hexoseaminodase A, required for the proper utilization of fats. Fatty deposits accumulate in cells of the central nervous system leading to their destruction. Tay-Sachs is another example of a genetic disorder in which the pathological effect stems from the absence of a product (the enzyme) and the consequent accumulation of materials (the improperly utilized fatty substances) which cause cellular degeneration by building up deposits. Afflicted children usually die before the age of 5 after experiencing severe nervous degeneration and suffering. The responsible recessive gene (*t*) causes a reduction in the level of the required enzyme. The heterozygous carrier (*Tt*) appears phenotypically normal but can be detected by chemical tests which show about one-half the enzyme level present in persons homozygous for the normal gene (*TT*). The heterozygote, however, has a sufficient amount of hexoseaminodase A to lead a normal existence. The incidence of the disorder is highest among Jewish persons whose ancestry traces back to eastern European populations. The possible significance of this is discussed in Chapter 8.

All the molecular disorders presented so far in this chapter are due to defective genes present on autosomes. However, the molecular basis is also known for certain defects caused by sex-linked genes. Actually, we became familiar with some of these in Chapter 4 in the discussion of blood clotting. The defects in hemophilia A and in Christmas disease result from deficiencies in the proteins AHF and Christmas factor (Fig. 4-11). These substances have enzymatic properties and act at specific points in the complex series of reactions terminating in blood clotting. The loci of the genes controlling these two required substances are located on the X chromosome.

An X-linked recessive gene which results in very bizarre effects is the one responsible for the Lesch-Nyhan syndrome. These babies exhibit an exceptional type of behavior which is self-destructive. Though irritability may be the only sign a few months after birth, by the age of 2 a severe nervous disorder is expressed. These infants are spastic and destructive; they grind their teeth and bite and chew their lips and fingers. Kidney damage, as well as nervous damage, terminates in an early death. Biochemical tests have shown that these babies are deficient in an enzyme which is involved in purine metabolism. All victims of this disorder are males, having inherited the defective gene from their mothers. No female victims can arise, since no male with the disorder survives beyond childhood. Consequently, it is impossible for any XX zygote to be homozygous for the recessive gene, since the gene cannot be transmitted by a male of reproductive age. (See Chapter 15 for more details of this disorder and its relation to behavioral studies.)

Mutant Genes and Hemoglobins

In 1957, an important discovery was made in human genetics which held great import for an understanding of gene action in all species. Dr. Vernon Ingram announced in that year his discovery of the difference between normal hemoglobin (hemoglobin A) and sickle cell hemoglobin (hemoglobin S). Sickle cell anemia is a fatal disease whose syndrome of effects is described in Chapter 1. This severe disorder arises when a person is homozygous for the gene responsible for the production of hemoglobin S (genotype Hb^S Hb^S). The gene acts as a codominant with its normal allele, Hb^A. Ingram was able to show that normal hemoglobin and sickle cell hemoglobin are identical to each other except for a single amino acid (Fig. 10-3). In any protein, it will be recalled from Chapter 9, the first amino acid is the one with the free—NH_2 group. This is free because it is not involved in a peptide bond. In normal hemoglobin A, position 6 in one of the polypeptide chains of the hemoglobin protein is occupied by the amino acid, glutamic acid. In the abnormal hemoglobin S, the amino acid, valine, is found at this position in the polypeptide chain. Ingram was able to determine this by a process known as *fingerprinting* (Fig. 10-4). In this technique, the hemoglobin (or any protein being studied) is broken into fragments by enzymes which can attack proteins. These protein fragments are then separated on a paper that is exposed to an electric current. The different fragments travel at different rates in the electric field due to the fact that they bear different electric charges. They may be further separated by being exposed to a flow of a chemical solvent in a direction at right angles to the direction of the current. Every kind of protein gives a characteristic picture or "fingerprint" when such a procedure is performed. Ingram found that a comparison of the fingerprints of hemoglobins A and S consistently shows a difference in just one fragment (Fig. 10-5). He was able to identify the sequence of amino acids in the various small fragments and to piece together the entire sequence of amino acids in normal and sickle cell hemoglobins. The only difference was the substitution at position 6 of valine in sickle cell hemoglobin for the glutamic acid in the normal hemoglobin A. As we will see a little later in this chapter, any one type of hemoglobin is actually composed of two different kinds of polypeptide chains. The substitution in sickle cell anemia occurs in the β chain of adult hemoglobin.

This discovery was significant for several reasons. It was the first demonstration that a gene mutation, a change at a point in the DNA, brings about a change in just one amino acid in a protein chain. The human was thus the first creature in which the exact relationship between the DNA and a protein was illustrated. Subsequent work with other organisms has verified the fact that a change in the DNA can be related to a change in an amino acid at a specific location in a polypeptide chain.

When the genetic code was finally deciphered in the 1960s, it then became possible to relate specific amino acids to specific codons in the DNA. Reference to Table 9-1 shows that only two DNA codons represent glutamic acid, CTT

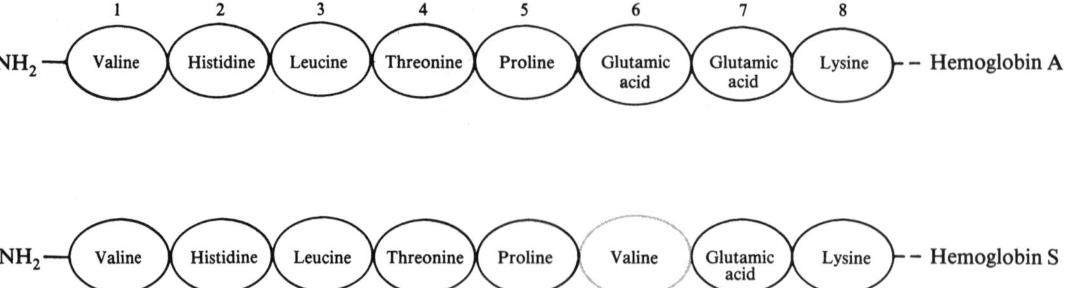

Figure 10-3. Hemoglobins A and S. The only difference between normal adult hemoglobin A and sickle cell hemoglobin (S) is an amino acid substitution at position 6 in one of the hemoglobin chains.

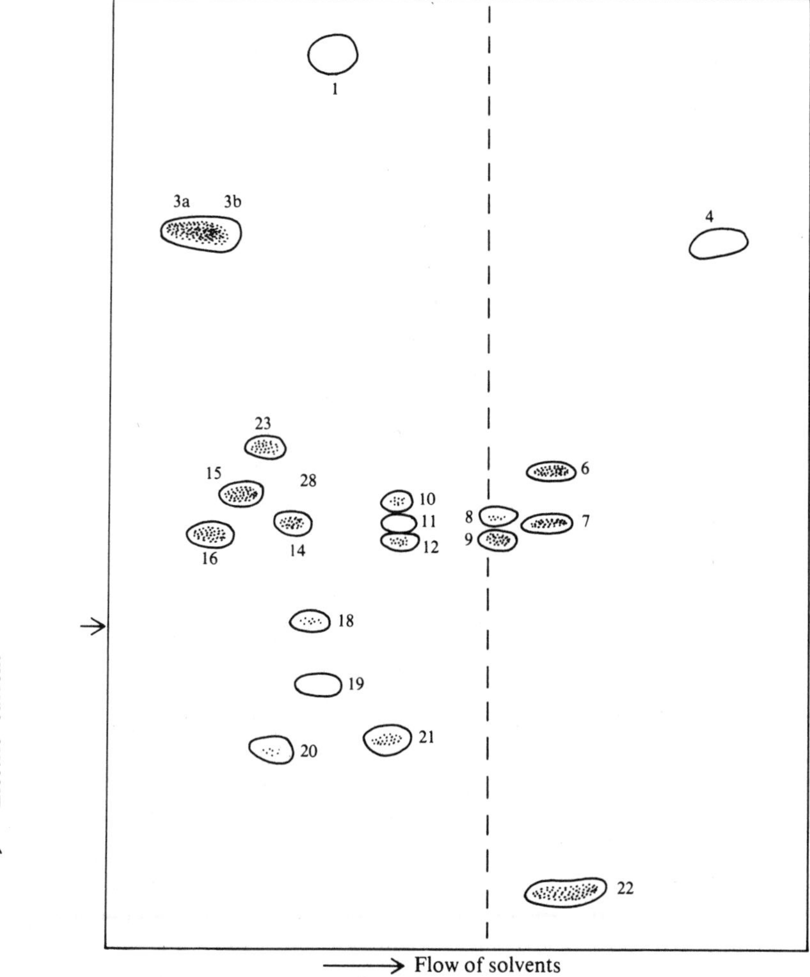

Figure 10-4. Fingerprinting. In this procedure, the protein to be analyzed is exposed to a protein-digesting enzyme, such as trypsin, in order to cleave it into smaller peptide fragments composed of just a few amino acids. The material is then applied to a spot (arrow) on a paper which is exposed to chromatography. When this is done, solvents are allowed to flow across the paper in one direction. This separates the fragments, since they will travel at different rates in a given solvent. After evaporation of the solvent, the paper is exposed to an electric current at right angles to the direction of the solvent flow. This achieves further separation of the fragments, since the fragments carry different electrical charges and will move at different rates in an electric field. The paper is then sprayed with a solution which stains the peptides. The stained regions may be cut from the paper and the peptides washed off each piece. The amino acid composition of a fragment can be determined by chemical analysis. The entire sequence of amino acids in a protein can be pieced together using the appropriate enzyme digestions, solvents, and stains.

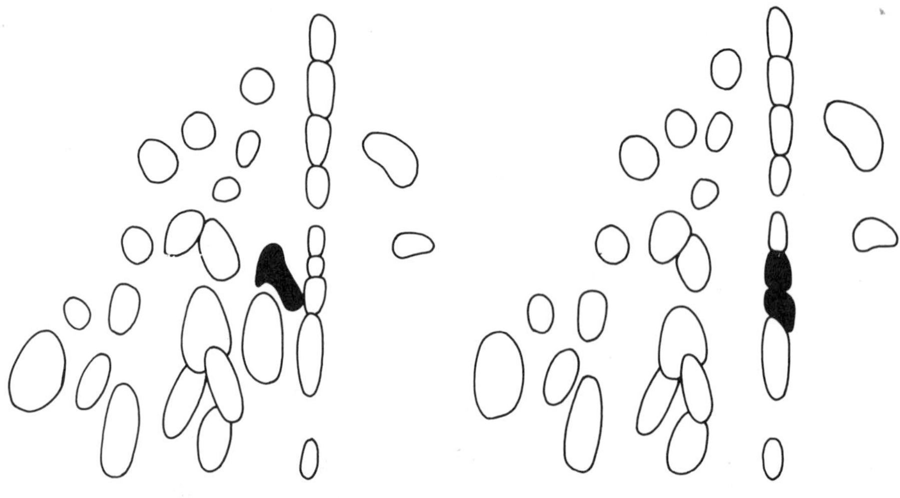

Figure 10-5. A comparison of the fingerprints of normal hemoglobin A and sickle cell hemoglobin shows them to be identical except for one fragment (shadowed). By piecing the fragments together, the difference was found to be the substitution of just a single amino acid, as indicated in Fig. 10-3.

and CTC. It is not known which of these two nucleotide triplets is actually the one in the hemoglobin gene which determines that glutamic acid will be placed at position 6 in the β chain. It can be seen (Fig. 10-6), however, that a change in one of the nucleotides in either of the two codons can give rise to a triplet which now specifies valine (CTT to CAT and also CTC to CAC). As will be presented in detail (Chapter 13), a gene mutation on the molecular level *is* just such a change in the DNA; one nucleotide replaces another and thus a new code word is formed. Noteworthy here is the fact that the syndrome of serious effects associated with sickle cell anemia can be related back to just two DNA codons, reducing to just two choices the exact alteration in the DNA which changed a normal gene to a defective one! This is a degree of precision in our understanding of the nature of a genetic disorder which would have been inconceivable 20 years ago.

The sickle cell story also illustrates that a single amino acid substitution can greatly alter the efficiency of a protein. Indeed, some proteins may become almost nonfunctional as a result of such a change. One reason for this is that a substituted amino acid may bear an electric charge which is very different from the one on the amino acid that is normally present in the protein. This change may alter the architecture or shape of the protein. The normal activity of most enzymes depends on a precise shape which permits them to interact with the substances in the reaction that is being catalyzed. A changed shape may prevent an enzyme from participating in the reaction which it governs to any extent. While hemoglobin is not precisely an enzyme, its efficiency in transporting oxygen in the red blood cells is undoubtedly dependent on the way the hemoglobin is arranged in the cell. A change in shape may decrease this efficiency. There is reason to believe that in sickle cell anemia the hemoglobin S molecules become stacked within the red blood cell, leading to changes which result in the formation of bizarre cell shapes when the oxygen level falls in the blood vessels. Aspects of the sickle cell story related to populations can be found in Chapter 8.

Sickle cell anemia is by no means the only disorder in which hemoglobin structure is altered. To appreciate the various types of diseases more fully, a little more must be said about normal hemoglobin. The hemoglobin molecule is not just one long polypeptide chain which becomes folded. Rather, each hemoglobin is composed of two different kinds of chains. Moreover, there are really different types of normal hemoglobin molecules which occur in the human (Table 10-1). In the adult, the commonest hemoglobin form, hemoglobin A, is composed of two different kinds of polypeptide chains, two identical alpha (α) and two identical beta (β) chains. These are fitted together along with non-protein components to form the hemoglobin A molecule. It is the β chain which is affected in sickle cell anemia.

Another type of hemoglobin, hemoglobin A_2, occurs in the red blood cells of an adult along with hemoglobin A. However, it makes up approximately only 1 part out of 40 of the total hemoglobin present in the adult. Hemoglobin A_2 is also composed of 4 polypeptide chains, 2 of the α variety associated with 2 of another kind, the delta (δ) type of chain. It can be represented as $\alpha_2\delta_2$, and the more abundant adult hemoglobin, hemoglobin A, as $\alpha_2\beta_2$.

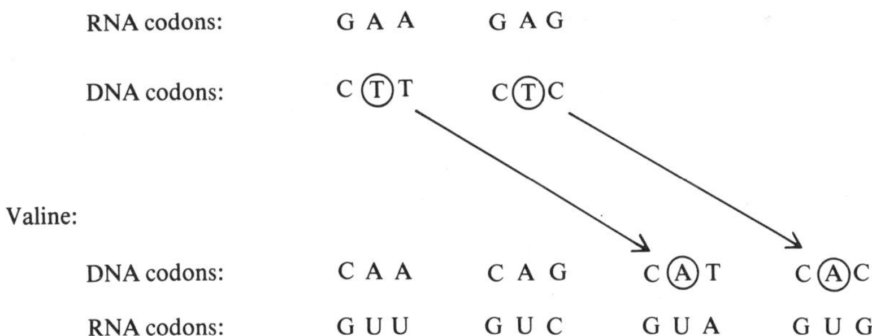

Figure 10-6. The exact codon change responsible for the production of sickle cell hemoglobin has been narrowed to just two, the two codons for glutamic acid. As shown here, a single nucleotide change at the second position in either one of them can give rise to a codon for valine. While the precise codon involved is unknown, it can be seen that the mutation certainly resulted in the replacement of T with A in the DNA codon which was affected.

Before birth, still other varieties of hemoglobin are present. Very early in the development of the embryo, the hemoglobin is composed of 2 α chains associated with 2 of another kind of polypeptide, 2 ϵ (epsilon) chains. This embryonic hemoglobin ($\alpha_2 \epsilon_2$) disappears after 12 weeks and is replaced by fetal hemoglobin, composed of 2 α chains and 2 gamma (γ) chains. This fetal hemoglobin ($\alpha_2 \gamma_2$) starts to decrease in amount before birth, as it is replaced by hemoglobins A ($\alpha_2 \beta_2$) and A_2 ($\alpha_2 \delta_2$). Normally, by 6 months of age, the fetal hemoglobin has almost disappeared. Table 10-1 summarizes the various normal hemoglobins. The β, γ, and δ chains of hemoglobin consist of 146 amino acids, whereas the α chain is composed of 141. The exact amino acid sequence is known for these chain types. The complete length and amino acid sequence of the ϵ chain of embryonic hemoglobin have not yet been fully determined.

Each kind of polypeptide chain is controlled by a different gene at a different chromosome locus. There is strong evidence that the gene locations of the β and δ chains are closely linked to each other on the same chromosome. Independent gene mutations can occur in the five genes which control formation of the different chains. These point mutations can in turn cause amino acid substitutions at various positions in the polypeptide chains. Many abnormal variants of the α and β chains are now known. In each of these forms, one amino acid difference is found between the variant hemoglobin chain and the normal one. Recall that the variant hemoglobin found in sickle cell anemia has a defect at position 6 of the β chain. The alpha chain is completely normal. There is another variant hemoglobin, called hemoglobin C, which is responsible for an anemia that is less severe than the sickle cell type. It is mentioned here because the amino acid substitution happens to affect the β chain at position 6, exactly the same position at which the variation occurs in hemoglobin S. The codominant gene responsible in this case causes a substitution by the amino acid, lysine, for the glutamic acid which is typical of normal β chains. This substitution alters the hemoglobin molecule, but no sickling occurs. We see from this that different changes at the same position in a protein can produce different degrees of inefficiency in the molecule. Actually, some protein changes in which an amino acid substitution occurs produce little or no effect. The hemoglobin molecule, however, appears in general to be adversely affected by any amino acid substitution.

Another very severe blood disorder is *thalassemia major*, sometimes called Cooley's anemia. This inherited condition is due to a variant form of the normal gene responsible for β chain production. Individuals homozygous for the aberrant gene develop an anemia in the first two years of life. The hemoglobin content becomes greatly reduced, and the red blood cells appear small and abnormal. A swollen abdomen, due to enlarged liver and spleen, is characteristic. This is accompanied by growth retardation and skin changes. Transfusions are required to offset the anemia. Death usually results from infection or hemorrhage. Blood analysis of victims of the disease shows that most of the hemoglobin is not the typical hemoglobin A but is of the fetal type, hemoglobin F ($\alpha_2\gamma_2$). The production of β chains has been greatly diminished. The gene responsible for *thalassemia major* is not a recessive strictly speaking, since the heterozygote may have a milder form of the disease, *thalassemia minor.* Others may show no apparent symptoms, but blood tests reveal the presence of a higher amount of hemoglobin A_2, meaning that more δ chains are being produced than in the normal person.

A very significant point in the story of the thalassemias is that the genetic change produces a decrease in the number of β chains, not a complete cessation of their production. Moreover, the β chains which are produced in the person homozygous for the defective gene are perfectly normal. Thalassemia is thus a *quantitative* rather than a *qualitative* disease. There is now evidence which makes it likely that the deficiency is due to insufficient amounts of that messenger RNA which undergoes translation and results in the formation of β polypeptide chains. The mutation involved here would thus appear to involve some sort of genetic change which influences gene regulation, a complex topic discussed at some length in Chapter 14.

In the person suffering from *thalassemia major,* there is a deficiency of adult hemoglobin A, as a consequence of the diminished amount of β chains. The amount of hemoglobin F which continues to be synthesized is not sufficient to compensate for this decrease. There is thus an accumulation of α chains. These tend to deteriorate, adding to the abnormalities of the red blood cell.

Actually, the term *thalassemia* is a general one, and we now know that there are various types of thalassemias due to the effects of different deleterious genes. The form of thalassemia just described is the one which has a high frequency among persons of Mediterranean origins. The frequency of heterozygotes among populations in this region may be as high as 15%. Again a relationship to the incidence of malaria seems to exist. More than one mutant gene form of the β gene seems to exist which can result in thalassemia. This means that there is probably a group of defective genes which, along with the normal one, comprise a multiple allelic series.

There is a less common genetic defect which can completely prevent the production of both the β and the δ chains. Hemoglobin F is found in place of any adult hemoglobin and is also present at an appreciable level in the heterozygote. There are even thalassemias which result from genetic defects which alter the α chain. These types of disorders have a higher incidence among populations of southeast Asia. Since the α chains may be almost completely absent, the embryo and fetus are severely affected because insufficient numbers of α chains are available to combine with ϵ and γ chains to form the prenatal hemoglobins. Consequently, there is a high risk of fetal death, more so than in other thalassemias. It is known that γ chains which are produced can, in the absence of α chains, associate with other γ chains to form molecules composed of four γ chains. The same pertains to β and δ chains in the absence of α chains.

The genetics of blood disorders has been shown to be exceedingly complex. Actually, the list of human hemoglobins in Table 10-1 is a bit of a simplification. Instead of five different types of polypeptide chains (and hence five different genetic loci), at least two more probably exist. In early embryos, another kind of hemoglobin has been recognized which is the only one that normally occurs without α chains. In this early embryonic form, two γ chains associate with still another kind of chain, the zeta (ζ) chain, under the control of still another genetic locus. Little is known about the chemistry of the ζ chain. There is also good evidence that there is more than one kind of γ chain and that two types of fetal hemoglobin, hemoglobin F, occur in all normal humans. Each of the two kinds of γ chains seems to be under the control of a different genetic locus. This means that a total of seven different genetic loci may be involved in the formation of normal human hemoglobin. We will continue to consider only the five which are most familiar. To complicate matters still further, many hemoglobin variants are known for all the major chains. It is likely that a multiple allelic series exists at each locus controlling the formation of a hemoglobin polypeptide chain. Blood genetics, while indeed a complex area of study, continues to produce information on gene action and the control of protein formation that is applicable to other systems.

Reference to Table 10-1 shows another remarkable feature about the normal hemoglobins. With the exception of the recently discovered early embryonic form, all possess α chains. The embryo does not have β or δ chains typical of the adult hemoglobin, but it has α ones in association with ϵ . These are then replaced, during intrauterine development, by γ chains which associate with the α to form the hemoglobin characteristic of the fetus.

Table 10-1. The major hemoglobin molecules which occur during the human life cycle are composed of four polypeptide chains. Two of these are α chains. The other two chains which may be associated with these are β , γ , δ , or ϵ . Two genes are thus involved for each kind of hemoglobin molecule.

Type of hemoglobin	Chains present	Controlling genes	Normal occurrence in red blood cell
Hb A ($\alpha_2\beta_2$)	$\alpha + \alpha + \beta + \beta$	α and β	Predominant component postnatally
Hb A_2 ($\alpha_2\delta_2$)	$\alpha + \alpha + \delta + \delta$	α and δ	Minor component postnatally
Hb F ($\alpha_2\gamma_2$)	$\alpha + \alpha + \gamma + \gamma$	α and γ	Early fetus until birth
Embryonic ($\alpha_2\epsilon_2$)	$\alpha + \alpha + \epsilon + \epsilon$	α and ϵ	Early embryo

Before birth, the γ chains begin to diminish as β chains are formed and hemoglobin A arises ($\alpha_2 \beta_2$). The δ chain type also arises so that, postnatally, the hemoglobin is of just two varieties, the major one, A ($\alpha_2 \beta_2$) and the minor component, A_2, ($\alpha_2 \delta_2$). We see here that some genetic mechanism must be operating at specific times in development to bring about the proper association of the α chains with the other chain types. Some regulatory device must exist to cause the disappearance of one kind of chain at a particular time and the synthesis of another kind. There must be some way in which expression of genes is controlled so that protein products are formed *only* when they are needed instead of being produced indiscriminately. This subject of control of gene expression will be a major topic in Chapter 14, which includes a discussion of genes whose function it is to regulate other genes.

Genetic Disease and Early Detection

The molecular natures of the genetic diseases discussed so far in this chapter are quite well understood. In each case, the presence of a defective gene has resulted in some detectable change in a specific protein, often an enzyme. Each of these disorders is inherited as a simple Mendelian trait; one gene or one genetic locus is primarily involved. This should not be interpreted to mean that *only* one gene influences the disorder. A disease, just as in the case of any other aspect of the phenotype, involves the interaction of several genes and the environment. There may be a primary gene responsible for the condition, such as the gene resulting in PKU, but other components of the genetic constitution may also play a role in its expression. Certainly the environment, as well illustrated by PKU, can alter the expression of a gene which could otherwise cause a severe malfunction. Knowledge of the primary molecular upset associated with a particular defective gene provides a direction from which the problem may be attacked. Steps can then be taken to permit early detection of the disorder or to help alleviate the unfortunate consequences it may entail.

An excellent illustration of the latter point is provided by an announcement from the National Institute of Health which reports some success in treating a rare genetic disease, Gaucher's disease. Approximately 1000 persons in the U.S. suffer from the disorder, which is classified as a lipid (fatty) storage disease. As in Tay-Sachs disease, there is an abnormal intracellular accumulation of fatty material due to the absence of a cellular enzyme. About nine different kinds of hereditary lipid storage diseases are known. In the recessively inherited Gaucher's disease, the genetic defect has been found to cause a lack of 1 of the enzymes needed in a chain of biochemical reactions leading to the normal breakdown of complex fatty substances. In the absence of the enzyme (glucocerebrosidase) there is an accumulation of the fatty material in the blood and certain other body tissues. An afflicted person suffers from an enlarged liver and spleen, weakened bones, bleeding, and anemia. Pain may be severe. Once the identity of the missing enzyme was discovered at the National Institute of Health, further research resulted in a method to obtain appreciable amounts of the normal enzyme. Sufficient quantities were produced to provide enough enzyme for single doses for 2 patients, an adolescent boy and a middle-aged woman. Following injections of the enzyme, the amount of fatty accumulations in the blood and liver of the patients dropped significantly. In just 1 day, the enzyme seems to have broken down and disposed of many years accumulation of fatty substances! While it is still unknown if this or further treatments will relieve crippling effects of the disease, the event is significant, since it demonstrates that an injected enzyme, when given to a person lacking it due to a genetic defect, can perform its normal role in the cells of the recipient. The results with the Gaucher's enzyme illustrate one practical use of knowledge of genetic disease at the molecular level and hold hope for the treatment of other genetic diseases whose molecular basis is understood. Unfortunately, the precise molecular defect caused by a specific gene is unknown in most cases. For example, the symptoms are well known for cystic fibrosis, one of the most serious diseases among American children and also a genetic disease with a high frequency (p. 201). However, the responsible recessive gene has not yet been related to a specific abnormality or biochemical block at the molecular level.

Not only is the primary molecular basis unknown for many Mendelian diseases, but some of them may also strike a victim with little or no warning. The fatal Huntington's disease is due to a relatively rare dominant gene, *H*. Affected persons, who occur in a frequency of 1 in 10,000, are heterozygotes, *Hh*. The presence of the dominant leads to progressive degeneration of nervous tissue and associated mental deterioration. The age of onset of the symptoms of the disease varies greatly, the average being 40, but they may appear as early as 15 or as late as 70. The disorder manifests itself in characteristic weaving movements of the body and twitching of the arms and legs along with facial grimacing. Death results after a number of years during which time the symptoms become more and more exaggerated. No cure has been devised for this fatal disorder. Perhaps most unfortunate is the fact that the heterozygote cannot be recognized for the presence of the dominant gene. An insight into the primary chemical upset caused by this gene could perhaps aid in early detection of heterozygous carriers and also provide some basis for the prevention of the fatal outcome.

An important point to bear in mind when discussing genetic disease is that a certain kind of disorder may have more than one kind of genetic basis. For example, muscular dystrophy is related to several different genetic loci. Among the more common forms of dystrophy is a sex-linked recessive type in which the symptoms of muscular weakness appear in early childhood. In an adolescent form of the disease, a dominant gene is responsible. Unfortunately, the exact primary biochemical defect in the several hereditary muscular dystrophies is unknown. However, certain enzymes are lost from muscle and appear in elevated levels in blood serum. In the X-linked recessive form, female carriers may show some mild symptoms, and a certain enzyme (creatine kinase) is found at higher than normal levels in their blood sera.

What seems to be the same condition or abnormality, as shown in the example of muscular dystrophy, may actually prove to stem from different gene defects and very different kinds of primary genetic blocks. What appears superficially to be one disease may actually encompass several whose primary defects are quite different, even though the clinical pictures of all may have many similarities.

Most of the disorders we have encountered in this and previous chapters have been the so-called "inborn errors of metabolism" and have been associated with one "main" gene with pronounced effects. Other rare conditions result from chromosome abnormalities (Chapter 11). Added together, all of these present a significant health problem, but each one is very uncommon. While a physician today must be aware of all of them, he may never see any one specific disorder during his entire practice!

However, certain more familiar disorders also carry a significant genetic component, but they do not show a simple Mendelian inheritance pattern. Instead, a polygenic basis is indicated. The inheritance of such a disorder would have the same kind of basis as the normal inheritance of skin pigmentation, height, etc., characteristics which show a continuous variation and which have no sharp, distinct traits (Chapter 8). No main gene may be responsible for the expression of the disorder, but rather several, each one adding a certain risk that the condition will develop. The effect of any one of these risk genes by itself may be small, but the effects are cumulative. A certain number of them may be required to reach a certain threshold, beyond which the condition can be manifested. Therefore, the effect of each risk gene can add to those of any other. The total may exert a pronounced influence on the phenotype. The severity of expression, however, will vary greatly from one person to the next, since individuals differ in the number of effective genes they carry and hence will vary in the degree to which they show the disorder. Moreover, the complexities of the environment, hormonal and external, can greatly influence this expression. An individual who is predisposed by his accumulation of risk genes may not express the potential effect if the environmental triggers are not there. On the other hand, a person with fewer risk genes, which just cause a slight push over the threshold, may suffer from pronounced disturbances due to exposure to an unfortunate set of environmental conditions. (Chapter 14 relates this matter to cancer expression.) While no definite conclusions can be reached, a polygenic basis has been suggested for cleft palate and lip. The degree of severity varies from one victim to the next and even among family members. This can be explained by differences in the number of effective genes each person carries plus differences in their environments, both before and after birth, both external and internal. That sex hormones may exert a pronounced influence on polygenic expression is seen by the fact that some disorders are as much as five times more common in one sex than in the other!

Perhaps the greatest number of afflictions of the human species have a polygenic basis. This likelihood emphasizes the magnitude of the importance of genetic factors in disease. Certain mental disorders may have a polygenic basis (Chapter 15). We will discuss at this point just three of the commonest physical disorders often attributed to the operation of many genes along with a strong environmental component. Hypertension is a chronic disease in which the systolic and diastolic blood pressures are elevated. It can develop in early or middle adult life and can persist over the course of many years. Death may eventually result from cerebral hemorrhage or heart failure. The primary biochemical defect or deficiency is unknown, but the kidney is thought by many to be the organ in which the derangement arises. Hypertension has a decided hereditary basis, showing a distinct tendency to run in families. An affected person frequently has brothers, sisters, or parents with the disease. However, the degree of severity among them may vary. This type of observation suggests that the genetic component is polygenic. Environmental factors involving stress are certainly involved in the expression of the disorder, but the exact nature of this is not clear. Common salt can play a significant role in hypertension, lower intake typically being followed by a reduction in the severity of the symptoms.

Hardening of the arteries (arteriosclerosis) is associated with an assortment of abnormalities. One of the best known is atherosclerosis in which fatty deposits accumulate on artery walls. These deposits can eventually cause destruction of a vessel. Moreover, they narrow it and provide surfaces upon which blood clots may develop. Atherosclerosis, whose incidence rises continually with increasing age, has attracted much publicity in the last few years, since it has been shown to be the most common cause of death in the United States and Europe. The genetic basis very possibly rests on many genes. However, the exact manner in which the cumulative action of such genes exerts an effect is unknown. The hereditary predisposition is closely interrelated with such environmental factors as stress and diet. Most informed persons today are aware of cholesterol, a fatty derivative present in many common foods such as eggs, liver, and shellfish and normally found in the body at a certain level. Elevated cholesterol levels are somehow associated with atherosclerosis, but the exact relationship continues to be a matter of debate.

Diabetes mellitus is a disease familiar to everyone. It may appear before puberty, but it is commonly a disorder of adulthood. It varies greatly in its expression, the greater severity generally associated with the earlier age of onset. The diabetic person cannot utilize simple sugar (glucose) properly. This defect in turn affects the utilization of fats, causing them to undergo improper metabolism. This leads to an accumulation of toxic products in the body. The heart and circulatory system and the kidneys may be seriously affected. The incidence of diabetes among Americans is quite high. Up to 5% of persons over 50 may have the disorder. The severity among them, however, is extremely variable. This is true even among diabetic members of the same family. A hereditary involvement is clearly indicated by the familial incidence, although the exact method of inheritance is uncertain. According to one idea, a recessive gene is the primary genetic factor responsible for the condition. However, there seems little doubt that many other genes adding to the severity of the disorder must be involved *if* one primary gene is truly the genetic basis of the disorder in some families. There are strong indications that diabetes usually has a polygenic basis, each gene producing just a small but cumulative effect. The possibility must also be considered that the genetic basis may be quite different from one family line to the next. Nonetheless, environmental factors play a role in the expression of any main gene or polygenes which are involved. Injections of insulin have dramatically increased the life expectancy of young diabetic persons, and certain oral drugs are used today in some cases to alleviate symptoms of the disorder. Diet plays an important role in the control of symptoms of diabetes. High caloric and high carbohydrate intake may even be involved in triggering expression of the disease.

The importance of an environmental agent in the onset of diabetes has been emphasized by a recent report from the State University of New York at Buffalo. It had been noted for years that diabetes often occurs in cycles, a feature that is typical of many infectious diseases. A study of case histories of about 120 diabetic children has now revealed that half of them had mumps about 4 years before the diabetes was diagnosed. About a dozen of the other children had been vaccinated against mumps by receiving a vaccine which is prepared from active mumps virus. The research team found no link between diabetes and several other infectious diseases of childhood, such as measles. A correlation was also made between the general incidence of diabetes in one New York State county and the incidence of mumps in the entire nation. For example, a high incidence of mumps was shown to parallel a high incidence of diabetes four years later. The mumps virus can apparently be harbored by the pancreas for several years. The observations raise serious questions about the advisability of injecting children with mumps vaccine prepared from active virus. They also point to the complexities which may exist between genetic factors which predispose an individual to a disease and the environmental factors to which he is subjected. The relative importance of each component, genetic and environmental, often becomes difficult or impossible to weigh.

Heredity and Environment in Expression of Disease

It is frequently asked whether heredity or environment is involved in a particular disease. The answer is that no disease is without both hereditary and environmental components. Just as heredity and environment are involved in the expression of any phenotypic feature (eye color, height, etc.), they are also involved in the expression of a disease, which is a kind of phenotype. The genetic component may play the more significant role in the case of one particular disease, whereas the environment may exert a more pronounced effect in the case of some other one. Sex-linked muscular dystrophy and Huntington's disease are examples of disorders in which the genetic component overshadows the environmental. Phenylketonuria is a disease with a very significant hereditary component, but the environmental factors can dramatically alter its expression. Diabetes, atherosclerosis, and hypertension are disorders in which the environmental and the genetic factors are so interrelated that their relative importance cannot as yet be evaluated. A disease may be known to be triggered by some infectious agent, such as a virus (smallpox, measles) or a bacterium (tuberculosis, some pneumonias). Certainly the environmental component overshadows the genetic in these cases in the sense that the microorganism must be present for the disease to manifest itself. However, it is well known that people vary in their susceptibilities to different infectious diseases. Surrounded by a germ-laden environment, some persons will develop a disorder; others will not. Those who do will experience many different degrees of severity. Studies of disease incidence among identical twins, unrelated persons, and comparisons of different populations and races leave little doubt that even infectious diseases have a genetic component which determines the relative susceptibility of the individual to the agent which precipitates the disorder.

The interrelationship of environmental and genetic factors in disease may be so complex that the very distinction between the words "environmental agent" and "hereditary factor" become blurred. Nowhere may this be more so than in the development of cancer, a topic reserved for special consideration in Chapter 14. Chapter 11 continues with the subject of genetic diseases, mainly those associated with chromosome abnormalities, and concludes with a discussion of procedures available for the detection and alleviation of several types of genetic afflictions.

Reference has been made throughout the chapters to specific human genetic disorders. This has in no way been an attempt to list the majority of the human afflictions known to have a significant genetic component. To list even a significant portion of those for which a genetic basis is known would be well beyond the scope of this book and its main goals, which are to provide an insight into an understanding of inheritance and an appreciation of the

Table 10-2. List of some metabolic disturbances associated primarily with single-gene defects. The grouping here is designed to emphasize the assortment of body parts or body functions which may become deranged by genetic defects. In some cases, a disorder could fall into more than one category. In others, the basic biochemical defect is unclear.

Disorder or variation of:	Mode of inheritance	Some major phenotypic effects
Amino acid metabolism:		
Albinism (oculocutaneous)	Autosomal recessive	Lack of melanin pigment in skin and iris
Albinism (ocular)	Sex-linked recessive	Lack of melanin pigment in iris
Alkaptonuria	Autosomal recessive	Darkening of connective tissues and urine; severe arthritis
Phenylketonuria	Autosomal recessive	Mental retardation; hyperactivity; seizures
Lipid metabolism:		
Gaucher's disease	Autosomal recessive	Liver-spleen enlargement; neurological disturbances; anemia. Usually affects the adult
Niemann-Pick disease	Autosomal recessive	Liver-spleen enlargement; severe central nervous system damage in infant
Tay-Sachs disease	Autosomal recessive	Severe central nervous system deterioration in infant; retardation in physical and mental development
Carbohydrate metabolism:		
Galactosemia	Autosomal recessive	Failure of weight gain in infancy; enlarged liver; mental retardation; cataracts
Mucopolysaccharide (protein-polysaccharide) metabolism:		
Hurler's syndrome	Autosomal recessive	Coarse facial features; mental retardation;
Hunter's syndrome	Sex-linked recessive	assorted skeletal defects; growth retardation; clouding of cornea (in Hurler's)
Hemoglobin:		
Thalassemia (various, including Cooley's anemia)	Autosomal recessive	Severe anemia; assorted bodily disturbances such as enlarged liver and spleen; skin changes; overall growth retardation
Polycythemia (various)	Autosomal recessive	Anemia due to increased affinity of hemoglobin for oxygen
Sickle cell anemia	Autosomal recessive	Severe anemia; jaundice; enlarged spleen; assorted body disturbances
Exocrine glands:		
Cystic fibrosis	Autosomal recessive	Repiratory disorders; digestive upsets; weight loss; ducts of mucous-secreting tissues distended with mucoid material
Endocrine glands:		
Pituitary dwarfism	Autosomal recessive	Short stature
Adrenogenital syndrome (various forms)	Autosomal recessive	Salt loss; hypertension; virilization

Table 10-2. Continued

Disorder or variation of:	Mode of inheritance	Some major phenotypic effects
Vitamin utilization:		
Methylmalonic aciduria	Autosomal recessive	Inability to synthesize vitamin B_{12} completely; developmental retardation
Cystathioninuria (various forms)	Autosomal recessive	Inability to bind normal amounts of vitamin B_6; anemia; mental retardation
Vitamin D resistant rickets	Sex-linked dominant	Failure of calcium deposition in skeletal tissue; various skeletal abnormalities
Drug sensitivity:		
Glucose-6-phosphate dehydrogenase deficiency	Sex-linked recessive	Lysis of red blood cells following administration of primaquine, sulfonamide, etc.
Succinylcholine sensitivity	Autosomal recessive	Pronounced reaction to drug used in anesthesia
Isoniazid inactivation	Autosomal dominant	Rapid inactivation of drug used in tuberculosis treatment
Cell membrane transport:		
Cystinuria (several forms)	Autosomal recessive	Defect in reabsorption of basic amino acids; kidney stones
Diabetes insipidus	Sex-linked recessive	Excessive urination; dehydration
Immune system:		
Agammaglobulinemia (several forms)	Sex-linked recessive Autosomal recessive	Decreased ability or complete inability to respond to antigenic stimuli
Porphyrins:		
Acute intermittent porphyria	Autosomal dominant	Abdominal pains; paralysis; psychosis
Cutaneous hepatic porphyria	Autosomal dominant	Abdominal pains; neurologic disturbances; sensitivity to light
Blood clotting:		
Hemophilia A	Sex-linked recessive	Internal bleeding; failure of blood to clot
Hemophilia B (Christmas disease)	Sex-linked recessive	
MISCELLANEOUS DISORDERS AFFECTING:		
Skeletal system:		
Marfan's syndrome	Autosomal dominant	Elongated limbs; defects of skeleton plus those of eye and aorta
Dwarfism (achondroplasia)	Autosomal dominant	Short leg bones; bulging cranium; associated skeletal defects; nose bridge depressed
Dwarfism (diastrophic)	Autosomal recessive	Short leg bones; assorted skeletal defects; clubbed feet; deformed external ear
Skin:		
Xeroderma pigmentosum	Autosomal recessive	Sensitivity to light; dilation of capillaries; freckling and thickening of skin; skin malignancies

Table 10-2. Continued

Disorder or variation of:	Mode of inheritance	Some major phenotypic effects
Muscular system:		
Myotonic dystrophy	Autosomal dominant	Progressive muscular weakening and atrophy; difficulty in relaxing grip; cataracts
Duchenne muscular dystrophy	Sex-linked recessive	Progressive muscular weakness and atrophy
Mesodermal and ectodermal outgrowths:		
Nail-patella syndrome	Autosomal dominant	Deformed fingernails and/or toenails; patella small or absent; kidney disorders
Nervous system:		
Huntington's disease	Autosomal dominant	Weaving body movements; muscular twitchings; impaired speech and intellect

advantages genetic knowledge can offer the human species. The student should be aware of the fact that, for almost every part or system of the human body, one or more genes are known whose allelic forms can lead to malfunction. For many of the derangements, the molecular basis is known; for others it is less clear. Table 10-2 is a very modest summary which lists only some disorders for which a genetic component is quite certain. The table attempts to communicate the diversity of the effects and the systems known to be under genetic control. The benefits which can be derived from this information to aid in the alleviation of human suffering should be apparent. This incomplete listing in Table 10-2 should also serve to impress us with the progress which has been accomplished by teams of outstanding investigators in a short period of time and which has increased our understanding of the genetics of the human, a creature once thought to be an unlikely candidate to yield an insight into the nature of gene action.

REFERENCES

Bach, G., Eisenberg, F., Jr., Cantz, M., and Neufeld, E. F., "The Defect in the Hunter Syndrome: Deficiency of Sulfoiduronate Sulfatase." *Proc. Nat. Acad. Sci.* (US), 70:2134, 1973.

Bergsma, D. (Ed.), *Birth Defects Atlas and Compendium.* Williams & Wilkins, Baltimore, 1973.

Brady, R. O., "Hereditary Fat Metabolism Diseases." *Sci. Amer.,* Aug: 88, 1973.

Capp, G.L., Rigas, D.A., and Jones, R.T., "Evidence For a New Hæmoglobin Chain (ζ-chain)." *Nature* (Lond.), 228: 278, 1970.

Carter, C. O., "Multifactorial Genetic Disease." In *Medical Genetics,* McKusick, V. A., and Claiborne, R. (Eds.), pp. 199–208. HP Publishing Co., N. Y., 1973.

Cerami, A., and Peterson, C. M., "Cyanate and Sickle-Cell Disease." *Sci. Amer.,* April: 44, 1975.

Conley, C. L., and Charache, S. "Inherited Hemoglobinopathies." In *Medical Genetics,* McKusick, V. A., and Claiborne, R. (Eds.), pp. 53–61. HP Publishing Co., N. Y., 1973.

Fitzpatrick, T. B., and Quevedo, W. C., "Albinism." In *The Metabolic Basis of Inherited Diseases,* Third edition, Stanburg, J. B., Wyngaarden, J. B., and Fredrickson, D. S. (Eds.), pp. 326–337. McGraw-Hill, N. Y., 1972.

Garrod, A. E., *Inborn Errors of Metabolism.* Reprinted by Oxford Univ. Press, London, 1963.

Huntsman, R. G., and Lehmann, "The Hemoglobinopathies." In *The Metabolic Basis of Inherited Diseases,* Third edition, Stanburg, J. B., Wyngaarden, J. B., and Fredrickson, D. S. (Eds.), pp. 1398–1432. McGraw-Hill, N. Y., 1972.

Knox, W. E., "Phenylketonuria." In *The Metabolic Basis of Inherited Diseases,* Third edition, Stanburg, J. B., Wyngaarden, J. B., and Fredrickson, D. S. (Eds.), pp. 266–295. McGraw-Hill, N. Y., 1972.

Knudson, A. G., Jr., *Genetics and Disease.* McGraw-Hill, N. Y., 1965.

Lorkin, P. A., "Fetal and Embryonic Hemoglobins." *J. Med. Genet.,* 10:50, 1973.

Macalpine, I., and Hunter, R., "Porphyria and King George III." *Sci. Amer.,* July: 38, 1969.

McKusick, V. A., and Rimoin, D. L., "General Tom Thumb and Other Midgets." *Sci. Amer.,* July: 102, 1967.

Mellman, W. J., "Human Cell Culture Techniques in Genetic Studies." In *Medical Genetics,* McKusick, V. A., and Claiborne, R. (Eds.), pp. 157–165. HP Publishing Co., N. Y., 1973.

Neufeld, E. F., and Fratantoni, J. C., "Inborn Errors of Mucopolysaccharide Metabolism." *Science,* 169:141, 1970.

O'Brien, J. S., "Ganglioside Storage Diseases." *Adv. Human Genet.,* 3:39, 1972.

Ohada, S., and O'Brien, J. S., "Tay-Sachs Disease: Generalized Absence of a Beta-D-N-acetylhexosaminidase Component." *Science,* 165:698, 1969.

Raivio, K. O., and Seegmiller, J. E., "Genetic Diseases of Metabolism." *Annu. Rev. Biochem.,* 41:543, 1972.

Stamatoyannopoulos, G., "The Molecular Basis of Hemoglobin Disease." *Annu. Rev. Genet.,* 6:47, 1972.

Weatherall, D. J., "The Thalassemias." In *The Metabolic Basis of Inherited Diseases,*

Third edition, Stanburg, J. B., Wyngaarden, J. B., and Fredrickson, D. S. (Eds.), pp. 1432–1448. McGraw-Hill, N. Y., 1972.

Witkop, C. J., Jr., "Albinism." *Adv. Human Genet.,* 2:61, 1971.

Zuckerkandl, E., "The Evolution of Hemoglobin." *Sci. Amer.,* May: 110, 1965.

REVIEW QUESTIONS

Select the correct answer or answers for each of the following:

1. In phenylketonuria: (1) phenylpyruvic acid accumulates in the blood; (2) tyrosine accumulates in the blood; (3) an autosomal recessive inheritance pattern is typical; (4) there is a genetic block in a metabolic pathway; (5) more than one biochemical pathway may be affected.

2. Tay-Sachs disease: (1) can be alleviated by the institution of proper diet at birth. (2) is a disorder in which a genetic block disrupts the proper metabolism of fats; (3) affects only male babies; (4) results from a genetic block in protein metabolism; (5) can be detected in the fetus by amniocentesis.

3. Sickle cell hemoglobin: (1) is identical to normal hemoglobin except for one amino acid; (2) does not contain the β chain due to a nonsense mutation; (3) results from the change in a single DNA codon; (4) gives a fingerprint in which most of the fragments differ from those seen in a fingerprint of homoglobin A; (5) is present in the heterozygote.

4. In the human: (1) a hemoglobin molecule is composed of four different kinds of polypeptide chains; (2) two different kinds of hemoglobin molecules are typical of the normal adult; (3) the hemoglobin is under the control of a single genetic locus; (4) α chains are not found in hemoglobin before birth; (5) genes for the five main kinds of homoglobin chains are called to activity in the embryo.

5. In the human: (1) muscular dystrophy may be autosomal in some families, X-linked in others; (2) Huntington's disease is typically expressed before puberty; (3) both heredity and environment are involved in the expression of any disease; (4) each of the common chronic diseases shows simple Mendelian inheritance patterns; (5) environment may trigger a disorder for which there is a significant genetic component.

chapter 11

CHROMOSOME ANOMALIES AND APPROACHES TO THE PROBLEM OF GENETIC DISORDERS

Inherited and Non-inherited Genetic Disorders

The genetic disorders discussed in Chapter 10 were all due to variant forms of normal genes. The stretch of DNA (the gene) concerned with a certain function may become altered or changed in some way so that another allele arises. We have noted that this change may be one which involves a codon (a code word of three nucleotides). A different amino acid may then substitute for the one typical of the protein whose structure that locus directs. Such changes in DNA are gene mutations, and their origin and significance will be the primary topic of Chapter 13. It is important here to understand that disorders such as phenylketonuria, sickle cell anemia, and alkaptonuria result from distinct changes in the normal or standard allele in each case. In those disorders termed *polygenic,* not one but many variant alleles at several genetic loci are involved. Certain allelic combinations may result in a susceptibility to a particular disorder. Whether or not the disorder is expressed and the degree to which it will manifest itself may depend on a complex of environmental factors which interact in different ways with the various allelic combinations. Nevertheless, as in the simpler Mendelian diseases, point changes in the DNA are involved. This means that specific sites within the genes have been altered so that new gene forms have arisen. Any new allele may not function as efficiently as the normal one, and so a phenotypic effect recognized as a disease results. In all the examples presented in the last chapter, the genetic disorder was inherited, in the sense that the defective allele could be passed down from one generation to the next. In the case of a recessive, the heterozygous carrier may exhibit no apparent effect of the variant allele but can pass it down to another individual. This transmission may continue until two such alleles come together in a zygote which develops into an individual who may express a serious disorder and who may even die from it.

Many genetic disorders, however, are not due to the presence of mutant alleles at one or even several loci. And many of them are not inherited in the usual sense of the word. Falling into this category are the disorders which are due to more gross changes in the hereditary material: actual additions or subtractions of one or more whole chromosomes, as well as changes in the very structure or shape of a chromosome. In these cases, many loci are involved, as is apparent if a whole chromosome is added or if part of one is deleted. However, there are not necessarily any mutant alleles involved. Instead, normal genes are present in excessive amounts in some cases or in reduced amounts in others. These departures from the normal are genetic in the sense that the hereditary material is definitely associated with the change, but they are not necessarily inherited, since many such abnormalities are not passed down from one generation to the next. While the abnormality may arise in the gamete of a parent and be transmitted to an offspring, it often stops there and goes no further. This is the case in the first few examples which will be presented shortly.

The chromosome changes we shall encounter in this chapter can be lumped under the general heading, "mutation," since in its broadest sense mutation refers to any sudden inheritable change, and these chromosome changes arise suddenly, just as do changes in a gene. However, the term *mutation,* unless otherwise qualified, is usually taken to mean a *point* or gene mutation. Those changes involving entire chromosomes or sets of chromosomes are designated *chromosome anomalies* or *chromosome aberrations.* The importance of chromosome anomalies can be appreciated from the following facts. At least 5% of the babies born have some kind of genetic defect. Actually, the number of genetic defects is much higher if embryos and fetuses are considered in addition to those babies who survive to birth. In fact, a quarter of the fertilized eggs will never result in full-term infants. Among those which are spontaneously aborted, chromosome abnormalities are very high. Moreover, chromosome aberrations are responsible for most of the congenital defects (defects present at birth) suffered by humans. As pointed out in Chapter 4, not only may they account for 20% of the spontaneous abortions in the U.S. alone, but 20,000 infants each year will incur some disorder due to a chromosome aberration. Since they present a definite problem both to the families concerned and to society, their nature must be understood so that steps can be undertaken to prevent their occurrence.

Aneuploidy Involving Sex Chromosomes

We have already encountered examples of chromosome aberrations in preceding chapters, particularly in Chapter 4, where anomalies involving sex chromosomes were discussed. Turner syndrome (XO) is an example of an abnormal condition resulting from the absence of a whole chromosome, a sex chromosome. Klinefelter syndrome is an example of a disorder caused by the presence of extra doses of entire sex chromosomes (XXY, XXXY). In such cases, the chromosome number is different from the characteristic 46 and departs from the normal diploid number by 1 or a few chromosomes. Persons with such chromosome anomalies are said to be *aneuploid* (Chapter 4). The origin of an aneuploid individual can be explained as a consequence of some abnormality arising at the time of meiosis. Reference to Figs. 4-4 and 4-6 will show how nondisjunction at the time of meiosis in a male or female can produce gametes with exceptional chromosome constitutions which give rise to persons with the Klinefelter or with the Turner syndrome. Figure 5-4 shows how accidents at both mitosis and meiosis can produce aneuploid individuals of a more complex type, such as mosaic persons, those who are mixtures of cell lines of different chromosome constitutions.

From the descriptions which were given of the Turner and Klinefelter syndromes, it is apparent that an addition or a subtraction of a sex chromosome can produce serious phenotypic upsets. The Turner condition is associated with critical disturbances of the heart and circulatory system, in addition to the defects in the sex organs. The severity of the absence of just one chromosome, as in the XO Turner female, is realized when it is noted that only about 2% of the zygotes which start out as XO manage to survive to full term. Actually, the XO aberration is believed to be the commonest of all the chromosome anomalies to arise, but almost all of them abort. So while many more XO zygotes arise than do XXY ones, the frequency of the Turner female is much lower than that of the Klinefelter male (about 1/3500 female births to 1/500 male births).

Any aneuploid chromosome complement or individual in which one chromosome is missing is termed *monosomic.* Monosomy in humans is a much less frequent type of aberration than the kind of anomaly in which one *extra* chromosome is present. Apparently, human development is less able to tolerate the absence of a chromosome than the presence of an additional one. Indeed, persons monosomic for an autosome are almost unknown. Fewer than half a dozen cases have ever been reported. Each involved one small chromosome in group G (Table 2-1). In the two cases where a small chromosome was clearly shown to be missing, the infants were very abnormal and died. The fact that autosomal monosomy is rarely if ever seen in abortuses supports the conclusion that a monosomic zygote is so unbalanced that it may not survive fertilization or may not be able to implant in the uterus.

Adding or subtracting an autosome seems to produce consequences which are distinctly more serious than those arising from addition or subtraction of a sex chromosome. An extra Y chromosome (Chapter 4) does not appear to affect bodily development in any adverse way. The presence of 1, 2, or 3 additional X's as in different Klinefelter males (XXY, XXXY, XXXXY) certainly causes bodily derangements, and the presence of extra X's in a female (XXX, XXXX) can result in mental retardation. Still, persons with these kinds of sex chromosome anomalies do manage to survive. Almost certainly, the ability to tolerate extra X chromosomes is somehow related to the fact that all X's in excess of 1 tend to become condensed or deactivated, as seen by the presence of Barr bodies (Chapter 5).

Aneuploidy Involving Autosomes

The extreme severity of extra autosomes in the chromosome complement is undoubtedly reflected by the fact that only a few of the autosomes have ever even been reported in extra doses postnatally. However, 1 extra chromosome has been reported in abortuses for every one of the identifiable chromosomes. It would seem, therefore, that an extra autosome tends to result in an unbalance so severe that survival to birth is prevented. This is supported by observations of the few individuals who have been born with chromosome #13 present 3 times and those with a triple dose of #18. These infants lived only a very short time. The extra chromosome in each case was associated with extremely severe abnormalities of the face, skeleton, and brain.

Any individual with 3 doses of a specific chromosome is called a *trisomic.* While trisomics (and even tetrasomics) for X chromosomes may be abnormal in some way, their abnormalities are less severe and their chances for survival are very much greater than those showing trisomy for an autosome. The only trisomy for an autosome in humans which permits survival well beyond infancy is that associated with an extra #21, the Down's syndrome which was briefly described at the end of Chapter 2. It was not until 1959 that the discovery was made (by Lejeune) that an extra chromosome was responsible for the well-known disorder. This was actually the very first time an association was made between any congenital malformation and a chromosome anomaly. The chromosome shown to be present in 3 doses in a Down's child is a very small one designated #21, a member of the G group, the smallest of the human chromosomes (see p. 270 for the precise identification of this chromosome). Approximately 1 out of 500 infants is born with this disorder, but the occurrence of the Down's trisomy is quite high in embryos as shown by the fact that the condition is observed in 20% of early abortuses. In addition to a number of serious bodily derangements stemming from a retardation of development, victims also have an increased chance of developing leukemia.

Survival beyond the age of sexual maturity is infrequent for these mentally retarded children, who tend to have a very affectionate nature. While Down's syndrome, or trisomy 21 as it is also called, is a severe disorder, the fact that survival is possible beyond infancy may be related to the fact that the extra chromosome has actually been found to be the smallest one of the human complement, not the second smallest as once believed.

Some Down's individuals survive past puberty, and female victims appear to be fertile. A few of them have actually borne offspring. About half of the children from Down's mothers also have the disorder, which is what we would expect on the basis of meiotic behavior. Since three chromosomes of one kind are present, an association of three chromosomes, #21, will form at meiotic prophase instead of the normal two. When anaphase takes place, about half of the time two of the #21 chromosomes will pass together to one pole, the third traveling by itself to the other. A nucleus derived from the latter will have the normal chromosome number; a nucleus derived from the former will carry the extra chromosome and can be the reason for the origin of a Down's child.

The rarity of encountering trisomy for any large autosomes may well be due to their size and hence the greater number of genes they must carry. Adding a larger chromosome would be expected to cause a greater upset than adding a very small one. No mechanism is operating in the case of autosomes to turn off part or all of the extra dosage, as is true for an X chromosome.

Parental Age and the Origin of Chromosome Anomalies

The incidence of Down's syndrome has been shown to be related to the age of the mother and to rise with increase of maternal age at time of conception. The father's age does not appear to be significant in trisomy 21. Women over 40 have a much higher chance of bearing a Down's offspring than do younger women. The incidence of Down's syndrome does not increase at just the same rate as maternal age. After the age of 40, there is a dramatic increase. While these older women bear only 4% of all the babies born, they producc 40% of the children who suffer from trisomy 21! The incidence of Down's syndrome rises from 1/3000 for the 20-year-old maternal age group to 1/40 for the 45-year-old group! The reason for this may be related to the fact that at birth the human female carries all the potential egg cells that she will ever produce (Chapter 3). These cells are already in first prophase of meiosis, and each remains in this stage until the age of puberty, at which time approximately 1 per month will resume meiosis. This means that a meiotic cell in a woman of 40 is twice as old as a comparable one in a woman of 20. The aging may somehow increase the chance for nondisjunction of chromosome #21 and hence the chance for the formation of an egg with 2 doses of that chromosome instead of a single one (Fig. 11-1). Following fertilization of such a gamete by a normal sperm carrying 1 chromosome #21, a Down's offspring could arise.

Maternal age also seems to exert an important influence in the XXY Klinefelter anomaly. The older the mother, the greater the likelihood that the extra X chromosome has been transmitted by an egg. While a male parent may contribute the extra X, the paternal age appears to wield no influence.

Various factors undoubtedly contribute to the higher incidence of nondisjunction in oocytes of older women. The oocytes of a woman over 40 have been suspended in first meiotic division for 4 decades. It is well known that cells in a meiotic state are especially vulnerable to the disturbing cytological effects of such environmental agents as viruses, certain radiations, and chemicals. These can damage the chromosomes or interfere with the meiotic spindle. The older a fertile woman, the longer the time for such factors to operate to the detriment of the meiotic cells. Moreover, a woman over 40 is approaching menopause, a time at which pronounced physiological changes

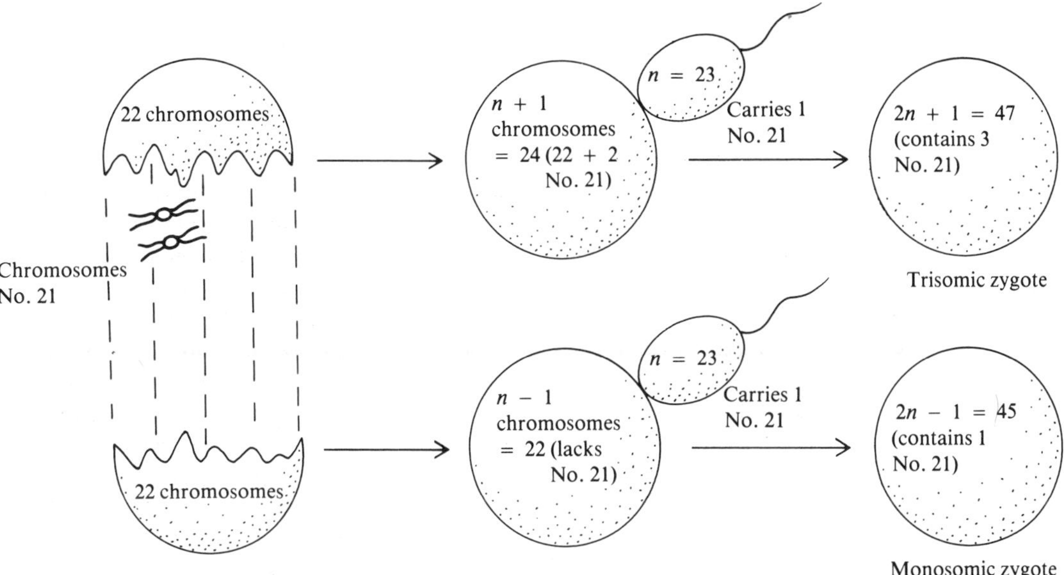

Figure 11-1. Nondisjunction of chromosome #21 at meiosis. If two #21 chromosomes fail to separate at anaphase I of meiosis, both will move to the same pole. A cell nucleus derived from this pole may become the nucleus of an egg. If it does, it will carry two #21 chromosomes. After fertilization by a normal sperm, the zygote would carry three #21 chromosomes, and a Down's individual would result. If the nucleus derived from the opposite pole became the egg nucleus which was later fertilized, a zygote with only one chromosome #21 would arise. Such a zygote would be so genetically unbalanced that it probably would not go on to full development.

are triggered by the internal hormonal environment. The ovaries will soon cease to function entirely. Such pronounced effects of normal aging of the ovary and surrounding tissues may contribute significantly to accentuate disturbances in the cell which have accumulated due to factors originating in the outside environment. Whatever the exact cause may prove to be, there is no doubt that age of the mother is correlated with several kinds of chromosome aberrations. Spontaneous abortions occur more frequently among older women, and among the abortuses, 1 out of 3 have been found to contain chromosome anomalies. Evidence has also been presented which suggests that the spontaneously aborted offspring of women 17 years old or younger have a significantly high incidence of chromosome aberrations. The possibility exists that either maternal age extreme may contribute for some reason to a higher incidence of chromosome anomalies.

Surprisingly, the age of the mother seems to play no role in the meiotic irregularities responsible for the Turner (XO) syndrome. For reasons which are still not clear, most of these cases seem to be related to mishaps in the male gamete!

Polyploidy

Another type of abnormal chromosome condition can result if a sex cell is formed which carries the unreduced diploid ($2n$) chromosome number instead of the haploid number composed of just 1 chromosome set. Formation of an unreduced gamete with the diploid number could arise from a complete failure of the meiotic process, so that an egg or a sperm forms carrying the chromosome number which is characteristic of the body cells. If such an unreduced sex cell fuses with a normal one having the haploid chromosome number, the zygote will contain 1 entire additional set of chromosomes. Any individual with 1 or more additional chromosome sets is said to be a *polyploid.* A polyploid with 3 sets instead of the normal 2 is a triploid; one with 4 sets is a tetraploid. Polyploidy is a very widespread feature of plants (Fig. 11-2) which are able to tolerate added chromosome sets, but it is rare among animals. In mammals, polyploidy produces a lethal effect. In the human, a few reportedly triploid babies having serious bodily defects have managed to survive for a few hours. No completely tetraploid baby has ever been reported. The killing effect of 4 chromosome sets is apparently even greater than that of 3. This is supported by the fact that the number of spontaneously aborted embryos containing tetraploid cells is very low; no completely developed tetraploid fetus has ever been found. In contrast, triploid cells are found in approximately 4% of human abortuses, some of which may approach full development. Most of the tetraploid zygotes or embryos apparently fail to survive early develop-

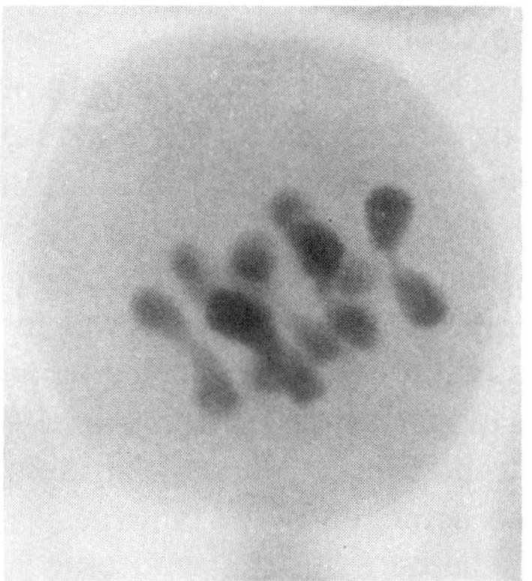
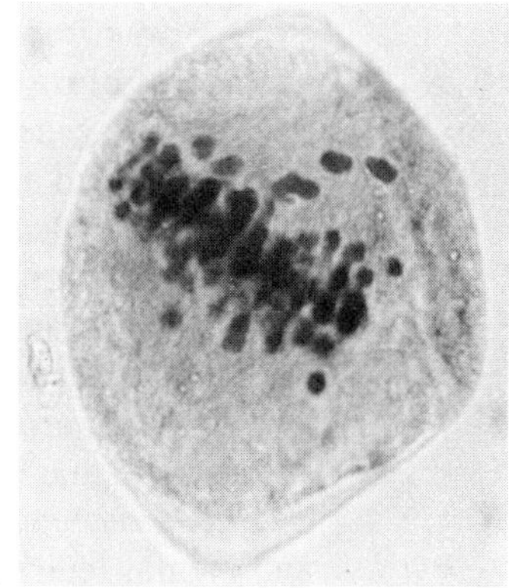

Figure 11-2. Polyploidy in a plant species, *Claytonia virginica,* the common Spring Beauty. First meiotic metaphase cells from some plants contain 16 chromosomes arranged as 8 pairs (left). Cells from other individuals have higher numbers, often multiples of 8. The cell at the right carries at least 48 chromosomes, some paired and others unpaired. Those plants with the higher numbers have undoubtedly been derived in nature from plants with lower numbers as a result of the fusion of unreduced gametes. In the Spring Beauty, both polyploidy and aneuploidy are commonly observed. Among plants in general, the polyploid types grow more slowly than the diploids. While polyploidy is a common feature of flowering plants, it is uncommon in animals and is lethal in mammals. (Photos courtesy of Robert F. Maller.)

ment. The unbalance which polyploidy evidently entails in mammals is so severe that death almost always occurs well before birth so that polyploidy is not a significant cause of any genetic disorders affecting the human.

The Deletion

Not only may the human chromosome constitution depart from the normal due to some modification in the chromosome number, it may at times contain a chromosome whose structure is abnormal in some way. Each chromosome of the normal human complement can be assigned to one of seven groups, since chromosome size and centromere position are very constant (Table 2-1). However, alterations in chromosome morphology do arise on rare occasions, and these may have serious consequences for the individual. The various types of structural chromosome changes can arise spontaneously following the breakage and rearrangement of chromosome segments. They may be induced by some radiations and chemicals (Chapter 13) and may also be caused by certain viral infections of the cell.

One way in which the structure of a chromosome may change is by the actual loss of a segment. This means that a portion of the DNA, and hence one or more genes, may be missing from the chromosome. Such a structural change is called a *deletion.* Some deletions may be so small that a size change cannot be detected with the aid of a microscope; others may result in an evident alteration in chromosome length. Deletions are known to occur in all groups of organisms, and as expected, they tend to result in abnormal phenotypes due to the unbalance caused by gene loss. In the human, a visible deletion is known which affects the longer arm of chromosome #5. An individual with a normal #5 and one with the deletion shows a group of effects known as the *cri-du-chat* syndrome, characterized by certain facial features, a small head, and mental retardation. The name of the syndrome is derived from the fact that affected children tend to have a cry which resembles that of a cat. Persons with *cri-du-chat* syndrome have been known to survive past the age of 30.

A condition is known which is similar to the *cri-du-chat* syndrome but without the cat-like cry. It is associated with a missing portion of chromosome #4. The origin of these deletions in disease is discussed a bit more on page 365.

A deletion of great interest is one associated with a form of leukemia, chronic myeloid leukemia. This deletion affects 1 of the 2 members of the G group, and until recently it was not known with certainty whether this was the same chromosome as that associated with Down's syndrome, #21. The 2 chromosomes of the G group are the smallest of all the human autosomes, and distinction between them on the basis of their general appearance is difficult. As noted in Chapter 2, advances in chromosome staining techniques have enabled the cytologist to bring out characteristic bands on the chromosomes, so that each chromosome within a group has now been distinguished from the other members. It is now certain that the chromosome with the visible deletion associated with the leukemia is not the same as the chromosome responsible for Down's syndrome. When classified, the human chromosomes are arranged and numbered according to decreasing size. As it turns out, the one involved in Down's syndrome is the smallest; so on the basis of numbering strictly on size, the extra chromosome in trisomy 21 is really chromosome #22! However, cytologists and medical persons have decided to keep the designation "trisomy 21" for Down's syndrome and to continue to call the smallest chromosome "number 21." The chromosome with the deletion will be considered chromosome #22. Two different chromosomes are therefore involved in these 2 disorders, classic Down's syndrome with the smallest of the entire complement (to be called #21 as has been done) and chronic myeloid leukemia with the next to smallest (to be considered #22).

It should be noted that another autosomal trisomy has been reported which apparently has been confused at times with the Down's syndrome. The techniques now available for precise chromosome identification using fluorescent dyes which bring out chromosome bands have shown that these cases involve an extra chromosome #22, the second smallest one. Some of the features of the trisomy 22 syndrome resemble those of trisomy 21. As in classic Down's syndrome, children frequently suffer heart defects and die early. If the latter are absent, survival for several years appears possible. The fact that any such survival is possible in trisomy 22 is very possibly related to that fact that it is one of the two smallest autosomes.

The deletion in chromosome #22 associated with the leukemia produces a visibly shorter chromosome, which has been designated the *Philadelphia chromosome,* after the city in which the deletion was first found. Approximately one-half of the long arm of the chromosome is missing. Unlike the other aberrations we have been discussing, the Philadelphia chromosome is found *only* in the leukocytes of persons with myeloid leukemia. This is the only specific chromosome aberration shown to be characteristically associated with a malignancy. The full significance of this deletion is not known. It is tempting to speculate that some sort of sudden change or mutation in the body cells, specifically those of the bone marrow, has resulted in the loss of the chromosome region and thus caused an unbalance which triggers the malignancy. However, the correlation between the Philadelphia chromosome and the leukemia is not 100%. In 5% of the leukemia victims, no structural chromosome abnormality can be detected. Moreover, it has not been demonstrated that the deletion brings about the disease. It is just as possible that the deletion results from some kind of derangement in the bone marrow that is the direct cause of the disease. The deletion may thus reflect some still unknown type of change which produces the malignancy. Another argument is that the deletion arises first and then renders the cells more vulnerable to the effects of certain environmental factors which actually stimulate the malignancy. A further complication is the more recent observation that the portion deleted from chromosome #22 has not been entirely lost from the cell but has been inserted, at least in part, into one of the larger chromosomes (p. 365). This would mean that the Philadelphia chromosome has resulted from a translocation, the subject of the following section. While the correlation between the Philadelphia chromosome and the leukemia is thus not clear, there is no doubt that an association exists between the two. This fact indicates the need for further research which may yield clues that hold significance for the general problem of cancer.

The Translocation and the Inheritance of Chromosome Anomalies

Other kinds of chromosome aberrations involve shifts of the genetic material in which chromosome segments are not lost but are placed in new locations. The one of greatest medical concern is the *translocation.* In a translocation, two nonhomologous chromosomes are usually involved, and segments are exchanged between them. Figure 11-3(A) diagrams two pairs of human chromosomes, pair #15 and pair #21. A break next to the centromere in the short arm of #15 and adjacent to the centromere in the long arm of #21 is followed by the insertion of most of chromosome #21 into #15. An abnormally large chromosome and a very tiny one arise as a consequence of the exchange. It is known that this large chromosome can definitely form in the human, but the tiny one has not been seen. It may actually carry very little genetic information and may usually become lost. It is also possible that most of chromosome #21 without the centromere may actually get inserted.

This translocation involving chromosomes #21 and #15 can be carried in either a sperm or an egg. As Fig. 11-3(B) shows, the union of a normal gamete (carrying 1 chromosome #21 and 1 #15) with a gamete carrying the translocation (1 large chromosome #15–21) produces a zygote which has, in effect, 2 normal doses of chromosome #21 and 2 of #15. An individual of this chromosome constitution would show no ill effects, since all the essential genetic material is present. However, a chromosome count on such a person would show 45 chromosomes, 1 less than normal. Close inspection would reveal that a chromosome larger than normal is present and that chromosomes #21 and #15 are represented only once. This condition, however, may remain undetected, unless a familial incidence of Down's syndrome becomes evident. The reason for the appearance of 1 or more Down's victims among the offspring of this type of person with the translocation is due to irregularities which take place at meiosis due to the presence of the translocation [Fig. 11-3(C)]. The abnormal chromosome, #15–21, can pair with both the normal #21 and the normal #15 which are also present in the complement. Instead of the normal pair formation, an association of 3 chromosomes forms at meiotic prophase. As a consequence of this abnormal grouping, the separation of the chromosomes during oogenesis or spermatogenesis will be irregular. More than 1 kind of gamete can arise. Some will be normal; some will be balanced but will pass down the translocation. Other gametes will be so unbalanced that they may not function at all or may produce a zygote which dies. However, one of the possibilities is the formation of a gamete which has chromosome #15–21, the translocated one, and also a normal #21. If this gamete combines with a normal gamete, the zygote has chromosome #15 represented twice, but chromosome #21 is present 3 times. The individual who develops from it will actually have trisomy 21 and will thus be a Down's victim. The chromosome number of such an individual, however, will appear normal, 46. Again close inspection will reveal the translocation. Two normal #21 chromosomes will be present, but a change affecting pair #15 will be evident. While one of the members is normal, the other is larger and does not match the normal-appearing one.

Most cases of the Down's syndrome result from nondisjunction, as described earlier (Fig. 11-1), and these victims have 47 chromosomes. However, a small percentage of the cases of the syndrome arise from the translocation as

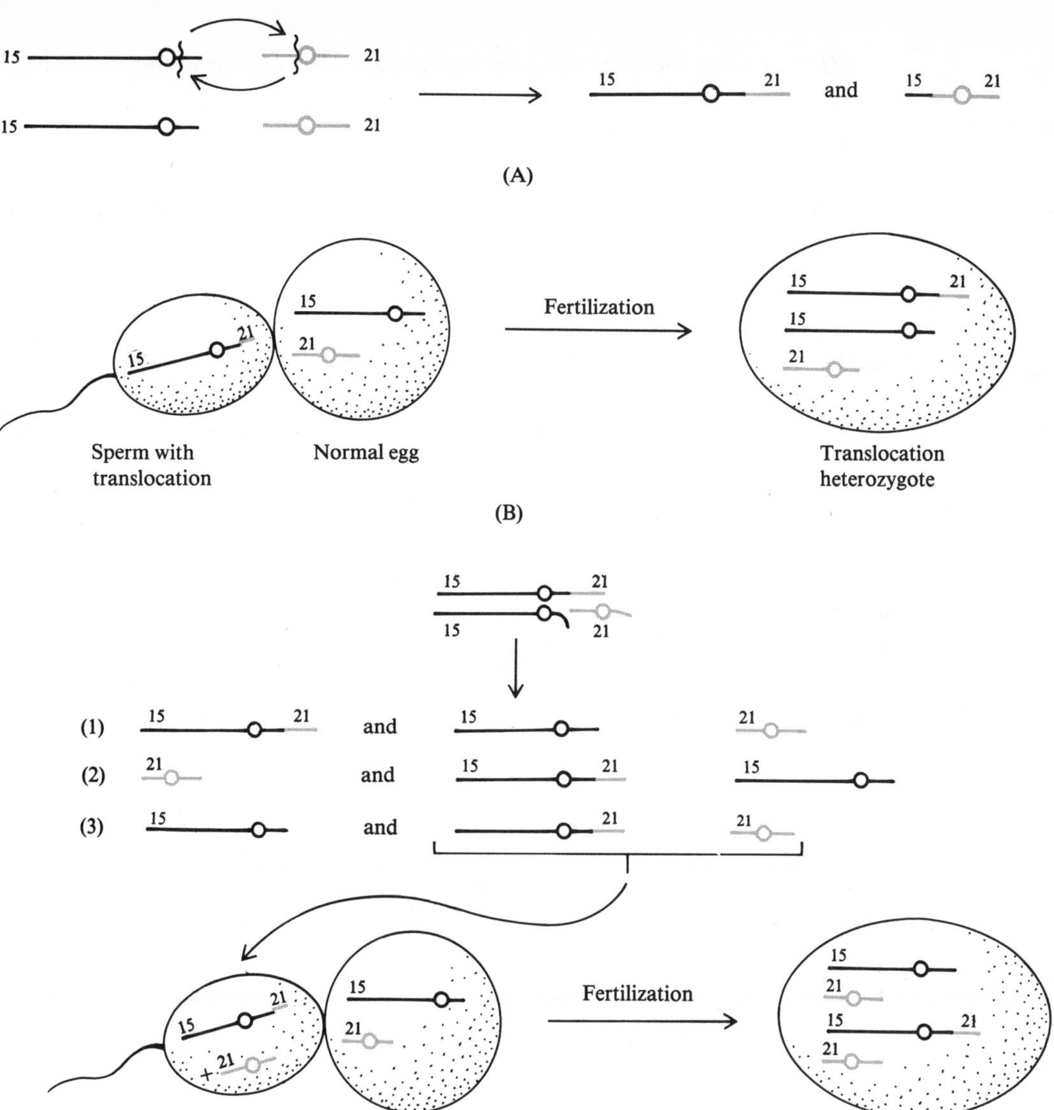

Figure 11-3. Translocation and the Down's syndrome. (A) Origin of a translocation involving the mutual exchange of portions of two nonhomologous chromosomes, #15 and #21. (B) Origin of a translocation heterozygote. (C) Meiotic irregularities occur in the translocation heterozygote, since three of the chromosomes (#15, #21, and #15-21) possess homologous regions, and an association of three chromosomes forms. As a result, different types of segregation to the poles can occur at meiotic anaphase, since any one chromosome may move to one pole and the other two to the opposite pole. In one possibility (1), the chromosome with the translocation, #15-21, moves to one pole. In this case, gametes which pass down the translocation will form, as well as gametes which are normal in chromosome content. In the second type of segregation (2), very unbalanced gametes will arise which will prove lethal. The third type of segregation (3) will result in gametes which lack chromosome #21 and will prove lethal, and also in gametes which can result in Down's syndrome (below).

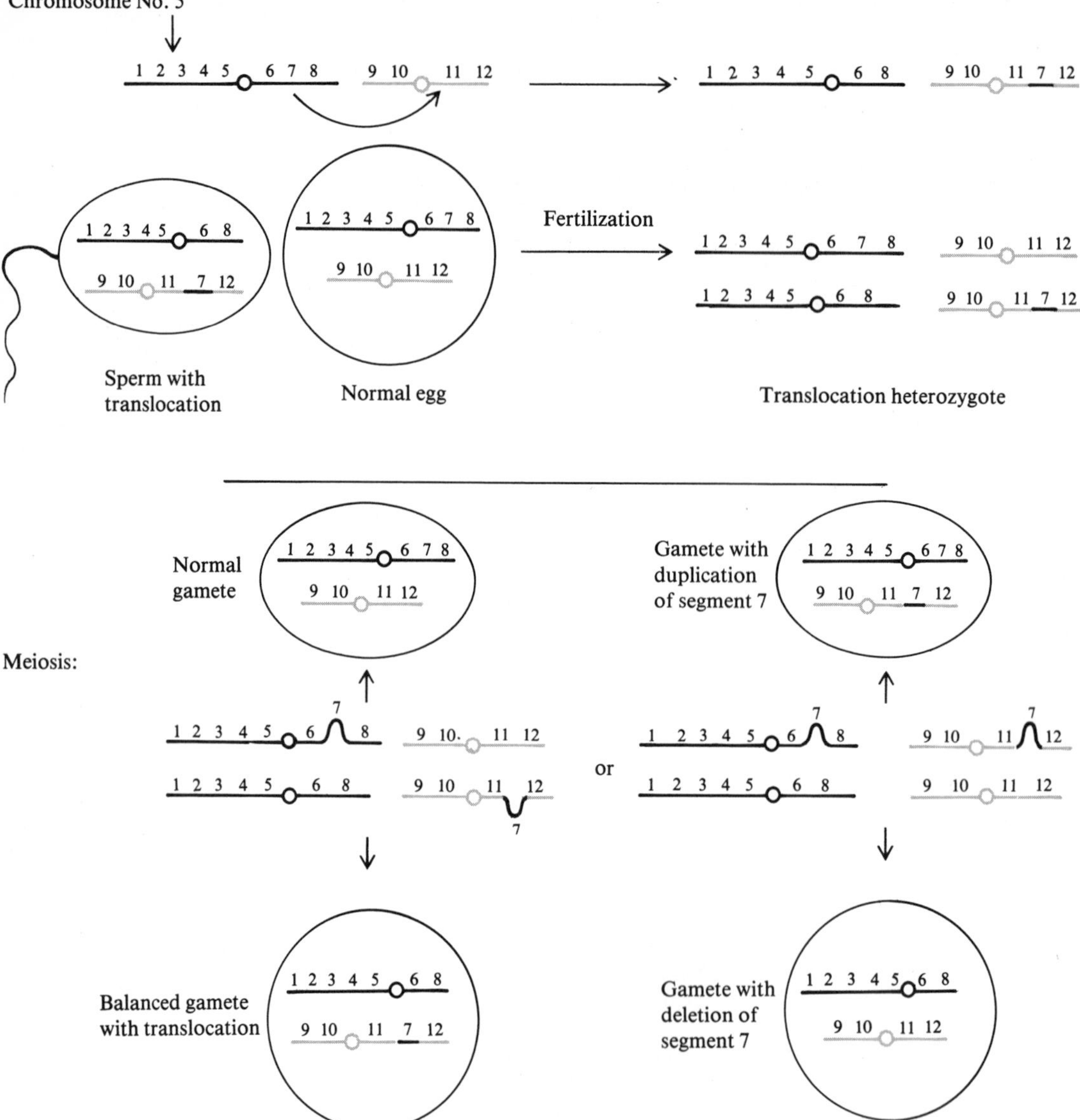

Figure 11-4. Consequence of a translocation. Assume that a portion of a chromosome, such as chromosome #5, is translocated to another non-homologous chromosome during gamete formation (above). If a gamete with the translocation combines with a normal one, the resulting individual would be a translocation heterozygote and would probably suffer no ill effects, since there is genic balance. Meiosis in such a person, however, may lead to the production of abnormal gametes (below). Some of these would have deleted material and some duplicated (segment #7 here). Such gametes might not survive or could produce offspring with abnormalities.

described here. It is important to realize that, in these cases, 1 parent is a carrier of the syndrome. Such a person can be a male or female and should be identified if possible, since he or she may produce more than 1 affected child. Moreover, these persons can pass the translocation down [Fig. 11-3(C)] to other children who in turn may produce victims of Down's syndrome. Therefore, parents of a Down's child who shows a translocation should have a chromosome analysis performed to determine if one of them is a carrier. Unlike the Down's cases due to nondisjunction, this form *is* familial. The responsible genetic condition *can* be inherited, since the abnormal chromosome may be passed from one generation to the next, just as any defective gene may be transmitted. The other chromosome anomalies discussed up to this point were all genetic defects, but they were not inherited in the usual sense. The chromosome aberration in each case arose as a rare event during gamete formation (as in the Klinefelter and Turner syndromes or as in the commoner form of trisomy 21). The affected offspring did not reproduce and so did not pass down any chromosome abnormality. Consequently, no pattern of inheritance or high incidence for these disorders is found in a family pedigree. (The Philadelphia chromosome occurs only in leukocytes and would not be carried through gametes.) While most known chromosome aberrations are not familial, the Down's translocation indicates the possibility that certain kinds of aberrations may be inherited. This must be recognized as syndromes related to still other types of chromosome anomalies become identified. This point is emphasized by the fact that there is some reason to believe that the deletion of material associated with the *cri-du-chat* and similar syndromes has arisen as a consequence of translocation. The person in whom the chromosome alteration occurred would carry the normal amount of genetic material, but a segment from the short arm of chromosome #5 would be inserted into another chromosome. This person would be a carrier of the translocation and could produce gametes with a defective chromosome #5 (Fig. 11-4). In several known cases of *cri-du-chat* children, one of the parents has been shown to carry a translocation which involves part of chromosome #5.

The exchange between chromosomes #21 and #15 is therefore not the only translocation that has been found in the human. Moreover, a very small number of Down's syndrome cases has been associated with translocations involving chromosome #21 and some other chromosome besides #15. Other translocations have been reported in which chromosomes #15 and #21 are not involved at all. In these cases, death in infancy and mental retardation accompany the translocation.

The Inversion

A type of chromosome change well known in many species but which has been identified in only a very few human cases is the *inversion*. As Fig. 11-5(A) shows, the inversion involves two breaks in a chromosome followed by a reinsertion of

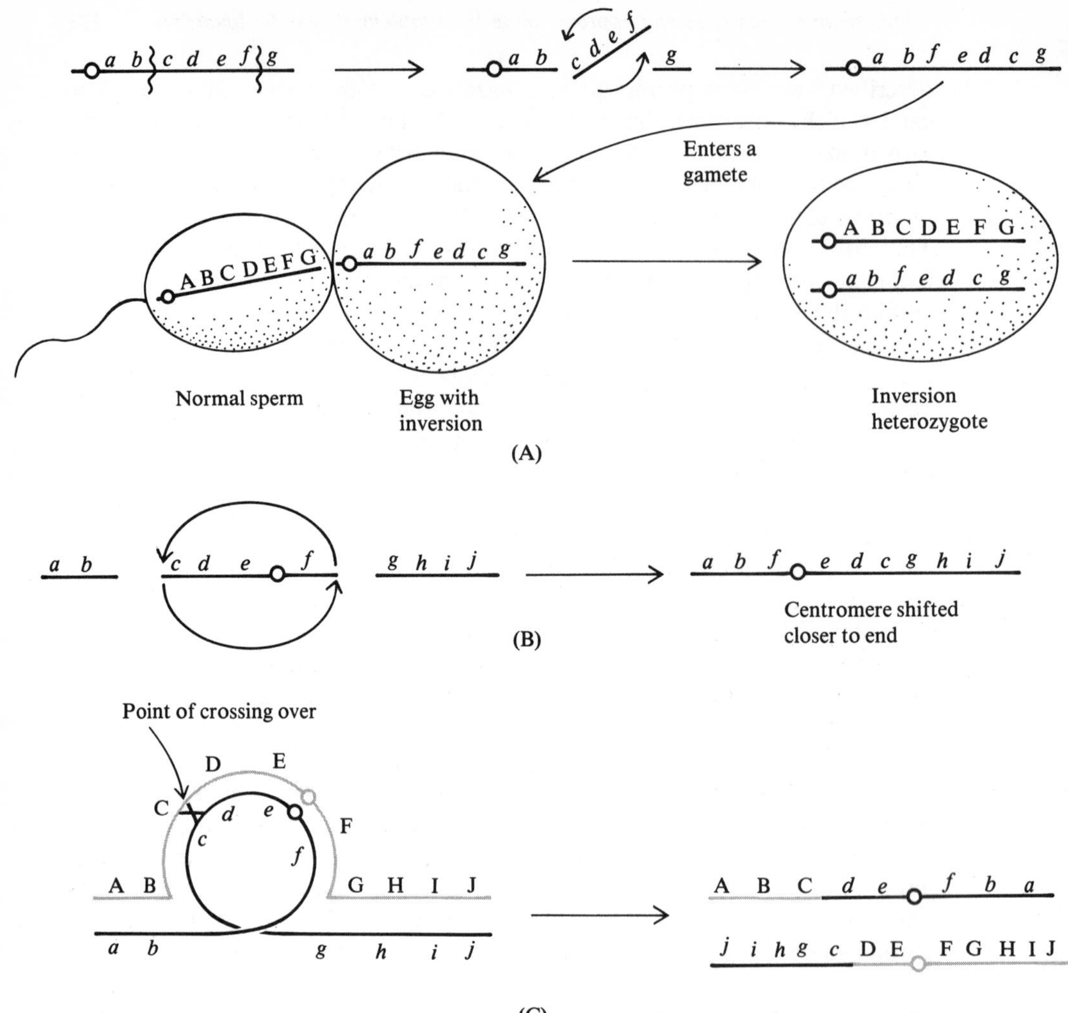

Figure 11-5. The inversion. (A) An inversion arises when two breaks occur in a chromosome and the piece between the ends becomes reversed. A gamete carrying the inversion can function, since no genes are missing. An inversion heterozygote is produced when a gamete with an inversion unites with a gamete carrying the normal gene sequence. The figure shows a paracentric inversion. Only one chromosome arm is involved. (B) In a pericentric inversion, the two breaks occur on each side of the centromere. If the breaks are not equidistant, the centromere position will shift. A person carrying such an inversion may be recognized by the abnormal position of the centromere in the affected chromosome. (C) While an inversion heterozygote possesses the normal balance of genes, a certain number of abnormal gametes will be produced by such an individual. This is a consequence of problems at prophase of meiosis. Since the inversion is present, one of the chromosomes loops so that the homologous regions of both chromosomes can pair. A crossover may occur in the region of the loops. If this takes place, chromosomes which are unbalanced will arise. These unbalanced chromosomes will have some genetic regions present twice and others which are missing. Gametes carrying these unbalanced chromosomes may not function or may result in the appearance of abnormal offspring.

a segment in a reversed order. No genes are lost or added, and an individual with the inversion is usually phenotypically normal. In many well-studied plant and animal species, the type of inversion illustrated in Fig. 11-5(A) has been well documented. Such an inversion is known as a *paracentric inversion,* one in which the two breaks occur to one side of the centromere so that only one arm of the chromosome is affected. The chromosome is not visibly different in appearance after the paracentric inversion, since the centromere position has not been involved. Paracentric inversions have not as yet been implicated in human disorders.

As Fig. 11-5(B) illustrates, if two breaks occur on opposite sides of the centromere, the position of the centromere may shift, giving the chromosome a different appearance, since the lengths of the two arms will have been altered. Such inversions, called *pericentric inversions,* have been detected in humans as well as in other life forms. The change in shape of one member of a chromosome pair due to an inversion can cause pairing problems at meiosis, as discussed in Fig. 11-5(C). The important point to realize here is that an individual carrying a chromosome with an inversion, along with the normal homologue, can give rise to a certain proportion of abnormal gametes. Some of these will be unbalanced due to deleted chromosome material; others will carry duplicated material. Aberrant gamete production due to an inversion can explain the case in which several types of abnormal children were born to a woman who was later found to have a pericentric inversion in chromosome #10. Inversions, while not believed to be a common cause of defects in the human, should nevertheless be considered in families with a high incidence of congenital disorders among the offspring.

Complications of Some Unusual Genetic Conditions

One type of chromosome aberration is known which, when undetected, can produce effects which seem inexplicable. This is *mosaicism,* which was mentioned briefly in Chapter 5 (p. 103). A mosaic individual possesses body cells of different genetic constitutions. Figure 5-4 illustrates how mitotic error during zygote development can generate two lines of cells differing in chromosome composition. Such errors can involve autosomes or sex chromosomes. The extent of the mosaicism depends on when the error arises. If it occurs late in development, a smaller population of aberrant cells will be present than if it occurs early. In the latter instance, the new cell line will have a greater opportunity to divide and produce more cells. In lower forms, such as the fruit fly, some of these errors occur at the first divisions of the zygote so that one-half of the body of the fly contains cells of the chromosome constitution present at fertilization, whereas cells making up the other half have a different constitution (Fig. 11-6). Such bilateral mosaics have not been reported in the human. However, persons have been described who have different degrees of mosaicism. Any body organ or tissue may be affected. The difficulty mosaicism can present in diagnosis is well illustrated by a case of two phenotypically normal young parents who produced Down's children. No explanation was immediately apparent from chromosome analysis, since no anomalies involving chromosome #21 were present in the cells examined. Cells

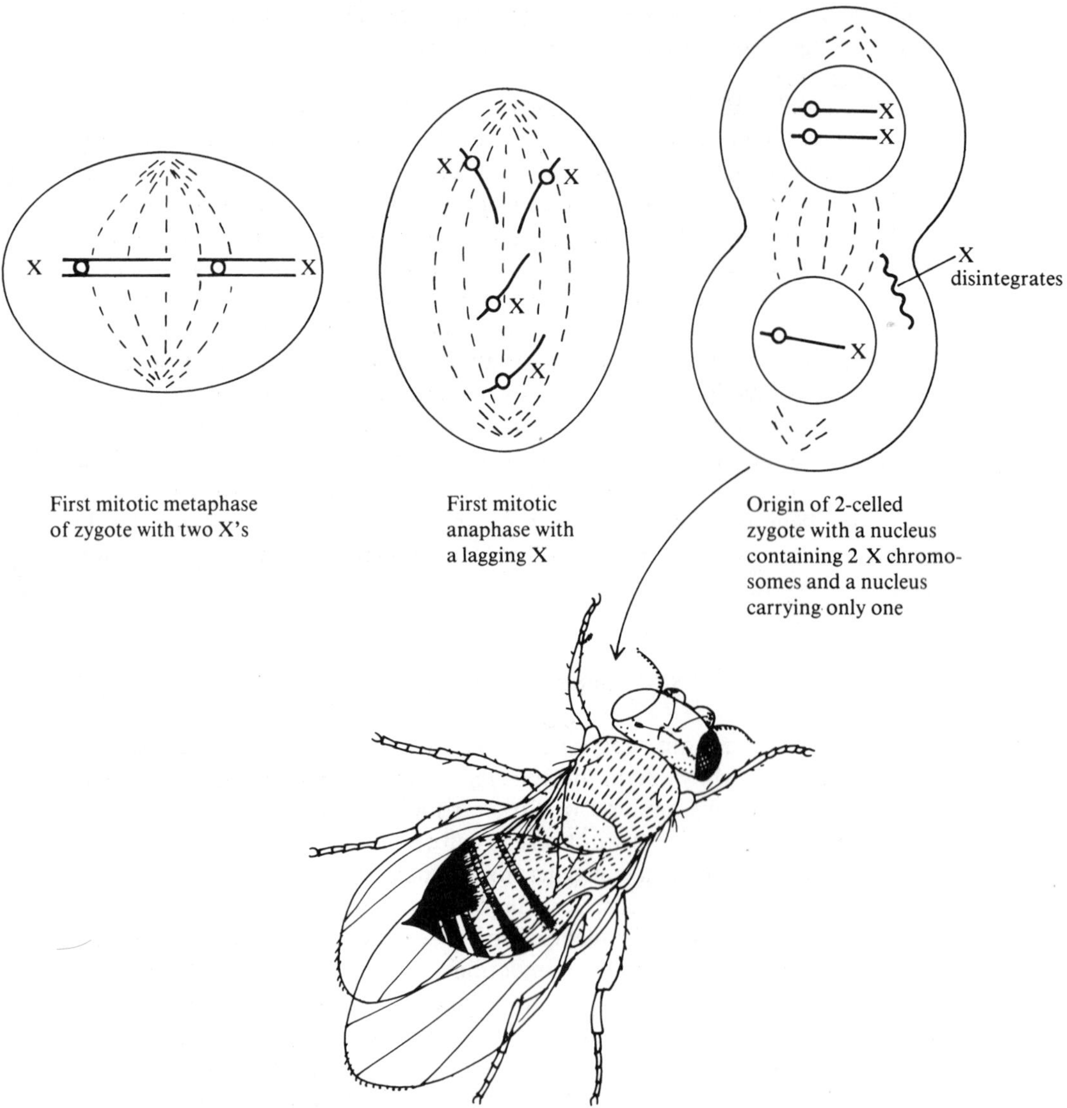

Figure 11-6.

were eventually taken from different body tissues and grown in culture. White blood cells were normal, but an extra chromosome #21 was found in cells taken from the skin and testicles of the father. This example again demonstrates the necessity to consider various alternatives in families with a high incidence of congenital defects among the offspring. Some medical geneticists suspect that mosaicism may be responsible for a significant number of the Down's offspring produced by young women in whom the disruption of oocytes due to age is minimal.

Three very unusual disorders are known which seem to combine the two major types of genetic defect—gene mutation and chromosome aberration. These are Fanconi's anemia, Bloom's syndrome, and the Louis-Bar syndrome. Each of these diseases is inherited as a typical Mendelian autosomal recessive. In addition to several distinct phenotypic symptoms characteristic of each condition, affected persons are more prone to develop leukemia and other types of malignancies than are those persons without the particular recessive gene (see Chapter 14 for more details). Indeed, otherwise normal-appearing heterozygous carriers of the recessive may actually be more disposed to malignancies than the average person. A very peculiar aspect of these syndromes is that the chromosomes of those with the particular disease appear to be very fragile and show a decided tendency to undergo breakage. Fanconi's anemia and the other two syndromes compose a unique group in which a particular gene in each case, inherited in a typical Mendelian pattern, is associated with chromosome changes. We do not expect to see visible effects on a chromosome when a certain gene is present. Exactly what this means in the case of these three conditions is unclear. The association of the disorders with malignancies is

Figure 11-6. Origin of a bilateral gynandromorph. In the fruit fly, occasional errors occur at the first mitotic division of the zygote. An XX zygote will normally give rise to a female fly. In rare cases, one X chromosome fails to reach a pole at anaphase of the division of the fertilized egg. A chromosome that does not get incorporated into a nucleus will disintegrate in the cytoplasm. In the fly, a cell with two X chromosomes and the normal number of autosomes produces a female, whereas a cell with only one X chromosome and the normal autosome number produces a male. In this example, the bilateral mosaic will be female on one side of the body and male on the other side (below). This fly has one red eye on the female side and a white eye on the male side, since the zygote carried an X with the dominant gene for red eye color and an X with the recessive for white eyes. The lost chromosome in this case carried the dominant gene, so that all the cells with one X (the cells in the male half of the body) have only the recessive. The eye on that side is thus white. Such sex mosaics cannot occur in mammals, since hormones would not permit distinct male and female tissue to grow in patches in the same individual. Because embryogenesis is much more complicated in the human, no bilateral mosaics of any kind are known in the human, although lesser degrees of mosaicism do occur.

suggestive, but as in the case of the Philadelphia chromosome, it is unknown whether the breakage may trigger a malignancy or whether some other factor is directly involved, the breakage being merely a secondary association. We will return to this important point in Chapter 14.

The Increasing Challenge of Genetic Impairments

Due to the combined efforts of many trained professionals and educated laymen, we can hope for an increasing public awareness of human genetics, especially in relation to human afflictions. The significance of genetic impairments to society becomes apparent when it is realized that of the admissions to pediatric hospitals, 10% to 25% of them are believed to be related specifically to genetic diseases. Well over 1000 conditions are known to be caused by single mutant genes. In addition, there are those disorders due to chromosome aberrations and those with a more complex genetic basis (diabetes, harelip, cleft palate). The somewhat staggering facts raise several questions. First of all, why does the number of genetic diseases seem to be increasing? The answer is that the true incidence of genetic disorders has probably remained constant but that the relative importance of genetic diseases has increased. Today infectious diseases do not pose the threat to mankind that they did 50 years ago. Infant mortality due to microorganisms has been dramatically cut in those countries where sanitary facilities have improved and where drugs, antibiotics, and vaccines are available. These same factors also enable more persons to survive to an age when certain kinds of genetic disorders may express themselves. Huntington's disease, for example, typically strikes after the age of 25. Decrease in the number of deaths due to infection preserves more individuals and thus increases the chance for the expression of genetic defects.

The conquest of infectious disease is presenting a new kind of challenge to medicine. Although an assortment of germ-transmitted diseases has plagued mankind into the present century, the number of these diseases becomes insignificant when compared to the number of different types of genetic afflictions. The skilled practitioner of yesteryear was alert to the diagnosis, treatment, and prevention of a comparatively small number of infectious disorders which attacked vast numbers of the population. Genetic diseases, on the other hand, compose an extremely large number of very different kinds of disorders, each afflicting relatively few persons in the population. This can be appreciated by the contrast between infectious diseases such as tuberculosis and smallpox, which once could affect thousands of persons at a time in the population, and genetic diseases such as phenylketonuria and Huntington's disease, which afflict only about 1 person in 10,000 or less. The commoner diseases known to have a significant hereditary component (hypertension, diabetes) present still other kinds of challenges in which the interaction of many hereditary factors must be untangled from factors in the environment. More and more, the medical person must be aware of an extremely large number of different types of genetic diseases, most of which he may never personally encounter. In addition, the role of the environment in the expression and prevention of many genetic disorders must be considered. By itself, this latter aspect of the problem may compound the difficulties as the human continues to be exposed to more stresses in the increasingly complex setting of modern life.

Detection of Genetic Defects

Detection plays a primary role in the management of genetic disease. Knowing what is wrong at a specific time may be critical in the prevention of fatal consequences or irreparable damage. Phenylketonuria has already been discussed in sufficient detail (p. 236) to illustrate this last point. In this disorder, no damage is done before birth, but soon afterward, the required enzyme must be present if phenylalanine is to be properly utilized. Feeding the infant a routine diet can doom him or her to a life of mental retardation. Another good example is the detection of the individual with a deficiency of the enzyme glucose-6-phosphate dehydrogenase (Chapter 5). A person who knows he carries the responsible sex-linked recessive and therefore lacks the enzyme is able to avoid the use of certain drugs which can evoke very serious reactions.

Prenatal detection of genetic defects can play a most important role in the prevention or management of some serious disorders. Amniocentesis (Fig. 5-9) has been a major breakthrough in this area and is an indispensable procedure in those cases where there is a high risk of a defective child. One example would be a situation where a mother who already has 1 offspring with Down's syndrome is found to carry the #15–21 translocation. Those cells of the fetus which are shed into the amniotic fluid can be cultured and their chromosome constitution examined. If the cells show 46 chromosomes, including 2 #21's plus 1 #15–21, there is then no doubt that the offspring will be a Down's child. Whether an abortion should be performed is an ethical matter left to the discretion of the persons involved, but nonetheless, the parents are well prepared for what might otherwise have been an unexpected traumatic event.

An increasing number of normal women aged 35 and over are opting to undergo amniocentesis during pregnancy to alert themselves to the possibility of a Down's offspring. Prenatal diagnosis of over 30 genetic disorders can now be made by studying cells obtained by amniocentesis. Another important use of this technique is seen in the prenatal detection of a Tay-Sachs child (p. 242). If 2 parents are known carriers of the responsible recessive gene, amniocentesis is certainly advisable. Cultured amniotic cells can be tested for the presence of the enzyme, hexoseaminodase A. A deficiency of this enzyme indicates an affected child who is certain to suffer from this dread disorder. The parents are given the opportunity to consider termination of the pregnancy. There are also several genetic disorders which can be detected by testing of the cell-free amniotic fluid alone. Unfortunately, many genetic diseases still cannot be detected *in utero.* A good example has been sickle cell anemia. While the disease can be diagnosed from just a minute sample of blood, techniques have allowed no safe way to obtain blood from the fetus. In cases such as this, 2 known heterozygous parents can be told that the chance of their having an afflicted child is 25% and that the chance is 2 out of 3 that any healthy child will be a carrier of the gene. They can then decide on the advisability of having children of their own. Fortunately in the case of sickle cell anemia, detection procedures are being highly refined. Success in prenatal diagnosis of a few sickle cell cases have been reported very recently, offering the hope that accurate prenatal detection may become a reality for this disorder.

A large number of the genetic diseases due to sex-linked recessives cannot be positively diagnosed prenatally. However, the study of cells obtained by amniocentesis can establish some guidelines. For example, a woman who has borne a baby with hemophilia may wish to have another child. Hemophilia, as well as other blood diseases, cannot be positively diagnosed prenatally, but chromosome study of amniotic cells can reveal the sex of the fetus. If the test indicates a boy, there is a 50% chance that the offspring will suffer from the disease. If the test shows the child is a girl, there is no chance, and anxiety will have been prevented. Any question of abortion in such cases can be left to the discretion of the parents. Without the use of amniocentesis, female carriers for sex-linked muscular dystrophy and agammaglobulinemia, two extremely serious sex-linked diseases, may decide to refrain from bearing children. However, prenatal determination of the sex of the offspring can afford them the chance to have healthy children if they so desire.

Genetic Counseling

Medical procedures such as amniocentesis, along with diagnostic techniques which include biochemical tests and chromosome analyses, are becoming familiar to more persons. Medical genetics units associated with certain hospitals and medical centers provide such services along with genetic counseling. "Genetic counseling" is a widely used expression but one which is often poorly defined. This important service is intimately related to medical and diagnostic procedures, but the counseling itself does not involve their actual performance. A genetic counselor may be a medical doctor or a well-trained professional who has specialized in human genetics. However, counseling in all of its ramifications requires the cooperative efforts of many trained personnel, medical persons, those in allied medical professions, and the educated layman. The qualified professional knows that teamwork and expertise from many areas are essential to the efficient performance of a counselor's duties. Primarily, a counselor is concerned with problems arising from the occurrence or the risk of occurrence of a genetic disorder. A counselor must try to explain to persons the way in which heredity operates in the case of a specific trait, if the genetic basis is known. The counselor must try to clarify the meaning of "chance" or "risk" in the appearance or reappearance of a genetic disease. He or she must be able to point out options and to give advice on the courses of action which are available under a given set of circumstances. Furthermore, the counselor must be available to those persons or families who will continue to require advice while making an adjustment in the face of a genetic disease or in the contemplation of its risk.

The effective accomplishment of these tasks makes several demands on the genetic counselor. For one thing, the counselor must have a good knowledge of the massive amount of literature which has accumulated on human genetic disorders. He or she must be aware of thousands of genetic afflictions, although in the course of any counselor's career only a very small percentage of them may be encountered. If the counselor is to benefit all those persons needing advice, it is essential that steps be taken to publicize the availability of counseling services. Educating the public is a most important aspect of counseling. Many who want or need such services are completely unaware of them. Extremely important is the need to alert persons who do not know that they have a high risk of transmitting a serious disorder. Others should be alerted to the fact that if they carry a certain gene, they themselves have a high risk of being harmed following exposure to certain environmental factors. In many cases, contacting a genetic counselor can prevent unnecessary anxiety. For example, a man whose brother suffers from sex-linked muscular dystrophy need not worry about transmitting the gene. Since the disorder is sex-linked, the counselor can assure him that he has no chance of transmitting it. The concerned man does not have the disease and thus must have received from his mother an X chromosome with the normal gene.

Besides locating and advising high-risk individuals, the counselor must also be ready to provide services to those groups or populations which have a high incidence of certain specific diseases. Persons of African descent should be made aware of their higher risk of sickle cell anemia; those of Mediterranean descent of their risk of thalassemia; east European Jews of their greater chance of carrying the Tay-Sachs gene. The counselor can then work with screening programs which provide diagnostic services. By educating a special high-risk group to the nature of the disease involved and how it can be detected, constructive steps can be undertaken to prevent its transmission and the unfortunate consequences it may entail.

To be fully effective, it is obvious that the counselor cannot work alone. He or she depends on an accurate diagnosis by a physician, since the counselor knows that similar symptoms are often associated with some disorders for which the genetic bases are very different. For example, is the muscular dystrophy in a certain case the sex-linked form, or is it the autosomal recessive type? The counselor does not offer advice based on a simple description of an affliction. He or she proceeds with assurance only when professional medical diagnosis has been made. The counselor depends on skilled persons to perform the chemical tests or the chromosome analyses which are so critical to detection and diagnosis. There is a dependence on informed and skilled community leaders who can publicize the need and availability of counseling services in high-risk groups. The counselor, in the follow-up of those cases where genetic disease has already struck, must often work along with the social worker to provide guidance and management of the problem.

Not the least of the requirements for a good counselor is a knowledge of psychology. The counselor must be able to appreciate the emotional trauma associated with the arrival of an afflicted child and also with a person's realization that he or she can transmit a defective gene. There are also psychological problems which arise when a person who seeks advice learns that he or she may run a high risk of expressing at a later date some gene with serious effects. For example, a person who learns that his parent has died of Huntington's disease may suddenly be faced with the realization that he runs a 50:50 chance of carrying the dominant gene and suffering its fatal consequences. The counselor must be prepared to explain in the most humane way that knowledge of a genetic disease, even when the exact molecular basis is understood, does not mean that medical science is necessarily now in a position to cure it or even to alleviate it. The psychological effects which knowledge of a genetic disease can generate in an individual or in high-risk communities cannot be overestimated. They involve many more situations and subtleties than have been mentioned in this brief treatment. The counselor must be alert to all of them and must be able to cope with specific ones as they arise.

It has been estimated that by the middle of the 1980s in North America, 1 person with training in medical genetics will be needed for every 200,000 persons. While 10% to 25% of cases admitted to pediatric hospitals are those which entail a genetic defect, as few as 10% of these may actually receive counseling. The figure is probably even lower when no genetics unit is available at all. It appears that many who could profit from genetic counseling are unable to obtain it. The need for counseling is probably greater than the demand for it. This follows, since the public is still largely unaware of human genetics and the significance of genetic afflictions. More awareness is required, and this calls for public education programs. Some advance has already been made through publicity given certain genetic diseases by concerned organizations, but a concentrated effort on the part of trained professionals in the field is required so that high-risk persons and groups can be reached. The demand for genetic counseling undoubtedly will continue to increase as progress continues to be made in these directions.

REFERENCES

Atnip, R. L., and Summitt, R. L., "Tetraploidy and 18-trisomy in a Six Year Old Triple Mosaic Boy." *Cytogenetics,* 10:305, 1971.

Carr, D. H., "Genetic Basis of Abortion." *Annu. Rev. Genet.,* 5:65, 1971.

Caspersson, T., Gahrton, G., Lindsten, J., and Zech, L., "Identification of the Philadelphia Chromosome as Number 22 by Quinacrine Mustard Fluorescence." *Exp. Cell Res.,* 63:238, 1970.

Francke, U., "Quinacrine Mustard Fluorescence of Human Chromosomes: Characterization of Unusual Translocations." *Amer. Journ. Human Genet.,* 24:189, 1972.

Fraser, F. C., "Genetic Counseling." *Amer. Journ. Human Genet.,* 26:636, 1974.

Friedman, T. F., "Prenatal Diagnosis of Genetic Disease." *Sci. Amer.,* Nov: 34, 1971.

Friedman, T., and Roblin, R., "Gene Therapy for Human Disease." *Science,* 175:949, 1972.

German, J., "Oncogenetic Implications of Chromosomal Instability." in *Medical Genetics,* McKusick, V. A., and Claiborne, R. (Eds.), pp. 39–50. HP Publishing Co., N. Y. 1973.

Hamerton, J.L., *Human Cytogenetics,* Vols. I and II. Academic Press, N.Y., 1971.

Hirschhorn, K., "Chromosomal Abnormalities I: Autosomal Defects." in *Medical Genetics,* McKusick, V. A., and Claiborne, R. (Eds.), pp. 3–14. HP Publishing Co., N. Y., 1973.

Hsu, L. F., Gertner, M., Leiter, E., and Hirschhorn, K., "Paternal Trisomy 21 Mosaicism and Down's Syndrome." *Amer. Journ. Human Genet.,* 23:592, 1971.

Lejeune, J., Gautier, M., and Turpin, R., "Étude des Chromosomes Somatiques de Neuf Enfants Mongoliens." *C. R. Acad. Sci.* (Paris), 248:1721, 1959.

Levy, H. L., "Genetic Screening." *Adv. Human Genet.,* 4:1, 389, 1973.

Licznerski, C., and Lindsten, J., "Trisomy 21 in Man Due to Maternal Nondisjunction During the First Meiotic Division." *Hereditas,* 70:153, 1972.

McKusick, V. A., "Human Genetics." *Annu. Rev. Genet.,* 4:1, 1970.

Nitowsky, H. M., "The Significance of Screening for Inborn Errors of Metabolism."

In *Heredity and Society,* Porter, I. H., and Skalko, R. A. (Eds.), pp. 225–261. Academic Press, N. Y., 1973.

O'Riordon, M. L., Robinson, J. A., Buckton, K. E., and Evans, H. J., "Distinguishing Between the Chromosomes Involved in Down's Syndrome (Trisomy 21) and Chronic Myeloid Leukemia Ph^1 by Fluorescence." *Nature,* 230:167, 1971.

Priest, J. H., *Human Cell Culture in Diagnosis of Disease.* Charles C. Thomas, Springfield, Ill., 1971.

Robinson, J. A., "Origin of Extra Chromosome in Trisomy 21." *Lancet,* 1:131, 1973.

Swanson, C. P., Merz, T., and Young, W. J., *Cytogenetics.* Prentice-Hall, Englewood Cliffs, N.J., 1967.

Swift, M., Cohen, J., and Pinkham, R., "A Maximum Likelihood Method for Estimating the Disease Predisposition of Heterozygotes." *Amer. Journ. Human Genet.,* 26:304, 1974.

Wahrman, J., Atidia, J., Goiten, R., and Cohen, T., "Pericentric Inversion of Chromosome 9 in Two Families." *Cytogenetics,* 11:132, 1972.

REVIEW QUESTIONS

1. Name a disorder or condition in the human associated with (1) trisomy for the sex chromosomes; (2) monosomy for the sex chromosomes; (3) a translocation; (4) trisomy for an autosome; (5) a deletion.

2. Assume that some animal has a haploid chromosome set which consists of five chromosomes designated A, B, C, D, and E. Give the chromosome complement of a body cell which happens to be (1) trisomic for E; (2) triploid; (3) monosomic for A; (4) tetraploid.

3. Give at least one way to account for the origin of humans with the following chromosome constitutions: (1) XXY; (2) XO; (3) XYY; (4) has only one normal metacentric chromosome #1 and an unusual chromosome the size of #1 but which is acrocentric; (5) three complete sets of chromosomes; (6) has only one chromosome #22 and only one normal #13 plus an unusually large chromosome; (7) XXX; (8) XX/XO mosaic; (9) XXY/XX mosaic; (10) XY/XYY/XO mosaic.

chapter 12

PEDIGREES AND PROBABILITY

The Geneticist and the Application of Probability

One very important aspect of genetic counseling is the evaluation of a family history. The person seeking genetic advice from a counselor is known as the *consultand.* Typically, this individual is concerned about his or her own genetic constitution and wishes to know the risk of transmitting a hereditary defect. The counselor will proceed to take as detailed a history as possible and to construct a pedigree from it. Genetic counseling, as mentioned in Chapter 11, is a team affair, and therefore the counselor relies on an accurate medical diagnosis provided by a physician. Working together, the counselor and the diagnostician may at times be able to arrive at a more precise diagnosis of a disease. For example, a pedigree constructed from the history of a family with a high incidence of muscular dystrophy may help to distinguish between types of the disease, a sex-linked form or one due to a gene carried on an autosome.

The counselor must be skilled in the recognition of different kinds of classic inheritance patterns which are associated with single-gene defects: sex-linked patterns, autosomal dominant and recessive ones. Frequently, the counselor may be unable to provide definite answers to questions presented by consultands. Often the clinical risks are unknown for certain diseases, even when a decided genetic component exists. This is particularly true in the matter of disorders such as diabetes and hypertension which involve many genes and unknown environmental factors. At other times, the consultand may lack the information required for the construction of a pedigree which lends itself to interpretation. A counselor must be able to recognize shortcomings in a pedigree which preclude the formulation of definite statements. Diagnosis at times may be in doubt, especially in the case of a disease which affected some ancestor now deceased. The counselor also realizes that an affliction in a particular family, which appears similar to a well-known affliction, may possibly have a unique genetic basis. For example, one cannot simply assume that all cases of albinism are due to an autosomal recessive gene, although they most commonly are (Chapter 4). An individual family may have its own peculiar genetic defect which resembles, but which is nevertheless genetically distinct from, a more common type.

The geneticist must have a good foundation in the rules of mathematical probability before he can state the chances that a consultand carries a defective gene and that the gene will be transmitted. A geneticist always deals with chance or probability. As pointed out in Chapter 1, genetic ratios such as 3:1 and 1:1 actually indicate the chances that a certain event either will or will not occur. Everyone continually deals with probability in the course of daily events. Our lives are lived according to the greater chance that this or that event will take place, although we usually do not think consciously of probabilities. We do, however, become aware of rules of chance and the involvement of probability in various kinds of common situations. We know that a tossed coin, when it falls flat on one side, has a probability of 1/2 of landing head side up and 1/2 of landing tail side up. The occurrence of a head and that of a tail are alternative events. Either alternative has a decided chance of happening, and in this case the probability of each alternative is 1/2 (or 50% or 0.5). It is common to use the letters p and q to represent the probabilities of any two alternative events. In this example, p can represent the probability for a head and q the probability for a tail (or the other way around, as long as one is consistent in a problem). Thus, $p = 1/2$ and $q = 1/2$. Obviously, $p + q$ must be 1, since there are just 2 alternatives. If one of these events does *not* occur, the other one certainly will. If the head does not show up, then the tail will. Therefore the probability is 1 or 100% that one *or* the other will occur once the coin is tossed and falls flat on one side.

Combining Simple Probabilities

This very elementary reasoning is almost instinctive, but at times it is forgotten when a question is stated in rather formal terms. Most people would answer that the chance for a couple to have a boy at a single birth is 1/2 and that it is also 1/2 for a girl. Again, we are dealing with 2 alternative events: p (the chance for a boy) $= 1/2$ and q (the chance for a girl) $= 1/2$. The chance that any baby who is born will be one sex *or* the other, *either* a boy *or* a girl, is 1 or 100% (the chance of the boy + the chance of the girl). This is an example of combining 2 probabilities. The combining of probabilities is a very important manipulation when one is dealing with chance, since it is often necessary to consider the chance of 1 event in relation to 1 or more others and the influence which 1 event may have on another. When we were asked the chance of *either* a boy or a girl arriving at a single birth (or either a head or a tail appearing on a fallen coin), we were considering 2 mutually exclusive events. Events are said to be *mutually exclusive* when the occurrence of 1 of them at a given time makes the occurrence of an alternative event impossible at that same time. If a boy is certain to be born, then obviously a girl cannot arrive, and vice versa. If the coin shows a head at 1 fall, it can't show a tail at the same time. The chance that either *one or the other* of the 2 events will occur at 1 time must be 100%, the sum of the 2 separate probabilities. So when we combine the probabilities of mutually exclusive events, we simply add them together as was done here with the coins and the births.

Probability values are obviously not always equal to 1/2. Consider 2 parents who are carrying the autosomal recessive for phenylketonuria. The genotype of each can be represented as *Pp,* and we are therefore dealing with a simple monohybrid cross: *Pp* × *Pp.* The chance that a normal child will arive at any single birth is 3/4, and the probability for an afflicted child is 1/4 ($p = 3/4$; $q = 1/4$). The chance that the first child will have 1 of the 2 phenotypes (either normal *or* PKU afflicted) is 1 or 100% (3/4 + 1/4), since any individual child must be either normal or afflicted.

Now suppose a question were phrased in a different way. This couple wishes to know the chance of their having a normal child, then a second normal, then a third normal, and finally an afflicted one. In this case we are dealing with *independent events.* Two (or more) events are independent when the chance of occurrence of 1 of them at a given time does not influence the chance of occurrence of the other. The arrival of a normal child at 1 birth does not determine what a second child will be. The chance remains 3:1 or 3/4 in favor of a normal *at every birth.* To answer the question, we simply multiply the separate probabilities for the normal and the afflicted: 3/4 × 3/4 × 3/4 × 1/4 = 27/256. Again we are combining probabilities, but the problem concerns independent events, not mutually exclusive ones, and so we *multiply* the separate probabilities instead of adding them. The probability that 2 or more independent events will occur together or in a certain prescribed way is equal to the product of the separate probabilities. A very familiar application of this principle concerns the arrival of a boy or a girl baby. The occurrence of a girl at 1 birth does not influence the sex of the next child. The probability of a girl continues to remain 1/2, as does that for a boy. And so the chance for a boy, a boy, a girl, a boy, and then a girl in that order is:

$$1/2 \times 1/2 \times 1/2 \times 1/2 \times 1/2 = 1/32 \text{ or } (1/2)^5$$

Once these elementary points on the combining of probabilities are understood, several types of questions can be easily answered. Suppose the 2 carriers for the PKU recessive gene wish to know the chance that their first child will be a normal boy. The answer requires no more than the combining of the probabilities of 2 independent events. The sex of the individual has no influence on whether or not the disorder will occur. The probability for a boy = 1/2 and that for a normal baby = 3/4. The chance for a normal boy (or girl) is thus 3/4 × 1/2 or 3/8. The chance for an afflicted boy (or girl) = 1/2 × 1/4 = 1/8.

What is the chance that the family will have children in this order: a normal boy, a normal girl, a normal boy, and an afflicted boy? To answer, nothing more is required than the multiplication of the separate probabilities:

$$3/8 \times 3/8 \times 3/8 \times 1/8 = 27/4096$$

At times, a counselor may be able to give a consultand a definite yes or no answer. In some cases the probability of a particular event is 100%; in others, it is 0. More frequently, the counselor can only state the mathematical probability as some value between 0 and 1. An example involving a sex-linked gene can illustrate different responses in a given situation. Suppose a young man and his sister wish to know the likelihood of their transmitting the gene for agammaglobulinemia (p. 167), a disorder which caused the death of their infant brother. The counselor, after being assured of the diagnosis by the proper medical persons, may proceed to evaluate the family history. The resulting pedigree indicates a sex-linked recessive [Fig. 12–1(A)]. The counselor is now in a position to supply some answers. He can assure the young man that there is no danger at all of his transmitting the gene. The probability is 0, since the gene is sex-linked. Any male receives his X from his mother. Since the male consultand is free of the disorder, he does not carry the X from his mother which has the defective gene. The young woman, however, cannot be so assured. A female receives 1 X chromosome from each parent. The father in this case obviously carries the normal gene on his X, but the probability that the young woman is a carrier is 1/2, since one-half of her mother's eggs will carry an X with the defective gene. Therefore, one-half of the daughters in such a family will be carriers, but no daughters will develop the disease. One-half of the sons will be afflicted, and the half who are not will not carry the gene at all. So while the counselor can give a definite answer to the young man in this case, he can only tell his sister that there is a chance of 1/2 that she is a carrier. This consultand may then wish to know the overall risk of having an affected child as her first offspring. The overall risk of an affected child is 1/8. This is so for the following reasons. Any husband of this woman will lack the sex-linked recessive, since the gene is lethal and causes death in infancy. There is a chance of 1/2 that the young woman is a carrier. If she *is* a carrier, the cross between her and any man is simply that shown in Fig. 12–1(B). So at any birth, the chance for an affected child is 1/4 *if* the woman is a carrier. But since this is uncertain and there is a probability of 1/2 that she is a carrier, we must modify the value of 1/4 by 1/2, giving a final answer of 1/8. This is again nothing more than combining the probabilities of 2 independent events.

To answer the woman's second question about the chance of an afflicted boy as the firstborn again entails the same reasoning. The chance that the first child will be a boy is 1/2. If the child *is* a boy, the chance is 1/2 that he will have the disease *if* the young woman is a carrier. The chance that she is a carrier is 1/2. The combined probabilities are therefore 1/2 × 1/2 × 1/2 or 1/8. As long as each probability is known for 2 or more independent events, their occurrence together may be easily calculated whether the case involves genes on the sex chromosome or those on autosomes.

To review the concept of mutually exclusive events, suppose the woman asks about the overall chance that the first child will be *either* a non-affected, non-carrier girl *or* a non-affected boy. It must be kept in mind that the chance is 1/2 that the woman is a carrier. Consequently, there is a chance of 1/2 that she is homozygous for the normal gene. If the woman is *not* a carrier, there is a chance of 1/2 that the first child will be a healthy, non-carrier girl and a chance of 1/2 that the baby will be a healthy boy [Fig. 12–1(C)]. But since there is just a chance of 1/2 that she *is* free of the gene, each figure must be modified by 1/2 to give 1/4 (a non-carrier girl) and 1/4 (healthy boy). But we must now consider the results if the woman is a carrier [Fig. 12–1(B)]. If she is a carrier, there is a chance of 1/4 that a healthy non-carrier girl will arrive first

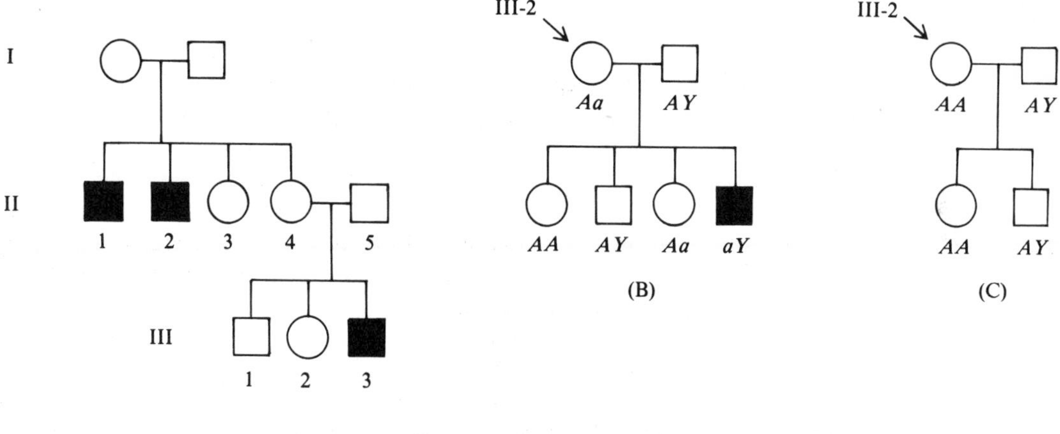

Figure 12-1. Use of pedigrees in counseling. (A) A diagnosis of a disorder that is typically inherited as an X-linked recessive, along with this pedigree, strongly suggests that the trait under consideration in this family is an X-linked recessive. Person III-1 can be certain he is free of the gene, since he would have the disorder if he received from his carrier mother (11-4) the X chromosome with the recessive gene. (B) Person III-2 may or may not be a carrier of the recessive gene. There is a chance of 1/2 that she is. Assuming she is a carrier, there is a chance of 1/4 that she will bear an afflicted child. (C) There is, however, a chance of 1/2 that person III-2 is not a carrier. In this case all the offspring will be normal. Both possibilities shown in (B) and (C) need to be taken into consideration.

and a chance of 1/4 for a healthy boy. Again each figure must be modified by 1/2, since the carrier state in the woman has this probability. This gives 1/8 and 1/8. The overall chance that the first child will be *either* a healthy boy *or* a healthy non-carrier girl is the sum of the separate probabilities for a healthy non-carrier girl and healthy boy, considering *both* the carrier and the non-carrier alternatives:

$$1/4 + 1/4 + 1/8 + 1/8 = 3/4$$

This figure agrees with the previously calculated figure of 1/8 for the chance of an afflicted son as the first child, since of the remaining 1/4 here, half will be carrier daughters and half afflicted sons (1/8 + 1/8). In this simple problem, we performed 2 types of operations. We first combined 2 independent events (multiplying the 1/2 probabilities of the carrier and non-carrier states by the separate chances for a normal boy and girl in each instance). We next combined mutually exclusive events (adding the chances of a normal girl and a normal boy in each instance).

Probability and Degrees of Relationship

The concepts of mutually exclusive events and independent events enable us to appreciate more fully degrees of relationship among family members, a subject treated briefly in Chapter 6. The degree of relationship can be expressed by a value, the *coefficient of relationship.* This figure indicates that proportion of genes which 2 related persons have in common on the average. Obviously, this is 1/2 for a parent and an offspring. Figure 12–2 shows, in general, the relationship between 2 first cousins. The coefficient value between them is 1/8. This can be appreciated from the following reasoning (refer to Fig. 12–2). Let us suppose that we are interested in learning the chance that any gene (assume it is a recessive *a*) is present in *both* of the first cousins, having passed to each of them from a common ancestor. Suppose the gene *a* was carried by the grandfather. The chance is 1/2 that the grandfather (person #1) passed it to his daughter (#4), and the chance would be 1/2 that the daughter would pass it to her daughter (#6). There is also a chance that *if* the grandfather has *a* he passed it to his son (#3), with a probability of 1/2. In turn, the probability is 1/2 that the son passed it to his son (#5). The chance, therefore, that the gene *a*, *if* present in the grandfather, went to both the first cousins, persons #5 and #6, is:

$$1/2 \times 1/2 \times 1/2 \times 1/2, \text{ or } 1/16, (1/2)^4$$

However, the gene *a* could have been present in the grandmother, and it could

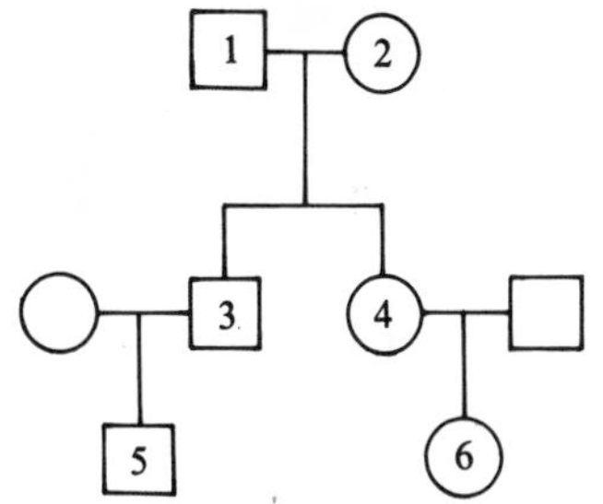

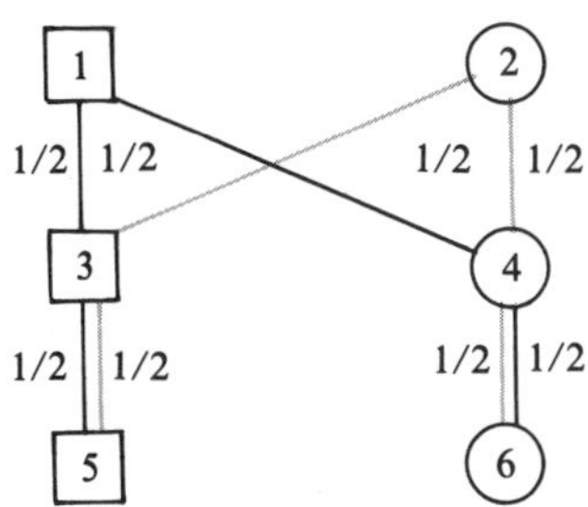

Figure 12-2. Relationship between first cousins. The family diagram on the left shows two first cousins (persons 5 and 6) and indicates by numbers those persons to whom *both* cousins are related. The diagram at right indicates only those persons related to both cousins, persons 5 and 6. The pathway by which a given gene, say *a*, could pass from the grandfather (person 1) to both cousins is indicated by dark lines. The path by which a given gene could be transmitted from the grandmother (person 2) is shown by light lines. The coefficient of relationship between the cousins is 1/8. (See text for details.)

be passed down from her with the same probability to the cousins as just explained *if* it had come from the grandfather: $(1/2)^4$. The chance that the cousins received it from *either* the one *or* the other thus becomes $(1/2)^4 + (1/2)^4$ or $1/16 + 1/16 = 1/8$. The events here are mutually exclusive. And so it would go for any gene that the first cousins carry. The overall chance is therefore 1/8 that they both have any 1 gene which has been derived from a common ancestor, the grandfather or the grandmother.

Now what is the chance that a child of the 2 first cousins is homozygous at a particular locus as the result of having received from both his parents some gene which was transmitted to them from a common ancestor? This is expressed by a value called the *coefficient of consanguinity*. Exactly the same logic used for the derivation of the coefficient of relationship is applied. Figure 12–3(A) shows person #7, the offspring of 2 first cousins. Since every person is diploid, any locus would be represented 2 times in a person. Let us number the alleles present at a locus in the grandfather A_1 and A_2 and the 2 alleles present at the same locus in the grandmother A_3 and A_4. We are simply placing a label on each allele regardless of what it is so that we can follow its transmission. Figure 12–3(B) shows how any 1 of the 4 alleles, in this instance allele A_1, can be transmitted from 1 of the grandparents. The chance is 1/64 that person #7 will be homozygous for the allele A_1, having received the allele from both his parents, who received it from a common ancestor. The chance would also be 1/64 that person #7 is homozygous for A_2 or A_3 or A_4. The chance that person #7 is homozygous for *any one* of them is the sum of the 4 separate mutually exclusive events:

$$1/64 + 1/64 + 1/64 + 1/64 = 1/16$$

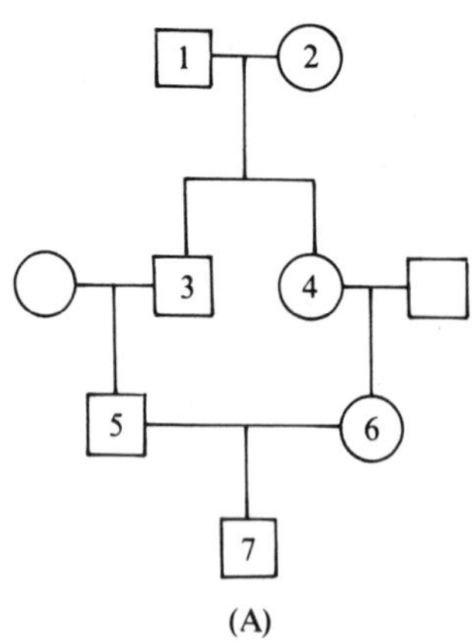

(A)

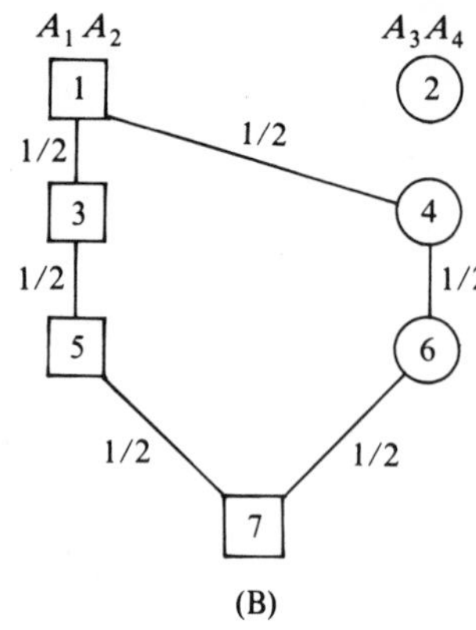

(B)

Figure 12-3. Offspring of first cousins. (A) Person 7 is the offspring of two first cousins. All common ancestors to his parents, persons 5 and 6, are indicated by the numbers. (B) The diagram shows how an allele (A_1 in this example) can be transmitted from the grandfather (person 1) to person 7. The chance is 1/64 that person 7 will have received it from *both* parents and will thus be homozygous for the A_1 allele. The same reasoning applies to A_2, A_3, and A_4, other alleles at the locus under consideration. The coefficient of consanguinity is 1/16. (See text for details.)

This value is the coefficient of consanguinity. It tells us what the probability is that the offspring from 2 related persons will be homozygous at a given locus for an allele transmitted by a common ancestor. The coefficient of consanguinity is equal to one-half the value of the coefficient of relationship. Table 12–1 gives the coefficients of consanguinity and of relationship for several degrees

Table 12–1. Coefficients of consanguinity and relationship for different kinds of parental relationships

Offspring Whose Parents Were Related as Follows:	Coefficient of Consanguinity of Offspring	Coefficient of Relationship of Parents
Father-Daughter	1/4	1/2
Uncle-Niece	1/8	1/4
Double first cousins	1/8	1/4
First cousins	1/16	1/8
Half first cousins	1/32	1/16
First cousins, once removed	1/32	1/16
Second cousins	1/64	1/32
Second cousins, once removed	1/128	1/64
Third cousins	1/256	1/128
Brother-Sister	1/4	1/2
Half-brother–Sister	1/8	1/4

of relationships. The student should try to derive the values in each case. These values indicate that the chance of bringing together in homozygous condition any particular gene is greater with related persons than with those who are not related by blood. One consequence of consanguinity (mating between blood relatives) is an increased chance or risk of homozygosity for an uncommon deleterious gene. The genetic counselor must keep this in mind if marriage between relatives has occurred in the pedigree.

Many questions of probability posed by a consultand are no more complicated than the ones discussed so far in this chapter. A great deal of information can be provided with the proper understanding of probability. One responsibility of the counselor is to be able when necessary to present concepts of chance and to make them intelligible to a consultand. This may be required when the risk has been calculated for the occurrence of a serious human problem. The counselor and consultand can then proceed to select the course of action which seems most appropriate. This may entail consultation with professionals skilled in medical management and in diagnostic and social services. Today genetic teams have available to them an increasing number of biochemical tests which allow more accurate detection of carriers of certain deleterious recessive genes. The reliability for some of these is low, but other tests permit the identification of the carrier with a high degree of certainty. For example, a chemical test with a reliability of 70–80% can be used to help determine if a woman is carrying the X-linked recessive gene responsible for the Duchenne form of muscular dystrophy. If diagnosis by a medical person verifies that this was the form of dystrophy which occurred in the family of a young woman seeking counsel, the counselor could acquaint her with the availability of the test. A woman concerned about carrying the X-linked hemophilia A gene can be informed that a highly reliable test is now available for the detection of heterozygotes. In the case of certain recessive disorders, a concerned person may learn from the counselor that further steps are available to clarify the nature of his or her genotype. Improved detection techniques can eliminate unnecessary anxiety for a person who is actually free of a deleterious gene but who would have been left in doubt without the knowledge a perfected test can provide. Unfortunately, tests are either not available or not accurate for the detection of most deleterious recessives in the heterozygous person. The risk must be estimated entirely from what is known about the family history.

Use of the Binomial Expansion in Genetics

Let us now turn our attention to another type of question which requires the application of another basic probability concept. If a couple asks for the probability of their having a boy, a girl, and then a boy in that order, we can quickly answer that the chance equals 1/8. However, a couple may ask the chance of their having 2 boys and a girl *in any order.* This is very different from the first question and must be handled in another way. Table 12–2(A) shows that the combination of 2 boys and 1 girl can arise in 3 different ways; each of these has a probability of 1/8. Therefore the probability of 2 boys and 1 girl is 3/8. Similarly, the probability for 2 girls and 1 boy in any order is 3/8. There are also 3 ways in which this combination can come about. The table summarizes all the ways that boys and girls may occur in a family of 3 children. Table 12–2(B) shows the same thing for a family of 4 children. The

Table 12–2. Various ways in which boys and girls may be born in families of 3 (A) or 4 (B) children.

(A) Order of Arrival*

3 boys:	$ppp = 1/8$, or p^3	
2 boys and 1 girl:	$ppq = 1/8$	$= 3/8$ or $3p^2q$
	$pqp = 1/8$	
	$qpp = 1/8$	
2 girls and 1 boy:	$qqp = 1/8$	$= 3/8$ or $3pq^2$
	$qpq = 1/8$	
	$pqq = 1/8$	
3 girls:	$qqq = 1/8$, or q^3	

(B) Order of Arrival

4 boys:	$pppp = 1/16$, or p^4	
3 boys and 1 girl:	$pppq = 1/16$	$= 4/16$ or $4p^3q$
	$ppqp = 1/16$	
	$pqpp = 1/16$	
	$qppp = 1/16$	
2 boys and 2 girls:	$ppqq = 1/16$	$= 6/16$ or $6p^2q^2$
	$pqpq = 1/16$	
	$pqqp = 1/16$	
	$qqpp = 1/16$	
	$qpqp = 1/16$	
	$qppq = 1/16$	
1 boy and 3 girls:	$pqqq = 1/16$	$= 4/16$ or $4pq^3$
	$qpqq = 1/16$	
	$qqpq = 1/16$	
	$qqqp = 1/16$	
4 girls:	$qqqq = 1/16$, or q^4	

*p = probability for a boy = 1/2
*q = probability for a girl = 1/2

chance of having a boy, a girl, a boy, and then a girl is 1/16, but Table 12–2(B) shows that this prescribed order is just 1 way in which 2 boys and 2 girls can arrive. There is actually a total of 6 out of 16. So if one simply asks the chance of 2 boys and 2 girls in a family of 4, the answer is 6/16, or 3/8. One question about a family may specify the order of occurrence of certain events, whereas another question may not. To answer the latter type, consideration must be given to the fact that 1 combination, such as 2 girls and 2 boys, may come about in more than 1 way.

Exactly the same reasoning applies when we are dealing with 2 alternative events whose separate probabilities depart from 1/2. We return to the 2 persons heterozygous for the PKU recessive. For 2 such parents, the risk of an affected child is 1/4 at each birth, whereas the chance for a normal is 3/4. If these people ask the chance of their having first a normal, then an afflicted, and then a normal, the answer is simply 9/64 ($3/4 \times 1/4 \times 3/4$). On the other hand, if they ask the chance of 2 normal and 1 affected in any order, we must consider the various ways in which this can happen. Table 12–3A illustrates this. This table is exactly the same as Table 12–2A, only now the chance for a normal child at any birth is represented by the letter p and that for an afflicted by q. There are 3 ways that 2 normal and 1 afflicted children may arrive, and each has a probability of 9/64. This means the overall probability for the combination of 2 normal and 1 afflicted is 27 out of 64 (the sum of the probabilities for the 3 ways).

Table 12–3. Various ways in which offspring may arise when $p = 3/4$ and $q = 1/4$.

3 normal:	$ppp = 3/4 \times 3/4 \times 3/4 = 27/64 = p^3$	
2 normal, 1 PKU:	$ppq = 3/4 \times 3/4 \times 1/4 = 9/64$	$= 27/64 = 3p^2q$
	$pqp = 3/4 \times 1/4 \times 3/4 = 9/64$	
	$qpp = 1/4 \times 3/4 \times 3/4 = 9/64$	
2 PKU, 1 normal:	$qqp = 1/4 \times 1/4 \times 3/4 = 3/64$	$= 9/64 = 3pq^2$
	$qpq = 1/4 \times 3/4 \times 1/4 = 3/64$	
	$pqq = 3/4 \times 1/4 \times 1/4 = 3/64$	
3 PKU:	$qqq = 1/4 \times 1/4 \times 1/4 = 1/64$	

It is obvious from Tables 12–2 and 12–3 that a problem can become very cumbersome if it is necessary to consider every type of combination possible in a given family of 3, 4, 5, or more. There is, however, a quick way of solving questions of this type which eliminates the need to consider every single possibility and allows us to concern ourselves with only the combination needed to answer a question. We simply expand the binomial $(p + q)^n$, where n represents the total number of trials, e.g., children in a family, tosses of a coin, etc. Anyone familiar with basic mathematics will recognize that the binomial expansion is applicable in the cases we have just been discussing. All the different combinations in Tables 12–2 and 12–3 form a binomial distribution. Actually, the diagram of a cross between any 2 monohybrids, such as $Pp \times Pp$, (Fig. 12-4) is really an expansion of $(p + q)^2$, as shown in Table 12-4. The ordinary ratio 1:2:1 reflects the fact that in a monohybrid cross, there is only 1 way in which each kind of homozygote can form, but there are 2 different ways in which the heterozygote can arise. In most problems dealing with human families, the binomial need not be expanded to any high power, since human families are small. Table 12–4 shows how the binomial can be expanded for any number of trials. Let us see how an expansion can be applied to a question in human genetics.

$Pp \times Pp$

	P 0.5	p 0.5
P 0.5	PP 0.25	Pp 0.25
p 0.5	Pp 0.25	pp 0.25

$= 25\,PP : 50Pp : 25pp = p^2 : 2pq : q^2$

Figure 12-4. The cross of any two heterozygotes gives a ratio of 1:2:1 and is really an expansion of the binomial $(p + q)^2$.

If the persons heterozygous for the PKU recessive ask the chance of their having 3 normal children and 1 afflicted, the binomial must be expanded to the fourth power (Table 12–4). We can allow p to stand for the probability of a normal and q for an afflicted. We next select the term $4p^3q$ from the expansion. This is the appropriate term, since the exponents in a term designate the numbers specified in the question for each of the 2 alternatives. Therefore, this term stands for the probability of 3 normal children and 1 afflicted. Substituting the known probabilities for p and q, simple arithmetic gives the answer:

$$4 \times (3/4)^3 \times 1/4 = 54/128$$

Note that we could just as well have allowed p to represent the afflicted and q the normal. Then we would have selected the term $4pq^3$ for 1 afflicted child and 3 normal. The answer again comes out to:

$$54/128\ (= 27/64, \text{ or approximately } 7/16)$$

The information given earlier on combining probabilities can also be used along with the application of the binomial expansion. For example, if the

Table 12–4. Expansion of the binomial to the fourth power. The expansion of the binomial to any power (n) is easily accomplished by following a few simple rules. In this example, $n = 4$. The number of terms always equals one more than n, so that five terms will be involved here. The exponents of these terms are easily derived. The first term will always be p to the power of n; so here the first term is p^4. The last term will always be q to that same power; so here the fifth term is q^4. The other terms will contain both p and q Starting with term 2, p decreases in value as q appears. This decrease of p and increase in the value of q continues until the last term is reached. The final step is to obtain the coefficient for each term. For the first and the last terms, it is always 1. For the second term, the coefficient is always equal to n, giving us 4 here for the second term. To obtain the value for the coefficient for the third term, we return to the second term. The exponent value of p of the second term is multiplied by the coefficient of that term. This product is then divided by the number of that term in the expansion. Since it is the second term, the value of the coefficient of the third term is $4 \times 3 \div 2 = 6$. For every succeeding term, the same process is followed. Using these rules, $(p + q)$ is shown below expanded to the fifth power.

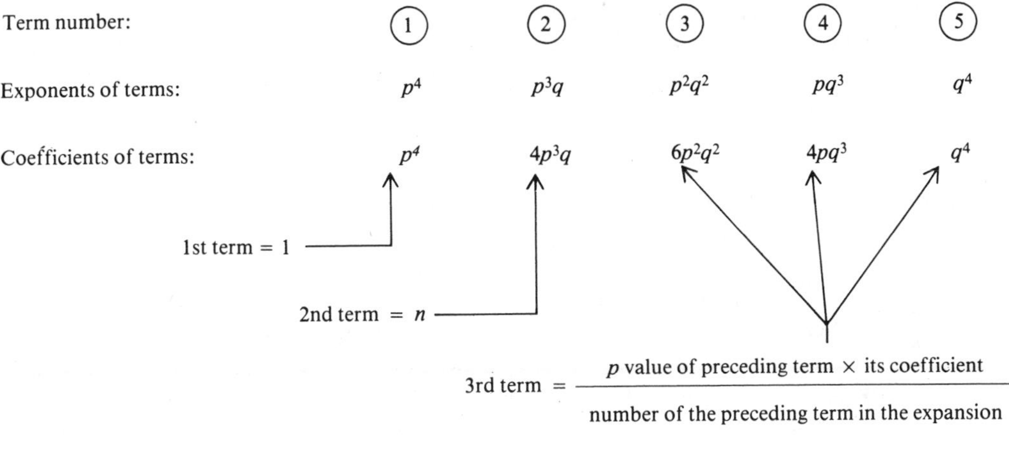

Term number:
1
2
3
4
5
Exponents of terms:
p^4
p^3q
p^2q^2
pq^3
q^4
Coefficients of terms:
p^4
$4p^3q$
$6p^2q^2$
$4pq^3$
q^4
1st term = 1
2nd term = n
3rd term = $\frac{p \text{ value of preceding term} \times \text{its coefficient}}{\text{number of the preceding term in the expansion}}$

$(p + q)^5 = p^5 : 5p^4q : 10p^3q^2 : 10p^2q^3 : 5pq^4 : q^5$

couple wished to know the chances of their having 3 normal boys and 1 afflicted girl, we could allow p to represent the probability for a normal boy at any birth and q the probability for an afflicted girl. The probability for the former is $3/4 \times 1/2$ or $3/8$, and that for the afflicted girl is $1/4 \times 1/2$ or $1/8$. The term $4p^3q$ is selected from the expansion of $(p + q)^3$ and the probabilities substituted: $4\,(3/8)^3 \times 1/8$.

The binomial is just as applicable to problems involving sex-linked genes. The example of the young woman concerned about transmitting the gene for agammaglobulinemia can illustrate this. Suppose she asks the chance of having 2 normal boys and 2 normal girls. Allowing p to stand for the boys and q for the girls, we select the term $6p^2q^2$ from the expansion (Table 12–4). If we take into consideration that she may or may not be a carrier (p. 290), the chance for a normal boy is 3/8 at any birth, and for a normal unaffected girl it is 1/2. Substituting in the selected term, we obtain the answer:

$$6(3/8)^2 \times (1/2)^2 = 54/256\ (= 27/128, \text{ or approximately } 7/32)$$

With the information presented so far, it is possible to give the probabilities for 2 alternative events occurring in a variety of ways. Many questions which can be asked in practical situations are of this type and can be handled with this kind of basic statistical knowledge. More complex questions could be presented, such as the chance that a woman carrying a harmful sex-linked recessive could have an afflicted boy, a normal boy, and 2 normal girls in any order. The binomial expansion cannot be applied here, since it is applicable to only 2 alternatives, and this question involves 3. Special mathematical formulas are applied to cases such as this which involve the occurrence of combinations with more than 2 alternatives. However, the same basic reasoning is used as that for the simpler questions. The handling of more complicated problems is somewhat beyond our present purpose, which is to introduce the student to basic probability concepts and their application to human genetics.

Representative Pedigrees and Genetic Counseling

Throughout the preceding chapters, we have encountered pedigrees involving a pair of alternative traits. Chapter 1 (Fig. 1–4) introduced some of the conventions used in the construction of a pedigree. The counselor must be able to evaluate a pedigree and to recognize any classic inheritance pattern which it may show. Certain patterns are so representative of single gene differences that the manner of transmission of certain traits can be recognized easily. When a pedigree is constructed, it is usually for the purpose of following a trait which is rather uncommon, at least one which occurs less frequently than the alternative trait. This will be assumed throughout the discussion of pedigrees which follows. The reason for stressing this point will become clear shortly.

If a counselor is presented with a pedigree such as that shown in Fig. 12–5, he would have little doubt that it reflects the transmission of a trait inherited as an autosomal dominant. The reason for this conclusion is that the trait never skips a generation. If a person expresses it, one of the parents did so as well. If two parents fail to show the trait, none of their offspring do, and the trait no longer appears in that family line. A complication that could enter the picture in the case of a dominant gene (or a recessive) is that of penetrance, where the trait may skip a generation (Chapter 1). The recognition of reduced

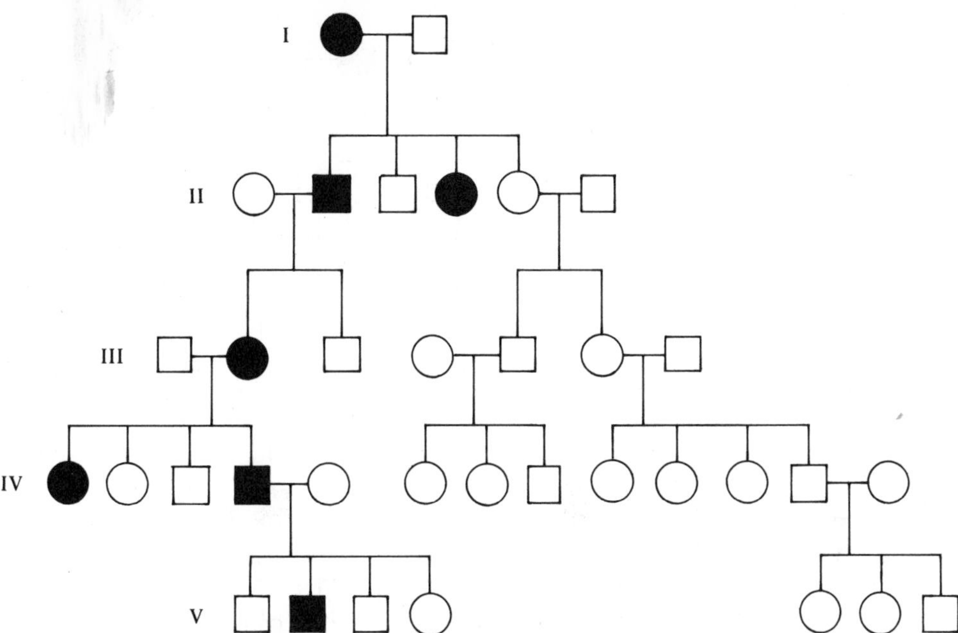

Figure 12-5. Pedigree showing the inheritance pattern typical for an autosomal dominant.

penetrance requires the skill of the counselor. We will ignore reduced penetrance in our discussion of pedigrees, but it must be kept in mind that it is a factor which can alter a dominant or recessive inheritance pattern and make the interpretation of a pedigree more difficult.

The pedigree in Fig. 12–6 is quite different from the preceding one. Most evident is the fact that the trait skips several generations and then reappears. This is typical of an autosomal recessive. Another very classic feature of the pedigree is that two first cousins have borne offspring who show the trait. The likelihood that an uncommon recessive gene will come to expression is greater, as we noted in this chapter, among the offspring of two family members. Two first cousins have in common a pair of grandparents, and therefore both have a certain proportion of their genetic material which traces back to the same relatively recent source. The chance is therefore much greater for them than for two unrelated people that they carry in the heterozygous condition certain identical recessives which can come to expression in their offspring. The pedigree shows that the trait does not appear whenever a mating takes place outside the family. A pedigree such as this one clearly indicates an autosomal recessive mode of inheritance.

The pedigree shown in Fig. 12–7 is typical of one for a sex-linked recessive. The distinguishing feature is that almost all the afflicted persons are

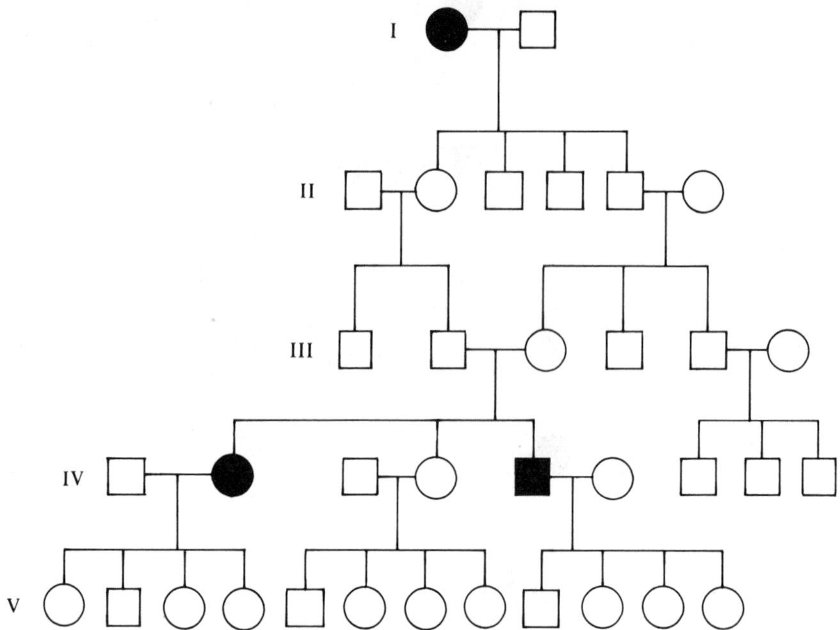

Figure 12-6. Pedigree showing the inheritance pattern typical for an autosomal recessive.

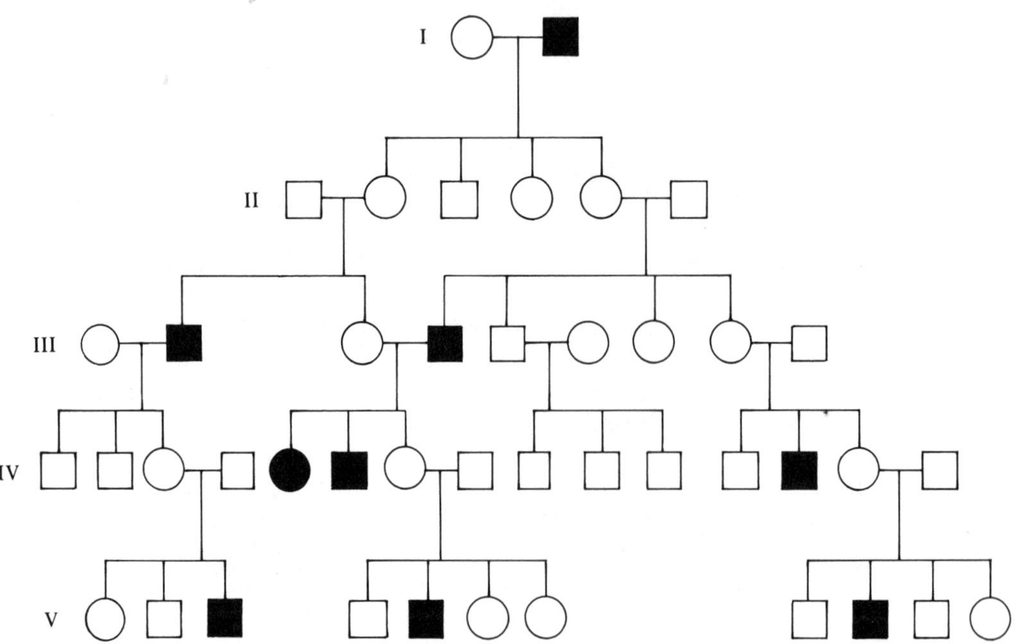

Figure 12-7. Pedigree showing the inheritance pattern typical for a sex-linked recessive.

males. The trait appears to skip a generation. When a male shows the trait, it does not appear in the next generation when he marries outside the family. It can, however, appear in the sons of his daughter. The pedigree shows one female with the trait. Note again that consanguinity is involved here. The father of the afflicted female expresses the trait, and the mother has an affected brother. The mother must be a heterozygous carrier. Any female expressing a sex-linked recessive must have a father who expresses the trait. Her phenotypically normal mother must be a carrier.

The pedigree shown earlier in Fig. 12–1(A) for agammaglobulinemia is typical of one for a rare sex-linked lethal. Since no male with the trait lives to reproductive age, only males can be afflicted. Obviously, no female in the case of such a gene can ever have an afflicted male parent. The trait is passed to sons by way of heterozygous mothers.

A gene on the Y chromosome would affect only males in a pedigree and would be transmitted directly from father to sons (Fig. 12–8). We will say no more about this kind of pattern, which certainly is the only one for the transmission of the Y, since there is no clear-cut evidence for any genetic factors on the Y which influence any characteristic other than those for masculinity. As noted (p. 81), there is some reason to believe that the genetic determinant for the hairy ear trait may be located on the Y chromosome; but the evidence is by no means conclusive. Conceivably, some Y-linked gene which affects some other characteristic will eventually be demonstrated. The

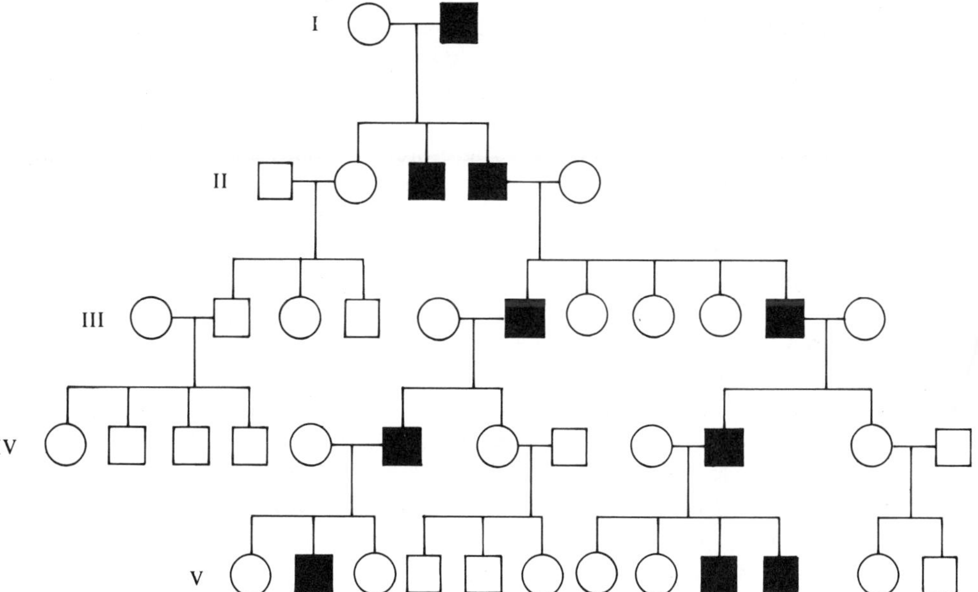

Figure 12-8. Pedigree showing the inheritance pattern typical for a Y-linked gene.

father-to-son transmission and the complete absence of affected females would be typical of the pedigree.

A few additional points must be considered in pedigree evaluation. Suppose a pedigree is submitted such as that shown in Fig. 12–9. Little information is provided other than the fact that a man is expressing a trait as do some

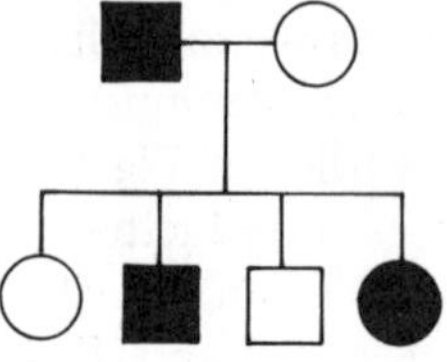

Figure 12-9. A pedigree such as this tells almost nothing by itself. Additional information is required if any definitive statements are to be made.

of his children. By itself, this pedigree offers no information on the inheritance pattern. One cannot identify the trait as an autosomal recessive, since either parent could be the heterozygous carrier. If the trait is an autosomal dominant, then the afflicted person would be the carrier, and the normal allele would be recessive to it. It is equally possible that the trait is due to a recessive, and the normal-appearing person is a carrier. So with no additional information, nothing can be said about the way in which this trait is inherited. If the trait is a rare one, and if it is known that the parents are unrelated, then the probability is great that the trait is due to a dominant gene. This would be likely, since the gene for the rare trait would obviously be rare in the general population. We would not expect the normal unrelated person in the pedigree (Fig. 12–9) to carry the same rare gene as the afflicted person. If we assumed, on the other hand, that the trait *is* due to a recessive, then the unrelated parent would have to carry the same rare gene as the afflicted parent in order to produce affected offspring. The likelihood of this combination of circumstances is improbable. Therefore, we could safely conclude that the trait more likely has a dominant gene basis. If, however, the parents are first cousins, then we can say nothing about dominance or recessiveness, since the trait, even if due to a rare recessive, could be carried by the normal-appearing parent. So without any additional information, a pedigree such as this one is extremely limited, and caution must be exercised before making any comments about it.

A pedigree by itself may therefore offer little or no aid in some cases. Human disorders which may be due to many genetic factors (hypertension, etc.) and/or which are greatly influenced by the environment will not tend to show the clear-cut patterns shown in Figs. 12-5 through 12-8. At present, all that can be said about most of them is that they have a familial tendency. The mode of inheritance of traits which depend on a single pair of alleles can usually be much more readily identified from pedigree analysis than can the pattern for those in which many genes are interacting in complex ways with modifying environmental influences.

Linkage, Probability, and Prenatal Diagnosis

Knowledge of linkage and chromosome maps in the human can, in some cases, be used along with pedigrees to aid in prenatal diagnosis. Chapter 6 included a discussion of the concept of linkage (occurrence of genes together on the same chromosome) and crossing over (the separation of linked genes), as well as a summary of the logic in chromosome mapping. While progress has been slow in the construction of human chromosome maps, enough information has been acquired to permit tentative mapping of some of the genes on the X chromosome and on chromosome #1. Well over 100 sex-linked genes are known, and over a dozen of these have been assigned positions on a map. This has been accomplished mainly by utilizing the data assembled from pedigree analyses. Recall that if 2 genes are close together on a chromosome (and hence map closely together), they will tend to be transmitted to the same gamete and thus be inherited together. The farther apart that 2 genetic sites are on the chromosome, the more likely they are to be separated at meiosis when exchange takes place between maternal and paternal chromosomes (Fig. 6–4). Indeed, if 2 genes are sufficiently far apart, the frequency of crossing over may be so high that they could appear to be unlinked, just as if their locations were on 2 different nonhomologous chromosomes. There are several human genes which have been assigned to the same chromosome on the basis of cell hybridization studies (p. 129) but which have not as yet been shown to be linked in their inheritance patterns. As noted (p. 130), the term *synteny* is used by many human geneticists to designate the association of genes on the same chromosome since distant loci may not show genetic linkage. As more detailed chromosome maps are constructed and as more complete linkage groups are established, linkage information will be used more routinely by genetic counseling teams.

The map in Fig. 12–10 shows that the locus for the gene responsible for the commonest form of hemophilia (hemophilia A) is closely linked to the one for production of G6PD, the enzyme whose absence can lead to drug sensitivity. The loci have been estimated to be about 6 map units apart on the X chromosome. Therefore there is the chance that 6% of the gametes will show new combinations between genes at these two loci. Suppose a woman is concerned because she has been identified as a carrier of hemophilia by a diagnostic test. Her father did not have hemophilia, but her mother was a carrier, and she is now certain that she received the X with the defective gene from her mother. This woman knows that the chance is 1/2 that any male child of hers will have hemophilia. Further information on this family reveals that the mother of this woman lacks the ability to produce the enzyme G6PD, whereas her father can produce it. Figure 12–11(A) shows what the arrangement of the genes must be on the X chromosomes of these 3 persons.

The woman, married to a normal man, becomes pregnant and does not

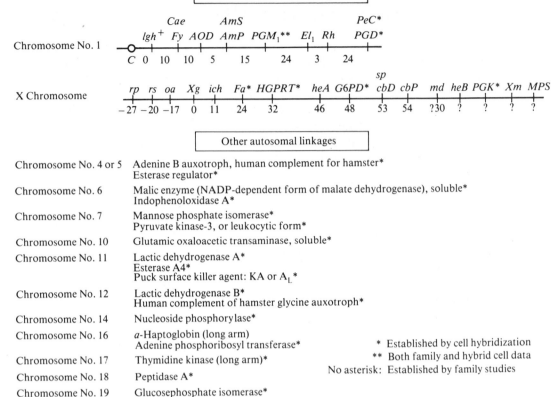

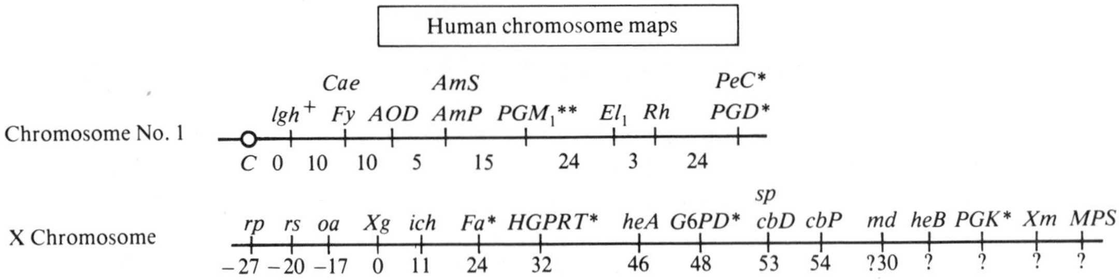

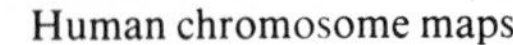

Chromosome No. 1

	Cae		AmS				PeC*
Igh+	Fy	AOD	AmP	PGM_1**	El_1	Rh	PGD*

C 0 10 10 5 15 24 3 24

X Chromosome

rp	rs	oa	Xg	ich	Fa*	HGPRT*	heA	G6PD*	sp cbD	cbP	md	heB	PGK*	Xm	MPS
−27	−20	−17	0	11	24	32	46	48	53	54	?30	?	?	?	?

Other autosomal linkages

Chromosome No. 4 or 5	Adenine B auxotroph, human complement for hamster* Esterase regulator*
Chromosome No. 6	Malic enzyme (NADP-dependent form of malate dehydrogenase), soluble* Indophenoloxidase A*
Chromosome No. 7	Mannose phosphate isomerase* Pyruvate kinase-3, or leukocytic form*
Chromosome No. 10	Glutamic oxaloacetic transaminase, soluble*
Chromosome No. 11	Lactic dehydrogenase A* Esterase A4* Puck surface killer agent: KA or A_L*
Chromosome No. 12	Lactic dehydrogenase B* Human complement of hamster glycine auxotroph*
Chromosome No. 14	Nucleoside phosphorylase*
Chromosome No. 16	a-Haptoglobin (long arm) Adenine phosphoribosyl transferase*
Chromosome No. 17	Thymidine kinase (long arm)*
Chromosome No. 18	Peptidase A*
Chromosome No. 19	Glucosephosphate isomerase*
Chromosome No. 20	Adenosine deaminase (?)*
Chromosome No. 21	Indophenoloxidase B* Antiviral protein A VP*

* Established by cell hybridization
** Both family and hybrid cell data
No asterisk: Established by family studies

Figure 12-10. Human chromosome maps. The tentative maps above are for chromosome #1 and the X chromosome.

Figure 12-11. Use of linkage information in prenatal diagnosis. (A) The genotypes of related persons may be deduced in a case such as this. The daughter knows she is a carrier for hemophilia. Her father is normal and can produce the enzyme G6PD. Since the enzyme and the hemophilia loci are linked, the daughter must carry both dominant genes on the only X chromosome she can receive from her father. Her mother is a non-enzyme-producer and a carrier. Since the woman concerned has been shown to be a carrier by a diagnostic test, the X she received from her mother must carry both sex-linked recessives. (B) The woman in (A) will produce four kinds of eggs in very unequal proportions. The recombinant types will make up only 6%. Any male offspring has only a slight chance of receiving an X which has arisen from a crossover event. Since the recessive for hemophilia was on the same chromosome as the recessive affecting enzyme production, lack of enzyme in the amniotic fluid when a male is being carried by such a heterozygous woman indicates that the defective gene is probably present. Presence of the enzyme indicates that the male is most likely not a victim of hemophilia.

wish to bear a child with hemophilia. The linkage information can be of aid in establishing the chances that the unborn child will be a hemophiliac. The woman's husband can be ignored, since only a male baby in this case runs the risk of hemophilia, and a father does not give his sons his X chromosome. Figure 12–11(B) shows the kinds of eggs this woman can form with regard to

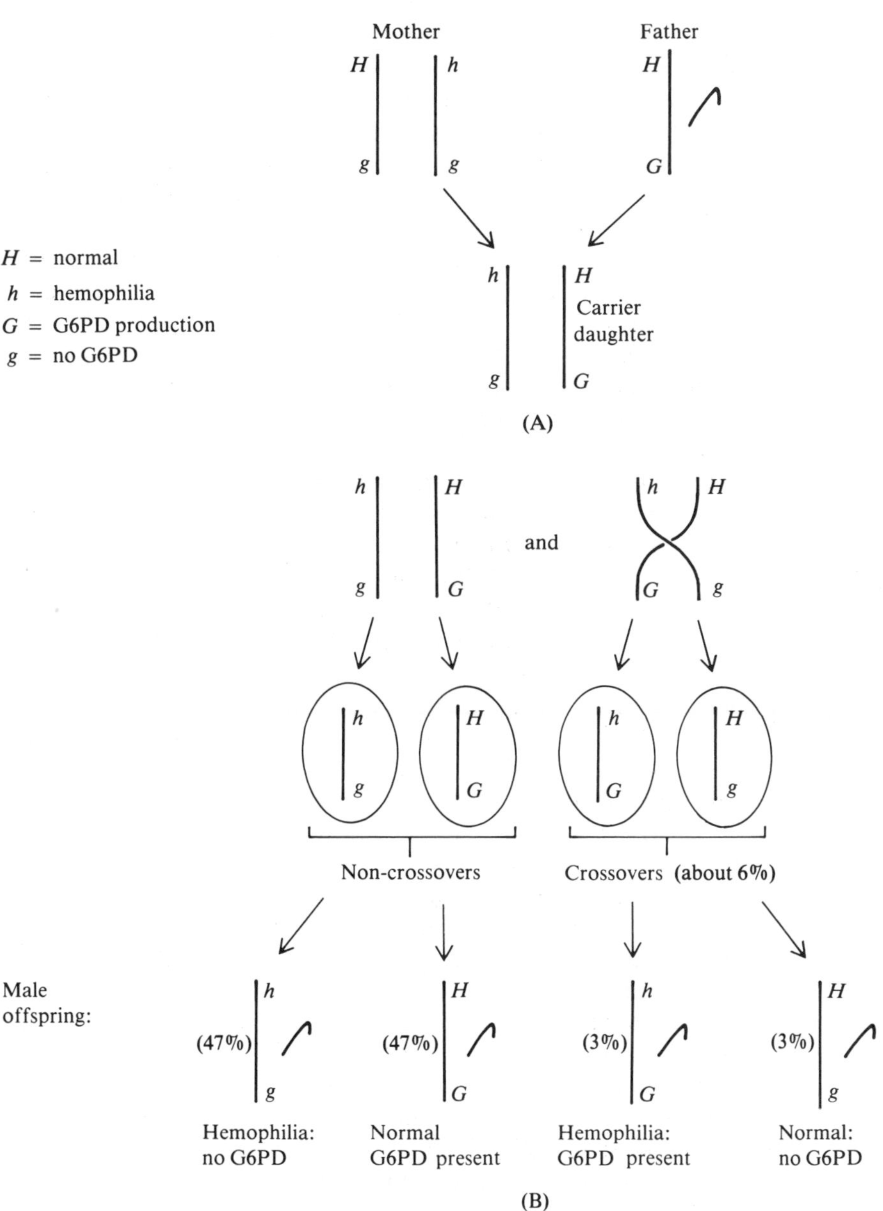

Figure 12-11

the X chromosome. Since the hemophilia locus and the enzyme locus are so closely linked, genes at these loci will tend to be inherited together in their original combination. Only a small percentage of the gametes, about 6%, would be expected to be crossover types carrying new combinations.

At the appropriate time in the pregnancy, the woman may opt to have amniocentesis. The cells obtained from this procedure may be cultured for study. If their chromosome constitution shows a Y chromosome, then it is positive that a male is being carried. The cells can then be tested for G6PD enzyme production. If the enzyme *is* present, the chances are very great (47/50) that the child will be normal. This is so because 1 of the woman's X chromosomes originally carried the normal gene along with that for enzyme production. Since the 2 genes are so closely linked, they will tend to stay together. If the enzyme is *not* present, then conversely, the child is most likely to be hemophiliac, since the defective gene occurs in combination with the gene for lack of the enzyme. The gene locus for G6PD production is being used in this case as a *marker.* The enzyme can be detected when the dominant gene *G* is present at the locus. When the recessive allele *g* is present instead, no enzyme will be detectable. Since the genotype of the woman was known, the marker can be followed to indicate the chance that a gene linked to it and which produces no detectable product *in utero* is also present. While it is possible that crossing over can take place, the probability is low (3/50) and therefore, it is most unlikely that the gene for enzyme production will be in a new combination with the defective gene. If no enzyme is detectable, the chance is very great (47/50) that the child carries the recessive for hemophilia. The woman, along with the counseling team, can then proceed to take the course of action which she finds suitable to the situation.

Similar reasoning utilizing information on linkage can be applied to prenatal detection of a few other human disorders. The dominant gene responsible for the serious disease, myotonic dystrophy, has been found to be on an autosome to which at least two other genes have been assigned. It is interesting to note that, although these genes are known to be linked together on one of the autosomes, the specific autosome on which they are found is not as yet known. (It should be mentioned that advances in our knowledge of human linkage are occurring so rapidly that we can expect the assignment of these loci to be made in the near future.) Nevertheless, as we shall see, the linkage information can be put to good use. The dominant dystrophy gene and its normal recessive allele do not produce any product which can be detected prenatally; so it is impossible to tell by following the dystrophy locus alone whether an unborn child will be afflicted or not. Moreover, the defective dominant gene does not necessarily express itself early in life; even after birth, no detectable product can be found in the body fluids. And so, the gene remains unexpressed until it strikes the person carrying it. In some cases, however, linkage information can aid in prenatal diagnosis. The marker locus here is the *secretor* locus which is very close on the chromosome to the dystrophy locus. The secretor locus is so-named, since the dominant allele *S* permits the secretion of antigens A and B into body fluids when a person is carrying genes *A* and *B* at the ABO blood group locus.

Suppose a woman is blood group A, but she is *not* a secretor. The A antigen is therefore absent in her saliva and other body fluids. Her husband has the dominant gene for myotonic dystrophy and carries it on the same chromosome as the dominant secretor gene (Fig. 12–12). The man is blood type AB, and he therefore secretes both antigens, A and B. The homologous chromosome in the man carries the normal recessive at the dystrophy locus in combination with the non-secretor trait. This information on genotypes was deduced from a family history. The man's normal mother was a non-secretor; therefore, he must have received a chromosome from her with the two recessives linked together. His afflicted father was a secretor, and since the

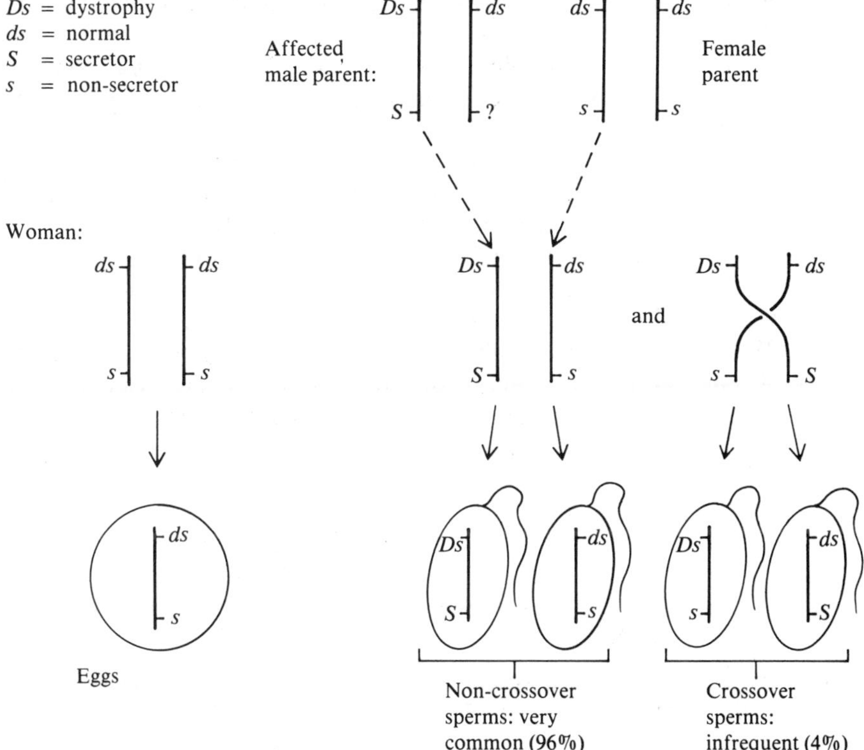

Figure 12-12. Use of linkage in prenatal detection of myotonic dystrophy. The distance between the *Ds* and the *S* loci has been estimated as 4 map units. Therefore, only 4% of the gametes will show new combinations of the genes at these two chromosome sites.

man himself is afflicted and a secretor, he must have received both dominants from his father. Since the loci are linked, the *Ds* and *S* must be together on one chromosome. This man and his normal wife are concerned about having an afflicted child and prefer to terminate the pregnancy if there is a high risk of having a child with the disorder.

The genetic information on the family can be put to good use in this case. Since the man is blood group AB and his wife is A, any offspring must be either A, B, or AB. This means that if the child also carries the dominant *S* gene for the secretor trait, then A or B antigens or both will be found in body fluids, including the amniotic fluid. As Fig. 12–12 shows, the man can produce four classes of sperms, but those with the new combinations between the dystrophy and secretor genes will be infrequent in comparison with the gametes carrying the old combinations. The secretor locus is being used as a marker for the dystrophy gene which produces no detectable product. During pregnancy, if withdrawn amniotic fluid contains any A or B antigens, then the unborn child is a secretor and most probably carries the dominant dystrophy gene as well. Should the fluid give a negative reaction for the antigens, then it is likely the child will not be afflicted. The counseling team and the parents can act accordingly once the prenatal information and the genetic facts are understood.

The last ten years have seen immense strides in our understanding of the inheritance of many human genes. Newer and more refined techniques will continue to provide more accurate methods for the detection of carriers, for the assignment of genes to chromosomes, and for the construction of maps. Counseling teams will undoubtedly be in greater demand as knowledge of the genetic component of many diseases becomes more clarified. Pedigree analysis and probability will continue to play very important roles in the alleviation of human genetic problems. Probability is a most valuable tool for any geneticist. Other aspects of probability will be discussed in Chapter 13 in relation to the calculation of frequencies of mutant genes in populations.

REFERENCES

Bailey, N. T. J., *The Mathematical Approach to Biology and Medicine.* John Wiley & Sons, N. Y., 1967.

Dixon, W. J., and Massey, F. J., *Introduction to Statistical Analysis,* Third edition. McGraw-Hill, N. Y., 1969.

Harper, P. S., Rivas, M. L., Bias, W. B., Hutchinson, J. R., Dyken, P. R., and McKusick, V. A., "Genetic Linkage Confirmed Between the Locus for Myotonic Dystrophy and the ABH-Secretion and Lutheran Blood Group Loci." *Amer. Journ. Human Genet.,* 24:310, 1972.

Jackson, C. E., Symon, W. E., Pruden, E. L., Kaehr, I. M., and Mann, J. D., "Consanguinity and Blood Group Distribution in an Amish Isolate." *Amer. Journ. Human Genet.,* 20:522, 1968.

Mayo, O., "The Use of Linkage in Genetic Counseling." *Human Hered.,* 20:473, 1970.

McKusick, V. A., *Human Genetics,* Second edition. Prentice-Hall, Englewood Cliffs, N. J., 1969.

Schull, W. J., and Neel, J. V., "The Effects of Parental Consanguinity and Inbreeding in Hirado, Japan, V. Summary and Interpretation." *Amer. Journ. Human Genet.,* 24:425, 1972.

Slatis, H. M., and Hoene, R. E., "The Effect of Parental Consanguinity on the Distribution of Continuously Variable Characteristics." *Amer. Journ. Human Genet.,* 1:28, 1961.

REVIEW QUESTIONS

1. In a family of 5 children, what is the probability that 3 girls and 2 boys will occur in (1) this order: a girl, a boy, a girl, a boy, a girl? (2) any order?

2. In a family of five boys, what is the chance that the next child will be a boy?

3. Let the allelic pair *A, a* represent, respectively, the autosomal factors for the dominant trait, normal skin pigmentation, and the recessive, oculocutaneous albinism. Assume that two married persons are known heterozygotes. *A.* What is the chance that the first child will be (1) a normal girl? (2) an albino boy? *B.* What is the chance that the first child will be either a normal boy or a normal girl? *C.* What is the chance that these people will have children in this order: an albino, a normal, and finally an albino? *D.* What is the chance the children will occur in this order: an albino boy, a normal boy, a normal girl, and finally an albino girl? *E.* What is the chance that in a family of four children, three of them will be albinos and one will be normal, in any order? *F.* If these people have three normally pigmented children, what will the fourth one be like?

4. A man with the sex-linked recessive trait, deutan colorblindness, is married to a nonrelated woman with no history of the trait in the family. What is the probability that (1) the first son will be colorblind? (2) that a daughter will carry the gene?

5. Suppose a woman with normal vision whose father was colorblind is married to a non-colorblind man. *A.* What is the probability that a son will be colorblind? *B.* What is the probability that the first child will be a colorblind boy? *C.* What is the chance that the first child will be a carrier daughter?

6. A pedigree shows that a male parent is expressing a certain trait, whereas the female parent is not. Of their six offspring, two boys and one girl show the trait, but three girls do not. Which of the following is (are) certain: (1) The trait is an autosomal dominant. (2) The trait is an autosomal recessive, and the female parent is a heterozygote. (3) The trait is an X-linked dominant. (4) The trait cannot be a sex-linked recessive. (5) The trait cannot be holandric.

7. Suppose that two normally pigmented persons are first cousins and that the grandparents they have in common are both carriers of the gene for the recessive trait, albinism. What is the chance if the cousins marry that their first child will be (1) an albino? (2) a normally pigmented girl? (3) What is the chance that one of the parents is a carrier? (4) What is the chance that the male parent received the recessive from his grandmother?

8. A normally pigmented woman has an albino brother. What is the chance that (1) she will have an albino son if she marries a normal unrelated man? (2) she will have an albino if the known incidence of carriers in the population is 1/70?

9. A normal man has a young brother suffering from the X-linked form of muscular dystrophy. He is married to a healthy woman. What is the chance that the man's first son will suffer from the disorder?

10. The fathers of a young man and a young woman are identical twins. What is the coefficient of relationship between the young people?

11. Suppose that the identical twins mentioned above are albinos married to normally pigmented women. If the young man and young woman mentioned should marry, what is the chance that the first child will be an albino?

12. Make a pedigree of the above family including the twins, their normally pigmented parents, wives, and the young man and woman. (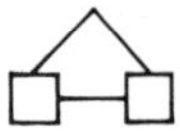 indicates monozygotic male twins.)

13. A man with hemophilia has a daughter who is concerned about the chance of bearing a child with the disorder. It is found that the man with hemophilia does not produce the enzyme G6PD but his wife does, and so does the daughter. The daughter is married to a man without hemophilia who can produce the enzyme. Amniocentesis and karyotype analysis show that the daughter is carrying a male child who can produce the enzyme. What is the chance that the baby has hemophilia? (A chromosome map shows the 2 loci concerned are X-linked and 2 map units apart.)

chapter 13

THE ORIGIN AND SIGNIFICANCE OF GENE MUTATION

General Effects of Gene Mutation

Mutation is any sudden inheritable change and includes alterations in chromosome number and shape (Chapter 11). The main topic of this chapter is *gene* or *point* mutations, which have played a much more important role during the evolution of life than have chromosome anomalies. From a long-term viewpoint, gene mutations are the most significant of the changes which affect the hereditary material. Gene mutations have a greater chance of being transmitted from one generation to the next than do most chromosome alterations. As noted in Chapter 11, chromosome aberrations are often not transmitted in this way, since an individual with an aberration frequently does not reproduce due to its upset of genetic balance. In contrast, a carrier of a changed gene has a good chance of reproducing and thus transmitting the new gene form. Mutant genes, those which depart from the standard, are the primary hereditary alterations associated with the vast number of genetic disorders which afflict the human. Mutant genes, some of them very harmful, continue to appear generation after generation. The student of human genetics must therefore understand the nature of gene mutations, their effects, how they arise and accumulate in populations, and how their frequencies may be estimated.

The gene is a stretch of DNA composed of nucleotides (Chapter 9). Different genes contain different kinds of coded information, since the sequence of nucleotides differs from one gene to the next. A primary function of genes is the control of polypeptide or protein formation. A point mutation within the gene can bring about a change in the code (Fig. 10–6) by substituting one nucleotide for another in the DNA. Such an alteration can change a code word or codon. Sixty-one of the 64 codons, composed of triplets (3 nucleotides in a specific sequence), represent specific amino acids (Table 9–1). If one codon is changed to another, the gene is changed at that position, and this alteration can result in the substitution of one amino acid for another in a protein. Such a sequence of events has undoubtedly taken place in the case of sickle cell anemia where the actual codon change in the DNA has been narrowed down to just 2 alternatives and to only 1 base substitution in 1 position (p. 247). Figure 9–15 gives a general summary of point mutation and its effect on amino acid content of a polypeptide chain. Many inherited metabolic disorders in the human (PKU, Tay-Sachs) undoubtedly have the same general type of molecular basis: a gene mutation produced a changed codon, which in turn resulted in an altered set of instructions. This change is then transmitted by way of the messenger RNA to the ribosomes, and a variant polypeptide is synthesized. This variant form may be less efficient than the standard one in its role in the cell, and it may bring about a serious alteration in cell or body chemistry.

The effect of most gene mutations is harmful; however, a certain amount of gene mutation is essential to the future of any living species (Chapter 8).

Table 13–1. Major classes of radiations and some examples: (A) Electromagnetic radiations; (B) corpuscular radiations. Angstroms (A) = 1/10,000 μ

(A)

Wave type	Wave length	
Radio	10^7–0.04 cm	Insufficient energy to excite or to ionize atomic electrons
Infrared	20,000 A–3800 A	
Visible light	7,800 A–3800 A	
Ultraviolet	3,800 A–150 A	Sufficient energy to excite
X-rays	150 A–0.15 A	Sufficient energy to excite and to ionize
rays	0.15 A–0.005 A	
Cosmic rays	0.005 A–0.0008 A	

(B)

Particle	Weight (atomic mass units)	Charge
Electron (beta particle)	0.00055	Negative
Positron (positive beta particle)	0.00055	Positive
Proton (nucleus of common isotope of hydrogen)	1.007	Positive
Deuteron (nucleus of heavy isotope of hydrogen)	2.013	Positive
Alpha (nucleus of helium)	4.002	Positive
Neutron	1.009	Neutral

This is so because evolutionary progress, which results in better and better adaptation of life forms to the environment, depends on gene mutation. In any species, a small percentage of the gene mutations in each generation may confer some type of advantage to those individuals carrying them. As a result of their advantage, these individuals will leave more offspring than other population members lacking the beneficial changes. Any mutant gene, if it confers a distinct advantage, may eventually increase to the point where it becomes the normal one. It must be kept in mind that all the hereditary diversity seen in different living forms has originated through gene mutations. Many very deleterious mutations arise and persist in the population, but the "good" ones play a major role in the progress of a species from simpler to more complex and better adapted. As discussed in Chapter 8, gene mutations account for all of the common hereditary variations we see in the human with respect to eye color, hair color, blood type, etc. Gene mutations are responsible for many of the variations which we accept as normal. If gene mutation were to cease, eventually no more variation would be possible. The only differences among individuals would result from different combinations of the gene forms already present in the population.

Spontaneous Mutation Rate and Some of Its Effects

Environment is not static but constantly varies. New genetic types, generated by gene mutations, may be better adapted to cope with changed conditions than were the standard types. However, since most mutations are harmful, a high mutation rate would be detrimental to any life form. Evidence indicates that approximately 5% of the sex cells produced by any species in one generation (approximately 25–35 years in the human) contains some kind of gene mutation which arose during that generation. The likelihood that any specific gene will undergo mutation is very rare, but since there are so many genes in a plant or an animal, the chance is appreciable that during a generation, a significant proportion of the total will have mutated. About 5% per generation is apparently not enough to harm most species but is sufficient to permit the origin of any gene mutations which happen to be beneficial. The mutation rate itself in a species is believed to be controlled by certain genes which keep it at the optimum rate for the species.

The frequency of gene mutations may be increased by certain environmental factors, such as some radiations and chemicals. Any agent known to raise the mutation frequency is called a *mutagen.* However, mutations continue to occur in any population for reasons which are unknown. These are called *spontaneous mutations,* since their origin is unpredictable and unassociated with any specific agent. It is quite possible that mutagens in the environment, such as natural radiations, *may* prove to play a role in the origin of spontaneous mutations. When a spontaneous mutation does occur, it is a random event, meaning that there is a likelihood that *any* gene or point on the chromosome has a chance to mutate at a particular time. Placing an organism in a certain kind of environment will not cause a specific gene or kind of gene to mutate. The application of a particular type of mutagen does not favor the mutation of one kind of gene over another. In experimental plants and animals, as well as in microorganisms, it has been demonstrated that the various genes mutate spontaneously at different rates; some may have a spontaneous mutation rate 2 or 3 times higher than that of others. However, this *same* difference in relative frequency will exist under different kinds of environments. These very important points can be summarized as follows: An environmental condition does not dictate which gene will change nor does it favor the mutation of one kind of gene over another. If a mutagen such as X-rays is applied, the entire mutation frequency will be elevated, and it will be elevated proportionately. This means that the mutation rate for all the genes will have been raised 2, 3, or more times, depending on the dosage. Mutations, then, are random events, unpredictable in their occurrence at a specific time.

Since gene or point mutations are changes in the DNA, they can occur in the hereditary material present in the body cells as well as in sex cells. A mutation which occurs in a body cell is called a *somatic mutation,* as opposed to a *germinal* one, which is carried in a gamete. Most somatic mutations probably remain undetected. This is so because most mutant alleles are recessive: the standard or normal gene is apt to be dominant to any variant allele which arises through mutation. This means that most somatic mutations will not be expressed in the body cells, since the chances are great that the individual will be homozygous for the standard gene [Fig. 13-1(A)]. Somatic mutations have been demonstrated in several plant species and in lower animals, as well as in mammalian cells grown in the test tube. They have also been

Figure 13-1. Somatic mutation. (A) Most commonly occurring forms of a gene in a population are dominant as a result of natural selection. Most persons carry dominant forms of those genes which affect basic physiological processes. Therefore, if a standard gene in a body cell undergoes mutation, such as the one for normal melanin production, the chances are that the mutant form of the gene will be recessive. Since the affected cell still carries one dominant allele, no effects of the mutation will be evident. All cells will exhibit the same phenotypic effect, pigment formation in this example. (B) In a heterozygote, it is possible that a gene mutation may affect the dominant allele in a body cell which is part of a developing organ, such as the eye here. All cells derived from the affected cell would be homozygous for the recessive. Two lines of cells would then compose the particular organ. This can explain the occurrence of a segment of blue in the basically brown iris of an eye.

proposed as factors in aging and the expression of certain pathological conditions such as cancer (Chapter 14). The full significance of somatic mutations to the human is still unknown. However, they can account for the occurrence of certain kinds of mosaic conditions. For example, a person may have a brown eye with segments of blue. This can be explained [Fig. 13–1(B)]

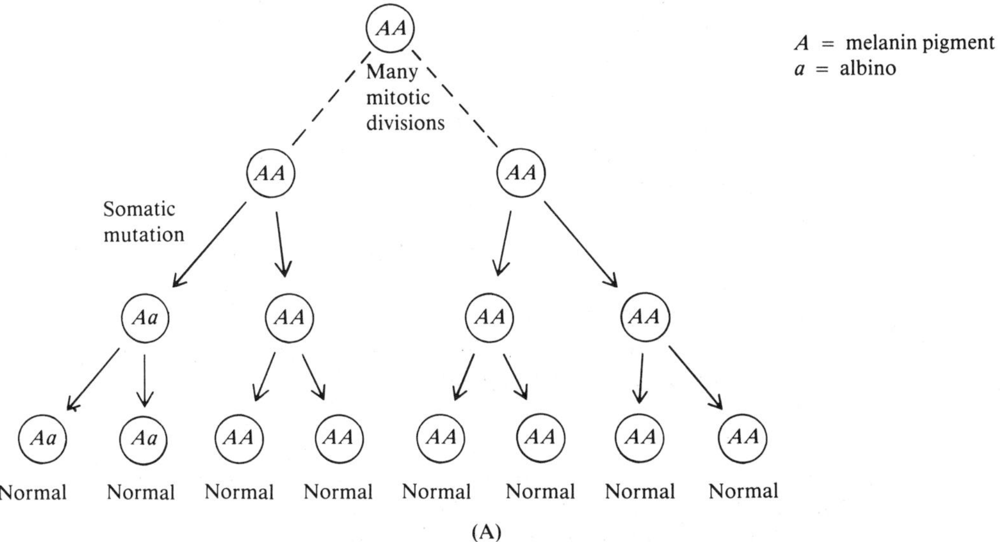

(A)

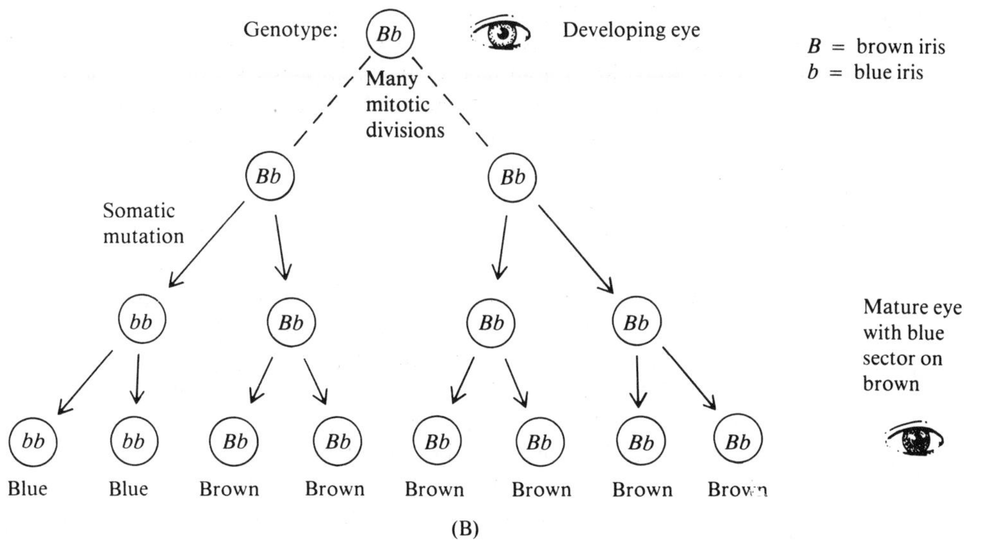

(B)

Figure 13-1

on the basis of a somatic mutation in the developing iris of the heterozygous individual. In the case of eye color, neither the allele for brown nor the one for blue color is abnormal, both gene forms being quite standard. There is more of a chance for mosaicism to occur and to affect two traits such as these than there is in cases where the allele associated with one of the alternative traits is very rare.

It is also possible that a recessive gene could mutate to the dominant allelic form, but this probability is much lower than mutation in the other direction, dominant to recessive. A mutant gene has the ability, therefore, to mutate back to the standard form, but the likelihood or rate of back mutation is much less than "forward mutation," mutation from standard to mutant.

Since spontaneous somatic mutations can affect the phenotype of an individual, it is conceivable that they could play some role in the origin of disorders or diseases. However, from the genetic point of view, more significant are those mutations which occur in the germ line, since the mutant allele can be transmitted from one generation to the next and affect many individuals. The mutant alleles may even accumulate to a point where they pose serious problems for the population, as the sickle cell story illustrates (p. 198).

If a mutant gene is beneficial, it will tend to spread throughout the population, since those carrying it will have some sort of advantage which enables them to leave more offspring than those without it. The once new mutant allele will supplant the original wild or standard allele and will eventually become the standard or normal. However, most mutant alleles are harmful, and it is these which are of more immediate concern to a population. The reason why most mutant genes cause some sort of deleterious effect is easy to understand, as a reexamination of Fig. 9–15 will illustrate. The fact that they are random and thus unplanned changes implies that they can occur in any gene and at any point within a gene. A spontaneous change within a normal gene can result in an amino acid substitution producing a variant polypeptide chain. A change in just 1 amino acid in 100 or more can seriously affect the efficiency of a protein, as demonstrated by the mutant hemoglobins. While it is conceivable that any random amino acid substitution in a long polypeptide chain will cause it to operate more efficiently, this is highly improbable. Many long polypeptide chains do not act alone but interact with other kinds of polypeptide chains or other substances in the cell. One or another of these interactions can become impaired by the single change stemming from the amino acid substitution. A gene mutation can be compared to a sudden, thoughtless change in a complex machine, such as an engine or a computer. It is most likely that the alteration will be a harmful one, since normal functioning of the system depends on the precise interaction of its parts. In a living organism, a most complex system, any sudden unplanned change in the hereditary material is almost certain to do more harm than good.

Mutation and the Degenerate Code

Reference to the coding dictionary (Table 9–1) shows that the genetic code is degenerate, meaning that 2 or more specific code words represent 1 specific amino acid. But note that it is always the same codons which specify 1 kind of amino acid. Actually, a benefit is derived from this. The genetic code has been perfected due to the operation of natural selection. The benefit imparted by a degenerate code is one which produces a buffering effect against the harmful effects of gene mutations. Note from the coding dictionary that 4 DNA code words designate the amino acid arginine. A particular gene may contain the codon CGA which would dictate that arginine is found at a specific location in a protein. If the genetic code were not degenerate, every functional codon would stand for only 1 amino acid. Only 20 codons would designate amino acids. All the others would be punctuation signals or stop signals. Assuming this to be the case for the moment, if the codon CGA mutated, it would most probably change to a mutant form which would terminate the protein. The fact that 61 of the codons mean amino acids gives some insurance that, when a gene mutation does occur, the chances will be that an amino acid substitution will occur and not the extremely serious effects caused by an incomplete protein or complete lack of a protein. However, as we have learned, most amino acid substitutions in a protein tend to be harmful in some

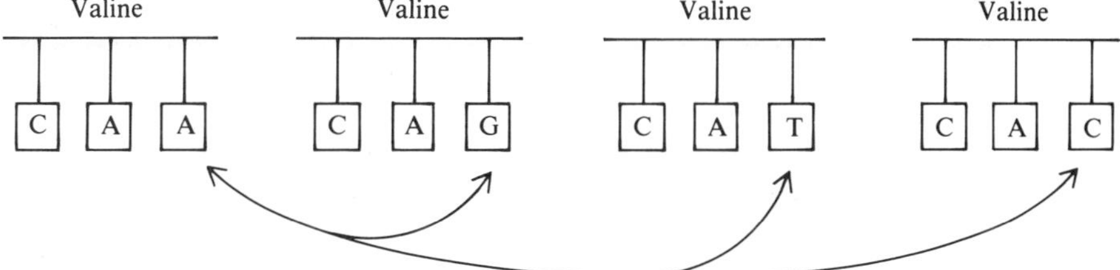

Figure 13-2. Silent mutations. The degeneracy of the genetic code affords an advantage in the buffering it provides in the case of gene mutation. As shown here, four DNA codons represent the amino acid, valine. A change in the third position in any one of them still results in a codon which represents valine. Therefore a mutation in any of these valine codons which alters the third position will not bring about an amino acid substitution in a protein nor will it cause chain termination. While a mutation has occurred in such a case, it is silent, since it produces no detectable effect.

way. If the genetic code is degenerate, there is then a chance that when a mutation occurs the codon will change to another codon which stands for the very same amino acid. As Fig. 13–2 shows, if CAA mutates to CAG, CAT, or CAC, the amino acid valine is still designated. Actually, a mutation would have occurred, but there would be no evidence of it, and no detriment would have been imparted. We call a mutation of this type a *silent mutation,* one in which one codon is changed to another which designates the same amino acid. The degenerate code thus affords some protection against what would probably be a very serious effect of almost every mutation. Without the buffering provided by a degenerate code, spontaneous mutation would undoubtedly slow down evolutionary progress, since most mutations would then assuredly be extremely detrimental to the population.

Harmful Genes and the Calculation of Their Mutation Rate

The degree of harm or detriment imparted varies from one gene mutation to the next. A mutation may produce a visible change in the phenotype; however, mutations which cause a readily observable change are far less common than those which produce no detectable effect. Those mutations which do produce visible effects also usually impart a certain amount of detriment at the same time. More common than the visibles are the *invisible lethals,* those mutant genes which remove the individual some time before the age of reproduction. Invisible lethals bring about death by causing serious derangements in the metabolism or chemistry of the organism. Lethals may act at different times in development, at the stage of the embryo or later. Much more common than visibles and the fully lethal mutations together are gene mutations which may be called *invisible detrimentals.* A *detrimental* does not necessarily kill the individual. However, each detrimental harms the carrier in some way and decreases the chances of survival. These harmful genes tend to act as recessives, and so it is possible for them to accumulate in the heterozygous condition. An increase in the number of heterozygotes within a population greatly increases the chance that two of them will come together in homozygotes where they exert their full deleterious effect. Many of the detrimentals actually cause some harm to the heterozygote as well. This is to be expected, since most genes designated as recessive probably express themselves to some degree even in the presence of their dominant alleles (Chapter 1).

Estimates have been made on the mutation frequency of certain specific genes in the human. Those genes which are X-linked or which produce a dominant phenotypic effect lend themselves more readily to studies of mutation frequency than do autosomal recessives. Suppose that a certain human disorder, as indicated by extensive pedigree analyses, is inherited as a dominant. If the condition suddenly appears in a family which has had no previous history of it, it is assumed that spontaneously a normal allele has mutated to the allele for the dominant trait. The total number of such occurrences among the newborn can be related to the total number of all births in a given period and the mutation frequency calculated. For example, 5 abnormal children born among a total of 100,000 births to normal parents would indicate that the rate at which afflicted persons are arising is 1 in 20,000. However, to determine the mutation rate on the basis of the number of germ cells per generation, it must be kept in mind that each individual represents 2 sex cells, the 2 gametes which united at fertilization. The new mutation could have occurred during either sperm or egg development. Therefore, the figure 1 in 20,000 represents 1 in 40,000 sex cells, and the mutation rate is thus 1/40,000 gametes per generation.

Similar reasoning can be applied to X-linked recessives, but the method is a bit more indirect, as will be seen from the example to be given. Since the male is hemizygous for all X-linked genes, a newly arisen sex-linked recessive has a much greater chance of expressing itself than does an autosomal one which may lie undetected in the heterozygous condition. The X-linked recessive will express itself when present in just a single dose on a male's X chromosome. The number of males with a certain sex-linked disorder who appear in families with no previous record of the disease can be counted. This figure is then related to the total number of births. Let us say that 1 male in 100,000 male births shows the disorder. It must now be kept in mind that any new sex-linked mutation can arise in an X carried by a male or in *either* of the 2 X chromosomes carried by a female. The origin of any new sex-linked mutation thus has 3 possible sources. Consequently, the figure 100,000 must be multiplied by 3 to account for the fact that the mutation could trace back to an X chromosome in a male ancestor or to 1 of the X's present in a female parent. It is evident that the affected male baby did not receive the X with the mutation from his male parent; so this method does not indicate the *immediate* source of the mutant gene. It could have occurred in the formation of germ cells in the affected male's mother, but being recessive, it could have been carried by her in an undetected state. The source of the mutation could have been a previous ancestor, *male or female,* hence the need to multiply the figure above by 3. This method is more indirect in this sense than in the case of the autosomal dominant.

The final estimates in these cases of both the sex-linked recessives and the dominant autosomal mutations entail many uncertainties, but even with their shortcomings, they do give some idea of mutation frequency in the human. While genes may differ in their rate of mutation, the overall likelihood that a gene at any human locus will mutate has been calculated to be approximately 1/100,000 gametes per generation. This figure is not at all out of the range which has been found for most other forms of life.

Though calculation of mutation rates for autosomal recessives generally involves additional difficulties, some estimates can be made for the origin of recessives which produce a fully lethal as well as a visible phenotypic effect.

Tay-Sachs disease is an example of a fully lethal disorder which kills in early childhood. Therefore, no homozygote can transmit the gene to the next generation. Nevertheless, this disorder, like others with lethal effects, tends to recur generation after generation with a characteristic frequency. The only way this can come about is through the origin of new lethal mutations. Without recurrent gene mutation to the recessive form, the disorder would become less and less frequent with each passing generation and would eventually disappear from the population. Since recessive, fully lethal disorders continue to arise, new gene mutations must continually originate to supplant those lethal recessive genes eliminated through the deaths of the afflicted persons. For example, suppose a fully lethal disorder is found in 1 newborn out of 100,000. This 1/100,000 directly reflects the mutation rate. A bit of thought tells us why this must be so. Since the disorder is always present in the population, the number of new victims must indicate the number of individuals dying from the affliction. Conversely, the number of deaths from the disorder must reflect the number of babies born with it. Otherwise the affliction would tend to decrease in frequency and eventually disappear. The figure 1/100,000 affected newborns represents the number of mutant births, and this represents 2 mutant recessive genes out of 200,000 gametes, since each human is diploid. This figure reduces to the 1/100,000 recorded for births or deaths; so either one gives the mutation rate per gamete per generation. The figure does not indicate the source of the mutant genes, which could have been carried for generations by unaffected heterozygotes.

Some Mutagenic Factors

Since the preponderance of new gene mutations is harmful in some way, any factor is of concern which can increase the mutation frequency above the spontaneous rate. Most studies relating to mutation induction have been conducted on species other than the human. However, there is no reason to believe that the human hereditary material, exposed to the same environmental factors, responds differently from that of other species. However, caution must be exercised when observations made on one life form are applied to another, and no conclusions on the human should be immediately drawn from them. For example, the first environmental factor found to influence the frequency of mutation was temperature. Fruit flies raised at temperatures below *or* above the normal range for the species were found to have a mutation rate double that of flies raised at ordinary temperatures. Evidence indicates that, in general, departures from the temperature range that is typical of the species group can increase the mutation rate. However, this must not be construed to mean that a person living in an extremely cold or an extremely warm environment will have a higher mutation rate than a person in more temperate regions. The human is well adapted to a wide range of temperature conditions, and the thermoregulation of the body goes a long way toward the maintenance of a constant internal body temperature, regardless of the temperature of the immediate surroundings. Moreover, the genetic material in the body and sex cells is protected in various ways from direct exposure to many factors in the environment. If some factor or specific mutagenic substance is to bring about a genetic change, it must somehow be able to bring about a change in the DNA, either by affecting it directly or by triggering some other factor which in turn causes the effect. This point will arise many times throughout discussions of mutation.

The mutation rate apparently can be increased by unknown factors involved in the aging process. This has been demonstrated by aging sperms of certain experimental animals and aging the pollen of plants. There is some reason to suspect that a similar aging effect pertains to the human. A few dominant disorders have been shown to arise unexpectedly (and hence are considered to reflect new gene mutations) when the father's age at the birth of the afflicted offspring is above that of the average male parent. One such genetic disorder in which an age factor is suspected is achondroplastic dwarfism.

A relationship between malnutrition and increased mutation frequency has been demonstrated with plants. Data in this direction on the human are needed and could yield important information.

Some Properties and Genetic Effects of Radiations

Among the better-known agents of mutation are certain drugs and radiations. Again, all of the experimental observations are from other species, but the implications of the assembled information are of profound significance to the human species. As noted above, a mutagenic agent must somehow alter the hereditary material by changing it either directly or indirectly through an intermediate. Certain kinds of radiations have the ability to penetrate almost all forms of matter. Since many radiations are highly penetrating, this means that most cells, even those composing a many-celled organism, are susceptible to their effects. The term *radiation* as used here refers to any phenomenon in which energy travels through space, even if the space is a vacuum. Not all radiations are mutagenic, since some of them are not associated with a high level of energy. Table 13–1(A) lists one of the two major categories of radiations, those known as *electromagnetic radiations.* Such a radiation does not have the general characteristics of a particle or small body of any kind and so cannot be assigned a weight or mass. Instead, the electromagnetic radiations behave primarily as waves which travel through space. Refined procedures have enabled the physicist to measure the lengths of these waves. As Table

13–1(A) shows, the size ranges all the way from the very long radio waves through the shorter light waves down to tiny X-rays and finally cosmic rays. On logical grounds, we would not expect visible light to be mutagenic, since life on earth is constantly exposed to it and would have been seriously hampered or eliminated millions of years ago.

The different wavelengths of light associated with the various colors of the visible spectrum, plus infrared and radio waves, are non-mutagenic. It will be noted from Table 13-1(A) that these wavelengths are longer than those of ultraviolet and the other electromagnetic radiations. As it happens, the *shorter* the wavelength, the *more* energy associated with that kind of wave. Therefore, visible light waves would carry with them much less energy than would X-rays. The amount of energy associated with a kind of wave is very important. The greater the amount of energy carried by the wave, the greater its ability to penetrate matter. Not only can X-rays and those radiations with still shorter wavelengths penetrate most common substances, they are able to disrupt the organization of the molecules which compose the particular material. X-rays, gamma rays, and cosmic rays carry so much energy with them that they can upset the stability of molecules. They usually do this by transferring their energy to electrons within atoms. Figure 13–3 summarizes some of the effects and shows that after application of high-energy radiations, there may be a separation of electric charges. Ultraviolet light is by no means as penetrating as those radiations of shorter wavelength. However, if ultraviolet is absorbed, it may cause a certain instability in atoms by imparting excess energy to some of the electrons. There is no separation of charges, but any atom with excess energy is more reactive than one whose energy level is lower.

Table 13–1(B) lists the commoner radiations in the other major category. These are the *corpuscular radiations* which may be emitted by radioactive materials. They do not behave as waves but rather as tiny bodies which have a definite weight. It is possible for them to disrupt matter by imparting excess energy to subatomic particles. The story is similar to that for the electromagnetic radiations; however, the corpuscular radiations can be envisioned as streams of particles striking electrons and causing them to become more energized and active. Again, separation of electric charges can occur.

The pioneer experiment which illustrated that radiation can be mutagenic was performed on fruit flies by H. J. Muller. For his demonstration that X-rays increase the rate of mutation, Muller was awarded the Nobel Prize in 1946. In the years following Muller's discovery, other radiations, those of wavelengths shorter than X-rays and the major corpuscular radiations, were shown to be effective mutagenic agents. Not only do X-rays and the other high-energy radiations raise the mutation rate of genes, they also increase the frequency of structural chromosome changes, such as deletions and inversions. Other kinds of chromosome anomalies are also more likely to occur after

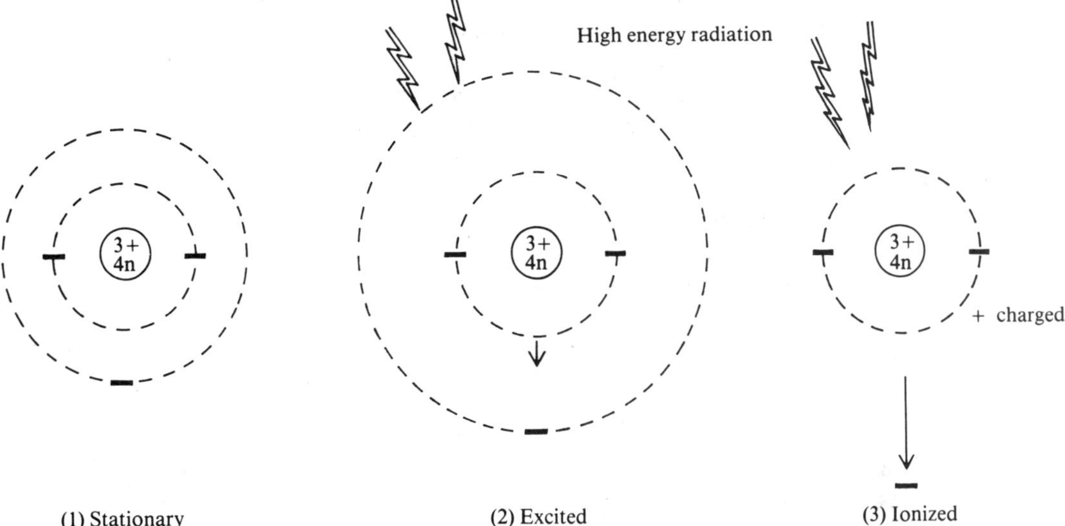

Figure 13-3. Radiation effects on matter. Certain radiations carry sufficient amounts of energy to disrupt the internal organization of matter. One way in which they do this is by imparting excess energy to single particles within atoms. Represented here is an atom of lithium (1) which has a nucleus of three protons (positive charge) and four neutrons (neutral). The electrons (negative charge) circle the nucleus of the atom. The atom is uncharged and possesses a certain basic level of energy in its stationary state. Certain radiations can impart excess energy to an atom, usually by affecting an electron in the outermost orbit. The electron may move farther away from the nucleus as a result and the atom is more reactive or excited (2). High-energy radiations may impart enough energy to cause an electron to leave its atom (3). An outright separation of electric charges results, and the atom is ionized, since it is electrically charged. The escaping electron may hit other atomic particles and excite or ionize them in turn, until it is captured by an atom which then becomes negatively charged.

exposure to radiations, such as an increase in the amount of nondisjunction and the failure of chromosomes to reach the poles at nuclear division. The general consequence of the various chromosome aberrations is the production of nuclei whose genetic content is unbalanced due to the loss of genes in some nuclei and the addition of extra gene doses in others.

Ultraviolet Light as a Mutagen

Ultraviolet light has also been shown to be an effective mutagenic agent, particularly in microorganisms. The wavelengths of ultraviolet are longer than those of the other mutagenic electromagnetic radiations. Ultraviolet is not able to penetrate matter with the effectiveness of the more highly energetic radiations. Therefore, it is unlikely that ultraviolet would reach cells of the sex organs of most many-celled organisms. However, it has been demonstrated that if ultraviolet *does* penetrate a cell, it can produce mutations. Moreover, it has also been shown that ultraviolet can cause changes directly within the DNA itself by altering its internal structure. The DNA actually tends to absorb certain wavelengths of ultraviolet when it is exposed to the radiation. This absorbed energy is then able to alter the DNA in such a way that it may not be able to duplicate itself properly or may not be able to act as a template for the formation of messenger RNA (Fig. 13–4). These sorts of changes in the DNA undoubtedly account for the strong mutagenic and killing effects which ultraviolet exerts in bacteria, cells which ultraviolet can penetrate.

Studies with bacteria have revealed some very interesting facts about ultraviolet which may hold significance for the human. It was discovered many

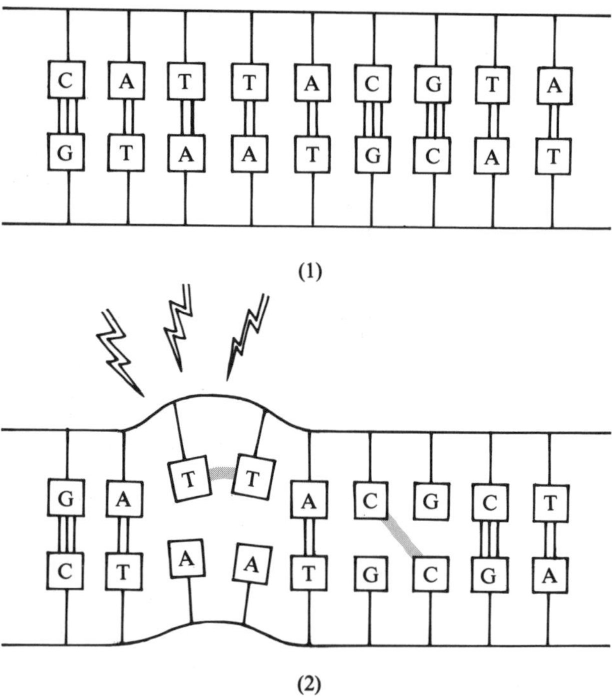

Figure 13-4. Effects of ultraviolet light on DNA. In the DNA molecule, the two chains (1) of the double helix are held together by chemical (hydrogen) bonds which are relatively weak and which can be eliminated at the time of DNA replication, permitting each chain to unwind and act as a template. Ultraviolet light may cause strong bonds to form (2) between adjacent bases on a chain or between two bases on opposite chains. Usually it is two thymines or two cytosines which are affected. Such bond formation can interfere with DNA replication or the formation of messenger RNA.

years ago that the killing effect of a dose of ultraviolet in bacteria can be reduced by as much as 40% if the cells are subjected to a strong source of ordinary white light following the ultraviolet exposure. This somewhat unexpected action of white light which reduces some of the lethal effect of ultraviolet was named *photoreactivation.* Later research revealed that the white light is responsible for the activation of a certain enzyme in the cell which is able to repair some of the damage caused by ultraviolet. This light-activated enzyme is now known to be just one of several cellular enzymes, called collectively *repair enzymes,* which can repair damaged DNA (Fig. 13–5). Not all repair enzymes require light to become activated. Recently one of the light-independent repair enzymes has been demonstrated in human cells. It can participate in the repair of DNA damaged by ultraviolet. A very significant observation has been made on persons with the inherited skin disorder, *xeroderma pigmentosum,* a condition in which skin malignancies may arise. Skin cells from affected persons have been shown to contain less of the light-independent repair enzyme than cells from other persons. This finding has several provocative implications. It is conceivable that repair enzymes are essential to all living things, since they may be required for the frequent repair of DNA damaged by some mutagenic agent. The amount of repair enzymes in cells may very well be under genetic control, as suggested by the observations made on the skin disorder. A certain normal amount may be important to keep the effects of spontaneous mutation at a minimum. If the normal enzyme level becomes decreased, due to an inherited disorder or to exposure to some environmental factor, damaged DNA may accumulate in cells. This could lead to cellular death or to the expression of an abnormality, especially in the body cells. It is also quite possible that the poorly understood process of aging may involve a decrease in the level of repair enzymes. The consequent aging effects may result in part from failure of the repair of damaged DNA.

Ultraviolet reaches everyone in the daily radiation received from the sun. Most of it is screened out in the upper atmosphere, but a sufficient amount arrives to cause sunburn or tanning. Tanning is a natural cellular response to screen out excess light through the production of more dark pigment. A tanned complexion is considered pleasing to the eye by some persons, and intentional exposure to the sun's rays is commonly practiced. That this may be unwise is suggested by the fact that excessive exposure to the sun by white persons can hasten the aging process of the skin. Moreover, the frequency of skin cancers among whites in the U.S. is found to increase significantly as populations farther and farther south (and hence those exposed to the more direct rays) are studied. The precise role of ultraviolet in these changes is unknown, but avoiding excessive exposure to the sun's rays would seem advisable. Further research on repair mechanisms in cells may give a greater insight into these matters as well as into factors involved in life span and certain inherited disorders.

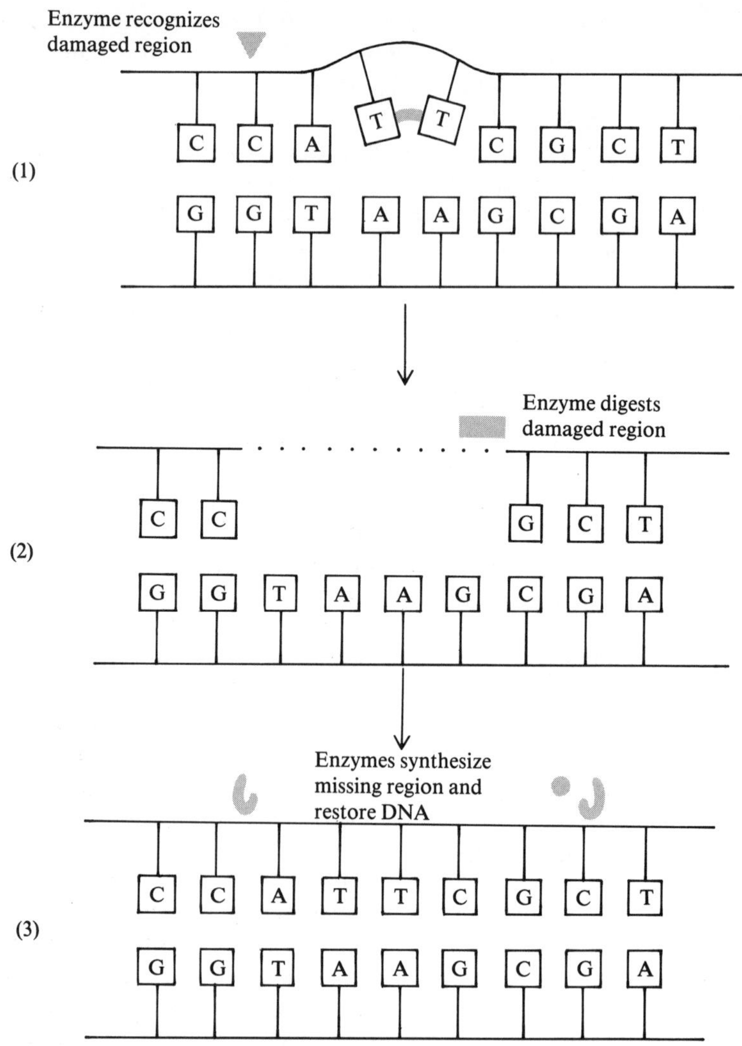

Figure 13-5. Repair enzymes. Cells are known to contain enzymes which can eliminate damaged sections of DNA, such as those caused by the union of two adjacent thymine bases (1). The damaged section is digested away (2). The missing portion is restored by enzymes which use the section on the intact strand as a template and which join the new piece to the free ends (3).

Action of Radiation in Genetic Change

While some of the ways in which ultraviolet can cause genetic change are understood, this does not mean that we know just how the high-energy radiations produce gene mutations or structural chromosome changes. Several ideas have been offered, and there are arguments for and against each of them. Actually, there may be no one way in which any mutagenic agent causes genetic change. Even ultraviolet is known to be able to cause the production of certain chemicals in the food medium upon which bacteria grow. These chemicals in turn can bring about genetic change, so that in some cases, the effect of the ultraviolet is not a direct one on the DNA. Radiations are known to be able to produce certain kinds of peroxides when they pass through a watery medium. By themselves, these are mutagenic, and so some effect of the radiations on the DNA may be through the production in a cell of mutagenic chemicals which in turn act on the DNA. Exposure to high-energy radiations is also known to hasten the aging process and to increase the risk of malignant diseases. The possible effect of these radiations on repair enzymes should be considered in this regard.

There is some good evidence that high-energy radiations such as X-rays or protons may actually strike the genetic material and in this way directly bring about a change. However, clarification is still needed on the exact mechanisms involved. Nonetheless, there is no doubt that ultraviolet and the high-energy radiations are effective mutagens and that their ability to produce gene mutations and structural chromosome alterations is increased as their dosage is increased. This knowledge should be kept in mind when dealing with radiations. Before World War II, radiation controls were often lax. Many persons can still recall the presence in stores of fluoroscopic machines to visualize the bones of the foot, supposedly to insure better fitting of shoes. When the inherent danger of radiation exposure became publicized, such frivolous use of radiation became legally prohibited. Medical and allied professionals have now become alert to the potential risk of radiation exposure. There is still an unsettled controversy regarding the level of a safe dose. Some argue that below a certain level there is little or no chance at all that the radiation will exert an effect on the genetic material. Others maintain that any dose, no matter how small, has a potential effect. The fact that the debate is still unsettled calls for public awareness but not overreaction to the matter. Intelligent, controlled use of radiations has very valuable applications in diagnosis and therapy and is instrumental in saving lives. The benefit of radiation exposure must be weighed against the risk involved in a specific case. Some routine procedures which are of great aid in diagnosis and detection of disease do not expose the individual to more than a fraction of the dosage generated by natural sources in the environment. However, use of high-energy radiations on the abdomen of a pregnant woman to obtain extra but unessential information would be ill-advised and would not be considered today by any competent medical person. The average individual should be sufficiently informed in order to avoid contact with high-energy radiations in those situations where authorized professionals are not in charge.

Chemical Mutagens

In 1941, the mutagenic properties of a chemical were reported for the first time. Mustard gas and related compounds were shown to be as powerful as high-energy radiations in their ability to increase the frequency of gene mutations and structural chromosome alterations. Like radiations, they can also cause serious burns and damage to cells of the body. A very unexpected finding was that following exposure to these compounds, an induced mutation did not necessarily appear immediately in the treated germ cell or body cell. The exposed cell may not contain the mutation, but one of its descendants may! The mutagenic effect of the mustard compounds may actually be delayed. This means that somatic mutation may occur later following exposure so that certain cells of the body may become affected. Extremely important is the fact that any genetic change due to the exposure may be delayed for several generations, the exposed individual showing no effect at all. It is now known that the mustards cause an instability to arise within the DNA by weakening the attachment of the base, guanine, to its sugar. If the guanine becomes deleted, it may be replaced by any of the bases. It is most probable that the delayed mutagenic effect of the mustards is due to this instability which can cause a change in the genetic code by permitting a substitution of one of the other bases in place of guanine. Perhaps other chemicals will be shown to act in a similar way. Such delayed effects are more insidious than those which become expressed soon after the exposure.

The list of mutagenic chemicals has increased tremendously, and the end is not in sight. Most of the chemicals are less potent mutagens than the mustards. They represent a wide assortment of different types of chemical substances, and there is no one common feature to their structures. The way in which they exert their action is unknown for many of them, and some present perplexing questions which remain to be answered. A chemical found to produce gene mutations or chromosome changes in one species may be completely inactive in another. Formaldehyde increases the rate of mutation in the fruit fly, but only if applied at a certain period in the development of the male. The female and any adults are completely immune to its effect! In some cases, a partial explanation is known for the differences in mutagenic response. The common substance, caffeine, is mutagenic in many lower life forms. In man and several other species, caffeine is denatured in the digestive tract before it can reach the internal environment of cells and exert any mutagenic action.

There is little question that the different kinds of mutagenic chemicals bring about genetic change in a variety of ways. Work with microorganisms has demonstrated that some mutagens such as nitrous acid can cause the DNA to make errors in one or more ways when it is in the process of replication (Fig. 13–6). The outcome of these mistakes can be the substitution of one base for another in the DNA, producing a change in the genetic code. Some mutagenic chemicals actually resemble the bases found in DNA to such an extent that they can substitute for the normal bases and in this way change the genetic code. No matter how it is effected, the eventual change in the DNA, whether due to radiations or chemicals, constitutes a change in the genetic message. As in the case of spontaneous and radiation-induced mutation, chemical mutagens cannot direct a mutational change but raise the overall frequency of mutation.

The subject of mutagenic chemicals poses several problems which must be considered. While radiations exert a pronounced damaging effect on the genetic material, human populations are not generally subjected to excessively large doses of them over extended periods of time. However, in today's world, the human contacts more and more chemicals in everyday life. The mutagenic potential of most of these is still unknown, but a few have been implicated as possible dangers to the hereditary material. One reason for the lack of clear-cut evidence in the matter of controversial food additives or substitutes such as the cyclamates is that the data are obtained primarily from experimental animals or from test tube situations. If a substance is shown to alter chromosome structure in animals given massive doses of the substance,

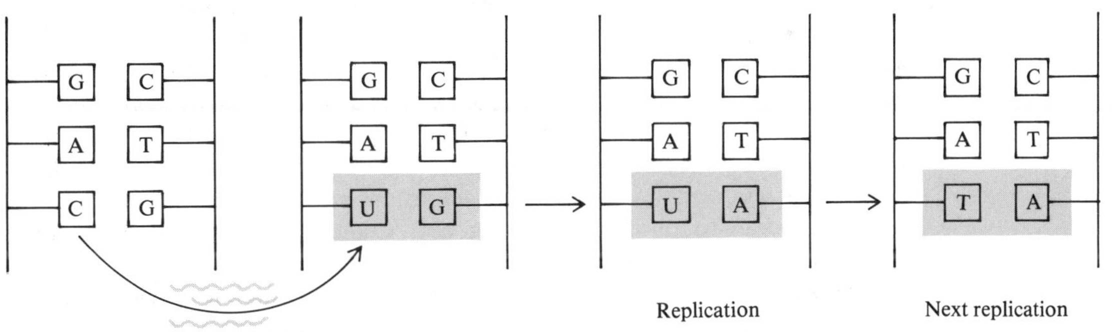

Figure 13-6. Action of nitrous acid on DNA. Nitrous acid can cause chemical alterations to occur in the DNA, such as changing cytosine to uracil. Uracil attracts adenine at the time of replication. At the following replication, adenine will attract thymine. The original C-G pair has been changed to T-A. Nitrous acid can also change adenine, which normally pairs with thymine, to a compound which pairs with cytosine. This results in the change of an A-T pair to a G-C pair. Chemical mutagens act in different ways, but the outcome is some alteration in the genetic message.

it is still not known if such effects pertain to the human in his normal environment who receives much smaller amounts. Unfortunately, gene mutations caused by a mutagenic agent cannot be expected to manifest themselves dramatically. They may remain hidden for many generations. And it must always be remembered that the commonest kinds of gene mutations are invisible recessives with only a slight deleterious effect on the carrier. We have no idea how many of these could accumulate in human populations before the little detriments add up sufficiently to afflict appreciable numbers of persons. Commonly used chemicals whose mutagenic actions are unknown should be subjected to as many critical tests as possible. Many chemicals are used for the immediate benefits they confer, such as food preservation, and could not be dispensed with without creating another major problem. In these cases, the present use of the chemicals can be justified as long as research continues to indicate no harmful effects. As with radiations, the wisest course seems to be the avoidance of unnecessary contact with chemicals whose potential danger has already been reported on the basis of reliable investigation. Where a controversy over genetic damage exists over a drug or other chemical, common sense would seem to dictate the individual's choice of action. The potential cancer-inducing effect of some chemicals will be pursued in Chapter 14.

Mutant Genes in Population

The geneticist who is concerned with entire populations instead of small groups or families is interested in the frequencies of rarer genes in the population, especially in the number of normal-appearing persons who may be carrying harmful recessives. This knowledge allows the genetic counselor to estimate more precisely the probability that a human carrier may have an afflicted offspring. A long-range study of populations enables the population geneticist to determine whether certain rare alleles are decreasing or increasing. From such information, he can gain an insight into the environmental factors which operate for or against certain genes. Studies of this type have been done on several plant and animal species and have revealed many points about the process of evolution. While the detailed study of population dynamics entails many mathematical concepts, there is one which is very easy to grasp and is fundamental to all population studies.

Early in this century, certain implications of inheritance patterns were appreciated independently by two men, Hardy and Weinberg. They were familiar with the basic laws of Mendel and recognized a very important fact which we have already noted (p. 11). This point is simply that Mendel's laws nowhere say that one gene will increase in frequency over another because of the mechanism of inheritance or gene transmission. A dominant gene does not become more abundant than its recessive allele just because it is dominant.

If 2 alleles are neutral in their effect on the reproductive abilities of those individuals carrying them, the frequencies of the 2 alleles will tend not to change generation after generation. A simple example will illustrate this concept. The gene for pigmented iris or brown eye color (*B*) acts as a dominant to the one for blue or lack of the pigment (*b*). Suppose a large number of heterozygous persons, ***Bb***, constitute the founders of a colony. The frequency of the gene for brown eye color (*B*) in this original population is 50% (0.5), and so is the frequency of the gene for blue, *b*. All these brown-eyed persons, when they mate, can produce brown-eyed children and also blue-eyed ones. Figure 13–7 represents a monohybrid cross, and this is how

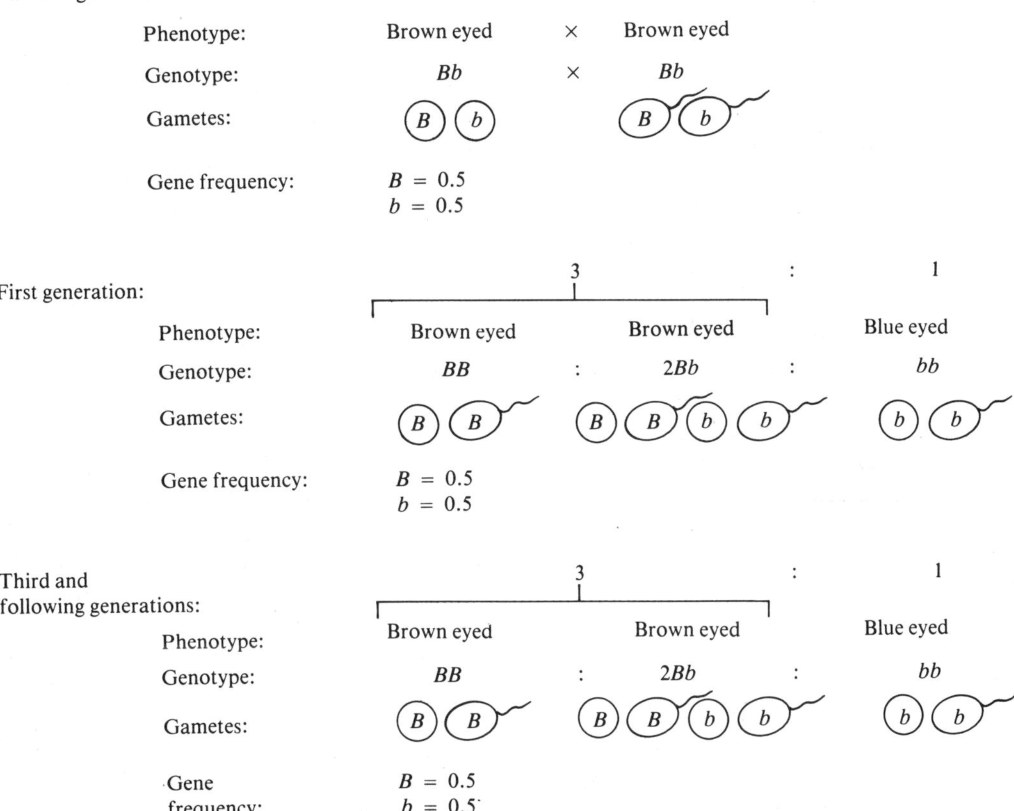

Figure 13-7. Genetic equilibrium. A pair of alleles in a population will tend to reach an equilibrium value in which the genotypes will occur in a frequency of 1:2:1. The frequency of the genes (0.50 and 0.50 in this case) will not change, nor will the ratio of genotypes and phenotypes, as long as mating is at random and several other factors pertain (see text for details).

the entire population can be treated, since *only* the 1 kind of mating is possible, that between 2 heterozygotes. From Mendel's first law, we expect the chance to be 3 out of 4 for a brown-eyed child at any birth. As Fig. 13–7 shows, the first generation of the new colony would occur in a phenotypic ratio of 3 brown-eyed to 1 blue-eyed. Note that the frequency of the allele for brown eyes and that for blue eyes has remained 50:50. Now if members of this first generation take mates only from within their own group and if mating is at random (without regard to eye color), then in the next generation the proportion of brown to blue eyes will again be 3:1, and the frequencies of the 2 alleles will still be 50% for each. In other words, starting out with the first generation in this example, the population came to equilibrium for phenotypes and genotypes. The 3:1 phenotypic ratio and the 1:2:1 genotypic (1*BB*: 2*Bb*: 1*bb*) will no longer change. They will remain constant from one generation to the next, as will the 50:50 relationship for the 2 alleles. The only ways it would change would be if an excess of persons of one kind of eye color arrived from other populations; if persons of one eye color tended to leave the group; or if a greater advantage were suddenly imparted by one of the genes over the other. This latter could come about if mating preferences change, for example if most persons suddenly wished to mate only with those having brown eyes. In the absence of such factors as these (selective mating, sudden advantages of 1 allele, migrations to the populations and emigrations from it), the frequencies of the phenotypes and genotypes and those of the alleles themselves will remain constant.

Hardy and Weinberg expressed this principle in a simple mathematical formula known as the *Hardy-Weinberg* law: $p^2{:}2pq{:}q^2$. This is nothing more than the expansion of a binomial to the second power $(p + q)^2$. Actually, each time we write the genotypes resulting from a monohybrid cross we are representing the expansion. The Hardy-Weinberg law tells us that populations in the case of a pair of alleles tend to reach equilibrium values for the frequencies of the three possible genotypes: $p^2{:}2pq{:}q^2$. If all the assumptions made above pertain, then the equilibrium values will continue to persist.

Of course, real populations are flexible, and therefore rigid adherence to these stipulations does not occur. Migration and emigration do take place; an advantage is often imparted by one of the alleles for some reason or other. Populations can also fluctuate in size. The Hardy-Weinberg law is applicable to large populations. If numbers become reduced, chance factors such as drift (p. 202) play a more important role, with the result that an allele may increase in frequency regardless of any advantage or disadvantage it imparts. Nevertheless, the law may be applied to most populations, since it does permit some estimate of the frequencies of two alleles and of the possible genotypes.

Suppose a very serious disorder which strikes adults is inherited as a recessive and is found to afflict 1 person in 90,000. We can allow L to represent the normal allele and l the harmful recessive. Once the frequency of afflicted persons in the population has been determined, the other frequencies can easily be derived by application of the Hardy-Weinberg law. The frequencies in the population of the 3 genotypes (*LL, Ll,* and *ll*) depend on chance combination of genes. The homozygous recessives whose frequency is known (1/90,000) represent the number of times 2 recessive genes have come together on the basis of chance. The number of homozygotes is therefore the product of 2 independent events and can be represented by q^2 in the Hardy-Weinberg formula (Table 13–2). Since we know q^2 (1/90,000), we can obtain the rest of the information by simple arithmetic. The frequency of the recessive gene must be the square root of 1/90,000 or 1/300. Consequently, the frequency of the dominant gene must be 299/300, since the number of dominant plus the number of recessive alleles must total 100% or 1 ($p + q = 1$). This is so for any locus. Knowing the frequencies of both alleles, we can easily estimate both the number of persons who are homozygous for the normal gene

Table 13–2. Calculation of gene frequencies and frequencies of genotypes applying the Hardy-Weinberg formula

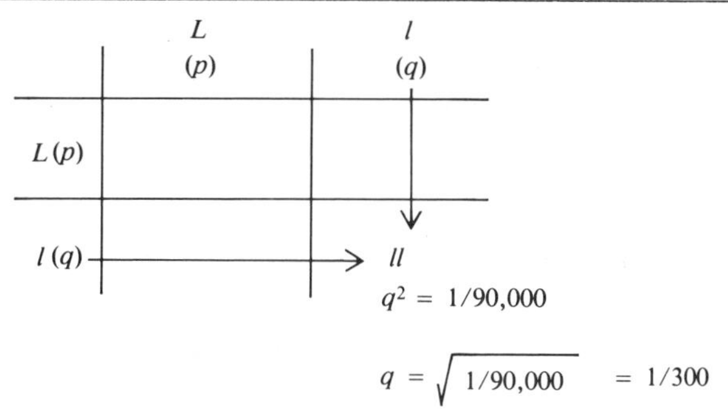

$$p + q = 1; \quad p + 1/300 = 1; \quad p = 299/300$$

Genotype	Frequency		
AA	299/300 × 299/300	=	89,401/90,000
Aa	2 × 299/300 × 1/300	=	598/90,000
aa	1/300 × 1/300	=	1/90,000
			90,000/90,000 = 1

and the number who are heterozygous carriers of the recessive (Table 13–2). The former is simply the square of the frequency of the dominant gene $(299/300)^2$, since the homozygotes represent the chance combination of 2 dominant genes. The frequency of the heterozygotes in the population is $2pq$, which represents the frequency to be expected from chance combination of 2 alleles. The number of these heterozygotes is thus:

$$2 \times 299/300 \times 1/300 = 598/90{,}000 \text{ or approximately } 0.7\%$$

Based on percentage, the frequency of the homozygous recessives, 1/90,000, is far below this figure (0.001%). Therefore, just by knowing the frequency of homozygous recessives, the Hardy-Weinberg formula permits the derivation of the rest of the information.

The preceding results illustrate another important point. Note that while the number of persons afflicted with the disorder is rare, the number of heterozygotes is appreciably higher. It is the case in populations that the number of individuals heterozygous for rare recessives exceeds by far the number of homozygous recessives. Therefore there is in populations a reservoir or pool of rare recessives which exist mainly in heterozygous carriers. Population members have arisen from genes in the pool and they return genes to it when they reproduce (Chapter 8). Every population has its mutant gene pool in which different recessives exist with different frequencies. The concept of the gene pool applies as well to family lines and becomes a matter of concern in cases of inbreeding and consanguinity (p. 294).

Once the geneticist knows the frequency of persons in a population heterozygous for a harmful recessive, he can give more precise answers to certain questions. Suppose a woman who has an albino brother wishes to know her chances of having an albino child if she marries a normal, unrelated man with no family history of the disorder. It is unknown whether or not the woman is a carrier, but the chance is 2/3 that she is. This is so, since the mating of any two heterozygotes such as her parents ($Aa \times Aa$) will produce offspring in the ratio of $1AA{:}2Aa{:}1aa$. The normally pigmented offspring therefore occur in a ratio of $1AA{:}2Aa$. Thus the chance that the woman is a carrier of the recessive is 2 out of 3. Suppose the geneticist knows that the frequency of albino carriers in the population as a whole is about 1 in 80. Elementary genetics tells us that *if* 2 carriers mate, the chance is 1/4 at each birth for an albino. The 3 separate probabilities here are then multiplied together: $2/3 \times 1/80 \times 1/4 = 1/480$. This figure is the chance for an albino at the first birth in this case.

Estimating the frequency of alleles for X-linked recessive traits and the number of female carriers in a human population is based on exactly the same reasoning and actually has some advantages over calculations for autosomal recessive traits. Since a hemizygous male will express an X-linked gene, whether the trait is dominant or recessive in the female, counting the number of afflicted males in the population tells directly the frequency of the detrimental gene. The incidence of deutan colorblind males in a population can illustrate how data on X-linked genes can be used. If a population shows that 3 males in 100 are colorblind, then 3 males in 100 carry the recessive gene and 97/100 carry the dominant allele. This is the frequency of the alleles among males, and it is also their frequency among females, since the whole population is sharing a gene pool and will come to an equilibrium for the 2 genes. The females are simply regarded as a separate population in order to calculate the number of female carriers. This is again $2pq$. Since we know the frequencies of the 2 alleles from observations of the males, the number of carrier females becomes:

$$2 \times 3/100 \times 97/100 = 582/10{,}000, \text{ which is close to } 6\%$$

The number of females who are colorblind is very rare: 3/100 × 3/100, or 9/10,000, a figure well under 1%. We see here again that rare recessives tend to accumulate in heterozygotes who act as a reservoir for them. We can also appreciate more fully the frequency difference between males and females who express sex-linked recessives.

Estimates of genes and genotype frequencies in a population will undoubtedly play an increasingly important role in human genetics and genetic counseling as more data are assembled on the numbers of deleterious genes in the population and as more expertise is achieved in the detection of heterozygotes.

REFERENCES

Auerbach, C., and Kilberg, B. J., "Mutation in Eukaryotes." *Annu. Rev. Genet.,* 5:163, 1971.

Bloom, A. D., "Induced Chromosomal Aberrations in Man." *Adv. Human Genet.,* 3:99, 1972.

Casarette, A. P., *Radiation Biology.* Prentice-Hall, Englewood Cliffs, N. J., 1968.

Drake, J., "Mutagenic Mechanisms." *Annu. Rev. Genet.,* 3:247, 1969.

Drake, J., *The Molecular Basis of Mutation.* Holden Day, San Francisco, 1970.

Green, E., "Genetic Effects of Radiation on Mammalian Populations." *Annu. Rev. Genet.,* 2:87, 1968.

Harris, M., "Mutagenicity of Chemicals and Drugs." *Science,* 171:51, 1971.

Miller, R. W., "Delayed Radiation Effects in Atomic Bomb Survivors." *Science,* 166:569, 1969.

Setlow, R. B., Regan, J. D., German, J., and Carrier, W. L., "Evidence That

Xeroderma Pigmentosum Cells Do Not Perform the First Step in the Repair of Ultraviolet Damage to Their DNA." *Proc. Nat. Acad. Sci.* (U.S.), 64:1035, 1969.

Sutton, H., and Harris, M. I. (Eds.), *Mutagenic Effects of Environmental Contaminants.* Academic Press, N. Y., 1972.

Wills, C., "Genetic Load." *Sci. Amer.,* March: 98, 1970.

Witkin, E. M., "Ultraviolet-Induced Mutation and DNA Repair." *Annu. Rev. Genet.,* 3:525, 1969.

REVIEW QUESTIONS

1. Which of the following is (are) true about gene mutations: (1) The commonest kinds produce no conspicuous visible effect. (2) They are usually beneficial, making possible evolutionary progress. (3) They represent changes in codons and therefore always cause amino acid substitutions in polypeptide chains. (4) They are usually recessive. (5) Mutant alleles which have a dominant effect will increase in frequency in a population.

2. Which of the following is (are) true about mutagens: (1) They direct the genetic material to change in a specific way or direction. (2) A substance found to be mutagenic in one species will prove to be mutagenic in another. (3) Mutagens usually cause mutations which produce very obvious effects. (4) They may cause changes in the structure of the chromosome. (5) All known mutagens cause their genetic effects in the same way by acting directly with the genetic material.

3. Select the correct answer(s). Ultraviolet: (1) is a corpuscular radiation; (2) is highly ionizing as it causes a separation of electric charges; (3) can cause damage directly within the DNA; (4) usually causes genes to become unstable, resulting in mosaicism; (5) includes longer wavelengths than do X-rays.

4. A certain rare disorder inherited as an autosomal recessive kills individuals by the age of 3. About 1/200,000 babies is born with it and the same number of infants dies. Give an estimate of the mutation rate for this particular genetic locus.

5. In a certain Indian tribe, the frequency of the MN blood types is found to be: M (36%); MN (48%); N (16%). What are the frequencies of the *M* and the *N* alleles in the tribe?

6. Assume that in a certain island population, the frequency of persons with oculocutaneous albinism is found to be about 1/900. *A.* What are the frequencies of the alleles *A* and *a* in the population? *B.* Suppose a normally pigmented woman in the population who has an albino brother wishes to marry a normally pigmented man from the island. The man has no history of albinism in his family. What is the chance that the first child will be an albino?

7. Assume that in another population about 15 males in 100 are found to carry the X-linked recessive *g* for inability to produce G6PD. What is the frequency of gene *g* and its allele *G,* and how many women in the population are heterozygous for them?

chapter 14

CONTROL MECHANISMS

Cell Specialization and Gene Expression

The information assembled from classical and molecular genetics has provided an insight into the nature of the hereditary material and some of the ways in which it acts to produce an effect. Gene expression involves transcription (the formation of an RNA, such as mRNA) and translation (the assembly of a polypeptide or protein using the information coded in the mRNA). A gene may therefore be recognized as a result of its association with a specific cellular product such as an enzyme or a particular type of protein which has a structural role in the cell. In multicellular plants and animals, all the body cells typically contain two entire sets of chromosomes, and consequently a double dose of each type of gene locus. Any fertilized egg or zygote contans this diploid number, since it results from the fusion of two haploid sex cells. The human zygote contains all the coded information required to direct the formation of a human being. The fertilized egg undergoes mitosis, forming an embryo of two cells, each with the identical genetic information. Mitotic divisions continue throughout embryology and produce the billions of cells of the human body. From the information in the fertilized egg, an individual can arise with the form, tissues, and organs typical of a member of the human species.

A consideration of these simple and obvious facts indicates something else about genes and their activity. The fertilized egg and the earliest cells derived from it are very similar, if not identical. As cell divisions continue, cells of the embryo begin to specialize or differentiate. Populations of cells assume differences which set them apart from other groups of cells, as they become associated with their specific roles. This is what happens as cells arise which are to become part of the skin, the liver, the brain, etc. These cells are differentiated and can be distinguished on the basis of appearance and the formation of specific products. Hemoglobin is made by red blood cells;

melanin is formed by certain skin cells; bile is manufactured by cells of the liver; and so on. While it is common knowledge that such specialization takes place, the factors which trigger and direct cell changes are largely unknown. The many different cell types which we see in a human contain exactly the same number and kinds of chromosomes, and hence the same genetic information. They all trace back to the one cell, the zygote. Although they carry the identical genetic information, cells of the embryo somehow manage to become different from one another and to undergo specific changes which result in cell differentiation. As development proceeds, this specialization of cell types continues in an orderly time sequence. Secretory cells of the pancreas do not suddenly arise in the embryo without any relationship to the development of another kind of cell type or organ. Cells of the heart are among the earlier of the specific cell types to arise and are essential to the establishment in the embryo of a separate circulatory system which will transport nutrients to specialized organs which arise later as the embryo becomes more complex. The developmental changes which occur to produce a human or any many-celled creature are coordinated; the normal completion of one part of a complex organ depends on the normal development of its component parts and even upon some other organ.

The orderly patterns typical of normal development must involve mechanisms which prevent haphazard cell proliferation and which insure the specialization of cells at the proper time and location in the embryo. One of the factors involved is the activation of specific genes in specific cells at specific times. It should be obvious that most of the genes in the zygote and cells of the early embryo must be inactive. Certainly genes for eye color or the production of gastric juice are not expressing themselves in these cells whose function is a more generalized one, largely concerned with the intrauterine establishment of the embryo. The specialization which does occur must involve the activation of certain genes, which up to a critical time, lie unexpressed. As a cell becomes specialized, however, only a small amount of the genetic information it carries will ever be expressed. Of course, there must be certain genes which remain active in all kinds of cells most of the time. These would be those genes coded with the information required to maintain the metabolic activities needed by any cell to survive: release of energy, construction and repair of cell parts, etc. As a cell specializes, it must retain the capacity to perform these basic tasks, but it acquires the additional ability to manufacture specific proteins. For example, a cell of the pancreas may become efficient in the secretion of certain digestive enzymes. Any specialized pancreatic cell will never form the blood protein, hemoglobin, nor the enzymes typical of a cell of the salivary glands. Since the human is such a diverse assortment of different cell populations, any one kind of specialized cell must express only a fraction of all the genetic information it carries. Indeed, most cells probably do not express more than 20% of the total. One example of a very highly specialized cell is the erythrocyte or red blood cell whose activities are concerned almost exclusively with hemoglobin production. Well under 5% of the entire amount of genetic information is ever called into activity in these highly differentiated cells.

It would appear, therefore, that most genes are never activated at all in a cell of a given type. A small percentage may provide vital information throughout the life of the cell, being called to activity as they are needed. Others may undergo transcription only at very restricted periods; still others may provide information for just a limited period and then become permanently inactivated. Table 14-1 lists several genes which are known to express themselves at characteristic times in the life span of the individual. The gene for polydactyly (extra digits) must act in the embryo at the time the limbs are forming. The presence of the dominant mutant allele causes some disturbance in the normal pattern of development. This indicates that the normal gene is necessary at this point of time in embryonic growth. Typically, the normal gene would be present and would be activated at this time. The effect of the recessive gene responsible for phenylketonuria becomes evident shortly after birth. The homozygous recessive condition results in a failure of production of phenylalanine hydroxylase, required for the proper conversion of dietary phenylalanine. It is at this time that the normal allele for the enzyme production would be called to expression. The gene which determines the deposition of melanin pigment in the iris of the eye is not activated until a few weeks after birth, at which time eye pigment is deposited. Certain genes express themselves

Table 14-1. Characteristic time of appearance of certain human traits

Trait or characteristic	Typical time for onset of expression of genetic factors involved
α Hemoglobin chain	Early embryo
ϵ Hemoglobin chain	Early embryo
Polydactyly	Embryo
Fingerprints	Embryo
γ Hemoglobin chain	12 weeks in development
β Hemoglobin chain	About time of birth
δ Hemoglobin chain	About time of birth
Phenylalanine hydroxylase production	Shortly after birth
Alkaptonuria	Birth
Pigments of iris of eye	Few weeks after birth
Lesch-Nyhan syndrome	Months after birth
Tay-Sachs disorder	4–6 months
Vitamin D resistant rickets	1 year
Sex-linked muscular dystrophy (Duchenne's)	2–5 years
Breast or beard development	Puberty
Pattern baldness	20–30 years
Huntington's disease	20–40 years

in later years, such as the dominant mutant gene responsible for Huntington's disease. Again this tells us that the normal alleles are required at these times to provide the information for some process essential to the normal maintenance of the individual.

Control of Gene Expression

The general picture which emerges for all life forms is one of gene control, the activation of particular genes in certain cells and the continued repression of others whose information would interfere with specialization. The factors responsible for the control of genes, their activation and repression, are required not only for normal development but also for the maintenance of a normal individual. The controlling factors themselves must be largely genetic ones, since the developmental patterns are typical for a given species. These factors must contain coded information which is transmitted from one generation to the next and which is needed to guide the orderly expression of other genes with other sorts of information. If something upsets a control, incorrect or superfluous information may be expressed in a cell at an inopportune time or the expression of the correct information may be completely prevented or delayed. Upsets such as these can have serious consequences which could kill the developing offspring, produce serious abnormalities to organs during development, or cause pathological conditions well after birth.

Controlling factors are therefore of the utmost importance to all living things. Without them, no matter how extensive the amount of genetic information, no organized creature could arise, since chaos would occur if any gene could express itself at any time. The nature of these controlling factors is central to any understanding of the normal process of development and also that of certain pathologies, among them malignancies, those disorders in which controlling mechanisms appear to have gone astray. Multicellular plants and animals are so complex that information on control in them is still quite scant, although as we shall see, certain factors have been implicated. Our clearest picture of genetic control comes from studies of lower forms, mainly bacteria and the viruses which attack them.

In the early 1960s Jacob and Monod, outstanding investigators in molecular genetics, proposed a model to explain the control of certain bacterial genes. In these one-celled organisms, as in any other species, only some of the genes are active at any given time. Some genes in bacteria are called to expression only under certain specific environmental conditions. Most of the information on this subject has been provided by work with *Escherichia coli,* the common colon bacterium which forms a normal part of the flora of the large intestine. Much of the bulk of the feces is composed of *E. coli.* This organism is routinely grown in the laboratory and has been the tool used in many genetic studies including those related to gene mutation, gene action, and genetic controls.

The normal *E. coli* cell carries its genetic information in one naked molecule of DNA, which is in the form of a circle (Fig. 14-1). This circular DNA molecule is often referred to as a "chromosome," but it is simpler by far than the chromosome of any higher life form. The normal *E. coli* can grow on a very simple food source in the laboratory. Supplied with the sugar glucose and a few other simple substances, the cell can manufacture all the other products required for its survival. The normal *E. coli* cell may also utilize the somewhat more complex sugar, lactose, for its growth, but it preferentially uses glucose

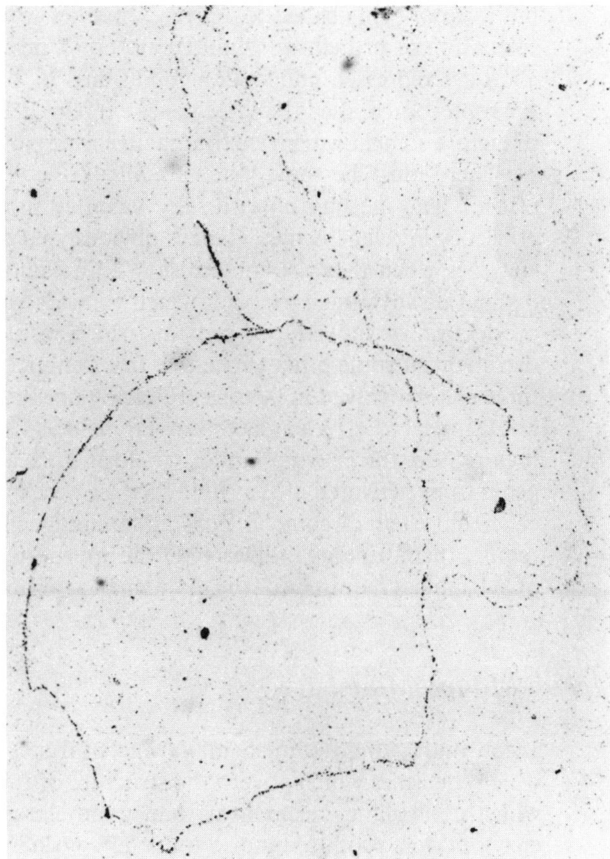

Figure 14-1. Autoradiograph of *E. coli* chromosome. The DNA is in the form of a closed circle which is shown undergoing replication here. It has been isolated from a cell grown on a food source containing a radioactive nucleotide (one with thymine). The DNA preparation was covered with a photographic emulsion which was then bombarded by the emissions of electrons from the radioactive material incorporated in the DNA. The result is an arrangement of spots which describes the appearance of the chromosome.

if supplied with both types. The cell, however, will be able to live on lactose if this is the sole sugar source. An interesting point is that when an *E. coli* cell is growing on glucose, it does not form the enzymes required to concentrate lactose and to break that sugar down. If the cell is given lactose, however, in place of glucose, the required enzymes for lactose utilization are produced in the cell. For the sake of simplicity, we will designate the three enzymes involved in lactose use as enzymes *a, b,* and *c.* It is very significant that these three proteins are not formed when a cell is using glucose exclusively. The genetic information is certainly there for the formation of the lactose enzymes, but it is not being called to activity. In other words, the genes which guide the formation of proteins *a, b,* and *c* are not expressed in the absence of lactose. They are repressed under such conditions. If, however, lactose is supplied as the food source, then the three genes express themselves. It is as if the presence of lactose somehow triggers the inactive genes to activity.

There must be some kind of control operating here to bring about these effects. The mechanism is a very valuable one for the cell and guarantees efficiency in cell activities. There is obviously no advantage for a cell to produce enzymes when they are unnecessary and cannot be used. There would be a distinct disadvantage to a cell which expends valuable energy in the wasteful production of superfluous proteins which would accumulate, perhaps to the detriment of some other processes. It is beneficial to the bacterial cell that the three genes involved in lactose utilization normally remain repressed when it is not using lactose. When lactose is supplied as the sugar source, the cell would then require the enzymes and it would then clearly be beneficial for the three genes to be activated at that time. We see here in *E. coli* one example of a very efficient system of control. Enzymes are induced only by the substances which require them. Energy is diverted at the appropriate time only to the production of substances required for the maintenance of normal cell activity.

The Jacob-Monod Model

Jacob and Monod studied many bacterial strains, among them mutants in which the genes which govern the structure of the enzymes were normal but in which controlling mechanisms had gone astray. In these mutant cells, the enzymes were produced indiscriminately, both in the absence and the presence of lactose. The enzymes themselves were perfectly normal, since the genetic change had not affected their structure in any way. Instead, it had interfered with the *control* of their formation so that they were being produced continuously, regardless of need.

From their observations of various kinds of *E. coli* cells in which the normal time and rate of enzyme production had become altered, Jacob and Monod presented a model of one type of gene control in microorganisms. Experimental work by others in the years which followed verified all the major premises of the model. The presentation of the model and the evidence to support it have been extremely significant to the biological world, since they opened up a pathway to approach the complex problems of cellular control, change and specialization. Jacob and Monod were awarded the Nobel prize in medicine for their many outstanding contributions. The model, though based on genetic analyses of bacteria and viruses, is pertinent to higher life forms including the human. It is therefore essential to know the basic features of the Jacob-Monod model which has provided a basis for an understanding of gene action in all life forms.

According to the model, two major classes of genes may be recognized. One includes all those genes bearing the information for the construction of the specific polypeptide chains which compose enzymes and other cellular proteins. These are the genes most familiar to us, and they have been designated *structural genes,* since they guide the structure (amino acid sequence) of polypeptide chains. The other class is composed of *regulatory genes.* These genes do not determine the amino acid sequence of a polypeptide which will play an enzymatic or structural role in the cell. Rather, they control the expression of the structural genes. A regulatory gene may repress a structural gene so that the polypeptide it governs is not formed. Another regulatory gene may activate a structural gene so that it is expressed at a particular time. A mutation in a regulatory gene can thus upset controls so that a structural gene is expressed at the wrong time, or perhaps not at all, even when the polypeptide it governs is needed. On the other hand, a mutation in a structural gene affects neither the time nor the rate of polypeptide production, but rather the assembly of the chain itself. The mutation responsible for sickle cell hemoglobin is a familiar example of a mutation in a structural gene which results in a structural change in a polypeptide due to an amino acid substitution.

Figure 14-2(A)-(B) brings out all the essential features of the Jacob-Monod model as based on the control of the lactose enzymes. As the model shows, certain structural genes concerned with the metabolism of the same substance (lactose utilization here) may be very closely linked on the DNA. The genes *a, b,* and *c* are found in sequence on the DNA [Fig. 14-2(A)]. They are also linked closely to a specific kind of regulatory gene, the *operator*, a genetic region involved with the expression of the structural genes closely associated with it. The operator and its structural genes compose a unit designated an *operon.* The operon interacts with still another kind of controlling gene. This is the *regulator*, which is not necessarily adjacent to the operon but which can be located some distance away on the DNA. The regulator gene is responsible for the production of a specific kind of diffusible substance. This regulatory substance can act as a repressor, as in the case of the lactose operon. The repressor is able to recognize and combine with the specific operator. This combination of lactose repressor and lactose operator prevents transcription of the structural genes in the operon. The reason for this is that the enzyme which assembles messenger RNA from the DNA (the enzyme RNA polymerase) cannot attach to the starting region of the operon because it is physically blocked by the repressor. The repressor thus blocks the site of attachment of the enzyme needed to construct RNA from the DNA blueprint found in the structural genes. If mRNA cannot be formed, the structural genes are in effect inactive. It is as if they were "switched off." No lactose enzymes can be manufactured.

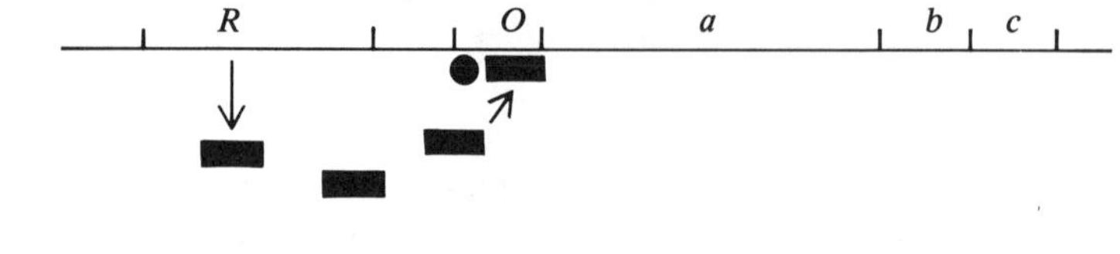

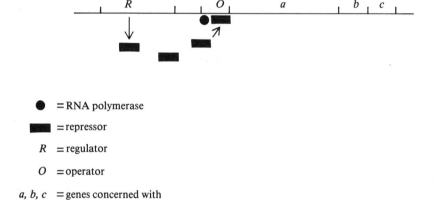

= RNA polymerase

= repressor

R = regulator

O = operator

a, b, c = genes concerned with lactose utilization

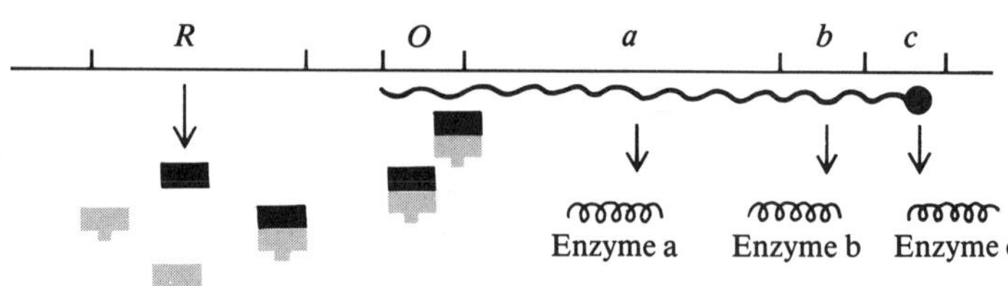

= lactose (inducer)

= mRNA

Figure 14-2. Model of the operon. (A) In the absence of the effector (lactose), the repressor produced by the regulator combines with the operator and blocks transcription of the structural genes, *a, b,* and *c.* (B) When the effector is added, it combines with the repressor, freeing the operator. The enzyme RNA polymerase is now free to transcribe genes *a, b,* and *c* into mRNA which is translated into enzymes *a, b,* and *c,* which are now needed by the cell for lactose utilization.

However, the situation will change in the normal cell if lactose is added [Fig. 14-2(B)]. The lactose acts as an *effector* or *inducer*, since it can combine with the repressor. By tying up the repressor substance, the lactose molecules prevent the combination of the operator and the repressor. The operator becomes freed of repressor, and the already-formed combination is dissolved. Consequently, the mRNA-forming enzyme, RNA polymerase, can attach to the proper initiating site in the vicinity of the operator and bring about transcription of the structural genes in the operon. The messenger which is formed carries the information for the construction of the three enzymes now needed in the cell for the breakdown of the lactose which is being supplied as a food source. If lactose is removed, the repressor is again able to accumulate and to form a complex with the operator. Transcription becomes blocked once more. No messenger RNA is formed from the structural genes and the three enzymes, now unnecessary to the cell, are not produced.

Figure 14-3(A) shows what may happen if a mutation occurs in the regulator gene. A defective repressor, or perhaps none at all, forms which cannot combine with the operator. Consequently, the RNA polymerase is free to attach to the beginning of the operon at any time and form messenger RNA with the information for the structure of the three enzymes. The enzymes will thus be produced indiscriminately, in the absence of lactose as well as in its presence.

Figure 14-3(B) shows another kind of mutation which has similar effects, but it is a mutation in the operator. In such a case, the operator may be unable to respond to the normal repressor, and so the structural genes are permanently switched on. Still other kinds of mutations in controlling genes are known, such as the one shown in Fig. 14-3(C). Here the mutation in the regulator produces a repressor which has a very strong affinity for the operator. Such a repressor may not be able to combine with the inducer, lactose. The outcome is that the enzymes are never produced, even when required. It should be apparent from a review of these basic features of the operon (Figs. 14-2 and 14-3) that controlling genes such as the regulator and operator are essential to normal gene expression and are critical to normal cell development and survival.

A number of different operons are now well known in many microorganisms. There is also evidence for them in higher life forms. Repressor substances have actually been identified for several regulator genes. The repressors have been found to be very large protein molecules which somehow are able to attach to specific operators, just as Jacob and Monod predicted. The description given here for control of the lactose enzymes is just one form of the operon model. There are variations which pertain to the regulation of some other kinds of structural genes. These need not concern us here, since all operons involve certain common features: structural genes, their operators and their regulators, and the regulatory substances which control them. Details of the control, however, are known to vary from one kind of operon to another.

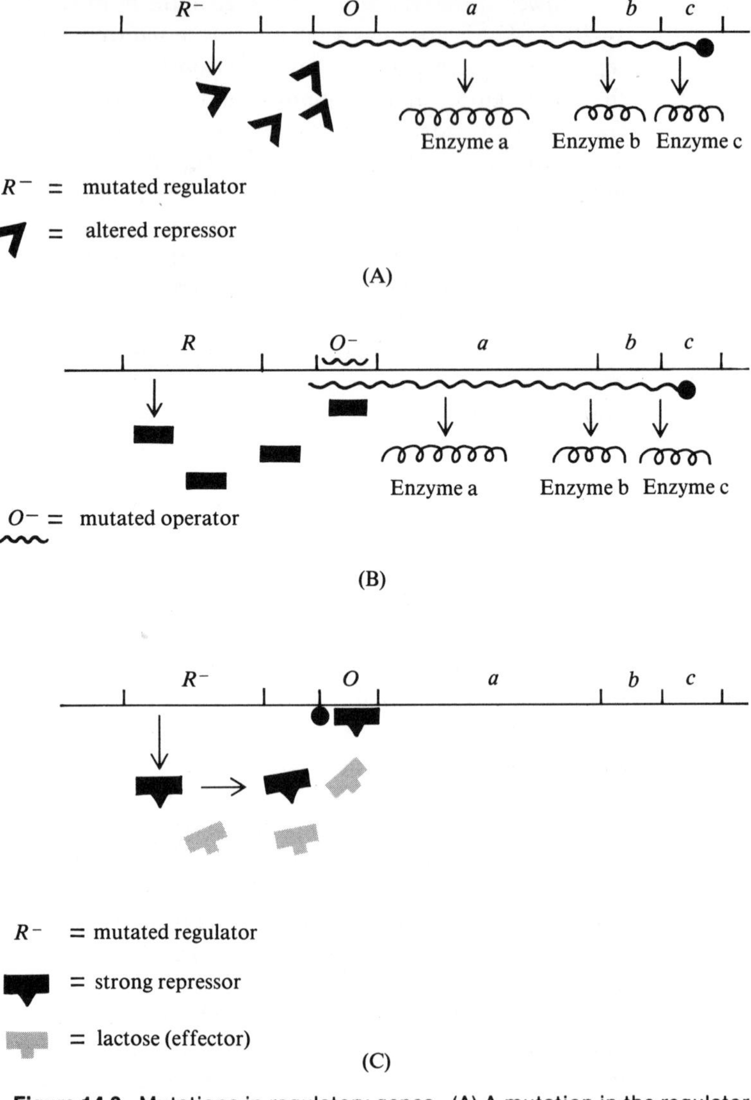

Figure 14-3. Mutations in regulatory genes. (A) A mutation in the regulator gene may result in the formation of a repressor which cannot combine with the operator, so that the structural genes *a, b,* and *c* undergo transcription indiscriminately, and the associated enzymes are produced even when they are not needed. (B) A mutation in the operator may render it incapable of combining with the normal repressor. Again the consequence is the indiscriminate transcription of mRNA and the production of enzymes. (C) A mutation in the regulator may affect it in such a way that the repressor which is produced has a strong affinity for the operator and cannot combine with the effector when it is added. In such a case, the mRNA and the enzymes are never produced, even when needed.

Control of Gene Activity in Higher Life Forms

The mechanisms which control gene expression in higher life forms must entail many more complexities than those in bacteria, which have only an exposed molecule of DNA instead of a complex chromosome. Bacteria, blue-green algae, and a few less familiar organisms are called *prokaryotes*. No nuclear membrane surrounds their naked DNA. Cell division in these groups does not involve a spindle or any of the features of a mitotic division, although the DNA is equally distributed. Other parts of these cells are much less specialized than those in higher species which include all the familiar plants and animals. The genetic system of a prokaryote is open in the sense that the regulatory products are released directly into the cell and interact directly with the DNA. On the other hand, the DNA of higher life forms is intimately associated with RNA and different kinds of proteins to form the typical chromosome. Species with complex chromosomes, nuclear membranes, mitosis, plus certain specializations of the cytoplasm are called *eukaryotes*. Any regulatory substances in a eukaryote are not free to interact directly with the genetic material but must encounter the components which are complexed with the DNA of the chromosomes.

Very active research in the past few years on the control of gene expression in eukaryotes has implicated the proteins of the chromosome as factors. One major group of proteins associated with the DNA is the *histone* proteins, also commonly referred to as *basic* proteins, since histones carry many amino acids which are more basic than acidic in nature. Another important category of chromosomal proteins is much more acidic. These are often designated *non-histone* or *acidic* proteins. Under laboratory conditions, chromatin (the assortment of DNA and proteins composing a chromosome) may be extracted from the nuclei of eukaryotic cells. This chromatin can then be studied *in vitro* for its ability to produce RNA. In these cell-free systems, RNA can be formed from isolated chromatin if the proper building substances are supplied in addition to those factors required for the maintenance of the chromatin under test tube conditions. The chromatin may be stripped in the test tube of some or all of its proteins (Fig. 14-4). When this is done, the degree of gene activity, as measured by the amount of RNA synthesis, is increased greatly. The assembled evidence indicates that the basic histone proteins somehow tend to keep genes in a state of repression. They may do this by binding tightly to the DNA, so that mRNA formation is prevented [Fig. 14-5(A)]. Recent data strongly suggest that some acidic proteins activate genes by somehow interacting with the basic proteins, causing the latter to bind less tightly to the DNA. In this way, the DNA becomes exposed to the enzyme RNA polymerase. The gene can now be transcribed and is thus activated. Certain hormones may act in this way by antagonizing repressor proteins, causing them to bind less tightly to the DNA [Fig. 14-5(B)]. It should be noted that these concepts which are being established for gene control in eukaryotes are based on those of the Jacob-

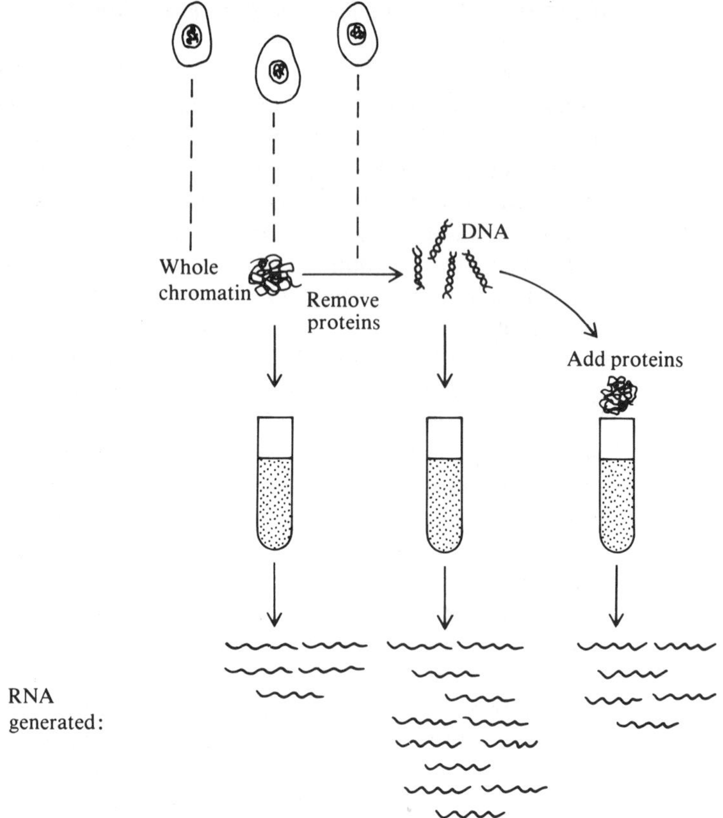

Figure 14-4. Chromatin and *in vitro* transcription. Chromatin may be taken from the nuclei of various kinds of plant and animal cells and then supplied to a test tube system which contains all the essentials needed for the synthesis of RNA. When chromatin from calf thymus is used, RNA is formed in the test tube which is characteristic of that organ. When proteins are removed from the chromatin, about 15 times as much RNA activity takes place. If the proteins are added back again to the denuded DNA, RNA synthesis becomes reduced. However, the RNA's that are made differ from the RNA characteristic of the native chromatin. The observations suggest that the proteins are suppressing much of the DNA, preventing it from being transcribed into RNA. This suppression is specific for a given tissue. When the proteins are removed, an increase in RNA formation occurs. When they are added back, there is again suppression, but due to conditions of the experiment, the genes being suppressed are not necessarily the same as those suppressed in the native chromatin.

Monod model. Structural genes are acting with repressors (histones). An operator region which must be exposed may also be involved in the binding of the RNA polymerase.

The controls of several genes in the human lend themselves to an interpretation according to the operon model. The expression of structural genes involved in the formation of the hemoglobin chains (Chapter 10) can be readily understood in this light. Figure 14-6 (which should be referred to in the following discussion) presents a summary of the genes and the loci involved in the formation of the five main types of hemoglobin chains.

The figure shows a close linkage between the locus for δ polypeptide chains and the locus for the β chains. Linkage between these two loci has been established primarily from studies of a large number of pedigrees and also by

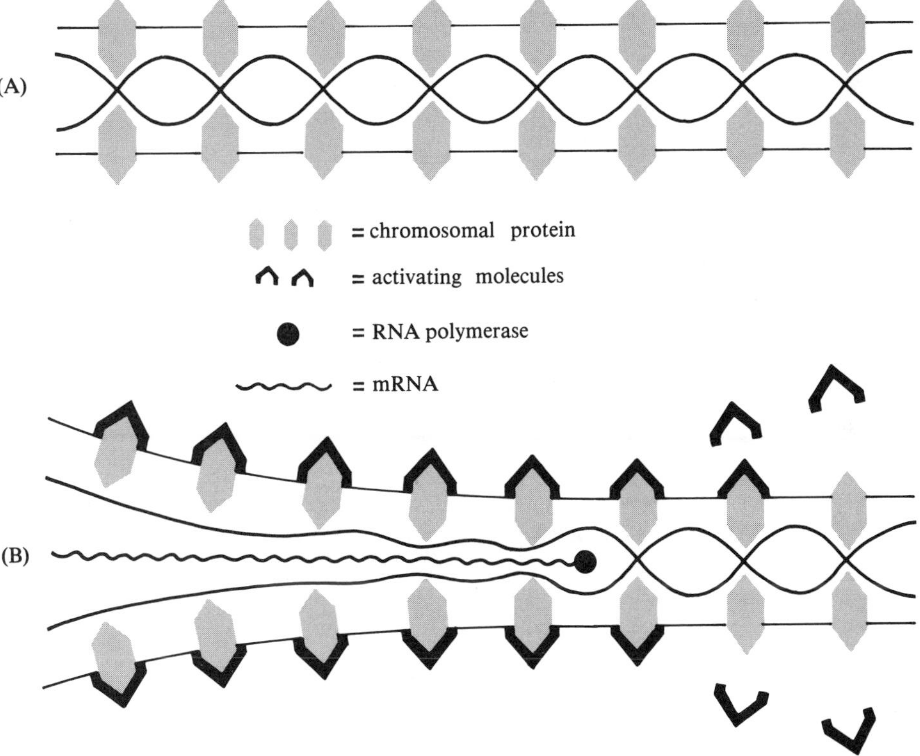

Figure 14-5. Histones and gene repression. (A) The histones may block the activity of genes by combining tightly with the DNA, thus preventing it from being used as a template for RNA synthesis. (B) Hormones and certain other substances may act as activating molecules by somehow causing the histone proteins to bind less tightly to the DNA so that the latter becomes exposed to RNA polymerase and transcription can occur.

the existence of certain rare and unusual forms of hemoglobin, the Lepore hemoglobins. In a person heterozygous for hemoglobin Lepore, approximately 10-15% of the hemoglobin consists of an α chain combined with a chain that has features of both the β and the δ chains. There is very good reason to believe that such a person carries on one of his or her chromosomes a genetic region in which part of the β gene has been combined with part of the δ gene, resulting in a single gene. This so-called Lepore gene will direct the formation of a hemoglobin chain with characteristics of both the β and the δ chains. More than one kind of Lepore allele has been identified, and each is believed to have arisen as a result of mispairing followed by crossing over in the $\beta \cdot \delta$ region of the chromosome during meiosis.

An operator gene is postulated for the δ and the β loci, so that the operator δ-β region composes an operon. There is some evidence that the γ locus, while not directly adjacent to the δ and β loci, is linked to them. It would

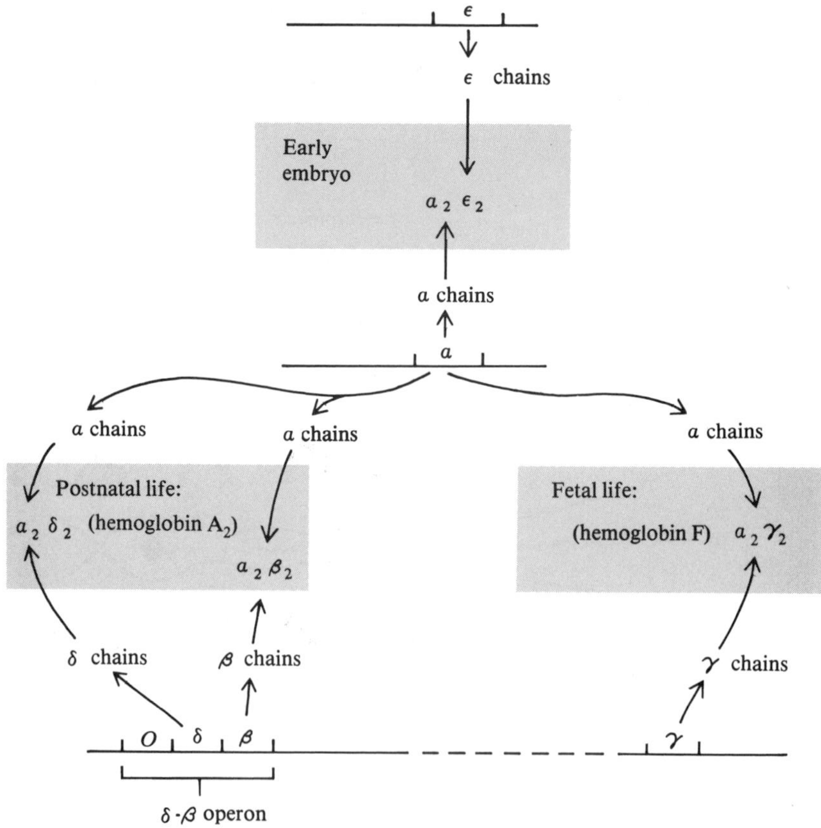

Figure 14-6. Summary of the genetic loci and their expression in the formation of human hemoglobins (see text for details).

not, however, be a part of the δ - β operon. However, it appears that the α and the ϵ loci are not linked to the other two nor are they linked to each other.

In the zygote and in the very early cells of the embryo, none of the genes at these five loci are expressed. But with the establishment of a circulatory system, gene activities will occur in the maturing red blood cells at different stages of the individual's life. In the embryo, the genes at the α and the ϵ loci can become activated or derepressed in the cell type specialized for hemoglobin formation. Exactly how this occurs is unknown. The α locus will now become activated in each cell which differentiates into a red blood cell. However, the ϵ locus, after being activated in maturing cells for a short span, no longer becomes "turned on." Instead, in maturing cells, the γ locus becomes activated. Gamma chains instead of ϵ chains now associate with the α chains to form the hemoglobin typical of the fetus. Toward the end of fetal life, the γ locus is no longer called to activity, as the δ and the β loci become activated. Fetal hemoglobin disappears, and the two types of adult hemoglobin are established $\alpha_2\beta_2$ (hemoglobin A) and $\delta_2\delta_2$ (hemoglobin A_2). The δ and the β loci appear to be activated in maturing cells at the same time, and they remain active in the differentiating red blood cells throughout the rest of the life span. The postulated operator gene may be partly responsible for the expression of the δ and β genes, and this may entail their release from chromosomal protein.

In the human, a recessive gene is known which prevents the formation of the δ and β chains, so that the adult possesses only fetal hemoglobin, α chains associated with the γ. The person who is homozygous for this high-fetal-hemoglobin gene suffers no apparent ill effects. It would appear that the mutation somehow affects the operator of the δ - β operon so that the δ and β loci remain repressed while the γ locus remains active. Those persons who are heterozygous and who carry one dose of the normal allele produce γ chains as well as the chains characteristic of the adult hemoglobins, A and A_2. This supports the idea (Fig. 14-7) that the normal operator region on one chromosome has been freed from repression at the appropriate time in development, whereas the defective operator on the other remains permanently associated with the repressor. Somehow as a result of the mutation, the γ locus is permanently switched on. The model presented here shows what is probably only a few of the complexities involved in the interactions of genes and regulator substances in higher species. The exact nature of the repressors and activators is unknown, but there is no doubt that such controlling factors are in effect. While certain modifications may be required in the Jacob-Monod model when it is applied to higher species, the model affords an excellent approach to an understanding of gene regulation in complex cells.

It should be appreciated here that special control mechanisms must be responsible for the random inactivation of most of one of the X chromosomes

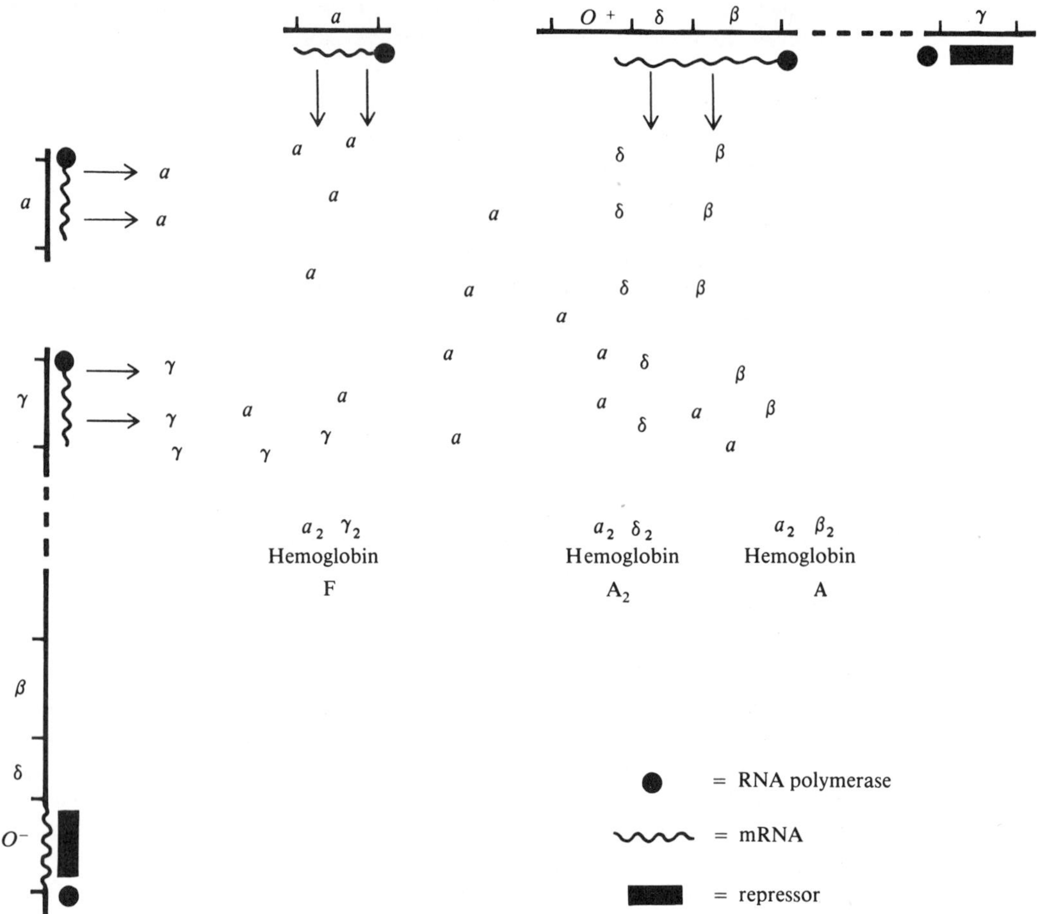

Figure 14-7. Heterozygote for high-fetal-hemoglobin gene. While the homozygote produces no δ and β chains at all and hence no adult hemoglobin (A or A_2), the heterozygote represented here produces hemoglobin A (α_2 β_2), A_2 (α_2 δ_2) and the fetal hemoglobin F (α_2 γ_2) in adult life. One explanation is that the high-fetal-hemoglobin effect is due to a defect in an operator (O) which is part of a $\delta - \beta$ operon. The defective operator remains permanently repressed. Since RNA polymerase cannot transcribe the genes in the operon, no δ or β chains can arise from the genes in that chromosome. The γ gene, however, remains switched on. In the other chromosome, the operator is normal (O^+) and is released from repression at the birth of the individual as the γ locus in that chromosome becomes repressed. The α genes on the other chromosomes continue to produce α chains as they did before birth and remain free from repression. Consequently, hemoglobins A, A_2, and F are present in the heterozygote.

during normal development of an XX zygote (Chapter 5). It is also most interesting that when one of the X chromosomes in an XX zygote carries some structural defect (such as a deletion), it is this defective X which tends to be inactivated in the body cells. This preferential inactivation of a defective X could be one way the cell's control mechanisms may operate to prevent abnormalities which could originate from defects in genetic material. It is conceivable that some such inactivation of defective parts of autosomes also occurs. Indeed, any clues to be discovered on factors which effect the repression of most of an entire X chromosome will be highly relevant to the problem of general genetic control in all eukaryotes. As yet, no experimentally sound ideas are available to explain the inactivation of an X chromosome.

Gene Control and Virus Activity

The subject of gene control is intimately related to the study of viruses. Virus particles attack not only animals, but plants and bacterial cells as well. Most of our appreciation of viral activity has emerged from work with bacterial viruses, also called *bacteriophages* or *phages.* A virus of any kind lacks cellular structure and is basically a package of nucleic acid surrounded by a protein envelope (Fig. 9-8). Certain phages which attack the colon bacillus, *E. coli*, have been extensively studied, and the events which follow viral infection of a bacterial cell are known in some detail. Viruses differ from one another in their host range, meaning that certain strains of cells are immune to one kind of virus but susceptible to another. A virus which attacks one strain of *E. coli* may be unable to infect another one. Any virus is typically restricted in its host range. A virus which can live in the cells of one species, say the cat, does not usually infect a very different one, such as the dog. Some bacterial viruses are virulent, meaning that once they infect the cell, they lead to its destruction. Figure 9-8 shows some of the typical events. When a bacteriophage (such as the well-studied one known as T/4) contacts a susceptible *E. coli* host cell, it attaches to the cell wall and injects its DNA while the protein coat remains outside. Once the viral DNA is inside the cell, it directs the formation of messenger RNA. This mRNA contains information for the formation of specific viral proteins, among them enzymes required for the construction of new viruses. The infecting viral DNA also succeeds in shutting down the normal synthetic activities of the infected cell. For example, all the processes involved in the maintenance of normal cell activities, as well as those needed for cell division and the synthesis of cellular DNA, come to a halt. The DNA of the host cell becomes degraded. New DNA is soon synthesized, but this is not the DNA typical of the cell. Rather, it is viral DNA. As the new viral DNA arises, proteins begin to appear. These are the proteins that compose the viral envelope. The viral DNA is packaged in the protein coats, and mature virus particles accumulate in the cell. Finally, an enzyme whose formation is guided by the virus causes the bacterial cell wall to rupture. The cell is destroyed and hundreds of viral particles are released which will repeat the cycle when other susceptible cells are encountered.

From this summary of viral infection of a bacterial cell, it can be seen that the viral DNA, once inside the cell, manages to gain control of the cellular

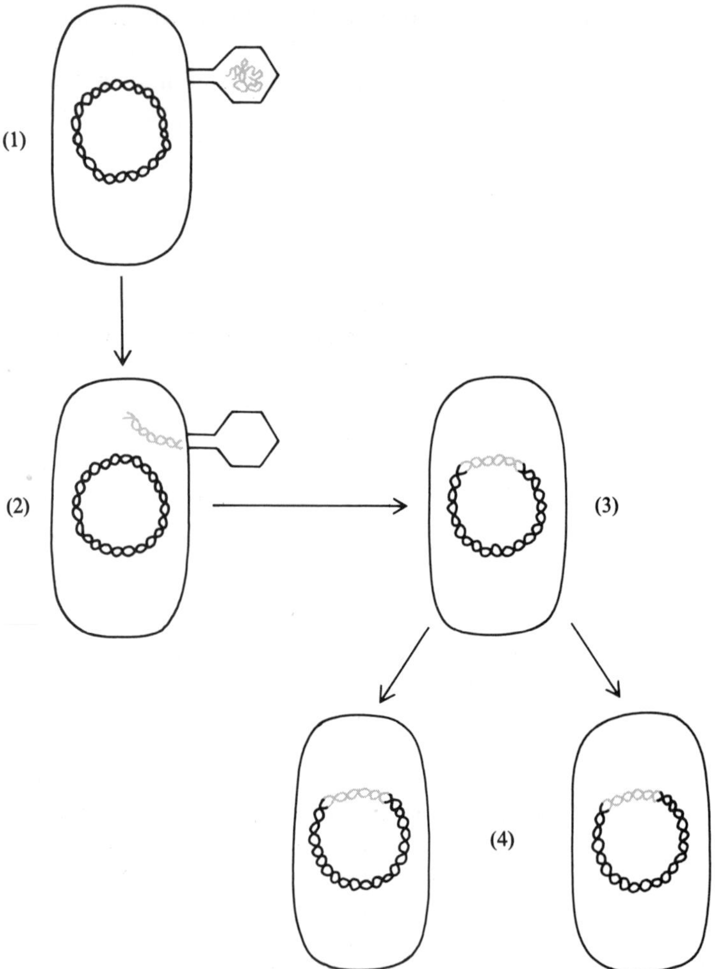

Figure 14-8. Temperate virus. When a symbiotic virus encounters the proper type of bacterial cell (1), it injects its DNA (2) leaving the protein coat outside. This DNA may become integrated (3) with the DNA of the bacterial chromosome. In this prophage state, it replicates in synchrony with the replication of the cellular DNA as a part of the cellular genetic material (4).

activities. All of those processes normal to the cell become shut down, and the only ones which persist are those concerned with the construction of new viral particles. In a very real sense, the control mechanisms of the cell have been disrupted by the presence of viral genetic information.

Not all viruses which infect bacterial cells are virulent. Some are known as *temperate* or *symbiotic* viruses or phages. When the DNA of a temperate virus enters the cell, it manages to insert itself into the DNA of the host (Fig. 14-8). Such a symbiotic virus does not necessarily harm its host. Instead, it behaves as part of the genetic composition of the cell and may remain inserted indefinitely. There may be no evidence whatsoever that the cell is harboring a virus, since no free mature virus with a protein coat can be detected in the cell. The virus is reduced instead to a piece of DNA integrated with the host DNA. In this reduced state, the bacterial virus is called a *prophage*. Bacterial cells which harbor prophage can often be shown to possess some new trait conferred by

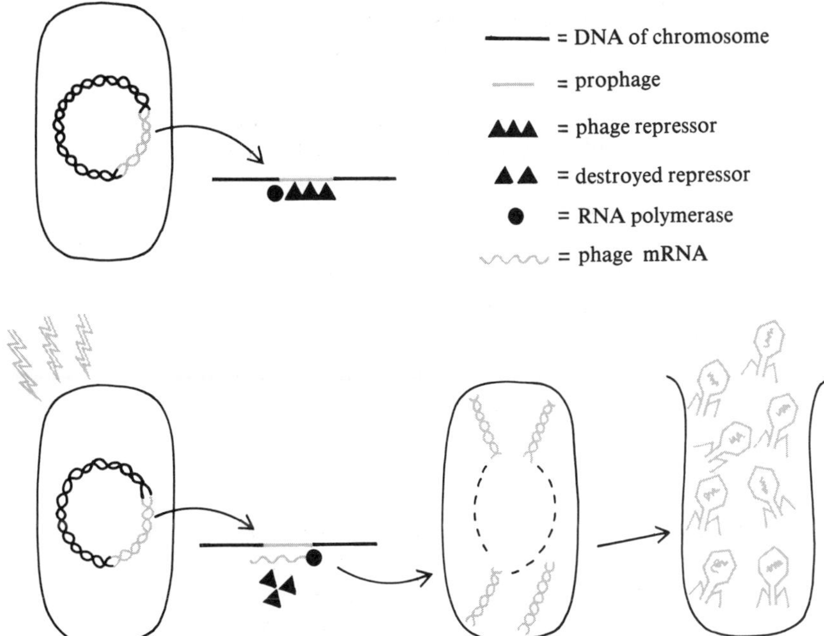

Figure 14-9. Maintenance and activation of temperate phage. The symbiotic phage is able to remain integrated in harmony with the cell as long as its genes are prevented from undergoing independent transcription. The viral genes remain repressed under the influence of a repressor substance controlled by the phage genetic material (above). This repressor can be destroyed by certain environmental agents such as ultraviolet light. The viral genes then undergo transcription (below). The viral mRNA which is formed directs the formation of viral DNA and viral protein. Mature virus particles arise, and the cell is destroyed.

the presence of the prophage. All bacterial cells with prophage are immune to further infection by a related virus. It is believed that the cell-virus association may actually be one which is beneficial to both the bacterial cell and the virus.

The integrated virus, however, does not necessarily remain inserted indefinitely in the host DNA. It can be stimulated to emerge by certain environmental factors, such as the presence of ultraviolet light. The virus may even remove itself spontaneously. A culture of bacterial cells which harbor prophage always contains a few cells in which the viral DNA will leave the host DNA. When this happens, the virus becomes virulent and sets off in the cell the chain of events described above leading to the formation of new mature viruses and the death of the cell. It is known that, in order for a virus to become inserted into the host DNA in the first place and to become a prophage, certain of its genes must be repressed. Otherwise the infecting virus will be virulent. In temperate phages, the repression of the viral genes which would otherwise lead to virus multiplication is accomplished by a repressor substance which is formed under the direction of a gene of the virus itself (Fig. 14-9). A failure to produce repressor results in the inability to become prophage, the assembly of new virus, and the destruction of the cell. Certain agents such as ultraviolet light destroy the repressor and thus cause a previously repressed virus to come to sudden expression. We see here in the simple bacterial-phage relationship events which depend on the activities of regulatory mechanisms. The activation of the prophage and the sudden changes in the cell which follow bring to mind changes which occur in cells of higher organisms which suddenly undergo pathological transformations. Cell abnormalities which involve control mechanisms are undoubtedly among the disturbances which lead to those derangements commonly grouped together as *malignancies* or *cancer.* The cancer problem is one of paramount importance today as more and more persons now live to an age when cancer is more likely to strike and as more and more environmental factors become implicated in pathological cell changes.

Viruses, Cancer, and Controlling Mechanisms

As pointed out by cancer researchers, there is probably no *one* way in which all cancers originate. The picture is a most complex one, and the term *cancer* embraces well over a hundred types of disorders. While changes affecting controls have undoubtedly become altered in some way, the nature of this change may also vary from one cancer type to another. Unlike normal cells, cancerous ones are characterized by unregulated growth. Studies of cancer cells in tissue culture demonstrate that, when a normal cell is transformed to a malignant one, various changes take place which affect the surface properties of the cell membrane. This is most clearly seen in a phenomenon known as *contact inhibition.* Figure 14-10(A) shows that when normal cells of a mammal (human, rabbit, monkey, etc.) are grown in plates supplied with appropriate factors, they divide and spread across the surface until a single compact layer is formed. Further divisions then cease. Figure 14-10(B) illustrates the behavior of cancer cells under the same conditions. The malignant cells spread over the plate, but they fail to exhibit contact inhibition when a solid layer is formed. They continue to grow and to pile on top of one another. The cancer cells fail to respond to the regulatory effects of contact with other cells.

The unrestricted growth of cancer cells at the expense of normal cells in the body implies an alteration in a controlling mechanism. In any normal plant or animal, all the differentiated tissues of the body interact and must be adjusted to one another; no one operates at the expense of another. The maintenance of harmony among all body parts requires restriction on growth. This mechanism of control seems to have been broken down in a cell which becomes cancerous. Search for the operation of factors which can bring about a transformation from the normal to the malignant state forms an active area of cancer research. Several viruses are known which are able to transform cells of certain mammals and birds from the normal to the cancerous state in tissue culture. Plant tumor viruses are also known. When the transformed cells are later implanted into experimental animals, they continue to grow and form malignant tumors. These tumor-inducing viruses are restricted in their host range, affecting cells of only one or a few species. A virus which can bring about tumor formation is termed an *oncogenic virus*; the induction of a tumor is known as *oncogenesis*. Among oncogenic viruses, two major categories may be recognized (Table

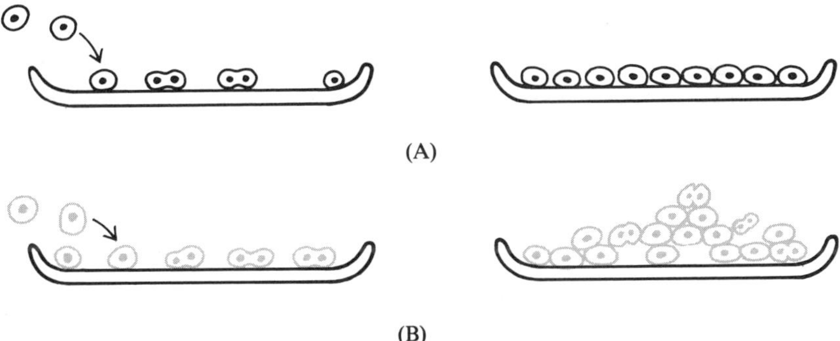

Figure 14-10. Growth of normal and malignant cells in culture. (A) Normal cells when added to a culture plate supplied with essential growth factors proceed to grow until a compact single cell layer covers the surface of the culture medium. Further growth of cells then ceases. The contact of the cells with one another eventually restricts the amount of cell growth. (B) When malignant cells grow across the surface of a culture plate, contact inhibition is lost, and cells grow in a disorganized fashion, piling on top of one another in the plate.

14-2). In one group, DNA is the genetic material; in the other, RNA composes the genes of the virus. Thus, certain viruses are exceptions to the almost universal occurrence of DNA as the genetic material. Viruses, it will be recalled, are not cellular entities. Only DNA is found as the substance of heredity in cellular forms. Some familiar viruses which are not oncogenic but which have RNA instead of DNA as their hereditary material are those which are causative agents of polio and influenza.

When an oncogenic virus induces transformation of normal cells in tissue culture to the malignant state, virus particles may no longer be detected in the cells. However, these may reappear at a later time. There is evidence that some oncogenic viruses may actually integrate their nucleic acid with that of the host. Observations of this type support the *oncogene theory* of malignant transformation. The concept is similar to that which was presented concerning the infection of a bacterial cell by a temperate virus, followed by the reduction of the virus to prophage (Fig. 14-8). According to the oncogene theory (Fig. 14-11), after entering a eukaryotic cell, the viral DNA becomes inserted into a chromosome where it proceeds to take up residence. The viral DNA would replicate along with that of the host cell and would in many ways behave as if it were a cellular gene as it continues to be transmitted from one cell genera-

Table 14-2. Some oncogenic viruses and their host cell types*

I. DNA VIRUSES (about 50 different viruses)
  A. Papilloma virus group
    Papilloma viruses of rabbit, man, dog, cows, and others
  B. Polyoma virus group
    1. Polyoma virus (murine) (Py)
    2. SV40 virus (simian)
  C. Adenoviruses
    1. Human adenoviruses—31 members, 12 members (at least) induce tumors in newborn animals and/or transform cells in vitro.
    2. Simian adenoviruses (6 viruses)
    3. Avian adenoviruses (2 viruses)
    4. Bovine adenovirus
  D. Herpesviruses
    1. Burkitt's lymphoma[a] (human)
    2. Lucké carcinoma[a] (frog)
    3. Marek's disease[a] (chicken)

II. RNA VIRUSES (about 100)
  A. Avian leukemia-sarcoma viruses (20 or more viruses)
  B. Murine leukemia-sarcoma viruses (several hundred isolates have been reported but the number of different types is not well established)
  C. Murine mammary tumor virus (3 types)
  D. Leukemia-sarcoma viruses of cat, hamster, rat, and guinea pig

*From M. Green, *Ann. Rev. Biochem.* 39:701 (Table 1).

[a]Recent evidence, not yet conclusive, associates these diseases with new members of the herpesvirus group.

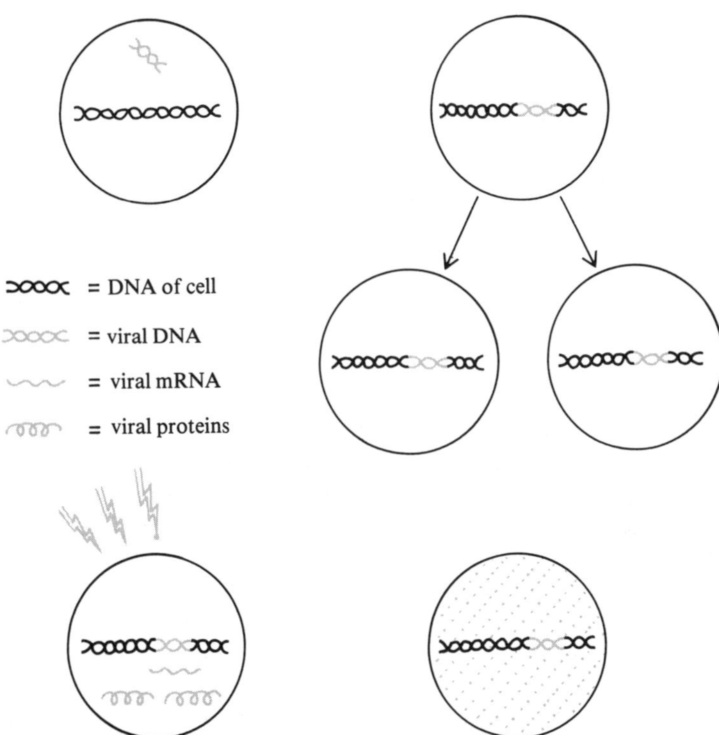

Figure 14-11. The oncogene theory. According to this concept of the origin of certain cancers, certain viral DNA may become integrated into the DNA of the chromosome of a cell (above). The presence of the viral DNA may cause no harmful effect as long as the viral genes remain repressed. The virus genetic material would behave as if it were genetic material of the host cell. The cell, however, is susceptible to cancer if subjected to certain environmental stimuli (below). The viral genes become released from repression and undergo transcription. Viral products are formed which can upset the normal metabolism of the cell and cause a malignant transformation.

tion to the next. Its presence in the chromosome would remain undetected, since the viral DNA might cause no perceptible change. The viral genes would be completely or at least partly repressed. According to this concept, however, the presence of viral nucleic acid is potentially harmful, since the viral genes can be released from repression and exert a deleterious effect. The trigger for release could be some factor in the environment, perhaps a chemical or some type of radiation. Recall that ultraviolet light can destroy the repressor of phage genes in bacteria. Perhaps carcinogenic agents (cancer inducers) alter or destroy the repressors of oncogenes and thus enable the viral DNA to express itself in some way which upsets the normal regulatory mechanism of the cell.

Since the incidence of cancer in a population increases with age, it is conceivable that some of the changes which accompany the aging process could similarly trigger the expression of oncogenic viral genes. According to the oncogene theory, genetic information for malignant transformation could thus be transmitted from one cell type to the next and even through germ cells from one generation to the next.

The action of those tumor viruses with RNA instead of DNA as the hereditary material would also involve the integration of viral nucleic acid with that of the host. However, the RNA of the virus does not insert itself directly into the chromosome. It is known that RNA oncogenic viruses carry with them into the cell an enzyme which can use the RNA of the virus as a blueprint for the manufacture of a DNA copy [Fig. 14-12(A)]. This DNA copy of the viral genes then becomes inserted into the DNA of the host where it may remain undetected. The RNA form disappears. According to the oncogene theory, it may effect a malignant transformation as described above and in Fig. 14-11. The RNA form of the virus may also reappear in the cell (as does occur) by a reverse of the process of integration. The inserted viral DNA would act as a guide for the formation of the same kind of viral RNA which originally infected the cell [Fig.14-12(B)].

Although strong experimental evidence exists to implicate viruses as agents which can bring about cancerous changes, it by no means follows that all cancers are virus-induced. One of the most active areas in medical research has been the pursuit of viruses which may induce cancer in the human. While various animal cancers are known to be caused by a virus, and while human cells in culture in the laboratory may be transformed from normal to malignant, this does not by itself mean that cancers which arise in people are caused by viruses. There has accumulated some strong evidence that certain viruses may indeed be involved in human oncogenesis. However, a recent report from the National Cancer Institute reveals that two of these are not responsible for causing cancer in humans after all. Doubt has also been cast on several others which have been suspected to be human cancer inducers. As of the moment, no human cancer-causing virus has been demonstrated.

From the discussions of the oncogene theory and the manner in which some viruses can integrate in the chromosome, it is apparent that great difficulties exist in studies designed to determine whether or not a virus has induced a certain cancerous change. It may be next to impossible to detect the virus by any known method, since the virus could possibly remain integrated while causing the cell disturbance. The question of whether a cancer, if it is caused by such an oncogenic virus, is genetic or not raises problems with the very definitions of the words "genetic" and "inherited." We noted in Chapter 11 that a disorder may be considered genetic if it involves some alteration in the genetic material, but it may not be inherited if it fails to be passed to the next generation. In the case of an oncogenic virus, a genetic alteration certainly occurs, since the host DNA is now integrated with viral genes and the viral genes may be transmitted. However, in such a case, the inheritance of the oncogene or possible cancer-causing nucleic acid involves foreign genes. It is not genetic material native to the individual nor to the species itself which is capable of being transmitted and producing the cancerous change. Instead, it is foreign genetic material which originally gained entry by infection. According to one idea, most persons are born with one or several kinds of viral-like DNA which

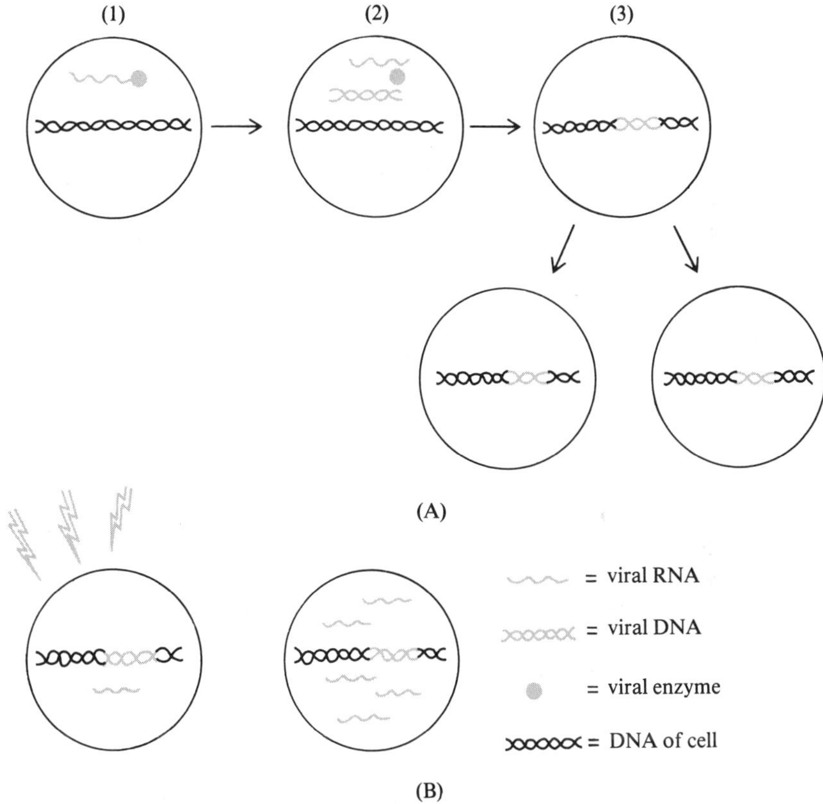

Figure 14-12. RNA tumor virus. (A) the RNA of the virus carries with it into the cell (1) an enzyme which can bring about the formation of a DNA form of the virus (2). The RNA form disappears, and the DNA form integrates into the DNA of the cell (3) and behaves as if it were cellular genetic material. It may also cause a pathological change if stimulated to undergo transcription as seen in Fig. 14-11. (B) The RNA form of the virus may reappear due to some environmental stimulus which causes the integrated DNA to undergo transcription. This also could possibly trigger some pathological cell change.

is integrated into their DNA and which may or may not be called to activity. Can the integrated viral DNA be considered genes of the host if it continues to be transmitted from one generation to the next, even though it originated by infection? Such possibilities make us pause to reexamine exactly what is meant by words such as "gene" and "virus."

A Consideration of Various Cancer-causing Factors

Since cancers are so variable in their nature and constitute *not* one entity but a group of diseases, the causative factors or agents may also prove to be very diverse. Induction by certain viruses may prove to be one of these, but in other cases, no virus at all may be involved. Some carcinogenic agents (certain chemicals and radiations) may upset cellular controls by destroying specific repressor substances which normally inactivate certain genes in differentiated cells. Released from repression, some genes could bring about metabolic unbalance leading to uncontrolled growth. The accumulation of certain products in a cell during the aging process could also act as the trigger in a similar way. This could account in part for the higher incidence of cancer among persons of older age groups. Supporting such an idea is the report that some persons show an inherited tendency to produce large amounts of a particular enzyme which can convert chemicals commonly found in the environment into other forms which can act as cancer inducers.

Somatic point mutation (gene mutation in a body cell) is another concept supported by many investigations. A point mutation could arise which affects some gene which in its changed form brings about a cancerous transformation. Certain radiations and many mutagenic chemicals are also able to induce cancer, lending support to the idea that changes in the gene itself can trigger malignancies. A cell carrying the mutated gene would then go on to divide and produce a malignant tumor.

Abnormal chromosome constitution has also been proposed as a cause of cancer. In most cases of human leukemia and different types of solid tumors, one or more kinds of chromosome abnormalities can be observed, such as aberrations in mitotic stages as well as those in chromosome number and structure. Agents such as high-energy radiations which are known to induce cancerous changes are also effective in producing chromosome abnormalities. However, there is no conclusive proof that an upset in chromosome number or structure causes a malignancy. It is very possible that the cells with the aberrations thrive better in the particular environment in which they are growing *after* the malignant change has occurred in a cell with the normal chromosome number and chromosome structure. Moreover, some malignant tumors show normal chromosome complements.

Nevertheless, certain well-known facts strongly implicate chromosome aberrations in the origin of malignancies. When cells are taken from many persons who are members of a group with a high cancer risk, due to group exposure to known carcinogens in the environment, an increase is found in the number of chromosome aberrations over that in persons not so exposed. In addition, we know of rare genetic diseases associated with a higher risk of cancer: Bloom's syndrome, Fanconi's anemia, and the Louis-Bar syndrome. The clinical picture characteristic of each syndrome is accompanied by chromosomal instability. Cultured cells from a person with one of the disorders, each of which appears to be due to an autosomal recessive, show a high incidence of chromosome aberrations of various types. Chromosomes from such persons appear to be abnormally fragile and susceptible to breakage. In addition, cells taken from these individuals and grown in tissue culture are much more likely than cells taken from normal persons to undergo malignant transformation when exposed to known carcinogenic agents.

The presence of minute deletions in the chromosome has also been suggested as a cause of cancer. Missing genetic information could bring about an upset in cellular controls in many ways. Certain oncogenic viruses are known to cause chromosome breakage, and the suggestion has been made that the production of tiny undetected deletions may cause the cancerous transformation. But again, other arguments can be offered. Any break caused by a virus may be the secondary and not the primary cause of the malignant change. Referance is frequently made to the association of chronic myeloid leukemia with the Philadelphia chromosome (chromosome #22 with a deletion; see p. 270). As we have pointed out, the association with the specific chromosome change, the deletion in the chromosome of the white blood cells in this case, does not indicate the nature of the association. The deletion may have caused the malignant change or it may itself have been brought about by the change. Moreover, it now appears that the Philadelphia chromosome anomaly actually represents a translocation rather than just a loss of material. Chromosome analyses in which banding techniques have been employed (Chapter 2) have shown that the portion missing from chromosome #22 in the cases so studied has been translocated to one of the larger chromosomes, usually #9. It is still possible, however, that some small amount of genetic material has been deleted during the translocation.

The kinds of chromosome abnormalities which are seen in cancers are very diverse. The occurrence of a high degree and assortment of chromosome aberrations seems to be characteristic of cell populations which are likely to give rise to cancer. However, although the chance is greater, by no means does a cancer always develop. The fact that persons with a higher cancer risk due to environment (group exposure to carcinogens) and persons with a higher risk due to genetic factors (Bloom's disease, etc.) both show a tendency for chromosome aberrations suggests that chromosome instability provides a background which makes it more likely, but not certain, that a malignant transformation will occur. The entire complex of factors, genetic and environmental, operating against the background of chromosomal instability is yet to be elucidated.

It must be emphasized at this point that most human cancers cannot be associated simply with any one mutant gene. Whatever genetic component is involved in their transmission, it appears most likely to be polygenic and therefore based on a large number of genetic loci with complex environmental interactions. The development of a cancer would depend on the accumulation of the effects of many contributing genes and environmental factors. When a certain level or threshold is reached, a cancer may develop. Such genes as those associated with the Bloom's and Fanconi's syndromes contribute significantly in some way so that the threshold level tends to be exceeded. However, these genes are most likely not the only ones involved.

Therefore, as seems to be the case, we would expect cancer to be very variable in its expression. A person with a large number of predisposing genetic factors may not develop cancer if he or she is not exposed to agents in the environment which may add to the contribution of the genetic component. On the other hand, the critical threshold for cancer development may be exceeded in a person who carries a relatively small number of cancer-predisposing genes but who is subjected to an excess of carcinogenic agents in the environment. In some cases, the effect of environmental factors on the genetic material could conceivably be the sole causation.

Present research into the cancer problem is also focusing on the role of the immune system of the body. Strong evidence suggests that cancer cells arise spontaneously in the normal person for one or more unknown reasons but that they are recognized and then destroyed by the antibody response. Failure of the immune response can allow a malignant cell to escape and then to divide and grow into a tumor. Accordingly, any chemical which suppresses the immune reaction could leave the body vulnerable, not only to infection by microorganisms, but also to cancerous growths which otherwise would have been destroyed. Since there is strong support for the idea that the immune system plays a role in the control of cancer cells, we should weigh the risk of exposure to any environmental factors which can depress the immune reaction. There is some evidence that the smoke of marijuana lowers the immune response, making the individual more susceptible to infection. Such information should be considered by anyone who is contemplating the use of any substance whose effects are still highly questionable.

While the discussion of cancer presented here is little more than a summary, it should illustrate that the cancer problem is a most complex one which necessitates a deep understanding of such areas as gene action, control mechanisms, cell physiology, and the genetics of microorganisms, to mention just a few. Whatever the causes and preventions of the many kinds of human cancers may prove to be, they will certainly in some way involve gene expression and control.

Factors in Aging

Any discussion of differentiation must encompass the subject of aging, a process which begins with the origin of the zygote and continues to the death of the individual. Very little is actually known of the causative factors of aging, though all of us accept it as the normal course of events. It is common knowledge that different species have different life spans which seem to be characteristic of the groups. No dog is expected to live much beyond 15 years, whereas a human can be expected to reach the seventieth year. It is also accepted that the chance of death increases as the individual ages. Fifty year olds are not expected to survive as long as their 20 year old children. Common observations such as these have led some to suspect that life span itself may be a hereditary characteristic, just as eye color or any other aspect of the phenotype. Perhaps every living creature is programmed to die by the genetic information carried in the body cells.

Support for some sort of inherent control of survival has been obtained from a series of fascinating investigations pioneered by Hayflick. In this work, human cells are taken and raised *in vitro* where they are allowed to divide. From the cultured cells, separate colonies of genetically identical cell populations can be derived. These colonies can be studied further in a variety of ways. Any such population of genetically identical cells is known as a *clone.* Actually the entire body of any individual comprises a clone, since the genetic content of all the body cells is identical, although gene expression differs from one cell type to another. From the study of clones, it has been found that the capacity for an embryo cell to divide appears to be no greater on the average than 50 divisions. By the time 50 doublings have occurred, most of the genetically identical colonies have died out. Some populations during the course of the study may be allowed to go through 5 cell divisions, 10, or 30 before being placed in a deep freeze to halt cell divisions. The clones can be kept suspended in this way for months or years. Of particular interest is the fact that, no matter how long any clone remains in the suspended state, when allowed to resume growth it still will not divide beyond the average total of 50 divisions (Fig. 14-13)! The clones seem to resume growth as if no interruption had occurred. A clone which had completed 10 divisions before the suspension will then go on through approximately 40 more; a population which had completed 30 will undergo approximately 20 more. In addition to these findings, cells taken at ages after birth go through fewer divisions than do those taken from an embryo. Moreover, cells from an older person (age 50) undergo fewer division than those from a young person (age 20).

Such findings support the idea that the longevity of a cell, as reflected in its capacity to divide, is set by its genetic factors. In its strictest interpretation, this would mean that all species have genetic restrictions imposed upon their life spans. The human life span has only seemingly been increased. In actuality, according to this theory, no true increase has occurred at all. Modern medicine

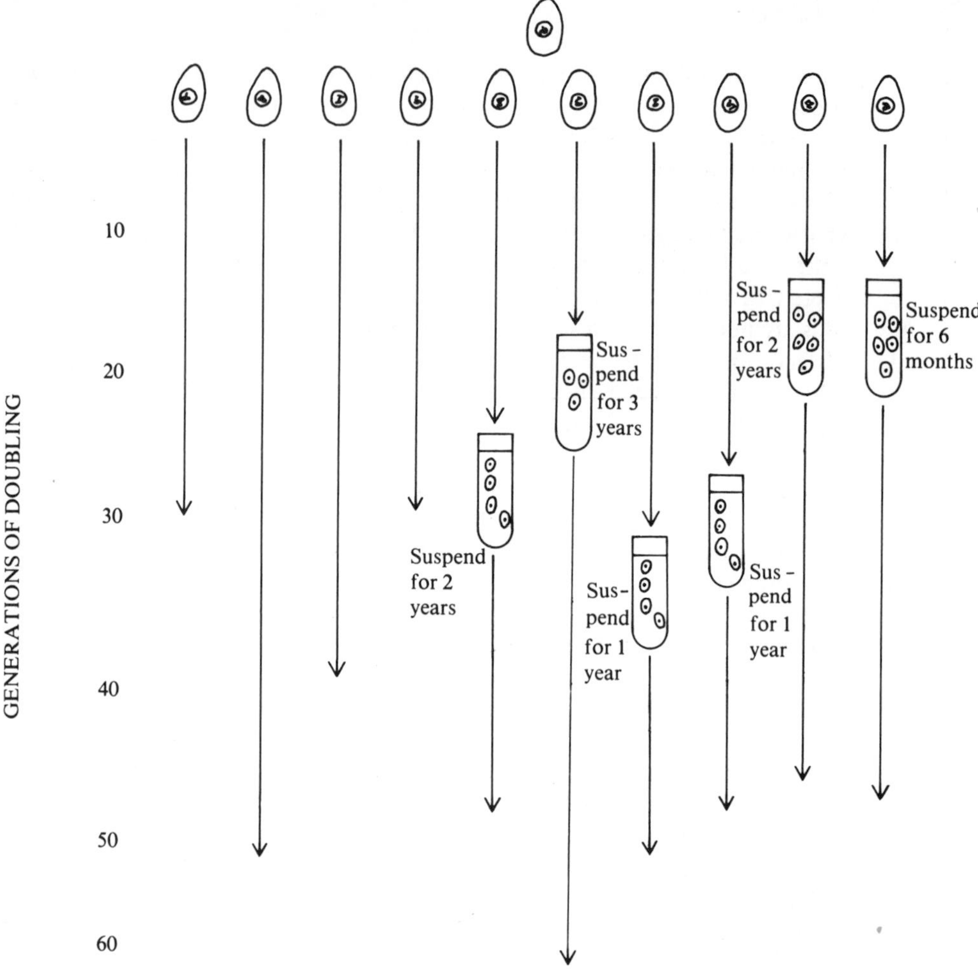

Figure 14-13. Life span of human embryonic cells. Separated populations of cells which trace back to one cell comprise a clone and can be studied separately. It is found that separate populations of identical cells tend to undergo 50 divisions. As they approach this number, more and more tend to die out. If some of the cell populations are prevented from dividing and are held in a state of suspended animation for months or years and then allowed to resume growth, they still tend to complete 50 divisions and then die out.

is enabling more persons to reach the upper limit. Beyond that limit, life expectancy cannot be increased. Man is thus not immortal.

It would be dangerous to draw conclusions on such an important topic strictly from studies of cell cultures. According to more recent reports, the longevity of the cell populations in culture has been greatly prolonged by the addition of vitamin E. If this and other additives can be shown to increase the capacity of the cells to divide, an argument can be made against the idea of a rigid limit to life span imposed by genetic factors. The subject of aging, like that of cancer, is certainly a complex one. Whether or not there exist distinct genes which limit the life span, the process of aging certainly entails many characteristic changes in cells which can be considered part of á sequence of development. For example, as an embryo develops, certain special cell types arise which enter into the formation of tissues and organs restricted to specific stages of embryology. After serving a particular function, these cells cease their mitotic activities and the prenatal organs derived from them disappear to be replaced by others. Such changes take place in the development of the human kidney, which is preceded by the formation and disappearance of simpler organs. It is also well known that in older age, several parts of the body, the tongue for example, contain fewer cells and weigh less than in younger years. This could be due to the expression of genes which place a limit on the number of divisions possible for a particular cell type.

However, other interpretations are possible. Perhaps no genes exist which specifically impose a limit on life span. Instead, aging may result from an accumulation of spontaneous alterations in the genetic material which arise with time. Some may be spontaneous gene mutations, others chromosome aberrations. Tissue culture studies show an increase in the number of chromosome anomalies with culture age, but this has not been demonstrated in people. Exposure to certain environmental factors may hasten the spontaneous process and add to those changes which can set the cell on the path of aging by altering cellular metabolism. Recall that the vulnerability of oocytes to environmental influences plus the physiological effects of ovarian aging are believed to account in part for the increased incidence of trisomies, such as Down's syndrome, among the offspring of older mothers.

Today a great deal of research is being directed to the area of repair enzymes, those enzymes which can undo damage to DNA (Fig. 13-5). Such enzymes are known to occur in microorganisms, and at least one of them has been found in the human. A deficiency of these enzymes is found in persons afflicted with *xeroderma pigmentosum,* a skin condition in which malignancies are frequent (p. 327). A lack of repair enzyme in any individual could conceivably lead to an accumulation with time of DNA damage which has occurred spontaneously or which has been induced. Such an accumulation of genetic alterations could hasten the aging process. It might also reach a point where a malignancy could be triggered as the result of disturbances in some of the DNA which is associated with control mechanisms. The very fact that *xeroderma pigmentosum* is an inherited condition associated with a deficiency of a repair enzyme lends additional support to the concept that gene mutation may induce a malignancy by bringing about a chemical change in the cell. Repair enzymes are very possibly involved in both the cancer problem and the aging process. If so, cell aging could entail mutations, spontaneous and induced, in the very genes which control repair enzyme formation and thus lead to inadequate amounts needed to repair damaged DNA. It is also possible that mutations may occur in genes whose products are necessary to activate these repair enzymes.

There is little doubt that heredity plays a role in the life expectancy of the human; however, it should be evident from our discussions that the relationship is not a simple one. No one or two genetic loci dictate the life span. Several disorders associated with aging (atherosclerosis, hypertension, diabetes, and cancer, among others) are known to have a genetic component, but it has been noted that the environmental component interacts with it in a complex way (p. 254). Of course, we recognize the role of certain single genetic factors which affect life expectancy in such disorders as Tay-Sachs disease, Huntington's disease, and several others, but these are not involved directly with the aging process. The role of genetic factors in the so-called degenerative diseases that are associated with aging is much more subtle. Part of this stems from the fact that polygenes are involved which operate in a quantitative fashion (Chapter 8). Some degenerative diseases are so typical of aging (atherosclerosis, for example) that one may debate if they are indeed diseases or normal developmental processes which accompany aging. Inherited genetic factors would appear to be involved, but irreversible genetic alterations incurred during the person's own life span (somatic mutations, chromosome changes) very probably contribute. While persons from families free of histories of diabetes or hypertension may have a good chance of escaping the effects of these disorders due to the genes they have inherited from their parents, the environments in which they find themselves may produce the same effects by causing genetic alterations in the body cells which are not inherited. A person from a family free of certain degenerative diseases has no guarantee that he or she will escape them, although the chances of doing so are better than those of a person whose pedigree shows a high incidence of relatives with hypertension, diabetes, atherosclerosis, and other degenerative diseases.

The complexities of the aging process are well illustrated by a very rare condition known as Werne's syndrome, which is characterized by progeria, premature aging. Young adults suddenly begin to exhibit changes associated with aging: graying of hair, balding, cardiovascular degeneration, increased risk of cancer, etc. The rare condition has appeared among brothers and sisters in a few families and also in families where first-cousin marriages have occurred. An autosomal recessive mode of inheritance is suggested. Another very rare syndrome described in some families has an onset in early childhood, so that by the age of 3, degenerative changes associated with old age have become very pronounced. A rare recessive is also believed to be the basis of this most unusual situation. The hastening of the aging process in these syndromes is not understood, but the existence of such rare, bizarre disorders suggests that much is yet to be learned about the genetic factors of aging and the control mechanisms involved.

REFERENCES

Brown, D., "The Isolation of Genes." *Sci. Amer.,* Aug: 20, 1973.

Dulbecco, R., "The Induction of Cancer by Viruses." *Sci. Amer.,* Apr: 28, 1967.

Gurdon, J. B., "Transplanted Nuclei and Cell Differentiation." *Sci. Amer.,* Dec: 24, 1968.

Hayflick, L. "Human Cells and Aging." *Sci. Amer.,* March: 32, 1968.

Hartman, P. E., and Suskind, S., *Gene Action,* Second edition. Foundations of Modern Genetics Series. Prentice-Hall, Englewood Cliffs, N. J.

Holland, J. J., "Slow, Inapparent and Recurrent Viruses." *Sci. Amer.,* Feb: 32, 1974.

Jacob, F., and Monod, J., "Genetic Regulating Mechanisms in the Synthesis of Protein." *Journ. Molec. Biol.,* 3:18, 1961.

Kappas, A., and Alvarez, A. P., "How the Liver Metabolizes Foreign Substances." *Sci. Amer.,* June: 22, 1975.

Koshland, D. E., Jr., "Protein Shape and Biological Control." *Sci. Amer.,* Oct: 52, 1973.

Leaf, A., "Getting Old." *Sci. Amer.,* Sept: 44, 1973.

Martin, R. G., "Control of Gene Expression." *Annu. Rev. Genet.,* 3:181, 1969.

Mazia, D., "The Cell Cycle." *Sci. Amer.,* Jan: 54, 1974.

Pastan, I., "Cyclic AMP." *Sci. Amer.,* Aug: 97, 1972.

Ptashne, M., "Genetic Repressors." *Sci. Amer.*, June: 36, 1970.

Rafferty, K. A., Jr., "Herpes Virus and Cancer." *Sci. Amer.,* Oct: 26, 1973.

Stein, G. S., Spelsberg, T. C., and Kleinsmith, L. J., "Nonhistone Chromosomal Proteins and Gene Regulation." *Science,* 183:817, 1974.

Stein, G. S., Stein, J. S., and Kleinsmith, L. J., "Chromosomal. Proteins and Gene Regulation." *Sci. Amer.,* Feb: 46, 1975.

Temin, H. M., "RNA-Directed DNA Synthesis." *Sci. Amer.,* Jan: 24, 1972.

Tomkins, G., and Martin, D. W., "Hormones and Gene Expression." *Annu. Rev. Genet.,* 4:9, 1970.

Wessells, N. K., and Rutter, W. J., "Phases in Cell Differentiation." *Sci. Amer.,* March: 36, 1969.

REVIEW QUESTIONS

Select the best answer(s) for each of the following multiple choices.

1. In an operon: (1) the regulator produces the effector; (2) the structural genes are transcribed by RNA polymerase on the same mRNA strand, (3) the repressor combines with the regulator and blocks transcription; (4) the operator must combine with the product of the regulator if transcription is to take place; (5) the effector can bind the repressor.

2. Suppose a cell begins to produce a related group of enzymes indiscriminately. Which of the following could account for this? (1) A mutation in the regulator has resulted in the production of a repressor which has a strong affinity for the operator and which cannot combine with the effector. (2) A mutation in the regulator has resulted in a defective repressor which cannot combine with the operator. (3) A mutation in the operator has rendered it incapable of combining with the repressor. (4) A mutation in the operator has produced an operator which now combines irreversibly with the repressor. (5) A mutation has arisen in one of the structural genes of the operon.

3. In eukaryotes: (1) the DNA is exposed directly to the regulatory products in the cell; (2) the histones act as stimulators of gene activity; (3) nonhistone proteins appear to be able to activate genes; (4) there is no evidence for the existence of operons; (5) hormones may activate genes by antagonizing repressors.

4. A bacterial virus: (1) injects protein into the host cell to provide a source of enzymes for the construction of new viruses; (2) injects only its DNA into the host cell; (3) causes the host cell to manufacture viral proteins; (4) requires a repressor of its structural genes if it is to take up residence in the cell; (5) causes the bacterial cell to synthesize increased amounts of bacterial DNA and protein.

5. Which is (are) true about viruses? (1) They are the simplest cellular entities. (2) Some do not have DNA as the genetic material. (3) Once integrated with the host genetic material, they remain permanently repressed. (4) They can bring about production of viral mRNA in a host cell. (5) They are prokaryotes.

6. Which of the following appear(s) to be true about cancer? (1) Cancer is not one single type of disease. (2) Cancer is a condition in which certain cellular control mechanisms have been altered. (3) Malignant cells have altered cell membrane properties. (4) Malignant cells are always associated with chromosome abnormalities. (5) Human cancers have been shown to be caused by viruses which integrate with the chromosome.

7. Which of the following genetic conditions is (are) associated with malignant transformation? (1) *xeroderma pigmentosum*; (2) sickle cell anemia; (3) the Philadelphia chromosome; (4) Bloom's syndrome; (5) *cri-du-chat* syndrome.

8. There is evidence that aging may entail: (1) exposure to certain environmental factors; (2) failure of repair enzymes; (3) cessation of mitotic activity; (4) certain inherited genetic information; (5) genetic alterations in somatic cells.

chapter 15

HUMAN BEHAVIOR AND OTHER PROBLEMS

Genetic and Environmental Components in Species-specific Behavior

Of all the organs of the body, by far the most complex is the brain. The evolution of the intricacies of the human brain has enabled *Homo sapiens* to become the dominant species on the planet Earth. The human is the animal which is best equipped to overcome challenges imposed by the environment. Largely as a result of his superior mental abilities, the human is less restricted than other species by environmental problems. In this sense, his greater independence of the environment, the human is the most highly developed of all creatures. Not only does he cope with existing environmental problems, but he is also capable of changing the environment to suit his needs. In so doing, he also influences the course of his own evolution as well as that of other species. The ultimate benefits to be reaped from tampering with natural selection poses many profound questions bearing on the very future of life on earth. How the human uses his superior brain to manipulate the environment is a direct reflection of human intelligence. Intelligence is reflected in the many facets of human behavior which has the potential to influence the very existence of all life forms.

Any information which can be obtained to provide an insight into the nature of human intelligence and behavior has a decided value. Since the human nervous system is so complex, it should come as no surprise that analysis of the hereditary and environmental components which influence human intelligence and behavior is the most difficult of all genetic investigations. However, many facts related to the problem are known. Sufficient data may in time accumulate to permit a breakthrough into a fuller appreciation of the human mind and the reasons why the human acts as he does.

Let us begin by noting several obvious facts about one aspect of human behavior, speaking a language. The ability to speak entails a decided advantage for the species. It enables members of one generation to pass the assembled knowledge, the culture of the society, to the next generation with no loss of time and no need on the part of the later generation to learn everything anew. Without this ability, each new generation of a species loses the advantage to profit from the knowledge accumulated by the efforts of previous ones. The ability to communicate through speech and the written words associated with oral language makes it possible for each generation to build on the culture of its ancestors, to utilize it, and improve it, thus becoming more and more independent of environmental problems.

These points about language indicate something about behavior in general. Just like any other feature of an individual, behavior is a phenotypic characteristic. We speak of *species-specific behavior* when we refer to the assemblage of behavior patterns which is considered typical of a given animal. We expect members of a given species to behave or conduct themselves in a certain way. A particular species of bird goes through its expected courting behavior, builds a nest, incubates its eggs, and feeds its young. The young, in turn, respond to the mother in a typical fashion. If the bird is a duck, the young will follow her shortly after hatching and will soon learn to swim.

The behavior of human parents in rearing their young has specific features which are expressed during the long period of development required by the very helpless human infant. While incapable shortly after birth of performing feats such as walking, swimming, or feeding himself, the potential of the human infant exceeds by far that of any other animal. The human's genetic endowment has provided him with a central nervous system of the utmost complexity. The countless nerve connections and relays in the brain itself provide him with the machinery to accomplish something no other animal can do. This is the ability to deal with abstractions, to receive symbols, arrange, and interpret them in a meaningful coherent fashion. There is evidence from studies with chimpanzees that these animals possess a certain ability to deal intelligently with symbols, but it does not approach the facility of the human in this respect. Moreover, the human is unique in the ability to communicate symbols freely through a spoken language. Again, chimpanzees have been taught after labor to utter a few simple words. The inability of the ape to speak freely may not be entirely a reflection of limited intelligence. It is to a large degree a consequence of the fact that the ape is not equipped anatomically to handle spoken words. Speech demands a voice box, palate, nose, and mouth cavities of a certain construction. No other animal possesses the type of anatomy required for the free formation of words.

Also essential for proper speech is a normal hearing apparatus. This is the receiving part of the system which delivers the sound stimuli to the brain. It is here that these stimuli are arranged and interpreted into a meaningful whole. Before any human can speak properly, he must be provided with these sound stimuli. These enable him in turn to form his own sound by association of symbols in the complexes of his brain. Ability to speak a language, therefore, is a complex characteristic. It depends in large part on anatomy, the hearing apparatus to receive the sound stimuli, the brain cells to interpret them, and the vocal apparatus to form words. The anatomical features necessary for language communication have a large hereditary component. Certainly, every fertile human must carry in the sperm or egg information required for the embryo to develop these anatomical requirements. A defective gene may cause upsets in the differentiation of one of these structural parts. Without sound reception, a human will never be able to speak clearly and freely, even though the brain and vocal apparatus have developed perfectly.

But proper hereditary information is by no means the sole requirement. An environmental trauma (viral attack during fetal development, birth accident) may deprive an infant of sound reception. Again, without the appreciation of sound, even though the proper genetic endowment is present, such a person will lack the facility with spoken words possessed by a more fortunate individual.

The ability to deal with symbols is a reflection of the intangible known as *intelligence.* The increase in intelligence which the human possesses over other animals is seen in the greater ability of memory retention, to think, and to arrange information stored in the brain in such a manner that foresight is possible. This foresight enables the human to act in ways which permit him to avoid potentially harmful environmental influences. With foresight, the human can plan and transmit his knowledge and culture. This thinking, learning and foresight depend, of course, on a brain with billions of neurons (nerve cells). Increase in intelligence thus depends to a large degree on increase in the number of brain cells and thus increase in the size of the brain. It is well known that differences in intelligence among different animal species is related to differences in brain size. However, *within* a species, there is no evidence to indicate that a larger brain reflects greater intelligence. Nevertheless, the species-specific behavior of the human in regard to the ability to think, plan, and speak depends on an important genetic component which provides the information to construct a brain of a certain size as well as the other anatomical features essential for the expression of the brain's full potential. However, the environment in which these genetic factors operate is of paramount importance to the final expression of that potential. Despite the fact that an individual may carry genetic factors for normal speech or above-average intelligence, the environment can deprive him of the full benefits which they provide. We see that behavior, like any other phenotypic characteristic, is thus the result of the interaction of many genes and the environment. It should also be evident that species-specific behavior has been subjected to the operation of the forces of evolution. Natural selection has provided each species with a particular behavior which enables it to cope best with its environmental challenges. Behavior, therefore, has an advantage just as any other characteristic of a species or individual. The human's behavior, as noted here, has enabled him to become less restricted than any other species, to be able to make use of the greatest number of environmental niches, and to manipulate the environment for his own end.

While much of human behavior depends on heredity, it must not be forgotten that a great deal of it depends on learning. Human behavior is distinguished by its plasticity, its ability to be changed and molded. A normal child is born with genetic factors which make it possible for him to receive environmental stimuli. Among the most important of these are the ones provided by other humans. An infant is associated with a family, and as he continues to develop, he is exposed to more and more members of his particular cultural group or society. Certain stimuli will determine what language he will speak. He carries the genetic information which provides the ability to speak, but whether he *will* speak and what language he will become expert in depends on other humans about him. The amazing inherent talent which a child possesses to learn a language is seen by the youngster's ability to become expert in several tongues if he receives the appropriate sounds during his formative years. No one must tell him what is English, Chinese, or French. He sorts them out automatically without being told a thing about rules of grammar. But alas, this genetic ability is lost about the age of puberty, so that no adult will ever have the facility of the child who can so effortlessly master two or more tongues. Nor will most adults ever speak a language perfectly if they have learned it after they have reached sexual maturity. There is nothing that dictates what language a child must speak. The great plasticity of this aspect of human behavior is quite apparent, as is the interaction of its hereditary and environmental components. Few human behavioral traits are based exclusively on genetic factors. The plasticity provided by the environmental interaction with the genetic component is seen in a child's general conditioning as he becomes integrated into his particular society. He learns the customs of his society and tends to behave according to the expectations of his specific cultural group. Reared in a completely different society with different values and expectations, another child with the same genetic potential as the first one will come to behave very differently as he acquires different habits, skills, and beliefs prescribed by the second group.

Human Sex Roles in Society

The relative roles of the hereditary and environmental components in the expression of human behavior has been a matter of heated debate in relation to the topic of human sex roles in society. Many persons expect a female to behave in a certain manner and a male to respond to a particular situation in a characteristic way which is distinct from that of a female. Generally, the female is expected to be more retiring, the male more aggressive. For a male or female to behave in a fashion considered inappropriate to his or her sex is often labeled "unnatural." According to the ideas of some students of behavior, characteristic male and female behaviors are fixed in the species. There is a genetic basis, according to this thesis, which is responsible for the expression of male social behavior and of female social behavior. The argument of the proponents of this idea is based mainly on studies of the fossil record, observations made on the behavior of various species of primates (the order to which the human belongs), and observations of behavior patterns in various human societies today. Proponents of the concept attribute the fixation of sex roles to the advantage this behavior imparted to the human species during its evolution in the relatively recent past. Social behavior in which sex roles are distinct would have given a decided benefit to the human when he became a hunter. Once the human species assumed hunting activities, so the argument goes, an assignment of roles to each sex would have given insurance to the success of these activities and hence would have increased the survival of the group. Any genes or genetic combinations which would magnify the differences in sex role behavior would be selected under the force of natural selection. As the hunter, the human male would require strength, endurance, speed, alertness, and prowess among other attributes. Such hunting efforts would benefit from improvement of these traits, and so any genetic determinants in this direction would be selected. However, such traits would benefit directly only the active hunter, the male. Interactions of genes for excellence in these traits with the hormonal environment would "fix them" in the male. On the other hand, any hormonal controls to make the female more different from the male would have value. The changes in the female during this part of human evolution would have occurred only in relation to the requirements of the male in his role as the hunter. In her role as childbearer and rearer, intelligence would not be at a premium. Cunning, strength, and swiftness would not be needed for food gathering activities. Genetic controls to make the female subordinate and docile would benefit the needs of the male after the demands of the hunt and would in turn make him a more efficient hunter. Hormonal regulations in the female to make her sexually receptive at any time would have value. They would have been selected in relation to the need of the hunter who would be assured that the female would be available to him whenever he so desired after return from the hunt. Strong bonds would form between males. Such bonding would be of advantage since it would assure cooperation during hunting activities and would reduce hostility among the males. No such bonding among females would be required, since the female's activities did not depend on major group activities. Therefore, sex roles became more and more distinct as the success of the hunt became more and more a certainty.

According to this idea, we see today in our human societies behavioral traits which characterize the male and those which are typically female. These exist because they are set in the genetic program where they became established by natural selection during the time the human emerged as a hunter. So today, mainly the men perform decision-making activities. Women accept these decisions. The male is aggressive, whereas the female is the homebody concerned primarily with child rearing and pleasing the male. All of the behavior patterns which we today consider male or female are largely a direct result of the genetic factors which control these activities and which became established in the past in relation to the hunting activities of *Homo sapiens.*

There are, of course, many arguments which can be offered to counter these ideas. Not the least is the demonstration that human behavior is highly plastic, a feature which distinguishes it from that of other species. There is good evidence to show that an individual, regardless of sex, may be just as intelligent, docile, aggressive, or strong as any other, depending on the experiences of the individual during development. No evidence exists for some sort of maleness or femaleness in behavior which is set and cannot be modified. There is no clear-cut evidence from the fossil record to reveal details of the social life of the human when the species emerged as a hunting society. It is possible that some of the participants in the hunt were actually females. The studies of behavior in other primate groups should be viewed with suspicion when conclusions are drawn from them and applied to the human. First of all, one should remember that the human is a human and the monkey today is a monkey. To draw conclusions on the basis of an animal's behavior today and to apply it to humans who once lived in the past are misleading tactics. All of our biological evidence indicates that no animal as it exists today is ancestral to any other which exists today. While they may have had a common ancestor in the distant past, this ancestor may have behaved very differently from either one of the modern groups today. In short, today's human did not arise from today's monkey or ape. Moreover, conclusions are often drawn from groups in which the male *happens* to be dominant. In some primate groups such as the gorilla, there are no sharp distinctions which set off the male as the strictly dominant sex. In addition, many conclusions are drawn on the behavior of animals in captivity, a factor known to distort an animal's behavior. Studies on so-called primitive human societies today provide no concrete evidence to the genetic assignment of sex roles. The primitive populations of the human species cannot simply be assumed today to be exactly like those populations in existence millions of years ago. We don't know that much about the history of these groups today nor of the cultural changes in them which are very recent adaptations.

While a genetic component certainly exists for species-specific behavior, we must recognize the advantage the human has in the flexibility which his genetic endowment permits in his behavior. It would seem less wise, on the basis of highly debatable arguments, to assert that certain types of behavior are programmed or inherent than to appreciate the *range* of human responses made possible by the very plasticity which the genetic endowment allows. A fuller understanding of the subtle interplay between genetic factors and environmental influences in the establishment of human behavior can lead to a deeper appreciation of differences among groups and individuals. It may also offer clues to the alleviation of serious problems which can arise from certain forms of human behavior.

Intelligence and Problems in Its Measurement

Before pursuing problems stemming from human behavior, some discussion is needed on the topic of intelligence. It would be impossible to arrive at a completely satisfactory definition of intelligence which takes into consideration all human populations in all their varied environmental settings. Most experts in the field would tend to agree, however, that intelligence does entail the ability to interpret symbols and to handle abstractions in a variety of ways. Many psychologists believe that the IQ test reflects this ability and hence intelligence. The IQ test was actually designed by Binet in France, early in this century, as one way to predict a student's success or failure in his studies in school. One reason the IQ test has become so well recognized is that performance on it does tend to indicate the degree of both academic success and occupational achievements later in life. IQ value is determined by obtaining the score which a person achieves on a test which supposedly denotes his ability to manipulate abstractions. The raw IQ test results for an age group in the population are then weighted in various ways (see below) to produce a so-called normal distribution or curve (Fig. 15-1). Some persons are found at the extremes of low score and high score. Most, however, fall somewhere in between. The peak or average for IQ has been established as 100 points for the Caucasian population. The score for any one person is determined by relating the individual score to the average for the age group (Table 15-1).

Whether or not the IQ test actually measures intelligence is a matter of

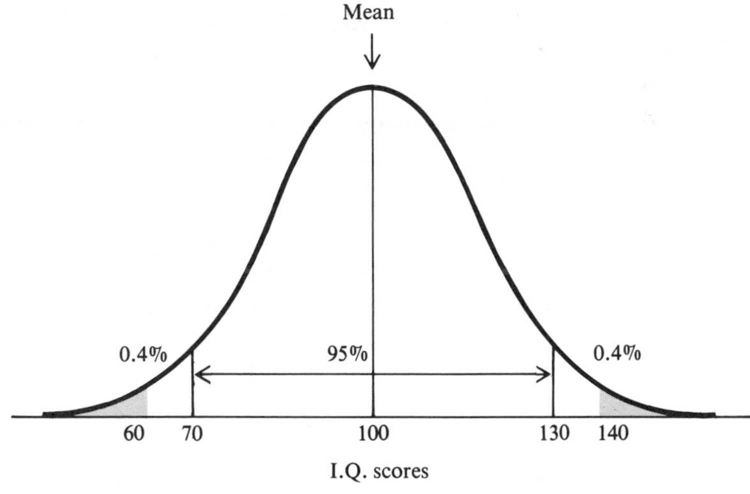

Figure 15-1. When IQ scores are plotted for Caucasians who have been sampled, they tend to fall into a normal (bell-shaped) pattern of distribution. The mean IQ score has been set as 100 points for an age group. Approximately 95% of the population have IQ's in the range of 70 to 130. Less than 1% have IQ's lower than 60 and over 140.

some debate, and space does not permit us to enter into details of the arguments. While IQ achievement serves somewhat to indicate accomplishments later in life, it must be noted that the very factors which determine academic achievement are correlated with those for occupational achievements in the society. It may not be intelligence itself which is being measured, but some ability to deal with factors upon which a particular society places great value. The question of just what is being measured relates to the very construction of the tests. A disturbing fact is that the IQ scores from the population do not just happen to fall automatically into a bell-shaped curve or normal distribution as shown in Fig. 15-1. Actually, the scores are forced to conform to this normal distribution by the very manner in which the examination is constructed.

Results have shown that boys perform better on test items dealing with mathematics, spatial relationships, and certain others, whereas girls surpass them in handling items dealing with language and memory. In order to prevent one sex continually scoring higher than the other on a test, let us say by the inclusion of more items dealing with mathematics which would favor boys, the items in the male-favoring category are balanced by a comparable number in the female-favoring group. The result of the selection of items eliminates high IQ performance which could be related to sex. A main point to note is that items can be selected if so desired to favor a certain group or to blur any intergroup differences which could be reflected in IQ test performance. We see that males and females of the same age group and from the same general environmental background can differ in IQ performance, depending on the items, but that this difference can be hidden by selection of the items in test construction. We have learned that different populations of humans differ in

Table 15-1. Calculation of IQ. The mean IQ for a given age group has been set as 100, since IQ's tend to fall into a normal distribution. The IQ for an individual is determined by dividing the mental age, as measured by performance on the IQ test, by the actual age and then multiplying by 100.

$$IQ = \frac{\text{Mental age}}{\text{Chronological age}}$$

Mental age	Chronological age	IQ
5	4	5/4 × 100 = 125
5	5	5/5 × 100 = 100
9	6	9/6 × 100 = 150
3	6	3/6 × 100 = 50

the frequency of certain genes and gene combinations (Chapter 8). Therefore, two persons from two different populations can be expected to have more genetic differences than two from the same population. Added to this would be very pronounced cultural differences. It would therefore seem that if two persons of different sex from the *same* population can differ on test performance, then two persons from very different populations should differ even more in IQ performance. Genetic differences which influence the IQ scores may have nothing actually to do with the intelligence of the two groups.

Attempts have been made to eliminate as much as possible the environmental effects which can influence IQ performance. The difficulties entailed are so overwhelming that no test has been demonstrated to be entirely culture-free. The intangible which we call "normal intelligence" is undoubtedly the result of the interactions in a genetic component composed of a vast number of genes which influence it in many different ways. Intelligence itself, moreover, is probably not just a single entity. It must in turn be composed of many separate parts, each influenced by heredity and environment. The IQ score is just a single numerical value which tells us nothing about the separate compartments, how they interact, or the relative importance of heredity and environment in each component. Therefore, while the IQ score does have some merit in predicting academic and later occupational success in a given society, it can still be asked: "What exactly is being measured, and what is the relative importance of hereditary and environmental factors in IQ performance?"

The Use of Twin Studies in Genetic Analysis

Whatever the full significance of IQ performance, there is no doubt that it does entail a hereditary component. To estimate that portion of IQ which rests solely on genetic factors, investigations have relied heavily on studies of identical twins. In Chapter 5 (p. 141), the value was briefly discussed of identical twins in studies designed to measure hereditary and environmental influences. Identical twins (monozygotic or MZ) have all of their genes in common; their genotypes are identical. Fraternal twins (dizygotic or DZ), like any other brothers and sisters, have on the average 50% of their genes in common. Since the genotypes of identical twins are the same, monozygotic twins provide a superb opportunity to help evaluate the relative roles of environment and heredity in the expression of a trait. A truly unique opportunity is provided by those rare identical twins who have been raised apart since infancy or early childhood. Since the environmental effects would be more varied on twins reared apart than on those raised together, the contribution of environmental factors to the expression of a trait can be revealed. A pair of identical twins thus gives the opportunity to study the expression of the same genotype in a similar environment (monozygotic twins reared together) and the expression of the same genotype in different environments (MZ reared apart).

Studies of dizygotic twins (DZ) and ordinary brothers and sisters reared together and reared apart are also related to the data from identical twin studies. Two identical twins should always show the same trait if that trait is based entirely on hereditary factors. Such a trait would be the blood type. Monozygotic twins always agree in their ABO, Rh, MN, or other blood groupings. We say that 2 members of a twin pair are *concordant* if they are alike with regard to a certain trait. They are thus concordant if they both exhibit a trait (such as the same blood type) or if they are both free of a trait (neither is an albino). Among identical twins, 100% would be concordant with regard to ABO blood type. Members of a twin pair are said to be *discordant* if they differ with respect to a given trait. If 1 pair of identical twins in a study is discordant, the operation of environmental factors in the expression of the trait under study is indicated. If MZ and DZ twins are very similar in concordance value for a trait, then the strong role of the environment is indicated, since the members of each dizygotic twin pair have only about 50% of their genes in common. If based largely on a hereditary component, the MZ twin pairs should show a much higher concordance value. We can thus use the extent to which twins, identical and fraternal, differ in their concordance values as a sort of measure of the effects of the environmental and genetic components on the expression of a characteristic or trait. Suppose 50 pairs of identical twins are studied for a trait and that 40 of the pairs (80%) show concordance. A study of 50 fraternal twins for the same trait shows that 10 pairs (20%) are concordant. This would provide evidence for a strong hereditary component. If both groups, monozygotic and dizygotic, showed approximately the same amount of concordance for a trait (or conversely, the same amount of disordance) such as 70%, then the greater importance of the environment would be indicated.

Figure 15-2 presents a list of conditions along with the percent of concordance. Note from the table that eye color shows a much higher concordance value between MZ than between DZ twins. This tells us that the genetic similarity of the MZ twins is highly significant in the expression of this condition and that eye color has a strong hereditary component. The fraternal twins show much lower concordance, since members of a pair will not always have the same genes for a trait because only about 50% of all their genes will be common to both. Note that susceptibility to measles is not so different when concordance is compared between the 2 classes of twins. This is a reflection of the greater magnitude of environmental influence on this condition. The values for some conditions, such as stomach cancer, suggest a hereditary component (significant difference between MZ and DZ twins in concordance value) but a large environmental component (73% of the identical twins are discordant).

Trait	Identical Twins (pairs)	Identical Twins	Fraternal Twins (pairs)	Fraternal Twins
Beginning of sitting up	(63)	82%	(59)	76%
Beginning of walking	(136)	68%	(128)	31%
Hair color	(215)	89%	(156)	22%
Eye color	(256)	99.6%	(194)	28%
Blood pressure	(62)	63%	(80)	36%
Pulse rate	(84)	56%	(67)	34%
Handedness (left or right)	(343)	79%	(319)	77%
Measles	(189)	95%	(146)	87%
Clubfoot	(40)	32%	(134)	3%
Diabetes mellitus	(63)	84%	(70)	37%
Tuberculosis	(190)	74%	(427)	28%
Epilepsy (idiopathic)	(61)	72%	(197)	15%
Paralytic polio	(14)	36%	(33)	6%
Scarlet fever	(31)	64%	(30)	47%
Rickets	(60)	88%	(74)	22%
Stomach cancer	(11)	27%	(24)	4%
Mammary cancer	(18)	6%	(37)	3%
Cancer of uterus	(16)	6%	(21)	0
Feeblemindedness	(217)	94%	(260)	47%
Schizophrenia	(395)	80%	(989)	13%
Manic-depressive psychosis	(62)	77%	(165)	19%
Mongolism	(18)	89%	(60)	7%
Criminality	(143)	68%	(142)	28%
Smoking habit	(34)	91%	(43)	65%
Alcohol drinking	(34)	100%	(43)	86%
Coffee drinking	(34)	94%	(43)	79%

Figure 15-2. Concordance and discordance. The shaded areas indicate the percentage of concordance among monozygotic (MZ) and dizygotic (DZ) twins. When both members express a certain trait or when both members of a twin pair are very similar with regard to a feature which does not have distinct traits, they are considered concordant. The percentage of discordance can be estimated by the unshaded areas and is equal to 100 minus the percentage of concordance. The numbers in the parentheses indicate the number of twin pairs studied for each item.

Heritability and the Intelligence Controversy

There are certain limitations in twin studies which must be recognized. For example, the operation and impact of prenatal environmental influences on both MZ and DZ twins is hard to evaluate, and these could operate to influence the expression of a condition even when twins are separated from birth. It could thus be a prenatal environmental factor which causes a concordance, and this would be measured as a genetic one. Nevertheless, valuable information can be gathered from twin studies on the relative roles of environment and heredity in the expression of a trait or characteristic. The information can be used to measure *heritability.* Heritability is a measure of that percentage of the variation shown in the expression of a condition or characteristic which is due to genetic factors. Any complex characteristic such as height or weight which involves many genes and environmental factors will exhibit variation. If all that variation could be shown to be due entirely to genetic factors, then the heritability would be equal to 1. This would mean that the variation seen from one individual to the next is due completely to genetic influences. If the heritability were 0.5, then half of the variation seen is due to genetic factors and the other half to environmental effects. A heritability of 0 would tell us that all the variation is due to the environment.

Intelligence is a most complex characteristic, and it is well known to vary greatly from one person to the next. While we do not know exactly what the IQ performance measures, it would appear that it does measure some kind of mental ability. IQ has been used as an index of intelligence in numerous studies, and the results leave no doubt that intelligence, as measured by IQ performance, has a significant hereditary component. IQ performance of identical twins, some reared apart, others together, is compared to that of fraternal twins and to that of related and unrelated persons. If heredity alone were involved, the correlation between monozygotic twins should be 1.0 and between fraternal twins and ordinary siblings, 0.5. The correlation between MZ's is significantly higher than that for the other groups. The difference between MZ's together as opposed to MZ's reared apart indicates an environmental influence. From analyses using this approach, some investigators have estimated that environment accounts for approximately 20% of the variation seen in a population in IQ performance. In other words, the heritability of intelligence as indicated by IQ is considered by some students of the subject to be in the vicinity of 80%. This would mean that 80% of the variation shown in IQ is the direct result of hereditary factors. The procedures used to arrive at a heritability value of 0.8 are somewhat complex and entail a number of assumptions. While there is certainly a significant genetic component involved in the variability seen in IQ performance in a population, it seems highly questionable to assign a rigid heritability value of 0.8 or 80%.

Relevant to this point is the concern which has been shown in some quarters concerning the lowering of intelligence due to unequal reproductive rates among the various economic groups in society. Some have feared that the higher rate of reproduction of persons in lower economic strata, who tend to score lower on IQ than those from higher socio-economic classes, will lead to eventual decrease of intelligence in the population. There have been few tests designed to test different generations in order to determine whether or not intelligence does vary from one generation to the next. Results from the few tests that have been conducted indicate a significant increase in test scores! Besides arguing against fears of lowering intelligence, the results seem to show a very rapid change in test score performance in just one generation or less. This can hardly be due to a genetic change of monumental proportions. Changes in environmental factors are strongly indicated, and these would appear to be instrumental in influencing the test scores and causing change in a short period of time.

Observations such as these lead us to be cautious when assigning a rigid value to heritability, a value which often rests on many assumptions. Heritability values for many characteristics have been shown to vary from one population to another similar population in a different environment. Heritability data on IQ have been obtained largely from studies of white populations in various regions of the United States and Europe. Since heritability for any characteristic is a figure derived from a population in a particular environment at one particular time, we may question the application of a fixed value for heritability of intelligence to one population when it has been calculated from data obtained from studies on another. Yet this is what appears to be the case in studies where conclusions have been drawn about racial differences in intelligence based on IQ test performance.

Few would deny that scores achieved by American blacks on IQ tests tend to be lower than those for American whites. Several studies suggest that the average IQ score for the white population is 100, whereas that for the black is 85. Even if this difference between mean IQ values should prove to be the case, it must be understood that many blacks score higher than the average for whites and many whites score lower in test performance than the average black person. No one can say that any black must score lower than a white person (Fig. 15-3). While a difference between an average score of 100 and one of 85 is mathematically significant, the interpretation of this difference has led to many heated debates. According to many, this difference reflects various disadvantages experienced by American blacks in educational and economic conditions. Little is known about the subtle effects which nutritional deficiencies, even those before birth, can have on a character as complex as intelligence. Perhaps all of the difference in IQ performance between the two groups stems from environment alone. Moreover, this performance may not be related to intelligence at all.

There are others who hold an opposing view, citing a heritability of 0.8 to which they rigidly adhere. These persons argue that 80% of the difference or variation seen in IQ achievement has a hereditary basis. Only 20% at the most is contributed by the environment. Jensen and other proponents of this viewpoint argue, therefore, that only 20% of this variation could be eliminated if all environmental differences were eliminated. Under exactly the same environment, there would still be a great variation between the black and white groups, and this would be entirely genetic. Assuming a heritability of 80%, it can be shown mathematically that elimination of the variation due to environment would raise the average IQ score for the black population only 1.6 points. The end result would be only a slight increase in IQ performance for American blacks. Therefore, the argument goes, the mean IQ value of the whites is truly higher than that of blacks and is largely due to a difference in hereditary factors which affect intelligence.

This sort of reasoning has been used by many well-intentioned investigators who are striving to improve educational and occupational achievements for all people. It has, of course, also been seized upon eagerly by those who wish to prove that one race is more intelligent or better than another for their own unscientific reasons. While the IQ controversy has led to many unfortunate debates and ill-founded statements, little can be accomplished by those persons of any persuasion who choose to avoid discussing intelligence entirely. Value for everyone can be derived from gaining insights into normal intelligence and the environmental and genetic factors which do mold it. Any differences between groups on some test such as IQ should not be construed to mean that any person or group is better in some way than any other. A test should be

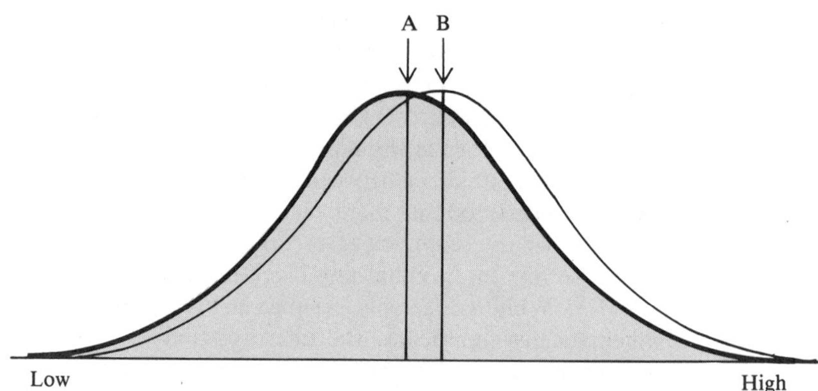

Figure 15-3. Theoretical mean IQ differences between two races. Assume that two races, *A* and *B*, are truly found to differ significantly in mean performance on IQ tests. In such a case, there would be a very large range of overlapping between the two races. Many persons of race *A*, the one with the lower mean IQ, would nevertheless score much higher than the average IQ of race *B*, even though the latter race has the higher mean score. The differences found within any one race are greater than those between the two races.

treated as one device which may enable the investigator to obtain a clue on some aspect of natural ability. It should not be taken to mean that it must reflect a person's innate intelligence, a characteristic composed of so many components. The subtle effects of environment must never be overlooked in the evaluation of such a complex attribute. Strict adherence to some mathematically computed value for a component of a complex character should be avoided until it can be demonstrated that such a value is constant under all conditions. A figure such as heritability can be considered truly reliable only when obtained from studies conducted on similar *and* very different populations under the widest assortment of environments. This has not been done. There is thus a complete lack of data on heritability differences seen within a group as compared to heritability differences which might pertain to separate groups of populations under a range of environmental backgrounds.

No true scientific investigation ever sets out to prove a point, nor does it use vague or ill-defined mathematical concepts to support a chosen argument. The most effective person is the one who gathers known facts, evaluates them, and tries to apply them constructively. He or she may use them to argue successfully against those who are ill-informed, unaware of pitfalls in their reasoning, and especially those blinded by emotion.

Genes, Environment, and Mental Capacity

Another point to bear in mind concerning any measure of intelligence is that any test score is a single numerical figure. As we have noted, intelligence is not just one entity. Rather, it is a complex of interrelated parts, some of which may have a high degree of independence. When we say that intelligence has a polygenic basis, meaning that many genes can influence intelligence, we do not mean that many genes contribute certain units of something called intelligence. It would be absurd to think of intelligence genes, each one adding a certain degree of intelligence in the way that genes which control melanin production influence pigmentation (Chapter 8). Instead, there are unknown numbers of genes which can influence development of the brain and other parts of the central nervous system in an unknown variety of ways. Normal intelligence must depend on a certain number of nerve cells and their spatial relationships to one another. Intelligence undoubtedly involves the physiological activities of these cells: their ability to carry on basic metabolic processes, to form messenger RNA, to carry on translation. Any of these aspects of central nervous system development could be influenced by genetic factors. Certainly it is influenced by environment as clearly seen in environment-heredity interactions in the case of certain mental disturbances. This is quite evident in phenylketonuria, a disorder we have cited frequently. Mental and behavioral disorders due to PKU clearly have a genetic component, but that is by no means all that is involved (Chapter 10). Environmental conditions can prevent many unfortunate consequences of the untreated disorder. The recessive PKU gene is thus one which can have an effect on intelligence, but we now have some understanding of what this means from the knowledge which has accumulated on the molecular basis of the syndrome. Certainly, the normal allele of the PKU recessive is important in the development of normal intelligence, but we see that its effect stems from the control of a specific biochemical step. It would be ridiculous to say that it contributes a unit of intelligence. Many normal genes are required for expression of normal intelligence, since a defective allele may cause a derangement in cellular chemistry which in turn may lead to mental retardation. An unfortunate victim of Huntington's disease may be a person of high intelligence who deteriorates mentally as a result of central nervous system deterioration later in life. The allele of the defective dominant gene is thus required for the maintenance of normal mental capacities, but again we see clearly that no unit of normal intelligence or unit of behavior is involved. Instead, human intelligence and human behavior are divided into many components. A gene or an environmental disturbance of some kind may influence one or more of them and operate against what is considered the normal expression of these human characteristics.

A lengthy list could be compiled of single-gene defects which are known to influence mental development and which can cause behavioral disturbances. No one of these produces more bizarre effects than the sex-linked recessive associated with Lesch-Nyhan syndrome, a kind of cerebral palsy mentioned briefly in Chapter 10; (p. 243). An early symptom of this disorder is vomiting. However, this may be associated with the aggressive behavior so typical of the syndrome. Performance on intelligence tests indicates that victims of this disorder are retarded, but many feel that these children possess a good level of intelligence. They are certainly bright in comparison to most persons who are institutionalized for retardation. Children with the disease develop a compulsion to mutilate themselves, biting the lips, tongue, and fingers. When upset, they may vomit, covering those about them. Speech is greatly impaired in the Lesch–Nyhan syndrome, and spastic movements are evident. Particularly poignant is the observation that affected children apparently desire to be protected against the mutilation they tend to inflict on themselves. They allow the fitting of restraints (mittens, splints) to protect them. The child tends to become relaxed with the restraints; when they are removed, he cries out, seeming to fear his own tendency to mutilate himself. There is some reason to believe that stress may trigger this compulsion. Besides harming himself, the Lesch-Nyhan child is extremely aggressive. He may kick, hit, and pinch persons caring for him. Still, he may apologize for his aggressive behavior!

There is variation in the expression of certain of the perverse behavior traits. Onset of self-mutilation was delayed until the age of 14 in 1 case, and 2 older children did not express this tendency. The variation may be a reflection of the amount of biochemical upset in a particular case. The molecular defect in the Lesch-Nyhan syndrome is known, and information on this very rare sex-linked disorder has great relevance to the study of aggressive and compulsive behavior in general. It raises the question of the role of genetic control in the expression of excessive aggression and in the manifestation of several types of compulsiveness. The recessive gene has been identified with a deficiency of a certain enzyme (hypoxanthine-guanine-phosphoribosyltransferase or HGPRT). This enzyme is involved in the metabolism of purines and is either absent or occurs in an unstable or inactive form in those children with the mutant gene. Lack of the normal enzyme leads to an accumulation of a substance which, when present in excess, causes an increase in the metabolic rate. How the enzyme defect is related to the aberrant behavior is unknown, but it apparently causes biochemical derangements which in turn are responsible for neurological damage. The biochemical upset in the Lesch-Nyhan syndrome also occurs in patients who suffer from gout, even though they exhibit no neurological damage. Like gout sufferers, Lesch-Nyhan children secrete excessive amounts of uric acid in the urine. This symptom may be apparent in a 6–9 month old child before some of the neurological disturbances become evident. Lesch-Nyhan victims eventually develop gout symptoms, and they also tend to suffer kidney damage, including the development of stones. Kidney damage is usually the responsible cause of death before reproductive age. Since the trait is sex-linked, no male can transmit the gene. Consequently, a female victim cannot arise, since such an individual must receive a defective gene from the paternal as well as the maternal parent. Attention undoubtedly will continue to be paid to this rare disease for the information it can provide on problems of arthritis, and especially for the light it can shed on behavioral disturbances and the biochemical upsets which may trigger it.

Other single genes are known (autosomal, sex-linked, dominant, and recessive) which exert their primary effects in very different ways but which can bring about mental retardation as well as certain behavioral disturbances. Table 10-2 includes several, but by no means all, of them.

Heredity and Environment in Mental Illness

Mental illness is a matter of great concern, since it takes such a toll in the burden it imposes on large numbers of the population. The psychoses are the most serious of the mental illnesses, and the most frequent psychosis is recognized as *schizophrenia*, characterized by withdrawal into the victim's personal world which is pervaded by very bizarre thoughts and gives rise to most unusual and most unexpected behavioral responses. Schizophrenia exists in several forms, making recognition and accurate diagnosis difficult. This has presented a problem in the elucidation of the genetic component associated with this psychosis. Some psychiatrists have held the view that the disorder stems solely from environmental causes. However, present data show that a significant genetic component exists. Concordance values for identical twins are much higher than those for fraternal twins or for unrelated persons (Fig. 15-2). Identical twins *do* show discordance; therefore, the environmental influence must be recognized as a significant factor in the expression of the psychosis. However, the importance of the genetic component is seen not only in the higher concordance between monozygotic than between dizygotic twins, but also from other kinds of observations. For example, children of schizophrenic parents have been separated from the mentally ill parents at birth and raised in normal families. The risk of developing schizophrenia, however, in such children is only slightly less than if they had not been separated. Comparison with other adopted children has shown that the act of adoption itself is not the causative stress which brings the schizophrenia to expression.

There has been much debate concerning the responsible genetic factors which can predispose toward schizophrenia. Some investigators favor a single-locus hypothesis in which a dominant gene is implicated; others favor a recessive-gene interpretation. According to some adherents of the latter, the predisposing recessive would produce an added risk of schizophrenia even in the heterozygote. Most schizophrenics would be homozygous for the recessive gene, but according to this hypothesis, unknown environmental factors can exert a strong influence on expression of the genotype. Carriers of the gene in both the homozygous and heterozygous conditions would thus run a greater risk of developing schizophrenia than would a person lacking the gene entirely. However, persons with the gene, including the homozygote, could be normal. On the other hand, even those lacking the predisposing gene could develop schizophrenia solely as a result of certain environmental stresses.

There are other students of the problem who support the concept of a polygenic rather than a single-locus hypothesis. Environmental influence is also recognized by adherents of this hypothesis. Whatever the exact nature of any defective genetic factors in schizophrenia, no molecular or biochemical disturbances have been identified, although some derangement in biochemistry which affects nerve hormones (neurohumors) is anticipated. The kind of environmental stress which can trigger the psychosis is also unknown.

Most of us have heard of the manic-depressive psychosis, another fairly common mental illness characterized by extreme changes in mood which sway from deep depression to great elation. A genetic component is also indicated for this condition, and it is a different one from that responsible for schizophrenia. Again there is a debate on the nature of the genetic component which is involved.

Studies of the genetic and environmental components in mental aberrations are very important, not only for the information they may yield to lead to successful treatment and prevention, but also for the information which may be provided in an understanding of other, more subtle aspects of behavior. Twin studies indicate that a genetic component may exist for personality type. The data show that personality can be greatly influenced by the environment, much more so than for certain other complex characteristics such as height, weight, or IQ achievement. But the possibility that heredity may be involved in the determination of personality type is also suggested. This has a great bearing on the study of criminal behavior, a very controversial topic which becomes very heated when discussed in relation to any hereditary predisposition. No definitive statements can be made at the moment, but the possibility cannot be dismissed that some genetic factors exist as well as environmental ones which predispose to criminal behavior.

Ethical Considerations in Genetic Investigations

We have already discussed the possibility of an extra Y chromosome as a predisposing factor toward criminal or aggressive behavior (p. 75). This topic leads us to another controversial matter, that of the ethics and morals in genetic investigations and procedures of a certain nature. Proposed studies of children with an extra Y chromosome have focused attention on the type of heated debate which can surround genetic research. More and more, we can expect to encounter problems which pose questions with serious ethical ramifications. The relationship of an extra Y chromosome to antisocial behavior is still unclear, and much more clarification is needed before positive statements can be made. The discovery at Boston Hospital that the XYY condition occurs with a higher frequency than formerly suspected (perhaps 1 male birth in 1000) indicated to some a rather urgent need to accumulate more information on the condition in order to settle the matter. A study was undertaken to follow up children with XYY and those with normal XY chromosome constitutions and to observe the two groups throughout a period of years. Any behavior differences between them would be noted. Such follow-ups, it has been argued, could yield several benefits. Any behavioral aberrations might be detected at an incipient stage when they could yield more effectively to treatment and guidance. Moreover, observations of those XYY children who happen to manifest no behavior problems could be related to studies of aggressive XY as well as to aggressive XYY children. The information obtained could cast light on those environmental influences which may mold a child's behavior or personality. Perhaps the extra Y may prove to contribute little or no effect.

The study has been challenged by at least one group of scientists who maintain that it is unethical and unscientific. Most important, it is argued, such a study is potentially harmful to the children involved. It immediately sets any child in the study apart by calling attention to him. Moreover, parents with an XYY child, knowing he has some kind of genetic anomaly, will inadvertantly treat him in a different fashion, which could distort his entire mental development. This in turn could set off a serious psychological disturbance which could or could not be related to the presence of the extra Y. Many who argue against the study feel that it focuses attention on very unusual genetic conditions and thus detracts from the socio-economic problems which are known to contribute to antisocial behavior and criminality.

Many proponents of the investigation admit that it may be better not to allow the XYZ person to know about his genetic constitution, since he might conceivably develop normally without this knowledge. If the design of the study could be altered to prevent this, it then raises the provocative question of a person's right to know and the right of parents to have such information if it is known to medical personnel. The controversy has been sufficiently heated to cause a cessation in further identification of XYY males. However, the study will continue on groups which are presently under investigation.

The XYY story illustrates the kind of moral dilemma which will continue to arise in genetic research as the procedures used in detection and treatment of genetic disturbances become more sophisticated. No one would argue that steps should be avoided which could prevent human suffering. The dilemma which arises often centers around the problem of whether the procedure will cause in the long run more problems for those it tries to treat, as well as greater problems for the human population in general.

Another type of research which is becoming more publicized involves research on the human fetus. Many have attacked such measures on ethical grounds, while others applaud it for the advances it can achieve in medicine. A study made for a national commission which will advise the Department of Health, Education, and Welfare on fetal research has reported that total abstention from this kind of study would have caused delays in our advance of medical knowledge and would have cost thousands of lives. One area which profited, so the study reports, is that of the treatment of babies suffering from blood disorders due to Rh incompatibilities (Chapter 7). Many thousands of infant deaths would have resulted with no research on fetal material, since animal research would not have been as satisfactory. Moreover, without the information obtained, many additional thousands would have become seriously incapacitated due to brain damage from Rh incompatibilities, placing burdens as well on society. As many as 12% of married couples may face some kind of risk due to Rh incompatibilities. The development of a vaccine which prevents the Rh negative mother from becoming sensitized by Rh^+ antigens (p. 160) has been accelerated greatly by research utilizing fetuses.

The report also pointed out that a complete ban on research involving fetuses would have prevented the development of amniocentesis, a procedure in almost routine use to diagnose fetal disorders as well as to detect genetic defects *in utero*. In some cases, the parents are presented with the option of abortion to prevent the birth of babies doomed to years of suffering.

Genetic Engineering and Problems for Society

The problems which will continue to confront society in areas of genetic research underscore the need to weigh carefully the benefits and the dangers entailed in a particular approach. The best course to take may, unfortunately, be unknown at the moment a decision is required. Urgency for some guidelines in certain types of genetic research is perhaps best illustrated by the decision reached at a conference of biologists assembled to debate research in genetic engineering, an area which involves the manipulation of genetic material by the insertion or elimination of DNA into a cell. An ideal solution to a genetic disorder would be the replacement of a defective gene by its normal allele in those cells in which activity of the normal gene is required. Accomplishment of such a feat is not in the immediate future and depends on research in various directions, among them the assignment of human genes to precise locations on the DNA of the chromosomes. Complete cure would require the insertion of the normal allele into germ tissue so that correct information would be transmitted to the next generation. While still a distance from effecting such complete cures, certain approaches have already been undertaken. It has been well established that some bacterial viruses can incorporate into their DNA bits of genetic information from a host cell and then transfer these to another bacterial cell which the virus later infects. Actual attempts have already been made to infect humans with viruses carrying genetic information missing from the humans' cells. Success has not been achieved, although enzyme-deficient human cells *in vitro* have been shown to acquire the ability to manufacture a specific enzyme after infection with viruses carrying the required functional gene. It is still unknown whether the viral genetic material with the added gene has been incorporated into the chromosome or whether it is residing elsewhere in the cell. Another question which arises is whether infection of a person with a virus would in the long run be detrimental, since the DNA of the virus may persist in the human cell and could very possibly undergo transcription leading to disturbances in the physiology of the cell (p. 361). A safer approach might be the injection of the enzyme known to be lacking in a disorder, as has already been reported with some success in the case of Gaucher's disease (p. 251). But this approach does not correct the genetic information and could therefore require continued treatments.

The isolation of specific genes has actually been accomplished in the laboratory (Fig. 15-4), as has the storage of particular genes from certain lower animals. Isolated genes can provide the geneticist with the opportunity to decipher the actual makeup of a gene and to study the mechanism of gene action and control. However, it is just this kind of approach to genetic engineering which has led to the conference of concerned biologists. Techniques have now been perfected which enable the transplantation of genes from one kind of animal into the DNA of another. This can permit the storage in a simple cell type of genes derived from a complex cell. This feat has been made possible by the use of some enzymes which were found to be capable of breaking DNA into segments and of others which can repair the breaks. Segments of DNA have been taken from animal cells such as those of the sea urchin or the South African toad. The segments have then been inserted into bacterial DNA through the action of enzymes which can recognize and break specific regions of the DNA. The foreign animal DNA can then be inserted into DNA of the simpler cell, and all breaks can be healed. Such bacteria then reproduce and form a storehouse of specific animal genes for study as needed. Large amounts of products of the inserted genes could pile up in the altered cells. Another advantage would then be the production in the simpler cells of large amounts of a valuable substance (insulin perhaps) required by persons with serious genetic diseases.

The potential dangers of this kind of genetic surgery for the human have caused alarm within scientific circles. It is conceivable that transfer of foreign genes from higher cell types to common bacteria such as *E. coli* could impart very infectious properties to the microorganism which could then run rampant, causing novel kinds of infections. There is even the danger of transfer to the

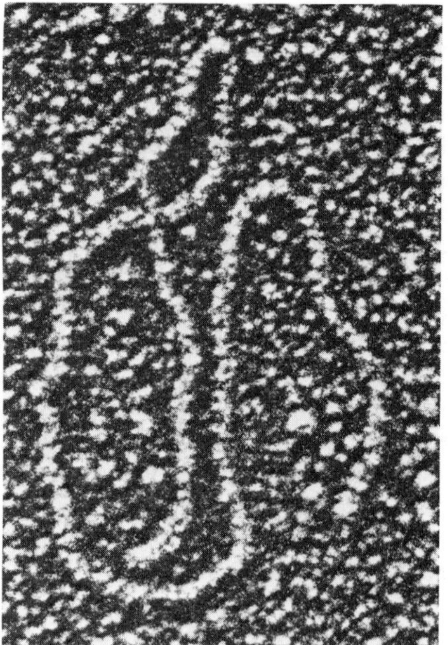

Figure 15-4. Electron micrograph of the *E. coli* lac gene which enables the cell to form the enzyme which breaks down the sugar lactose. Isolation was achieved utilizing phages which had picked up the lac gene and which could carry it from one cell to another.

human of known cancer-inducing animal viruses from cells in which virus genetic material may be stored. The potential for harm in this field of research has prompted the scientific world to recommend guidelines in the pursuit of such studies. The guidelines have no legal basis, but the moral compunction to follow them should be strong, since so many illustrious scientists have recognized the need for them. The proposal of safeguards and guidelines is very unusual in the scientific world, which has tended to avoid the imposition of restraints on investigations along certain pathways of inquiry. We see in this case complete intellectual freedom being weighed against the dangers inherent in a particular approach. To many, restraints appear preferable during the pursuit of an investigation whose capacity for overwhelming harm is at the moment unknown.

Our understanding of human genetic disorders will continue to present no end of ethical dilemmas. There are many who seriously question the morals of any kind of abortion, even when the fetus is known to be defective and the victim of an unfortunate disorder, such as Tay-Sachs disease or Down's syndrome. Perhaps even more disconcerting are problems which relate to the successful treatment of genetic disease. The life of an affected person may be spared, but the treatment required may be lifelong and expensive. It may impart an excessive burden to the victim, the family, and the medical teams which are required. There are those who ask what benefits are gained by saving from early death those children with genetic defects which require expensive treatment that permits survival for just a few additional years. Could the efforts be better utilized in another direction even though death in certain cases would be a certainty if the disorders are left untreated? Remember that many genetic disorders due to single-gene defects are very rare. As more and more continue to be understood and treated, will this divert attention to an ever-growing list of disorders, each affecting few people? Will this shift attention, thus restricting the resources and efforts needed to solve health problems which affect very large numbers of people?

Not to be overlooked is the problem presented by saving persons with genetic disorders who will later reproduce and pass on the defective genes. Phenylketonuria illustrates more than one problem of this kind. No one would question the value of treatment of PKU victims. However, when a PKU woman, saved from mental deterioration, becomes pregnant, additional steps must be taken to protect the child she is carrying, (Chapter 1). This is so, since the PKU adult will very likely be on a normal diet. Procedures must be undertaken to control the diet of PKU mothers in order to avoid fetal damage as a result of high levels of phenylalanine in the maternal blood.

There are those who point to possible dangers to the population as more and more persons with defective genes reproduce and pass the defective genes into the population gene pool (Chapter 8). Many geneticists consider it possible that a crisis will eventually arise in future generations if freedom to reproduce is left unchecked. Others answer this by saying that, while increase of a dominant gene would be quite rapid (about 6 times the original frequency in 5 generations), the frequency of the more common defective recessives would rise very slowly (only about 1% in 5 generations). Man's knowledge in the long run, so this argument goes, would enable him to devise ways throughout the years to cope with the increase in the frequency of affected persons. In the future, persons who are now considered defective in some way would impose no burden to the population and incur no disadvantage as a result of the perfection of procedures and treatment which are at the moment unimaginable.

In the long run, it becomes a question of human behavior, that multisided characteristic composed of so many intangibles. The human's individual behavior can certainly result in harm to himself. Should restrictions be imposed upon individual behavior if it affects only the individual who practices it? The use of certain drugs is known to be harmful to the person. Even certain drugs known to be beneficial to the average person in the treatment of illness may cause violent reactions in some individuals, as seen clearly in the case of G6PD-deficient persons to primaquine and other substances. Such people are normal in all respects other than their drug response. Another example is seen in the violent reaction to barbiturates experienced by those with one or another form of porphyria, a group of disorders with different genetic bases. Some of these people are essentailly normal but may exhibit varying degrees of sensitivity to light. Exposure to barbiturates, however, can result in death. Many more examples of drug anomalies could be given. It should at least be appreciated that normal persons may respond differently to a given drug, in their obvious physical reactions to it and even in their ability to respond in any way at all. More and more drug anomalies are certain to appear as more therapeutic agents are designed.

Since harmful drug responses with a genetic basis are well known, what should be practiced to regulate personal use of substances whose effects are poorly defined? Should a marijuana user be punished because some evidence indicates that the immune response is lowered with habitual marijuana use? Perhaps just the dangers of usage should be publicized, as in the case of cigarette smoking, and personal choice permitted. But what about evidence that marijuana may depress the level of the male hormone in habitual male users of this substance? In mice, rats, and even in some primates, there is experimental evidence which indicates that interference with male hormones during fetal development can interfere with the completion of normal male development, including behavior. While still an unknown area, the question arises whether marijuana use by a pregnant woman could adversely affect the development of the male fetus. Other reports suggest that marijuana use interferes with the assembly of RNA and even of amino acids into proteins. Such possibilities raise the point that personal use of a substance may do more than harm the individual user alone. The potential for harm to future generations by any mutagenic action of drugs in current use is also unknown but cannot be disregarded.

It must be kept in mind that the personal behavior of each human can affect the welfare of others. The behavior of the human species collectively can alter the existence of other species on the planet. Research into the genetic and environmental components of the human's most intricate characteristic, behavior, may hold significant bearing on the future of *Homo sapiens* and all other forms of life. Whether the human's genetic potential can be put to its best use for the survival and advancement of all life forms will finally yield an answer to the tantalizing question, "Is today's human truly a higher form of life?"

REFERENCES

Allen, M. G., Cohen, S., and Pollin, W., "Schizophrenia in Veteran Twins: A Diagnostic Review." *Amer. Journ. Psychiat.*, 128:939, 1972.

Avers, C. J., *Evolution.* Harper & Row, N. Y., 1974.

Bodmer, W. F., and Cavalli-Sforza, L. L., "Intelligence and Race." *Sci. Amer.,* Oct: 19, 1970.

Bok, S., "The Ethics of Giving Placebos." *Sci. Amer.,* Nov: 17, 1974.

Cattell, R. B. *The Scientific Analysis of Personality.* Penguin Books, London, 1965.

Cohen, S. N., "The Manipulation of Genes." *Sci. Amer.,* July: 24, 1975.

Connery, A. H., and Burns, J. J., "Metabolic Interactions Among Environmental Chemicals and Drugs." *Science,* 178:576, 1972.

Dahlstrom, E. (Ed.), *The Changing Roles of Men and Women.* Beacon Press, 1971.

Eichenwald, H. F., and Fry, P. C., "Nutrition and Learning." *Science,* 163:644, 1968.

Erlenmeyer-Kimling, L., and Jarvik, L. F., "Genetics and Intelligence: A Review." *Science,* 142:1477, 1963.

Eysenck, H.J., *The IQ Argument.* Library Press, N.Y., 1971.

Falek, A., "Differential Fertility and Intelligence: Current Status of the Problem." *Social Biology* (supplement) 50, 1971.

Fox, J. R., "The Evolution of Human Sexual Behavior." In *What a Piece of Work is Man,* Ray, J. D., and Nelson, G. E. (Eds.), p. 273. Little, Brown, Boston, 1971.

Gardner, L. I., "Deprivation Dwarfism." *Sci. Amer.,* July: 76, 1972.

Geschwind, N., "Language and the Brain." *Sci. Amer.,* April: 76, 1972.

Hilton, B., Callahan, D., Harris, M., Condliffe, P., and Berkley, B. (Eds.), *Ethical Issues in Human Genetics.* Plenum Press, N.Y.

Jensen, A. R., "How Much Can We Boost IQ and Scholastic Achievement?" *Harvard Educ. Rev.,* 39:1, 1969.

Jensen, A. R., "Race and the Genetics of Intelligence: A Reply to Lewontin." *Bull. of the Atomic Scientists,* May: 17, 1970.

Kandel, E. R., "Nerve Cells and Behavior." *Sci. Amer.,* July: 57, 1970.

Kidd, K. K., and Cavalli-Sforza, L. L., "Analysis of the Genetics of Schizophrenia." *Social Biology*, 20:254, 1973.

LaDu, B. N., "Pharmacogenetics: Defective Enzymes in Relation to Reactions to Drugs." *Ann. Rev. Med.,* 23:453, 1972.

Levine, S., "Stress and Behavior." *Sci. Amer.,* Jan: 26, 1971.

Lewontin, R. C., "Race and Intelligence." *Bull. of the Atomic Scientists,* March: 2, 1970.

Lipman-Blumen, J., "How Ideology Shapes Women's Lives." *Sci. Amer.,* Jan: 34, 1972.

Morris, D., *The Naked Ape.* McGraw-Hill, N. Y., 1967.

Premack, J., and Premack, D., "Teaching Language to an Ape." *Sci. Amer.,* Oct: 92, 1972.

Seegmiller, J. E., "Lesch-Nyhan Syndrome and the X-Linked Uric Acidurias." In *Medical genetics,* McKusick, V. A., and Claiborne, R. (Eds.), pp. 101–119. HP Publishing Co., N. Y., 1973.

Thompson, J. S., and Thompson, M. W., *Genetics in Medicine.* W. B. Saunders, Philadelphia, 1973.

Tiger, L., *Men in Groups.* Random House, N. Y., 1969.

Weiss, J. M., "Psychological Factors in Stress and Disease." *Sci. Amer.,* June: 104, 1972.

REVIEW QUESTIONS

1. Which of the following is (are) true about behavior? (1) Species-specific behavior is a phenotypic feature. (2) Species-specific behavior is innate and cannot be influenced by environment. (3) Species-specific behavior generally entails an advantage to the species. (4) Human behavior is more rigidly fixed than that of other species. (5) The ability to speak a language is a purely cultural behavior trait.

2. The following is (are) true concerning IQ. (1) Performance on an IQ test entails some genetic component. (2) Achievement on an IQ test has provided some indication of future academic and occupational achievement. (3) The tests are purposely designed to produce a bell-shaped curve for the population. (4) A mental age of 8 and a chronological age of 10 indicates an IQ of 180. (5) The heritability value of 0.8 shows

clearly that 80% of the variability among persons in IQ performance rests on a genetic basis.

3. Name a human disorder or condition in which (1) a dominant gene causes progressive mental deterioration, usually after age of 20; (2) the children exhibit behavior characterized by compulsive self-mutilation; (3) a person withdraws into a private world of fantasy; (4) diet must be carefully controlled in certain pregnant women to prevent mental retardation of the offspring; (5) an extra autosome is associated with mental retardation; (6) a chromosome deletion is associated with mental retardation.

4. From the information presented on twin studies shown in Fig. 15-1, what appears to be the relative importance of the genetic and environmental factors in (1) alcohol drinking; (2) diabetes; (3) schizophrenia; (4) rickets; (5) beginning of sitting up; (6) feeblemindedness; (7) right- or left-handedness?

GLOSSARY

acrocentric. A chromosome in which one arm is very short since the centromere is close to one end.

adenine. One of the nitrogen bases, a purine, which occurs in DNA and RNA.

albinism, ocular. An X-linked trait in which melanin pigment is absent from the eyes but not from the skin or other body parts.

albinism, oculocutaneous. Absence of melanin pigment from eyes, skin, and hair.

alkaptonuria. A metabolic disorder involving a block in the chain of reactions by which amino acids are converted to simpler compounds. As a result of the block, an intermediate compound collects and darkens the urine and various soft body tissues.

allele. One form of a gene which occupies a specific locus on a chromosome. Alternative gene forms may reside at a specific locus, and these are said to be alleles or to be allelic to one another.

amino acid. One of the building units of a polypeptide or protein. An amino acid has the general formula :

$$R{-}\overset{\displaystyle H}{\underset{\displaystyle NH_2}{\overset{|}{\underset{|}{C}}}}{-}COOH.$$

R represents anything from a hydrogen atom to more complex groupings. It is the R group which distinguishes one amino acid from the next. (Of the 20 commonly occuring amino acids, only proline deviates from this general pattern.)

amino group. The—NH_2 chemical group, typically basic in nature and found in all amino acids. (Proline is a slight exception to this.)

amniocentesis. A procedure in which a sample is obtained of the amniotic fluid surrounding the fetus. This is accomplished by the insertion of a hypodermic needle through the woman's abdomen and the wall of the uterus.

anaphase. Stage of nuclear division in which the chromosomes move to opposite poles.

androgens. A general expression for the male sex hormones.

aneuploidy. Chromosome anomaly in which the number of chromosomes in a cell departs from the normal diploid number by less than a haploid set. (In the human, a haploid set = 23; the diploid number = 46. Aneuploid numbers would be 45, 47, 48, etc.)

antibody. A protein molecule which is able to recognize and react with a specific antigen. Antibodies are also called *immunoglobulins* and are produced in response to specific antigens.

anticodon. A sequence of three nucleotides in tRNA which pairs with a specific codon in mRNA. It is the part of the tRNA responsible for giving identity to an amino acid in the cell and for positioning that amino acid in its proper place in a polypeptide chain.

antigen. Any large molecule or substance which, upon entrance into the tissue of a vertebrate, can stimulate the production of antibodies. Antigens are extremely diverse in their nature (foreign animal cells from the same or a different species, bacteria, viruses, and chemical substances of various kinds).

autosomes. All the chromosomes other than the sex chromosomes.

backcross. The cross of a member of the F_1 generation with one of the parental lines or parental types.

back mutation. (*See* reverse mutation.)

bacteriophage. (*See* phage.)

Barr body. (*See* sex chromatin.)

binomial. An expression consisting of two terms (such as a and b) separated by a + or a − sign. A binomial may be expanded to some power indicated by an exponent, such as $(a + b)^4$. Expansion of the binomial to a particular power has great value in the determination of the probability that a certain combination of events will occur.

bivalent. An association of two intimately paired homologous chromosomes at first prophase of meiosis. Each of the chromosomes in the bivalent is double, composed of two chromatids.

blood serum. Blood minus the formed elements (red and white cells and platelets) and the protein, fibrinogen, which can be converted to a clot.

cancer. A term given to a group of diverse diseases which have in common the characteristic of uncontrolled cellular growth.

carboxyl group. The —COOH chemical grouping, acidic in nature and found in all amino acids.

carcinogen. Any agent which is able to induce a transformation in a cell to the cancerous state.

catalyst. A substance which influences the rate of a chemical reaction but which is not itself changed permanently by the reaction.

cell fusion (hybridization). Formation of a cell with the nuclei and cytoplasm of two different cells.

centromere. That part of the chromosome responsible for its movement and to which the microtubules of the spindle attach.

chain-terminating codon. A sequence of three nucleotides which indicates the end of the information for the assembly of amino acids into a polypeptide chain. There are three chain-terminating codons, each one of which can bring translation to a halt. Also called *nonsense codons.*

characteristic. A general feature or attribute of a living organism. The variation seen in some characteristics is recognized as distinct, contrasting traits. The variation in other characteristics is continuous, showing a gradual transition from one extreme to the other.

chiasma. A cross-like figure seen at prophase of meiosis. One or more *chiasmata* usually occur in a bivalent, and each is believed to represent the actual point at which homologous chromosomes have undergone a reciprocal exchange, resulting in a crossover event.

Christmas disease. Hemophilia B.

chromatid. A strand of a chromosome which has replicated. At prophase and metaphase of mitosis, a chromosome is composed of two chromatids. These separate at anaphase, with the result that each chromosome is then single until replication takes place again before the next prophase.

chromatin. A term which refers to the substance in the nucleus which stains following treatment with certain dyes. Chromatin is generally used to indicate the material composing the chromosome. It is thus not one kind of substance but an assortment of macromolecules.

chromosome. A structure in the cell nucleus composed mainly of DNA and associated proteins and which stores and transmits genetic information. A chromosome may be single or may be composed of two chromatids held together at the region of the centromere.

clone. Any group or population of cells which has been derived from a single cell following mitotic divisions and which thus carries identical genetic material.

codominance. The expression of both alleles independently when present in a heterozygote. Codominance is generally associated with the presence of two products, each associated with a specific allele. *Example:* Both antigens A and B are present on the red blood cells of the type AB person who carries one allele for each antigen.

codon. A sequence of three adjacent nucleotides (a triplet) which designates a specific amino acid or which indicates that translation is to be terminated. Sixty-one codons designate amino acids; three are chain-terminating (nonsense) codons.

coefficient of consanguinity. A value which expresses the chance that an individual has received from both parents a specific gene which was transmitted to them by an ancestor which they have in common.

coefficient of relationship. The value which expresses the chance that two related persons are both carrying a specific gene which traces back to an ancestor which they both have in common. The value also indicates the proportion of all the genes in any two related persons which *on the average* is identical. The value increases with the closeness of the relationship. It is exact (1/2) only for a parent and child relationship, since any child must carry one-half of his or her genetic component in common with a parent.

colinearity. The corresponding relationship which exists between the order of the amino acids along a polypeptide chain and the order of the codons in the DNA of the gene which specifies that polypeptide.

concordant. An expression generally used in twin studies to indicate that both members of the twin pair express a given trait. (*See* discordant.)

congenital. Present at birth. A congenital trait may be due to genetic factors or solely to environmental ones.

consanguinity. A mating between two persons having a common ancestor.

consultand. The person who consults a counselor for advice.

contact inhibition. The property possessed by normal cells which places a restriction on cell proliferation in response to contact with neighboring cells.

continuous variation. That phenotypic variation seen in many characteristics which does not fall into distinct, sharply defined categories but which grades from one extreme to the other.

corpuscular radiations. Streams of subatomic particles which can travel through space and dissipate their energy upon contact with matter.

cri-du-chat syndrome. A collection of clinical features characterized by mental retardation and often a cat-like cry of the infant. Associated with a deletion in the short arm of chromosome #5.

crisscross inheritance. The pattern of inheritance typical of a mating involving sex linkage in which a trait expressed in a male parent is found only among the daughters, whereas the contrasting trait expressed in the female parent is found only among the sons. Clearly seen in the human in the mating between a woman with deutan colorblindness and a non-colorblind male.

crossing over. The separation of two genes associated with the same chromosome.

cystic fibrosis. An autosomal recessive condition in which certain glands do not function normally, resulting in a collection of symptoms, among them excess secretion of thick mucous which obstructs the air passages of the lungs.

cytoplasm. That part of a cell outside the nucleus.

cytosine. One of the nitrogen bases, a pyrimidine, found in DNA and RNA.

degenerate code. The genetic code in which two or more codons may represent the same amino acid.

deletion. A chromosome anomaly in which a segment of a chromosome has been lost.

deoxyribonucleic acid. (*See* DNA.)

detrimental gene. A gene which imparts a disadvantage to the individual possessing it. The carrier of a detrimental is hampered in some way by the gene so that the chances of reproduction or survival are less than those of the individual free of the gene.

diabetes mellitus. A disease in which the body is unable to metabolize sugar properly, resulting in an assortment of bodily derangements.

differentiation. The process in which cells derived from a common, generalized cell type become specialized in structure and function for the performance of more specific tasks.

dihybrid. Heterozygous at two genetic loci.

diploid. having two sets of chromosomes, one set having been contributed by each parent. The diploid chromosome number typically occurs in the body cells of higher plants and animals. In the diploid state, each kind of chromosome (except for the sex chromosomes in the heterogametic sex) is represented twice.

discontinuous variation. That phenotypic variation in some characteristics which can be recognized by two or more distinct traits or clear-cut differences.

discordant. A term generally used to indicate that two members of a twin pair differ with respect to the expression of a given trait. (*See* concordant.)

disjunction. The normal separation of chromosomes at anaphase of mitosis or meiosis.

dizygotic twins. Those twins derived from two separate eggs, each fertilized by a separate sperm. DZ twins may be of the same sex or different sexes. Also known as fraternal twins.

DNA. The genetic material in all cellular forms. A molecule composed of two long complementary polynucleotide strands held together as a result of the preferential pairing of its component purine and pyrimidine bases and twisted in the form of a double helix.

dominant. A gene which expresses itself in some way in the heterozygote (in the presence of its allele). Also a trait that is expressed phenotypically when the responsible gene is present singly in the heterozygote or in double dose in the homozygote. The phenotype of the heterozygote may or may not be detectably different from the homozygote. (Contrast with recessive.)

Down's syndrome. A type of mental deficiency associated with a variable collection of traits such as short stature, a large fissured tongue, short fingers and toes, and a characteristic fold of the skin of the eyelid. Attributed to three doses of chromosome #21.

drift. Random or chance fluctuations in gene frequencies as a result of chance factors and independent of natural selection or mutation. The smaller the population, the greater the effect of changes in gene frequency due to chance fluctuations.

drumstick. The sex chromatin seen in the polymorphonuclear leukocyte and which appears as an appendage on the lobe of the nucleus. (*See* sex chromatin.)

DZ. Dizygotic.

effector. A small molecule that can combine with a repressor substance and influence the activity of an operon.

electromagnetic radiation. A form of energy which can travel through empty space and which has properties which are wave-like and also those resembling particles. Electromagnetic radiations of wavelengths shorter than those of visible light may act as mutagens.

enzyme. A cellular protein that acts as a highly specific catalyst.

epistasis. A type of genic interaction in which one gene is able to mask the expression of another gene which is at a different locus and which is thus not its allele.

erythroblastosis fetalis. A condition of a fetus or newborn in which red blood cells are destroyed, resulting in anemia, jaundice, and various severe effects.

erythrocyte. A cell which contains hemoglobin. A red blood cell.

estrogen. Any one of a group of hormones secreted by the ovary and required for feminization.

eukaryote. A cell containing a distinct membrane-bound nucleus and various cytoplasmic organelles such as mitochondria, Golgi, etc., which are specialized for the performance of certain specific cell functions. Eukaryotic cells compose all living things except bacteria and blue-green algae. (*See* prokaryote.)

evolution (organic evolution). The genetic changes which occur with time in populations of living things.

expressivity. The variability in the phenotypic expression associated with a particular genotype when that genetic constitution is penetrant. (*See* penetrance.)

fallopian tube. An oviduct. One of the two tubes which can receive the egg and conduct it to the uterus. Fertilization in the human typically occurs in the upper portion of a fallopian tube.

fertilization. The union of two sex cells.

first filial generation (F_1). The generation of individuals derived from the first parental generation (P_1), the first set of parents under consideration in a mating or pedigree.

fetus. An unborn individual in the later stages of prenatal development.

fingerprinting. A procedure used to separate the component parts of a protein following enzyme cleavage of the protein into fragments and their separation by a combination of electrophoresis and paper chromatography.

first parental generation (P_1). The original set of parents under consideration in any pedigree or experimental mating.

F_1. (*See* first filial generation.)

follicle. A fluid-filled cavity with a surrounding jacket of cells in which an immature female reproductive cell may mature into an egg.

fluorescent stains. Those stains which can combine with chromosomes and render them luminous as a result of their ability to emit some of the light which is absorbed.

founder effect. The establishment of certain gene frequencies in a small population due to the origin of the group from a few founders who happened to possess a certain genotype. The gene frequency in the parental population from which the founders migrated will differ from that of the smaller one.

fraternal twins. (*See* dizygotic twins).

fungi. Those plants which lack chlorophyll and are thus unable to manufacture their own food.

gamete. A mature sex cell, such as a sperm or egg. In some plants and lower animals, the gametes may appear identical and cannot be designated sperms or eggs.

gene. A unit of inheritance composed of a segment of DNA and carrying coded information associated with a specific function.

gene flow. The transfer of alleles from one population to another as a result of migration and ensuing mating.

gene mutation. An alteration at a site within the gene in which one kind of nucleotide has been substituted for another, resulting in the formation of a different codon or code word. Also called point mutation.

gene pool. The total of all the genes of all the members of a population.

genetic death. Failure of an individual to reproduce due to sterility, failure to mate, or elimination from the population before reproductive age.

genetic drift. (*See* drift.)

genetic marker. (*See* marker gene.)

genotype. The genetic constitution of a cell or individual.

germinal mutation. A mutation which occurs in a reproductive cell and which can thus be transmitted to the offspring.

glucose-6-phosphate dehydrogenase (G6PD). An enzyme whose absence can result in red blood cell breakdown following the ingestion of certain drugs or substances such as primaquine and fava beans.

glycogen. The principal storage form of carbohydrate in animals. Composed of units of sugar linked together. Also known as animal starch.

gonad. An organ specialized for production of sex cells. A testis or ovary.

guanine. One of the nitrogen bases, a purine, which occurs in DNA and RNA.

haploid. Having one set of chromosomes. In the haploid condition, typical of sex cells, each kind of chromosome is represented only once.

Hardy-Weinberg law. A mathematical expression which illustrates that the frequency of alleles and the resultant genotypes in a population will remain constant from one generation to the next and will not change due to dominance or any mechanism of transmission. The forces of evolution (mutation, natural selection, and drift) operate in nature and disturb the Hardy-Weinberg equilibrium so that, in reality, genes and gene frequencies do tend to change in populations.

hemizygous. Having particular loci present in just a single dose as is the case for all the X-linked genes of a human male.

hemophilia. One of several disorders in which the blood does not clot normally. In hemophilia A, there is a lack of antihemophilia factor. In hemophilia B, the deficiency is in plasma thromboplastin component.

heritability. That proportion of the phenotypic variation in the expression of a characteristic which is due entirely to the genetic component.

hermaphrodite. An individual possessing both male and female reproductive organs.

heterogametic sex. That sex which produces two kinds of gametes with respect to the sex chromosomes. In mammals, the male is heterogametic and produces both X-bearing and Y-bearing gametes.

heterozygous. Having alternative gene forms (alleles) at a particular genetic locus.

hexosaeminidase A. The enzyme deficient in Tay-Sachs disease.

histocompatibility loci. Those genetic loci whose resident genes influence the ability to accept foreign tissue such as a graft or transplant.

histones. Those proteins basic in nature and of low molecular weight which are associated with the DNA of the eukaryotic chromosomes.

holandric gene. Referring to any gene whose locus is on the Y chromosome.

homogametic sex. That sex which produces a single type of gamete with respect to the kind of sex chromosome. In mammals, the female is homogametic, since all the gametes bear an X chromosome.

homologous chromosomes. Chromosomes which correspond in size, shape, and genetic loci.

homosexuality. The performance by an adult of a sexual act with a member of the same sex.

homozygous. Having the same allele present at a particular locus on both homologous chromosomes.

hormone. A chemical which is secreted in one part of the body and which affects other parts. Also called a chemical messenger.

Huntington's disease. A nervous disorder generally manifested well after puberty and characterized by involuntary movements of the face and limbs along with a gradual loss of mental faculties.

hybrid. Heterozygous at one or more loci. Also applied to an individual derived from the crossing of two different genetic lines. A cell hybrid is a cell resulting from the fusion of two somatic (body) cells from different ancestries, often from different species.

identical twins. (*See* monozygotic twins.)

incomplete dominance. The expression of both alleles in a heterozygote. Generally used in reference to those situations where the phenotypic effect of one of the alleles in the heterozygote appears more pronounced than the other. The distinction between incomplete dominance and codominance is often arbitrary, and the two are frequently used interchangeably.

inbreeding. Mating between related individuals.

independent assortment. The segregation at meiosis of a pair of alleles independently of other pairs of alleles which are located on different chromosomes.

independent events. Two or more alternatives whose separate occurrences do not influence one another in any way.

inversion. A chromosome anomaly involving two breaks in a chromosome followed by a reversal of the segment and a consequent reversal of the sequence of the genes in the segment.

in vitro. Referring to any biological process taking place under experimental conditions away from the intact organism.

karyotype. The chromosome constitution of a cell or individual.

Klinefelter syndrome. Abnormality due to the presence in the body cells of one or more extra X chromosomes along with a Y chromosome. Individuals are males with nonfunctional testes and often some breast development.

kinetochore. The region of the chromosome to which the spindle fibers attach at nuclear divisions. Often used synonymously with centromere, although some consider the kinetochore a separate entity located in the region of the centromere.

law of independent assortment. (*See* independent assortment.)

law of segregation. The separation at meiosis of the two genetic factors which occur at the same locus on homologous chromosomes. Gene segregation is a result of the separation of homologous chromosomes which move to opposite poles at meiotic anaphase I.

Lesch-Nyhan syndrome. An X-linked trait characterized by an elevated level of uric acid, spastic cerebral palsy, and compulsive self-mutilation.

lethal gene. A gene which kills the individual who possesses it sometime before the age of reproduction.

leukemia. A malignancy of the blood-forming tissue resulting in excessive production of white blood cells.

linkage. The association of genes on the same chromosome. Genes that occupy loci which are close together will tend to be transmitted together. Those which are found farther apart may be separated frequently by crossover events, making the demonstration of linkage difficult through genetic analysis.

linked genes. Genes which are associated on the same chromosome and thus tend to be transmitted together in a cross. Since linked genes occur at different loci, they are not allelic.

locus. The specific position occupied by a particular gene on a chromosome. Any one of the alternative forms of a gene may be found at a given locus.

Lyon hypothesis. The concept that, in any given cell of a mammalian female, only one X chromosome is entirely active, the other one being inactive completely or in part. The inactivation occurs at random so that the paternally derived X is active in some cells, the maternally derived one active in others.

Lyonization. (*See* X-inactivation.)

lysis. Destruction or disintegration of a cell, as lysis through bursting.

malignant transformation. A change from a noncancerous to a cancerous state.

map unit. A figure which measures the distance between two linked loci as inferred from the amount of crossing over which occurs between them. For loci which are close to each other, the map units are about equal to the number of new combinations which can be detected, so that 1 map unit = 1% recombination. For loci more distantly situated with reference to each other, the amount of multiple crossovers cuts down on the number of new combinations which can be detected, so that the map units and the percentage of recombination are not equivalent.

marker gene. Any specific gene or genetic trait which can be used to follow the transmission of a specific chromosome or chromosome region in a mating.

mean. The sum total of the values of a group of quantities divided by the number of individual measurements in the group. Also called the arithmetic mean.

meiosis. The nuclear event involving two divisions and resulting in the reduction in the number of chromosomes from diploid to haploid. Meiosis occurs sometime before gamete formation in all sexual life forms.

melanin. The brown pigment found in the cells of the skin, hair, and iris of the eye.

messenger RNA (mRNA). The transcript of DNA which carries the information for the amino acid sequence of a polypeptide.

metabolism. The sum of all the chemical reactions which occur in a cell including the synthetic reactions (anabolism) and the energy-releasing processes (katabolism).

metacentric. A chromosome with a centromere located in the middle, producing a chromosome with two arms which are equal in length.

metaphase. That stage of nuclear divisions at which chromosomes are arranged at the equator or midregion of the spindle.

mitochondria. The cytoplasmic bodies in the eukaryotic cell which are the sites of aerobic respiration and energy release.

mitosis. The nuclear division which results in the accurate distribution of the genetic material of the chromosomes from the parent cell to the two cells derived from it.

modifier (modifying gene). A gene whose expression affects that of other genes at other loci. Often modifiers have no other known effects.

monohybrid. An individual carrying a pair of alternative gene forms (alleles). Heterozygous at a single locus under consideration. Also a cross in which a single pair of alleles is being followed.

monosomy. A chromosome anomaly in which only one of a given kind of chromosome is present instead of the normal two.

monozygotic twins. Those twins derived from the same single fertilized egg. MZ twins are always of the same sex. Also called identical twins.

mosaicism. The occurrence of two or more genetically different cell types within the body of an individual.

multiple alleles (multiple allelic series). Three or more gene forms associated with a given locus. Any one form can occur at the locus, but being diploid, any individual can carry no more than two of the members of a multiple allelic series.

multiple-factor inheritance. Inheritance which involves many genes at different loci and the complex interactions of environmental factors, many unknown.

mutation. A sudden inheritable change. In its broadest sense, includes gene (point) mutation and chromosome anomalies.

mutation pressure. The spontaneous rate of gene mutation which occurs continuously in a population.

mutually exclusive events. Two alternatives, the occurrence of one precluding the occurrence of the other.

myotonic dystrophy. A disorder characterized by progressive atrophy and weakness of muscles, the development of cataracts, and also abnormalities of the heart. Associated with a dominant autosomal gene.

MZ. Monozygotic.

nail-patella syndrome. A hereditary disease in which the fingernails are abnormal and several skeletal defects are present, among them an abnormal or missing kneecap.

natural selection. Differential reproduction which results in an increase in the frequency of certain genes and a decrease in others. Natural selection tends to raise a population to the highest level of efficiency in a specific environment and is the major force of evolution.

nondisjunction. The failure of chromosomes or chromatids to separate at a mitotic or meiotic anaphase. Nondisjunction is responsible for several human chromosome anomalies.

nonsecretor. A person with A or B antigens on the red blood cells but not in any body secretions.

nonsense codon. (*See* chain-terminating codon.)

nonsense mutation. A mutation which results in the change from a codon specific for an amino acid to a chain-terminating codon which will bring translation to a premature halt.

normal curve (distribution). A symmetric, bell-shaped curve which is often approximated when a large number of measurements is taken with respect to some quantitative attribute which shows continuous variation. The peak of such a curve is the average or mean value.

nucleic acid. An organic acid composed of many nucleotide units.

nucleolus. A dense body within the nucleus, associated with a specific region of a specific chromosome and concerned with the assembly of the ribosomes.

nucleotide. A building unit of a nucleic acid, composed of a 5-carbon sugar associated with a phosphate and a nitrogen base.

nucleus. That part of a eukaryotic cell bounded by a membrane and containing the chromosomes and the nucleolus.

oncogenic virus. Any virus which can induce cell proliferation and tumor formation.

oogenesis. The process in a female which results in the formation of a gamete from the maturation of an immature germ cell.

oogonium. A primordial cell in the female which can give rise to a primary oocyte.

operator. A DNA region which interacts with regulatory substances and thus influences the activity of genes adjacent to it.

operon. A unit of DNA consisting of an operator site and one or more structural genes whose activity is influenced by the operator.

organelle. A subcellular structure specialized for a certain role.

organic compound. A molecule containing carbon and hydrogen.

ovum. A cell of a female which can combine with a male gamete.

paracentric inversion. An inversion involving a single chromosome arm, both breaks being to one side of the centromere.

pedigree. A record of inheritance for two or more generations, often presented as a diagram.

penetrance. The ability of a genotype to be expressed in any way at all when present. If a gene or genotype is penetrant, that expression may vary (be of variable expressivity).

peptide. A compound composed of two or more amino acids. Peptides combine to form polypeptides.

pericentric inversion. An inversion involving both arms of a chromosome, the breaks being on opposite sides of the centromere. As a result, centromere position can be shifted.

phage. Any virus which has a bacterial cell as a host.

phenocopy. An environmentally induced, non-genetic imitation of the effects of a specific genotype.

phenotype. Any of the detectable attributes of a living thing. The phenotype is the result of the interaction of the genetic and environmental components.

phenylketonuria (PKU). A recessively inherited condition characterized by the inability to convert the amino acid phenylalanine to tyrosine, resulting in the accumulation of toxic products which may cause mental retardation unless dietary precautions are instituted at the proper time.

photoreactivation. The reversal of the killing effect of ultraviolet light by white light.

placenta. In mammals, the organ composed of tissues derived from the offspring and the mother and which makes possible the exchange of materials between them during pregnancy.

pleiotropy. The multiple phenotypic effects of a single allele.

P_1. (*See* first parental generation.)

point mutation. (*See* gene mutation.)

polar body. One of the small cells receiving little cytoplasm following the division of a primary and a secondary oocyte. The polar bodies disintegrate.

polygenic inheritance. Inheritance in which many genetic loci are involved which interact with the environment. ''Polygenic'' refers specifically to the many genetic factors as opposed to the environmental ones.

polynucleotide. An assembly consisting of many nucleotide units joined together.

polypeptide. A single, long chain composed of many amino acid units. Some proteins are composed only of one polypeptide chain.

polyploidy. Presence in a cell of 3 or more sets of the basic haploid number of chromosomes. In the human, the haploid number is 23; the diploid is 46. Polyploid numbers are 69 (triploid) and 92 (tetraploid).

population. An assemblage in a more-or-less defined area of individuals who tend to mate with one another more than with members of other groups.

primary oocyte. In the female, the cell type derived from an oogonium and in which the first prophase of meiosis takes place.

primary spermatocyte. In the male, the cell type derived from a spermatogonium and in which the first prophase of meiosis takes place.

prokaryote. A cell type lacking a nucleus bounded by a membrane, as well as certain specialized cell organelles such as mitochondria. The genetic material in a prokaryote is not associated with histone or nonhistone proteins. Bacteria and blue-green algae are prokaryotes.

prophage. A bacterial virus which exists as nucleic acid integrated with the nucleic acid of the host cell.

prophase. That phase of nuclear division in which the chromosomes become more and more evident. In prophase of meiosis, the chromosomes pair and engage in crossing over.

protein. A large molecule composed of one or more polypeptide chains, each of which is composed of many amino acid units.

purine base. A nitrogen base composed of a double ring. Adenine and guanine are examples.

pyrimidine base. A nitrogen base composed of a single ring. Examples are cytosine, thymine, and uracil.

quantitative inheritance. Inheritance involving many alleles at different loci, each allele producing some measurable effect. Often used interchangeably with polygenic and multiple-factor inheritance.

race. A population of individuals in which the frequencies of certain genes in the gene pool differ from that in other populations.

radiation. (*See* electromagnetic and corpuscular radiations.)

recessive. A gene which does not express itself in the heterozygote (in the presence of its allele). Also a trait which is expressed phenotypically only when the responsible gene is present in the homozygous condition. In actuality, some effect can often be detected in the heterozygote. Absolute dominance and recessiveness may not truly exist, and the distinction between the two is often artificial and used for convenience.

reciprocal crosses (matings). Two crosses which are similar to each other in which the same traits are being followed but in which the sexes of the parents are interchanged. *Example*: In one cross, a female parent shows trait *A* and the male parent trait *B*. In the second cross, the female shows trait *B* and the male trait *A*.

recombination. The formation of new combinations of genes following meiotic events. Often restricted in use to those new combinations between linked genes.

reduced penetrance. Failure of a genotype when present to express itself in any way at all in an individual. (*See* penetrance.)

regulator. A gene which produces some substance such as a repressor which can influence the transcription of an operon by interacting with an operator.

regulatory genes. Genetic regions concerned primarily with the control of the activity of other genes through their influence on transcription.

repair enzymes. Those enzymes which are able to remove and replace segments of damaged DNA, as seen in the repair of damage incurred following exposure to ultraviolet light.

replication. The construction of a large molecule (such as DNA) under the guidance of a template, resulting in two identical molecules, the copy and the template.

repressor. A protein substance produced by a regulatory gene which can combine with an operator and regulate the activity of an operon.

reverse mutation. A mutation from a mutant gene form back to the form considered standard or wild. The rate of reverse mutation is generally lower than mutation from wild to mutant.

Rh negative. Having red blood cells lacking D antigen and which do not react with anti-D antibodies.

Rh positive. Having red blood cells which carry D antigen and which react with anti-D antibodies.

ribonucleic acid. (*See* RNA.)

ribosome. A cellular organelle which interacts with mRNA and tRNA and is able to join together amino acid units into a polypeptide chain.

RNA. A large molecule composed of nucleotide units in which the nitrogen base, uracil, occurs in place of thymine and in which the sugar portion contains one less oxygen than the sugar in DNA. RNA is single-stranded and is more widely distributed in the cell than is DNA. Occurs as messenger RNA (mRNA), transfer RNA (tRNA), and ribosomal RNA (rRNA).

schizophrenia. A serious mental disorder characterized by a withdrawal from normal relationships into a personal world of the victim's making.

secondary oocyte. The larger of the two cells formed following the division of a primary oocyte.

secondary sexual characteristics. Those bodily features other than the sex organs themselves which distinguish members of one sex from the other.

secondary spermatocyte. One of the two cells resulting from the division of a primary spermatocyte.

secretor. An individual who possesses a water-soluble form of antigens A and/or B which can be detected in certain body secretions such as the saliva and the amniotic fluid.

selection pressure. The operation of natural selection on the frequency of genes in a population. Selection pressure interacts with mutation pressure and differs from one environment to the next, with the result that certain genes increase in frequency in some populations but are kept at a lower frequency in others.

selective medium. A source of food or growth substances which will maintain only specific cell types. Those cells which lack the traits required to grow on the selective medium die out. Hence such a medium effectively screens a population of diverse cell types by permitting the growth of only those cells with certain traits.

semiconservative replication. The synthesis of two new DNA molecules utilizing each of the two single DNA strands composing the original molecule as a template for the assembly of complementary strands.

sex bivalent. The association of an X and a Y chromosome in the primary spermatocyte.

sex chromatin. A densely staining body visualized in cells containing more than one X chromosome. Sex chromatin represents an inactivated X (or part of an X). The number of sex chromatin bodies equals one less than the number of X chromosomes in the cell. Barr body is synonymous with sex chromatin. The drumstick represents sex chromatin in a polymorphonuclear leukocyte.

sex chromosomes. The chromosome or chromosome pair which shows a difference in kind or amount between the two sexes of a species. The X and Y are the sex chromosomes of all mammals.

sex-influenced genes. Genes which can be located on any chromosome, an autosome or sex chromosome, but which are expressed as dominant in one sex and recessive in the other.

sex-limited genes. Genes located on autosomes or sex chromosomes which express themselves only in one of the sexes.

sex-linked. (*See* X-linked.)

sexual reproduction. Reproduction which involves the union of the nuclei of cells set aside for reproduction.

Siamese twins. Conjoined twins connected by some portion of their bodies as a result of the failure of complete separation of the early embryo into two unattached embryos.

siblings (sibs). Brothers and sisters in the same family.

sickle cell anemia. A severe hereditary disorder in which the red blood cells assume a crescent shape in situations of low oxygen level and are destroyed, resulting in severe anemia and an assortment of derangements of the body.

silent mutation. A point mutation which results in a change from a codon specifying an amino acid to another codon specific for the same amino acid.

somatic cell. A cell of the body, in contrast to a germ cell (reproductive cell).

somatic mutation. A genetic change which arises in a body cell and which is not transmitted to the next generation.

species. The largest population unit whose component members can freely exchange genetic material with one another. Two species are recognized when there is a decrease in the ability of members from the two groups to mate and produce fertile offspring.

spermatid. One of two cells resulting from the completion of second meiotic division by a secondary spermatocyte. Undergoes transformation to a sperm without further changes in the genetic content.

spermatogenesis. The entire series of events leading from the origin of a primary spermatocyte through the development of mature sperms. The process responsible for the generation of male gametes.

spermatogonium. A cell in the testis from which a primary spermatocyte is derived.

spermiogenesis. That portion of spermatogenesis concerned with the transformation of the spermatid to a sperm and involving cytoplasmic rather than genetic changes.

spindle fibers (spindle). The microtubules composed of protein subunits which are involved with chromosome movement and distribution at nuclear divisions. The visible spindle represents the aggregation of the separate microtubules.

structural gene. A gene which contains information for the assembly of a polypeptide or protein which will act as an enzyme or structural component of the cell. As opposed to a regulatory gene, a structural gene is not concerned with the synthesis of regulatory substances or with gene regulation.

suppressor. A gene which is able to prevent the expression of a mutant gene found at a different genetic locus.

symbiotic phage. (*See* temperate phage.)

synapsis. The pairing of homologous chromosomes at meiotic prophase.

syndrome. The group of symptoms or features associated with a particular disease or abnormality.

synteny. The association of genes on the same chromosome. Often used synonymously with linkage. Some prefer to reserve the term "linkage" for only those genes which have been shown by genetic analysis to fall into a linkage group. For example, two genes shown by cell hybridization to occur on the same chromosome would be considered syntenic. They would be called linked only if genetic analysis showed that they do not assort independently. (*See* linkage.)

Tay-Sachs disorder. A condition inherited as an autosomal recessive in which an infant develops loss of muscular function, paralysis, blindness, deafness, and other derangements, usually resulting in death before the age of 4.

telocentric. A chromosome with a centromere so close to one end that one arm is tiny or inconspicuous.

telophase. The stage of nuclear division during which the nuclear membrane reforms.

temperate phage. A symbiotic bacterial virus which is able to integrate with the host DNA without causing apparent harm to the host cell.

template. The large molecule which acts as a mold or blueprint for the synthesis of another identical or closely similar molecule.

testcross. A cross of an individual to one expressing a specific recessive trait.

testosterone. A hormone secreted by the testes and required for the development and maintenance of male secondary sex characteristics.

thalassemia. An inherited blood disorder in which the homozygote suffers a severe anemia (*thalassemia major*) and the heterozygote a milder form (*thalassemia minor*). Thalassemia is actually a general term which includes various anemias due to the effects of different genes.

thymine. One of the nitrogen bases, a pyrimidine, which occurs in DNA and is not found in RNA.

thymus gland. An organ in the upper part of the chest required for the development of various immune responses.

thyroxine. A hormone secreted by the thyroid gland which can stimulate the rate of metabolism.

trait. Alternative or variant form of a characteristic.

transcription. The assembly of RNA on a DNA template.

transfer RNA (tRNA). A small RNA molecule which is able to recognize a specific amino acid, transport it to a ribosome, and position it properly in a growing polypeptide chain.

translation. The assembly of a polypeptide chain using the information in the mRNA to direct the amino acid sequence of the chain.

translocation. A chromosome aberration in which a segment of a chromosome is shifted to a new location in the same or a completely different chromosome.

transsexual. An individual who considers himself or herself to be a person of the other sex. A transsexual is not considered a homosexual, since the latter person identifies with his or her anatomical sex.

triploid. Possessing three complete haploid sets of chromosomes.

trisomy. A condition in an otherwise diploid cell in which three doses of a certain chromosome are present instead of the normal two.

Turner syndrome. An abnormal condition resulting from the presence of only one X chromosome unaccompanied by another X or a Y. This is the only monosomic condition known in the human. The individual is a female characterized by absent or non-functional ovaries in addition to circulatory and other bodily disturbances.

uracil. One of the nitrogen bases, a pyrimidine, which occurs in RNA but not in DNA.

uterus. The organ in which the embryo undergoes development. The womb.

variable expressivity. (*See* expressivity.)

virus. A noncellular parasitic particle composed of a core of nucleic acid surrounded by protein.

wild type. The normal or standard form of a gene or of an individual, such as the form found typically in nature.

X-inactivation. Condensation and resultant inactivation of all or part of any X chromosome in a cell in excess of one.

X-linked. Any gene having a locus on the X chromosome. Also called sex-linked.

zygote. The cell resulting from the fusion of two gametes. In most animals, the fertilized egg.

REVIEW QUESTION ANSWERS

CHAPTER 1

1. Reduced penetrance. **2.** Pleiotropy. **3.** First female: *Bb*. Puppies *Bb* and *bb*. Second female: *BB*. Puppies *Bb* in each litter. **4.** Gene *E* shows variable expressivity and reduced penetrance. Both offspring of the parent with the extra toe are genotype *Ee*. **5.** *A*. (1) *Ff*; (2) *ff*. *B*. (1) 1 mildly frizzled, 1 normal; (2) 1 frizzled, 1 mildly frizzled. **6.** *A*. 1 out of 4. *B*. 1 out of 4. *C*. 2 out of 3. **7.** $Hb^S Hb^S$. Yes, since the gene for normal hemoglobin is absent. **8.** (3) and (4).

CHAPTER 2

1. Centromere. **2.** Two at prophase; one at telophase. **3.** (1) Metaphase; (2) Prophase; (3) Telophase; (4) Prophase; (5) Anaphase. **4.** *A*. (1) 92; (2) 46. *B*. (1) 46; (2) 46. *C*. 44. **5.** 47. **6.** 47 and 45. **7.** Before onset of prophase, commonly called *interphase.*

CHAPTER 3

1. (1) First prophase; (2) Second anaphase. **2.** (1) Synapsis; (2) Bivalent. **3.** At mitosis, single chromosomes are arranged at the midregion of the spindle; at meiosis bivalents are oriented at the equator. **4.** (1) 46, (2) 46, (3) 23, (4) 23, (5) 23, (6) 23. **5.** Primary oocyte. **6.** *A*. 4; *B*. 1. **7.** (1) *AaF?* (2) *Aaff*; (3) *aaff*; (4) *AaFf.* **8.** (1) *AF, Af, aF, af;* (2) *AF, Af;* (3) *AF;* (4) *AF, aF.* **9.** Equal chances for: normal pigmentation, free lobes; normal pigmentation, attached lobes; albino, free lobes; albino, attached lobes. **10.** One out of four.

CHAPTER 4

1. 44 in both. **2.** Bar males and non-bar females. **3.** *A*. (1) Both are *aa;* (2) *ii* and *i*Y. (3) *aaIi*; (4) *Aai*Y. *B*. (1) *AI, Ai, aI, ai;* (2) *Ai, A*Y, *ai, a*Y. **4.** (1) *mmDd*; (2) *Mmdd;* (3) *M?d*Y. **5.** *A*. Non-colorblind female with 3 X chromosomes (*DDd*); non-colorblind Turner female (*DO*); non-colorblind Klinefelter male (*Dd*Y). The YO zygote would not survive. *B*. Non-colorblind Turner female (*DO*); non-colorblind Klinefelter male (*DDY*); colorblind Turner female (*d*O); non-colorblind Klinefelter male (*Dd*Y). **6.** Man—*R*Y. Woman—*rr*. Offspring—normal sons (*r*Y) and daughters with rickets (*Rr*).

CHAPTER 5

1. (4) and (5) **2.** (4) and (5) **3.** (1) female—*Bb*; (2) could be either female, *bb*, or male, *b*Y; (3) female—$\underline{W?}$. **4.** The male carries genes influencing milk yield, but these sex-limited genes are expressed only in the cow who receives genes for milk yield from both parents. **5.** *A.* (1) *LL;* (2) *ll. B.* Daughters with long index finger (*Ll*) and sons with short index finger (*Ll*). **6.** (1) All; (2) All; (3) Half; (4) None; (5) None.

CHAPTER 6

1. *A.* (1) 8; (2) 4;(3) 32; (4) only one type. *B.* (1) 8 phenotypes, 27 genotypes; (2) 4 genotypes, 9 phenotypes. **2.** (1) $\underline{Mn}$ and $\underline{mN}$; (2) $\underline{MN}$ and $\underline{mn}$. **3.** (1) $\underline{Op}$ (40%), $\underline{oP}$ (40%), $\underline{OP}$ (10%), $\underline{op}$ (10%); (2) $\underline{OP}$ (40%), $\underline{op}$ (40%), $\underline{Op}$ (10%), $\underline{oP}$ (10%). **4.** The original combinations: $\underline{Op}$ and $\underline{oP}$. **5.** $\underline{d\text{-}4\text{-}m\text{-}10\text{-}t}$. **6.** Approximately 20%. **7.** (1) $\frac{dP}{dP} \times \frac{Dp}{\mathrm{Y}}$; offspring: $\frac{dP}{Dp}$ (non-colorblind daughters) and $\frac{dP}{\mathrm{Y}}$ (deutan colorblind sons). (2) $\frac{Dp}{dP} \times \frac{DP}{Y}$; offspring: $\frac{Dp}{DP}$ and $\frac{dP}{DP}$ (daughters, normal vision); $\frac{Dp}{Y}$ and $\frac{dP}{Y}$ (sons with protan colorblindness and sons with deutan colorblindness); (3) offspring: $\frac{Dp}{DP}, \frac{dP}{DP}, \frac{DP}{DP}, \frac{dp}{DP}$ (daughters all with normal vision); $\frac{Dp}{\mathrm{Y}}, \frac{dP}{\mathrm{Y}}, \frac{DP}{\mathrm{Y}}, \frac{dp}{\mathrm{Y}}$ (sons: protan colorblind, deutan colorblind, normal vision; protan and deutan colorblind). **8.** $\frac{Do}{dO}$ (daughters phenotypically normal); $\frac{Do}{\mathrm{Y}}$ (sons with normal color vision and ocular albinism). **9.** Daughters phenotypically normal: $\frac{DO}{DO}$ and $\frac{DO}{do}$; sons: $\frac{DO}{\mathrm{Y}}$ (normal) and $\frac{do}{\mathrm{Y}}$ (colorblind and ocular albinism). **10.** Offspring: Daughters—all enzyme producers with normal vision. Sons—enzyme producers and colorblind, 47%; nonenzyme producers, normal vision, 47%; enzyme producers and normal vision, 3%; nonenzyme producers and colorblind, 3%.

CHAPTER 7

1. (1). **2.** (1), (4) and (5). **3.** A, B, AB, and O. Each may be Rh positive or negative. **4.** The man carries the gene for A antigen but it is not expressed. He is probably double-recessive for gene *h.* He could also be carrying the gene for B antigen. **5.** Colored to white in a ratio of 9 to 7. **6.** (3) and (5). **7.** Children cannot accept from the parents or donate to them. There is about a 25% chance that two children in the family would be compatible with respect to these loci. Possibilities are:

$$\frac{A^1S^2B^1D^2}{A^3S^4B^3D^4} \quad \frac{A^2S^1B^2D^1}{A^3S^4B^3D^4} \quad \frac{A^1S^2B^1D^2}{A^4S^3B^4D^3} \quad \frac{A^2S^1B^2D^1}{A^4S^3B^4D^3}$$

CHAPTER 8

1. 69.5 inches. **2.** (1) *AABb* or *AaBB*; (2) *AaBb*, *AAbb*, or *aaBB*; (3) *Aabb* or *aaBb*. **3.** The light-skinned parent is *aaBb* or *Aabb*. The dark parent could be *AaBB* or *AABb*. **4.** The white person is *aabb*. The intermediate parent is probably *AAbb* or *aaBB*, since genotype *AaBb* would make additional shades probable. **5.** The white person is *aabb*; intermediate is *AaBb*. **6.** Three gene pairs. Effective genes each contribute 1/2 lb. **7.** Four gene pairs. Effective genes each contribute 1/4 in. **8.** Four gene pairs are involved. The extremes are *AABBCCDD* (96 in.) and *aabbccdd* (32 in.). The original parents were not extremes. **9.** Parents could have genotypes *AABBccdd* and *aabbCCDD*. Other genotypes are also possible. **10.** While the defective gene occurs in low frequency in the European population, some of the few founders who left it and migrated to Canada happened to be carrying the gene in the heterozygous condition. Those founders of the Peruvian population happened to be free of it. **11.** While there are three phenotypically distinct populations, only two species can be recognized. Populations 2 and 3 have not accumulated genetic differences which limit free genetic exchange. They may be considered varieties of the same species. Population 1 is genetically isolated from 2 and 3 and respresents a distinct species.

CHAPTER 9

1. *A.* A-T-C-A-T-A-G-T-A-T-A-C. *B.* A-U-C-A-U-A-G-U-A-U-A-C. *C.* 4. *D.* 4. *E.* U-A-G U-A-U C-A-U A-U-G. *F.* Translation would halt at that point. **2.** (2), (3), and (4). **3.** (3) and (4). **4.** (5). **5.** (2), (3), (4), and (5) **6.** *A.* UUU is one. *B.* Adenine and guanine in both. *C.* Thymine and cytosine in DNA; uracil and cytosine in RNA. *D.* Phosphate, sugar, and a nitrogen base. *E.* tRNA and RNA of the ribosome. *F.* All possess an -NH_2 group, a -COOH group, and an R group. *G.* It has a free -COOH.

CHAPTER 10

1. (1), (3), (4), and (5). **2.** (2) and (5). **3.** (1), (3), and (5) **4.** (2). **5.** (1), (3), and (5).

CHAPTER 11

1. (1) Klinefelter syndrome; (2) Turner syndrome; (3) Down's syndrome; chronic myeloid leukemia (the latter has long been associated with a deletion but recent evidence indicates a translocation); (4) Down's syndrome; (5) *cri-du-chat* syndrome; chronic myeloid leukemia (recent studies indicate Philadelphia chromosome associated with the latter involves a translocation.) **2.** (1) AABBCCDDEEE; (2) AAABBCC-CDDDEEE; (3) ABBCCDDEE; (4) AAAABBBBCCCCDDDDEEEE. **3.** (1) nondisjunction of X's at first anaphase of oogenesis giving an XX egg which becomes fertilized by Y-bearing sperm; nondisjunction of X and Y at first anaphase of spermatogenesis giving XY sperm which fertilizes an egg; (2) nondisjunction of sex chromosomes at spermatogenesis or oogenesis producing gametes with no sex chromosomes; loss of one X chromosome at first mitotic anaphase of an XX zygote; (3) nondisjunction of Y at second anaphase of spermatogenesis; (4) pericentric inversion; (5)

complete failure of anaphase at meiosis giving an unreduced sperm or egg which unites with a haploid gamete; (6) translocation; (7) nondisjunction of X's at oogenesis producing an XX egg which becomes fertilized by an X-bearing sperm; (8) loss of an X during mitotic divisions of an XX embryo; (9) loss of a Y during mitotic division of an XXY embryo; (10) nondisjunction of a Y during mitotic divisions of an XY embryo.

CHAPTER 12

1. (1) 1/32; (2) $10(1/2)^3 (1/2)^2 = 5/16$. **2.** 1/2. **3.** *A.* (1) 3/8; (2) 1/8. *B.* 3/4. *C.* 3/64. *D.* 9/4096. *E.* 3/64. *F.* 3/4 for a normal; 1/4 for an albino. **4.** (1) Little chance; mother probably doesn't carry the recessive; (2) 100%. **5.** *A.* 1/2. *B.* 1/4. *C.* 1/4. **6.** (5). **7.** (1) 1/16 (1/4 chance both are carriers; 1/4 for an albino if they are carriers); (2) 3/8 (chance is 3/4 that parental combinations are either *AA* × *AA* or *Aa* × *AA* or *AA* × *Aa*. Chance is 1/2 for a girl); (3) 1/2; (4) 1/4. **8.** (1) Little chance; unrelated man probably not a carrier; (2) 2/3 × 1/70 × 1/4 = 1/420. **9.** No chance. **10.** 1/4. **11.** 1/4. **12.** **13.** 1/50.

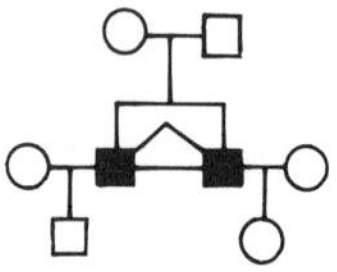

CHAPTER 13

1. (1) and (4). **2.** (4). **3.** (3) and (5). **4.** 1/200,000. **5.** M = 0.6; N = 0.4. **6.** *A.* $A = 29/30$; $a = 1/30$. *B.* 2/3 × 58/900 × 1/4. **7.** $g = 3/20$; $G = 17/20$. Heterozygous women = 2 × 3/20 × 17/20.

CHAPTER 14

1. (2) and (5). **2.** (2) and (3). **3.** (3) and (5). **4.** (2), (3), (4). **5.** (2) and (4). **6.** (1), (2), (3). **7.** (1), (3), (4). **8.** All.

CHAPTER 15

1. (1) and (3). **2.** (1), (2), (3). **3.** (1) Huntington's disease; (2) Lesch-Nyhan syndrome; (3) schizophrenia; (4) phenylketonuria; (5) Down's syndrome; (6) *cri-du-chat* syndrome. **4.** Stronger hereditary component in (2), (3), (4), (6); stronger environmental component in (1), (5), (7).

Index

A

ABO blood grouping, 148, 149-151, 156, 161, 162
  population differences and, 193-194
ABO locus, tissue compatibility and, 164-165
Acidic proteins (*see* Nonhistone proteins)
Adenine, 209, 210
Agammaglobulinemia, 167
Age, parental effect of, 266-268
Aging, 367-371
  cancer and, 362
  genetic and environmental components of, 370
  mutation rate and, 323
  premature, 370-371
Albinism, 22, 242 (*also see* specific types of)
  ocular, 85, 90, 92, 94
  oculocutaneous, 90-92, 94, 240
Alkaptonuria, 18, 241
Alleles, 9
Amino acids, 219
  structure of, 222, 223
Amniocentesis, 113-115, 281-282, 308, 392
Anaphase, mitotic, 34
Aneuploidy, 80
  of autosomes, 265-266
  of sex chromosomes, 264-265
Antibody, 148
Anticodon, 227
Antigen, 148
Arteriosclerosis, 254
Arteriosclerosis obliterans, 162
Arthritis, 241
Asexual reproduction, 131
Atherosclerosis, 184, 254, 256
Autosomes, 39, 68

B

Bacteriophage (*see* Phage)
Baldness, 99-102
Barbituates, 396
Barr body, 197, 108, 109
Beard growth, 98
Behavior, 387, 388-389, 396-397
  PKU and, 239-240
  sex roles and, 376-378
  social, 376-378
  species specific, 374, 375, 378
  XYY constitution and, 75, 77, 391-392
Binomial expansion, 295-300
Bivalent, 50
Blending theory of inheritance, 5
Blood clotting, 86, 87, 110, 111
Blood groupings (*see* specific groups)
  biological significance of, 162
  minor, 161-162
Bloom's syndrome, 279-280, 365
Bombay phenomenon, 152-154
Breast development, 98

C

Cancer, 167, 271
  aging and, 362
  chromosome anomalies and, 364-366
  control mechanisms and, 358-364
  environment and, 365, 366
  genetic factors and, 366

Cancer (*cont.*):
immune system and, 366
repair enzymes and, 369-370
skin, 327
Carcinogens, 361, 365
Cell-free systems, 228, 229
Cell fusion, 127-130
Centromere, 32
Characteristic, 2
Chemicals, as mutagens, 330-332
Chiasma, 120
Chimpanzee, 374
Christmas disease, 89, 243
Chromatid, 32
Distribution at meiosis, 49, 52-53
Chromatin, 32, 35
Chromosome anomalies, cancer and, 364-366
incidence of, 71, 263
Chromosome map, use of, 305
Chromosome mapping (*see* Mapping of chromosomes)
Chromosomes, 30-31, 32-35
banding pattern of, 40, 41
homologous, 37
human complement, 36-45
morphology of, 37, 42
number of, 36-37, 40, 45, 47-48
sex (*see* Sex chromosomes)
Cleft palate, 253-254
Clone, 367
Code (*see* Genetic code and Triplet code)
Codominance, 19-20
Codon, 229
sickle cell anemia and, 244, 246, 247
Coefficient of consanguinity, 293-295
Coefficient of relationship, 144-145, 292-293, 294
Colinearity, 230-231
Colorblindness, 127
deutan, 81-84
protan, 84
total, 85
tritan, 85
Concordance, 382
Contact inhibition, 358-359
Continuous variation, 170-172
Control, cancer and, 359-361
gene expression and, 340-348
hemoglobin chains and, 351-354
in higher forms, 349-355
Jacob-Monod model of, 344-348
Cooley's anemia, 200, 248-249
Cori, C. F., 235
Cori, G. T., 235
Cosmic rays, 324, 325
Cousin matings, 293-295, 304
Cri-du-chat syndrome, 270, 275
Crisscross inheritance, 83-84
Crossing over, 119-122, 122-124, 126
biological significance of, 135-136
multiple, 124, 126
variation and, 135-136
Cystic fibrosis, 201
Cytosine, 209, 210

D

Davenport, C. B., 173
Deletions, 269-271
cancer and, 365
in X chromosome, 113
Deoxyribonucleic acid (*see* DNA)
Detrimental genes, 294, 320
in populations, 198-204
Diabetes, 184, 254-255, 256
Diet, PKU and, 239-240
Differentiation, general aspects of, 339-341
Diploid, 48
Discontinuous variation, 170
Discordance, 382
Disease, hereditary and environmental components of, 256
single gene defects and, 257-259
DNA (deoxyribonucleic acid), 35, 207-208
composition of, 208-209
structure of, 209-214
Dominance, defined, 11, 13
Down's syndrome, 45, 265-266, 395
chromosome involved in, 270
familial, 272, 273, 275
maternal age and, 266-267
mosaicism, 278-279
prenatal detection of, 281
Drift, 202-204
Drunstick, 108, 109
Dystrophy, myotonic, 308-310 (*also see* muscular dystrophy)

E

Environment, behavior and, 373-376
disease expression and, 256
influence on gene expression, 24-27, 240

Environment, behavior and (*cont.*):
IQ and, 384-387
mutation and, 315, 322, 323
polygenic inheritance and, 175
Enzymes, nature of, 216, 217
relation to genes, 216-217
Epistasis, 154
Erythroblastosis fetalis, 157, 162
Escherichia coli, chromosome of, 343
control mechanisms in, 342-348
infection by phage, 355-358
Ethics, genetic research and, 391-392, 395-397
Evolution, variation and, 131, 136
Expressivity, 20-22, 24

F

Fanconi's anemia, 279-280, 365
Fingerprinting technique, 244, 245
Fingerprints, 192-193
Fluorescence, human chromosomes and, 40, 41
Follicle, 56
Founder effect, 202-204

G

Gamma rays, 324, 325
Ganglioside lipidosis (*see* Tay-Sachs disease)
Garrod, A. D., 216-217, 235, 241
Gaucher's disease, 251-252
Gene, DNA and, 214, 216
enzyme control and, 216-217
isolation of, 393-394
protein and, 217
replication of, 215, 216
Gene pool, 186
Genetic blocks, 235-243
Genetic code, 227-230, 319-320
Genetic counseling, 92-93
nature of, 282-285
pedigrees and, 287-288, 300-305
probability and, 288-295
Genetic defects, detection of, 281-282
incidence of, 280
Genetic engineering, 393-395
Glucose-6-phosphate dehydrogenase (G6PD), 111-112, 281, 305-308, 396
Glycogen, breakdown of, 235, 236
Glycogen storage disease, 235
Grafting, 163-168
Guanine, 209, 210

H

Hairy ear trait, 81
Haploid, 48
Hardy-Weinberg law, 332-337
Height, 180, 182-184
Hemoglobin, types of, 247-248, 250, 251
linkage and, 351-354
Hemoglobin A, 243-244
Hemoglobin S, 243-244
Hemoglobins Lepore, 352
Hemophilia A, 86, 89, 305-308
detection in heterozygote, 295
Hemophilia, autosomal, 89-90
in royal families, 86, 88-89
Heritability, 384-387
Hermaphroditism, 103
Heterogametic sex, 81
Heterozygote, selection for, 199
Heterozygous, 11
High fetal hemoglobin, 353, 354
Histocompatibilty loci, 163-167
Histones, regulation and, 349, 351
Holandric genes, 81
Holley, R. W., 228
Homogametic sex, 81
Homosexuality, 105-106
Homozygous, 11
Hormones, gene expression and, 98-101, 349, 351
Huntington's disease, 252, 256, 284, 388
Hybridization, cell (*see* Cell fusion)
Hypertension, 184, 254, 256

I

Immune response, 163-168
Immune system, cancer and, 366
Inbreeding, 142, 143
Incomplete dominance, 19-20
Independent assortment, variation and, 131, 132, 133-134
Independent events, 289
Ingram, V. M., 243
Intelligence, genetic factors and, 387-389
measurement of, 379-381
Inversions, 275-277
IQ, 384-386
racial differences and, 385-387
test, 379-381

J

Jacob, F., 342, 344, 345, 347, 349
Jacob-Monod model, 344-348, 349, 351
Jensen, A. R., 386
Jervis, G., 236

K

Karyotype, 37, 40, 45
  analysis, 41
Kell factor, 161
Khorana, H. G., 228
Kidney transplant, 166
Kinetochore (*see* Centromere)
Klinefelter syndrome, 71-75, 80, 264, 265
  maternal age and, 267

L

Landsteiner, K., 148, 154, 155, 156
Language, 373-376
Law of independent assortment, 60-65
Law of segregation, 9, 13
Lepore hemoglobins, 352
Lesch-Nyhan syndrome, 243, 388-389
Lethal genes, 320, 321, 322
  calculation of mutation rate of, 321, 322
Leukemia, 270, 279, 364
Levine, P., 154
Linkage, 117-119, 119-122, 122-127, 130, 135-136
  hemoglobin chains and, 351-354
  prenatal diagnosis and, 305-310
Locus, 61, 63
Louis-Bar syndrome, 279-280, 365
Lymphocytes, immune response and, 163
Lyon hypothesis, 109-112
Lyonization (*see* X-inactivation)

M

Malaria, 199, 200
Malignancy (*see* Cancer)
Manic depressive psychosis, 390
Mapping of chromosomes, 122-127, 127-130, 305
Marijuana, 396
Meiosis, 48
  in the female, 56-59
  in the male, 48-56
  Mendel's laws and, 60-65
Meiosis (*cont.*):
  variation and, 59-60
Melanin, 173, 237
Mendel, G., 5, 8, 9, 13, 15, 18, 119
Mendel's laws, 9, 13, 15-18, 60-65, 92-94, 332
Mental illness, 254, 389-391
Mental retardation, 74, 75, 79, 236, 265
Messenger RNA (*see* mRNA)
Metaphase, mitotic, 34
Mitosis, 31-35, 36
MN blood grouping, 154-156
Modifiers, 23-24
Monod, J., 342, 344, 345, 347, 350
Monosomy, 264
Mosaicism, 103-105, 109, 277-279
mRNA, 219-221, 224, 226, 227, 228
Muller, H. J., 324
Multiple alleles, 150-151
Multiple factor inheritance (*see* Polygenic inheritance)
Muscular dystrophy, 252-253, 256, 295
Mustard gas, 330
Mutagens, 315 (*also see* Radiations and Chemicals)
Mutant genes, in populations, 332-337
Mutation, 263
  amino acid substitution and, 318
  back (reverse), 318
  effect on enzyme, 216-217
  general effect of, 313-315
  genetic code and, 319-320
  point, 313
  in regulatory genes, 347-348
  somatic, 316-318, 330
Mutation rate, calculation of, 320-322
  spontaneous, 315-319
Mutation pressure, 188
Mutually exclusive events, 289

N

Natural selection, 189, 190, 204
  behavior and, 375
  for heterozygote, 199
  skin pigmentation and, 197-198
Nirenberg, M. W., 228
Nondisjunction, 75, 77, 79
Nonhistone proteins, 349
Nonsense mutation, 229
Nucleic acid (*see* specific types of)
Nucleolar organizer, 227

Nucleolus, 227
Nucleotide, 208, 209, 210, 211

O

Oncogene theory, 360-364
Oncogenesis, 359
Oncogenic viruses (*see* Viruses, tumor inducing)
Oogenesis, 56-59
Oogonia, 56
Operator, identification of, 345
Operon, adult hemoglobins and, 351-354
  identification of, 345
Osteogenesis imperfecta, 20-22
Outbreeding, 142, 144

P

Particulate theory of inheritance, 5-8
Pedigrees, construction of, 7
  genetic counseling and, 287-288, 300-305
  linkage and, 125, 127
  mapping and, 127, 130
  prenatal diagnosis and, 305
  representative types of, 300-305
Penetrance, 22, 24
Peptide, 222
  linkage, 222, 224
Phage, 217, 218
  infection of *E. coli* by, 355-358
  symbiotic (temperate), 357-358
Phenocopy, 25-27, 102, 106
Phenylketonuria (*see* PKU)
Philadelphia chromosome, 271, 365
Phocomelia, 26-27
Photoreactivation, 327
Pigmentation (*see* Skin pigmentation)
PKU, 24-26, 256, 314, 387-388, 395
  biochemical basis of, 236-240
Pleiotropy, 18, 22, 237, 241
Polar body, 57
Polydactyly, 24
Polygenes, disease and, 253-255
Polygenic inheritance, 175, 192-193, 262
  problems in analysis of, 177, 179-184
Polypeptide, 222
Polyploidy, 268-269
Populations, 184-194
  detrimental genes in, 198-204
  mutant genes and, 332-337
Porphyria, 396
Prenatal detection, 281-282 (*also see* Amniocentesis)
Primary oocyte, 56
Primary spermatocyte, 49
Probability, combining of, 288-292
  degree of relationship and, 292-295
  genetic counseling and, 288-300
  prenatal diagnosis and, 305-310
Progeria, 370
Prophage, 357-358
Prophase, mitotic, 32
Psychosis, 389-390
Punnett square, 14
Purine bases, 209-211
Pyrimidine bases, 209-211

Q

Quantitative inheritance (*see* Polygenic inheritance)
Queen Victoria, pedigree of, 86, 88-89

R

Race, 192, 194-197
Radiations, corpuscular, 324
  electromagnetic, 323-324
  mutagenic effect of, 329-330
  nature of, 323-324
Ratio, numerical, 1:1, 15
  1:2:1, 14
  3:1, 14
  9:3:3:1, 63
Recessiveness, defined, 11, 13
Regulator, identification of, 345
Regulatory genes, identification of, 345
Relationship, coefficient of, 144-145
Repair enzymes, 327, 328
  aging and, 369-370
  cancer and, 369-370
Replication, 34, 215-216, 219, 220
  semiconservative, 216
Rh blood grouping, 156-161
  incompatibility, 392
  population differences and, 194
Ribonucleic acid (*see* RNA)
Ribosome, 219, 224, 225, 226, 227
Rickets, 85
RNA, 35, 219, 221 (*also see* specific types of)
  ribosomal, 227
RNA polymerase, 346, 349, 351

S

Schizophrenia, 389-390
Secondary oocyte, 57
Secondary spermatocyte, 52
Secretor locus, 152, 309-310
Selection pressure, 190
Selective medium, 128
Sendai virus, 128
Sex bivalent, 50
Sex chromatin, 107-109, 113
Sex chromosomes, 37, 39
  anomalies of, 71-80
  Mendel's laws and, 92-94
  sex determination and, 68-71
Sex determination, 68-71
Sex influenced genes, 99-101, 102, 103
Sex limited genes, 97-99
Sex-linked genes, 80-86 (*also see* X-linked genes)
Sex preferences, 105-106
Sex roles, 376-378
Sexual reproduction, variation and, 130-134
Sickle cell anemia, 18-20, 198-200, 243, 246-247, 283, 314
Sickle cell trait, 199
Skin pigmentation, 173-177, 178, 179, 197-198
Somatic mutation, cancer and, 364
Species, 185-194
Speech, 373-376
Spermatid, 53, 56
Spermatogenesis, 54-55
Spermatogonium, 48
Spermiogenesis, 56
Spindle, 34
Structural genes, identification of, 345
Suppressor genes, 23-24
Synapsis, 49
Syndrome, 18
Synteny, 130, 305

T

Tay-Sachs disease, 201, 242-243, 283, 314, 395
  prenatal detection of, 281-282
Telophase, mitotic, 34
Temperature, mutation rate and, 322, 323
Testcross, 15, 122, 123
Testosterone, 70
Thalassemia, 200, 248-249, 283
Thalidomide, 26-27
Thymine, 209, 210
Thymus gland, 163
Trait, 2
Transcription, 219-221
Transfer RNA (*see* tRNA)
Translocation, 271-275
Transplants, 163-168
Transsexuals, 106, 113
Triplet code, 228
Trisomy, 21-22, 270-271, (*see also* Down's syndrome)
  autosomal, 265-266
tRNA, 222, 224, 225, 227
Turner syndrome, 77-79, 80, 109, 113, 264, 268
Twinning, 138-141
Twins, in genetic analysis, 141, 381-384
  mental illness studies and, 390
Tyrosinosis, 241

U

Ultraviolet light, 324, 325-327
  phage activation and, 358
  skin pigmentation and, 197-198

V

Variable expressivity (*see* Expressivity)
Variation, significance of, 130-137
  types of, 170-171
Viruses, bacterial (*see* Phage)
  cancer and, 359-364
  genetic engineering and, 393
  RNA containing, 362, 363
  tumor inducing, 359-361
Vitamin D, 197-198

W

Watson-Crick model, 209-216
Wiener, A. S., 156
Wilkins, M. H. F., 209
Wild-type genes, 102, 133-134

X

X chromosome, deletions in, 113
  extra doses of, 79, 265
  height and, 180, 182
  inactivation of, 109, 113, 353, 355

X chromosome, deletions in (*cont.*):
  mapping of, 127
  sex determination and, 70-71
Xeroderma pigmentosum, 327, 369-370
X-linked alleles, calculation of in populations, 336-337
X-linked dominant genes, 85-86
X-linked genes, 80-86 (*also see* Sex-linked genes)
X-linked recessives, calculation of mutation rate of, 321
XO condition (*see* Turner syndrome)
X-rays, effect on matter, 324, 325
XXY condition, 74, 80 (*also see* Klinefelter syndrome)
XYY condition, behavior and, 75-77, 80, 391-392

Y

Y chromosome, height and, 180, 182
  sex determination and, 69, 75